Undergraduate Texts in Mathematics

L. E. Sigler

Algebra

Springer Science+Business Media, LLC

L. E. Sigler
Bucknell University
Department of Mathematics
Lewisburg, Pennsylvania 17837

AMS Subject Classifications: 13–01, 15–01

Library of Congress Cataloging in Publication Data

Sigler, L(aurence) E(dward)
Algebra.
(Undergraduate texts in mathematics)
Bibliography: p. 409
Includes index.
1. Algebra. I. Title.
QA152.2.S56 512′.02 76–21337

The cover design is based on John Croker's medal of Newton. The figure is that of winged Science holding a tablet upon which appears the solar system. More about it in D. E. Smith, *History of Mathematics*, Vol. I, (Dover).

Originally published by Springer-Verlag, Inc. in 1976
Softcover reprint of the hardcover 1st edition 1976

ISBN 978-3-540-90195-2 ISBN 978-3-662-26738-7 (eBook)
DOI 10.1007/978-3-662-26738-7

Preface

There is no one best way for an undergraduate student to learn elementary algebra. Some kinds of presentations will please some learners and will disenchant others. This text presents elementary algebra organized according to some principles of universal algebra. Many students find such a presentation of algebra appealing and easier to comprehend. The approach emphasizes the similarities and common concepts of the many algebraic structures. Such an approach to learning algebra must necessarily have its formal aspects, but we have tried in this presentation not to make abstraction a goal in itself. We have made great efforts to render the algebraic concepts intuitive and understandable. We have not hesitated to deviate from the form of the text when we feel it advisable for the learner. Often the presentations are concrete and may be regarded by some as out of fashion. How to present a particular topic is a subjective one dictated by the author's estimation of what the student can best handle at this level. We do strive for consistent unifying terminology and notation. This means abandoning terms peculiar to one branch of algebra when there is available a more general term applicable to all of algebra. We hope that this text is readable by the student as well as the instructor. It is a goal of ours to free the instructor for more creative endeavors than reading the text to the students.

We would have preferred to call this book *College Algebra* because this was the name of the standard algebra course for undergraduate students in the United States for many years. Unfortunately, the name "College Algebra" now seems firmly attached to a body of material taught in the 1930's. Perhaps in time the name "College Algebra" will once again describe the algebra studied by college students. Meanwhile we have names like "Modern Algebra" and "Abstract Algebra" using inappropriate modifiers.

Included in the first half of the text and providing a secondary theme are a development and construction of the number systems: natural numbers (Sections 3.1–3.4), integers (Sections 3.5–3.6), fractions or rational numbers (Section 4.5), and complex numbers (Section 5.8). The construction of the real number system is properly a topic in analysis; we refer the reader to reference [10, p. 234] for an algebraically oriented presentation. The use of the integers as exponents and multiples as in secondary school algebra is covered in detail (Section 4.4). All of the material on number systems can be stressed advantageously by those students preparing for school teaching.

As in all texts, size considerations eventually begin to exercise influence. Group theory is not stressed in this text although there is a respectable amount of material for an elementary text in Chapter 9. There is no Galois theory in this text. Although lattice theory is a central concept for universal algebra we have pretty well omitted study of that area. For that reason and others, this cannot be considered to be an elementary text in universal algebra.

Considerable attention has been paid to the algebraic properties of functions and spaces of functions. One of the primary uses of algebra for an undergraduate is in his analysis courses. We hope that the attention we have paid to functions will be found rewarding by the student in his analysis courses and in turn we hope that the somewhat concrete nature of spaces of functions helps illuminate some of the algebraic structures by being tangible examples of those structures.

Chapters 1–5 are devoted to rings, Chapters 6, 7, and 10 to linear algebra, Chapter 9 to monoids and groups, and Chapter 8 to algebraic systems in general. We envision the text being used for a year's course in algebra, for a one semester course not including linear algebra, or for a linear algebra course. A shorter course in algebra might consist of Chapters 1–5, omitting possibly Section 3.8, Section 4.6, and parts of Chapter 5, supplemented by Sections 9.1–9.4. A course in linear algebra for students already familiar with some of the topics included in the first five chapters could concentrate on Chapters 6, 7, and 10 after reviewing Sections 5.1–5.6. Ideally we envision the book for a one-year course covering all the chapters.

The questions at the end of each section are to help the reader test his reading of the section. Certainly the section ought be read carefully before attempting to answer the questions. Many of the questions are tricky and hang upon small points; more than one of the answers may be correct. The exercises provide for practice and gaining a better knowledge of the material of the section. It is our practice to use in examples and in the exercises some material on an intuitive bases before the material is treated in the text more formally. Provided one guards against circular reasoning this provides for a more immediate illustration of the principles the student is trying to understand.

Algebra as an undergraduate course is frequently the subject in which a student learns a more formal structure of definitions, theorems, and proofs.

The elementary calculus is often a more intuitively presented course and it is left to the algebra course to institute the more formal approach to mathematics. For this reason the student should be very aware of what are definitions, what are theorems, and what is the difference between them.

We now make several comments on style. In this text sentences are frequently begun with symbols which may happen to be small letters. *e* is a transcendental number. We consider such symbols proper nouns and beg forgiveness; we have found the practice of avoiding such sentences too limiting. Secondly, we use a number of run-on sentences connected with *if and only if*. They are too logically appealing to avoid. We leave to the reader without comment all other perversions of the queen's English.

It is our opinion that one of the most rewarding things a student of mathematics can learn is some of the history of the subject. Through such knowledge it is possible to gain some appreciation of the growth of the subject, its extent, and the relationships between its various parts. We cannot think of any other area where such a little effort will reap such a bountiful harvest. For reasons of length and cost we have not included historical material in this text despite the opinion just expressed. We recommend the purchase, reading, and rereading of at least one mathematical history book: see the references for suggestions.

The author wishes to thank Bucknell University for a sabbatical leave to work on the manuscript for this book, the editors of Springer-Verlag for their encouragement and advice, and several readers for their suggestions. All errors I claim for myself.

Lewisburg
March, 1976

L.E.S.

Contents

1 Set theory

This chapter on sets establishes the language of set theory in which this book on algebra is written. The elementary operations on sets such as subset, union, and intersection may well be familiar to the reader. In this case he should certainly move on to matters which are new. We suggest that this chapter first be read hurriedly to see what topics are familiar and what topics are new. The reader can then spend the time necessary to master whatever is new for him and can avoid spending time on material he knows.

Concepts of this chapter such as set, subset, quotient set, function, and the fundamental isomorphism theorem will be used repeatedly for each separate algebraic structure in later chapters.

Since this is not a textbook for a course in set theory the treatment of sets is abbreviated. We have tried, as do all authors of elementary texts in algebra, to compromise between too much material on sets and too little. The goal is to give the necessary amount to support the algebra presented later. Certain other concepts of set theory not included in Chapter 1 will be presented later in the text as needed.

In Section 1.8, after a discussion of composition of functions, we introduce the algebraic concept of group by example. Commutative groups occur again in Chapter 6 as modules; groups not necessarily commutative are treated in detail in Chapter 9.

1.1 Sets

In this first section we discuss the fundamental concepts of set membership, the relations of equality and subset.

A set is a collection of objects. Some examples of sets are

the set of all letters of the English alphabet,
the set of letters a, b, and c,

the set of states of the USA,
the set of all teams of the National Basketball Association,
the set of numbers $-2, 0, 2, 4$.

We use $x \in S$ as an abbreviation for x is a member of the set S. Some alternate expressions are x is an element of S, x belongs to S, x is in S and S contains x.

A set can be denoted by listing its members between a pair of braces:

$$\{a, b, c\},$$
$$\{-2, 0, 2, 4\}.$$

Several true statements about these two examples are: $a \in \{a, b, c\}$, $2 \in \{-2, 0, 2, 4\}$.

We use the symbolism $x \notin S$ to mean x is not a member of the set S. For example, $d \notin \{a, b, c\}$, $1 \notin \{-2, 0, 2, 4\}$.

We also denote sets by means of some defining property:

the set of all letters of the English alphabet.

In order to work principally with numbers we now name five sets:

$\mathbb{N}$ = the set of natural numbers,
$\mathbb{Z}$ = the set of integers,
$\mathbb{Q}$ = the set of rational numbers,
$\mathbb{R}$ = the set of real numbers,
$\mathbb{C}$ = the set of complex numbers.

These symbols will be used consistently throughout the entire text for the sets indicated. By the natural numbers we mean all positive whole numbers and zero. By the integers we mean all whole numbers, positive, negative, and zero. By rational numbers we mean all fractions, quotients of integers with nonzero denominator. By real number we mean any number representable by a decimal, terminating or nonterminating. Examples of real numbers are 2, 3.00000 . . . , 3.14159 . . . , 0.33333 . . . . At this point we are relying upon previous knowledge and acquaintance with these concepts. We will develop these number systems more fully later in this text. Finally a complex number is a number of the form $a + bi$ where a and b are real numbers and $i = \sqrt{-1}$.

The symbol $\{x|x \text{ is an integer}\}$ is read "the set of all x such that x is an integer." This is a stilted way of saying "the set of integers," but what is lost in euphony is more than compensated for by the gain in utility. The symbol is called a classifier. For example, $\mathbb{N} = \{x|x \text{ is a natural number}\} = \{x|x \in \mathbb{Z}$ and x is nonnegative$\}$. $\{-2, 0, 2, 4\} = \{x|x \in \mathbb{Z}$ and x is even and $-3 < x < 5\}$. In general if $p(x)$ is some statement involving the letter x and if there is a set consisting precisely of all objects x such that $p(x)$ is true then we denote that set by $\{x|p(x)\}$.

We shall consider two sets to be *equal* when they have the same members:

Definition. $X = Y$ if and only if every member of X is a member of Y and every member of Y is a member of X.

The definition of equality makes irrelevant the order of listing of members in any set: $\{a, b, c\} = \{b, c, a\}$. Given an object and a set, the object is either a member of the set or it is not a member of the set. There is no question of duplicate membership. We must regard the sets $\{2, 8, 8\}$ and $\{2, 8\}$ as equal; 2 and 8 are members of the sets and no other objects are. If we consider the set $\{x, y, z\}$ and it is given that x, y, and z are integers we have many possibilities. Some possibilities are $\{1, 0, 3\}$, $\{0, 4, 2\}$, $\{7, 12, 13\}$. Another possibility, however, is $\{5\}$ which is obtained by setting $x = 5$, $y = 5$, and $z = 5$.

Equality of sets enjoys the three properties of *reflexivity*, *symmetry*, and *transitivity* which are listed in order in the next theorem.

Theorem. *$X = X$ for all sets X. $X = Y$ implies $Y = X$ for all sets X, Y. $X = Y$ and $Y = Z$ imply $X = Z$ for all sets X, Y, Z.*

PROOF. X has the same members as X. Every member of X is a member of X and vice versa. $X = X$.

If every member of X is a member of Y and every member of Y is a member of X then $Y = X$ as well as $X = Y$.

If X and Y have the same members and Y and Z have the same members then X and Z have the same members. □

We now take up the concept of *subset*.

Definition. $X \subseteq Y$ if and only if every member of X is a member of Y. $X \subseteq Y$ is read "X is a subset of Y."

EXAMPLES. $\{a, b\} \subseteq \{a, b, c\}$. $\mathbb{N} \subseteq \mathbb{Z}$. $\mathbb{Z} \subseteq \mathbb{Q}$. $\mathbb{Q} \subseteq \mathbb{R}$. $\mathbb{R} \subseteq \mathbb{C}$. $\{a, b, c\} \subseteq \{a, b, c\}$. $\{-2, 0, 2, 4\} \subseteq \mathbb{Z}$.

$A \nsubseteq B$ means A is not a subset of B. $A \subset B$ means $A \subseteq B$ and $A \neq B$. For A not to be a subset of B there must be some member of A which is not contained in B. For A to be a proper subset of B ($A \subset B$) we must have A be a subset of B and also there must be some member of B which is not a member of A.

Theorem. *$X \subseteq X$ for all sets X. $X \subseteq Y$ and $Y \subseteq X$ imply $X = Y$ for all sets X, Y. $X \subseteq Y$ and $Y \subseteq Z$ imply $X \subseteq Z$ for all sets X, Y, Z.*

These three properties are called respectively: reflexivity, antisymmetry and transitivity.

PROOF. Every member of X is a member of X and therefore $X \subseteq X$. Because $X \subseteq Y$ every member of X is a member of Y. Because $Y \subseteq X$ every member of Y is a member of X. We have $X = Y$.

Let a be a member of X. Because $X \subseteq Y$ we have $a \in Y$. Because $Y \subseteq Z$ it follows that $a \in Z$. Thus $X \subseteq Z$. □

We must distinguish carefully between the two symbols $\in$ and $\subseteq$. Here are some examples which require this. We denote the set of teams of the National Basketball Association by N, the Boston Celtics team by C and a player Bill Russell of the Boston Celtics by R. All of these statements are true: $C \in N$. $R \notin N$. $\{C\} \subseteq N$. $R \in C$. $\{R\} \subseteq C$. $N \notin C$. $\{R\} \nsubseteq N$. $N \nsubseteq C$.

1.2 Operations on sets

In this section we find the operations of union and intersection defined, the empty set introduced and the complement defined.

We now give means of producing a set from two given sets.

Definition. The *union* of the sets X and Y, written $X \cup Y$, is defined to be the set $\{x | x \in X \text{ or } x \in Y\}$. The *intersection* of the sets X and Y, written $X \cap Y$, is defined to be the set $\{x | x \in X \text{ and } x \in Y\}$.

EXAMPLES. $\{a, b\} \cup \{b, c, d\} = \{a, b, c, d\}$. $\{a, b\} \cap \{b, c, d\} = \{b\}$. $\mathbb{N} \cap \mathbb{Z} = \mathbb{N}$. $\mathbb{N} \cup \mathbb{Z} = \mathbb{Z}$. The two Venn diagrams in Figure 1.1 illustrate the intersection (a) and union (b).

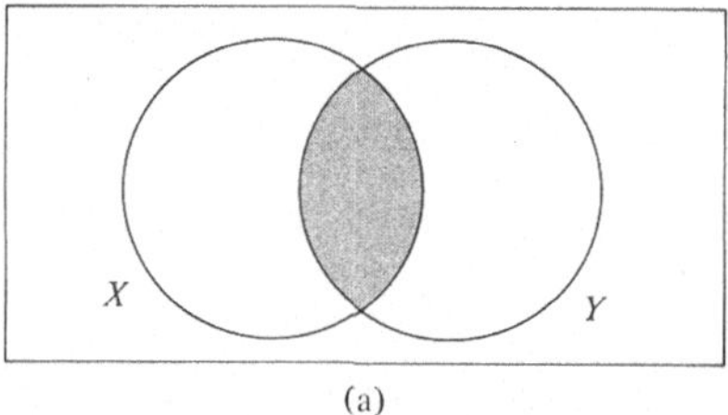

(a)

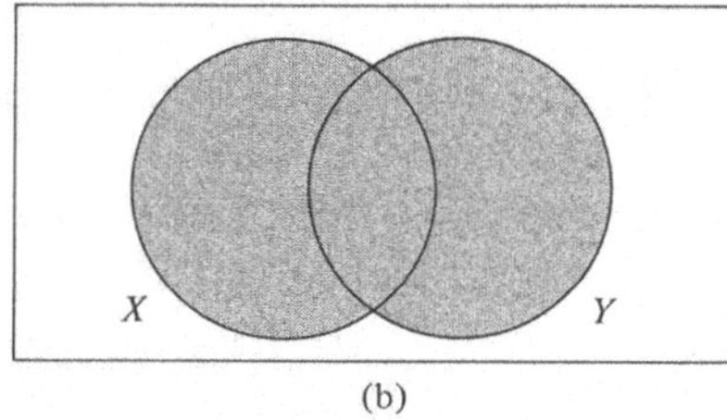

(b)

Figure 1.1 (a) $X \cap Y$ is shaded. (b) $X \cup Y$ is shaded.

The following theorem states that both union and intersection of sets are commutative.

Theorem. *$X \cup Y = Y \cup X$ for all sets X, Y. $X \cap Y = Y \cap X$ for all sets X, Y.*

PROOF. If x belongs to $X \cup Y$ then x belongs to X or x belongs to Y. If x belongs to X or x belongs to Y then x belongs to Y or x belongs to X. Then x is a member of $Y \cup X$. Likewise, every member of $Y \cup X$ is a member of $X \cup Y$. $X \cup Y = Y \cup X$. The proof for intersection is entirely similar. □

The next theorem states that both union and intersection of sets are associative.

Theorem. *$X \cup (Y \cup Z) = (X \cup Y) \cup Z$ for all X, Y, Z. $X \cap (Y \cap Z) = (X \cap Y) \cap Z$ for all X, Y, Z.*

Proof. Let $x \in X \cup (Y \cup Z)$. Then $x \in X$ or $x \in Y \cup Z$. $x \in X$ or $(x \in Y$ or $x \in Z)$. This means the same as $(x \in X$ or $x \in Y)$ or $x \in Z$. This is to say the word "or" is associative. $x \in X \cup Y$ or $x \in Z$. $x \in (X \cup Y) \cup Z$. This demonstrates that $X \cup (Y \cup Z) \subseteq (X \cup Y) \cup Z$. In like manner we can show $(X \cup Y) \cup Z \subseteq X \cup (Y \cup Z)$. Therefore $(X \cup Y) \cup Z = X \cup (Y \cup Z)$. The second equation for intersection is proved by substituting the word "and" for the word "or" in the proof just given. □

If the intersection of two sets is always to be a set, such as in the case $\{a, b\} \cap \{c, d, e\}$ then the need for a set with no members is clear.

Definition. *We use the symbol $\varnothing$ to represent the empty set, a set with no members.*

Examples. $\{a, b\} \cup \varnothing = \{a, b\}$. $(a, b\} \cap \{c, d, e\} = \varnothing$. $\varnothing \cap \{a, b\} = \varnothing$. $\varnothing \cup \varnothing = \varnothing$. $\varnothing \cap \varnothing = \varnothing$. $\varnothing \subseteq \{a, b\}$.

Theorem. $X \cup \varnothing = X$ *for any set* X. $X \cap \varnothing = \varnothing$ *for any set* X.

Proof. We show $X \cup \varnothing \subseteq X$ and $X \subseteq X \cup \varnothing$. Let $x \in X \cup \varnothing$. $x \in X$ or $x \in \varnothing$. Since x cannot belong to the empty set then x must belong to X. $X \cup \varnothing \subseteq X$. On the other hand, if $x \in X$ then certainly the statement $x \in X$ or $x \in \varnothing$ is true. Therefore, $x \in X \cup \varnothing$. $X \subseteq X \cup \varnothing$. $X \cup \varnothing = X$. The second statement of the theorem is proved using similar techniques. □

Theorem. $X \subseteq (X \cup Y)$ *for any sets* X, Y. $(X \cap Y) \subseteq X$ *for any sets* X, Y.

Proof. Again we prove only the first statement leaving the second as an exercise. If $x \in X$ then the statement $x \in X$ or $x \in Y$ is true. Therefore $x \in X \cup Y$. $X \subseteq (X \cup Y)$. □

Theorem. $\varnothing \subseteq X$ *for any set* X.

Proof. Suppose the empty set were not a subset of some set X. Then there would be some element c of the empty set which failed to belong to the set X. But this cannot be since the empty set cannot contain any element c. We conclude $\varnothing \subseteq X$. □

When we prove $\varnothing \subseteq X$ we prove the statement "every member of the empty set is a member of the set X" to be true. This argument is often a bit tricky for the reader unaccustomed to arguing vacuous cases. In order for the statement to be false there would have to exist some member of the $\varnothing$ which was at the same time not a member of X. For another example of this kind of reasoning take the statement "every pink hippo in this room weighs precisely 47 grams." This statement does not assert the existence of any pink hippos; the statement merely asserts that if there are any then they weigh 47 grams. If there are no pink hippos at all in the room then the statement is true. For the statement to be false one must demonstrate the existence of at least one pink hippo in the room which does not weigh 47 grams. In mathematical usage a universal statement never asserts existence.

For a bit more practice with proving theorems about sets:

Theorem. *$X = X \cup Y$ if and only if $Y \subseteq X$ (for all sets X, Y). $X = X \cap Y$ if and only if $X \subseteq Y$ (for all sets X, Y).*

PROOF. First assume $Y \subseteq X$. We must prove $X = X \cup Y$. From an earlier theorem we know $X \subseteq X \cup Y$. To show $X \cup Y \subseteq X$ let $x \in X \cup Y$. Then $x \in X$ or $x \in Y$. If $x \in Y$ then $x \in X$ because Y is a subset of X. Therefore in either case $x \in X$. $X \cup Y \subseteq X$. Both subset relations imply $X \cup Y = X$.

For the converse begin with $X = X \cup Y$. Now we must prove $Y \subseteq X$. Let $x \in Y$. If $x \in Y$ then $x \in X \cup Y$. Since $X \cup Y = X$ we have $x \in X$. $Y \subseteq X$.

The second statement involving intersection can be proved in a similar manner. □

Definition. We define the *relative complement* $X - Y$ to be the set $\{x | x \in X \text{ and } x \notin Y\}$.

The Venn diagram of Figure 1.2 illustrates the relative complement.

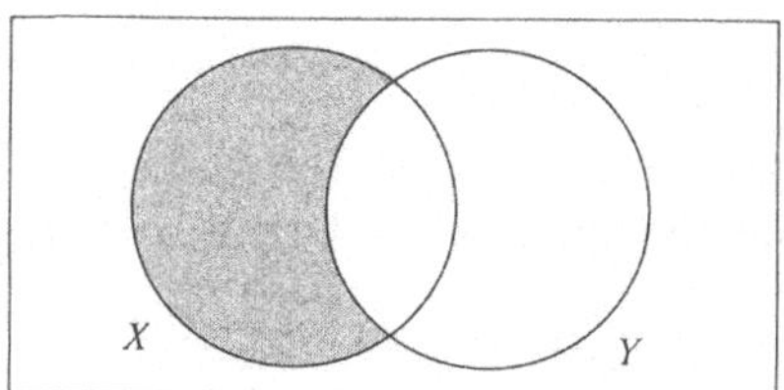

Figure 1.2 X-Y is the shaded area.

EXAMPLES. $\{a, b, c\} - \{a, b\} = \{c\}$. $\{a, b, c\} - \{c, d\} = \{a, b\}$. $\mathbb{N} - \mathbb{Z} = \varnothing$. $\mathbb{Z} - \mathbb{N} = \{x | x \text{ is a negative integer}\}$. When the first set X in the relative complement $X - Y$ is understood from context then the set is often indicated by $Y^{\sim}$ and is simply called the complement of Y.

Theorem. *Let S be a given set and let $X^{\sim}$ denote $S - X$ where $X \subseteq S$. Then*

$$X^{\sim\sim} = X, \qquad S^{\sim} = \varnothing, \qquad \varnothing^{\sim} = S, \qquad X \cup X^{\sim} = S, \qquad X \cap X^{\sim} = \varnothing.$$

PROOF. $x \in X^{\sim\sim}$ if and only if ($x \in S$ and $x \notin X^{\sim}$) if and only if ($x \in S$ and not ($x \in X^{\sim}$)) if and only if ($x \in S$ and not not ($x \in X$)) if and only if ($x \in S$ and

$x \in X$) if and only if $x \in X$. The proofs of the remaining parts can be supplied by the reader. □

The following results are usually called *De Morgan's identities.*

Theorem. *Let X and Y be subsets of some set S.*

$$(X \cup Y)^{\sim} = X^{\sim} \cap Y^{\sim}. \qquad (X \cap Y)^{\sim} = X^{\sim} \cup Y^{\sim}.$$

PROOF. Let $x \in (X \cup Y)^{\sim}$. Then $x \in S$ and x is not a member of $X \cup Y$. x is not a member of X and x is not a member of Y. $x \in X^{\sim}$ and $x \in Y^{\sim}$. $x \in X^{\sim} \cap Y^{\sim}$. This proves $(X \cup Y)^{\sim} \subseteq X^{\sim} \cap Y^{\sim}$. We reverse the steps to prove the other inclusion and have $(X \cup Y)^{\sim} = X^{\sim} \cap Y^{\sim}$. The proof of the second De Morgan identity is left entirely for the reader. □

QUESTIONS

1. Which of these statements are true?
   (A) $\mathbb{Q} \subseteq \mathbb{Z}$
   (B) $\mathbb{R} \subseteq (\mathbb{Z} \cap \mathbb{R})$
   (C) $\mathbb{Z} \subseteq (\mathbb{R} \cap \mathbb{N})$
   (D) $\mathbb{N} \subseteq \mathbb{Z}$
   (E) None of the four statements is true.
2. $X \subseteq Y$ and $Y \subseteq X$ fail to imply
   (A) $X = Y$
   (B) Y contains some member that X does not contain
   (C) $X \subseteq Y$
   (D) X and Y have the same members.
   (E) $X \subseteq Y$ and $Y \subseteq X$ imply all four statements.
3. Which of the following is not a subset of $A = \{a, b, c\}$?
   (A) $\varnothing$
   (B) a
   (C) A
   (D) $\{a, b\}$
   (E) All four listed sets are subsets of $\{a, b, c\}$.
4. Which of these statements are false?
   (A) $\varnothing \in X$ for all sets X
   (B) $\varnothing \subseteq X$ for some set X
   (C) $(\varnothing \cap X) \subseteq X$ for all sets X
   (D) $X \subseteq (\varnothing \cup X)$ for all sets X.
   (E) All four statements are true.
5. Which of these sets have no proper subset?
   (A) $\mathbb{R}$
   (B) $\mathbb{R} \cap \mathbb{Z}$
   (C) $(\mathbb{N} \cup \mathbb{R}) \cap \varnothing$
   (D) $\mathbb{N} \cap \mathbb{N}$
   (E) All four listed sets have at least one proper subset.

6. If A is a nonempty set and B is any set whatsoever then the union of A and B
- (A) may be empty
- (B) must be nonempty
- (C) may strike for higher pay
- (D) is a subset of B.

(E) None of the four possibilities completes a true sentence.

7. Equality of sets is not
- (A) symmetric
- (B) reflexive
- (C) antisymmetric
- (D) transitive.

(E) Equality of sets has all four listed properties.

8. Which of these statements are true?
- (A) $(X \cap Y) \cup Z = X \cap (Y \cup Z)$ for all sets X, Y, Z
- (B) $(X \cap Y \cap Z)^{\sim} = X^{\sim} \cup Y^{\sim} \cup Z^{\sim}$ for all sets X, Y, Z
- (C) $\{x, y\} \not\subseteq \{x\}$ for all $x, y \in \mathbb{R}$
- (D) $(X \cap Y) \cup Y \subseteq X$ for all sets X, Y, Z.

(E) None of the four statements is true.

9. Which of these statements are true?
- (A) $9 \in \mathbb{Z}$
- (B) $9 \in \{x | x \in \mathbb{Z} \text{ and } x \leqslant 0\}$
- (C) $\{a, b\} \in \{a, b, \{a, b\}, d\}$
- (D) $\{a, b\} \in \{a, b\}$
- (E) $\{a, b\} \subseteq \{a, b\}$
- (F) $(\{a, b\} \cap \{b, c\}) \subseteq (\{a, b\} \cup \{b, c\})$.

Exercises

1. Prove $X \cup X = X$ for all sets X.

2. Prove $X \cap X = X$ for all sets X.

3. Prove $(A \cap B \cap C) \subseteq (A \cap B) \subseteq A$ for all sets A, B, C.

4. Show that $A \cap (B - C) = (A \cap B) - (A \cap B \cap C)$.

5. Show: If $A \subseteq C$ and $B \subseteq D$ then $(A \cap B) \subseteq (C \cap D)$ and $(A \cup B) \subseteq (C \cup D)$.

6. Is it true that $A \not\subseteq B$ and $B \not\subseteq C$ imply $A \not\subseteq C$? (You should either prove the statement true or give an example illustrating that the statement is false.)

7. Is it true that $A \subset B$ and $B \subset C$ imply $A \subset C$?

8. Prove $A \subseteq \varnothing$ implies $A = \varnothing$.

9. Letting $X^{\sim} = \mathbb{Z} - X$ find $X^{\sim}$ in all these examples.
- (a) $X = \{x | x \in \mathbb{Z} \text{ and } x \geqslant 0\}$
- (b) $X = \{x | x \in \mathbb{Z} \text{ and } (x \geqslant 4 \text{ or } x \leqslant -3)\}$
- (c) $X = \{x | x \in \mathbb{Z} \text{ and } 0 \leqslant x \leqslant 1\}$
- (d) $X = \{x | x \in \mathbb{Z} \text{ and } x = 2n \text{ for some } n \in \mathbb{Z}\}$.

1.3 Relations

In this section we treat Cartesian products and relations: equivalence relations and orders.

Definition. $X \times Y$, the *Cartesian product* of X and Y, is defined to be the set $\{(x, y) | x \in X \text{ and } y \in Y\}$.

The Cartesian product of two sets, X and Y, is the set of all ordered pairs (x, y) in which the left member of the pair is selected from X and the right member from Y.

EXAMPLE. $\{a, b\} \times \{c, d, e\} = \{(a, c), (a, d), (a, e), (b, c), (b, d), (b, e)\}$.

The following chart gives some intuition for the Cartesian product.

		Y		
		c	d	e
X	a	(a, c)	(a, d)	(a, e)
	b	(b, c)	(b, d)	(b, e)

It is to be understood that two ordered pairs (r, s) and (u, v) are equal if and only if $r = u$ and $s = v$. $(2, 1) \neq (1, 2)$. It is quite possible to have an ordered pair with left and right side equal: $(2, 2)$.

EXAMPLES. $\mathbb{R} \times \mathbb{R} = \{(x, y) | x \in \mathbb{R} \text{ and } y \in \mathbb{R}\}$ is the set of all ordered pairs of real numbers. This is a model for the Euclidean plane. $\mathbb{Z} \times \mathbb{R} = \{(n, y) | n \in \mathbb{Z}$ and $y \in \mathbb{R}\}$ is the set of all ordered pairs with left side an integer and right side a real number. This is a subset of the Euclidean plane consisting of all lines parallel to the Y-axis and having integral abscissae.

It was seen earlier that equality of sets enjoyed the properties of reflexivity, symmetry, and transitivity. Starting with these properties we define any relation which is reflexive, symmetric, and transitive to be an equivalence relation. Equality of sets is one such relation. Before we investigate more formally the concept of relation we offer one more example of an equivalence relation. This example depends upon the reader's prior knowledge of $\mathbb{Z}$ and arithmetic.

Beginning with $\mathbb{Z}$ we define xRy if and only if $x - y$ is divisible by 2. For $x - y$ to be divisible by 2 simply means $x - y$ is an even number. This is to say, $x - y = 2n$ for some $n \in \mathbb{Z}$. $5R7$ because $5 - 7 = -2 = 2 \cdot (-1)$. $12R2$ because $12 - 2 = 10 = 2(5)$. $8R8$ because $8 - 8 = 0 = 2(0)$. Clearly all odd numbers are equivalent to each other and all even numbers are equivalent to each other. We now argue that the R in this example is reflexive, symmetric, and transitive. xRx because $x - x = 0 = 2(0)$. Suppose xRy. $x - y = 2n$ for some $n \in \mathbb{Z}$. $y - x = 2(-n)$. yRx. Finally suppose xRy and yRz. $x - y = 2n$ for some $n \in \mathbb{Z}$ and $y - z = 2m$ for some $m \in \mathbb{Z}$. $x - z = x - y + y - z = 2n + 2m = 2(n + m)$. xRz. R is transitive.

In fixing a formal definition of a relation what is essential is that given any two members x, y of the set S on which the relation is defined we must be able to decide whether or not the relation holds. This can be done precisely by means of a set of ordered pairs. We simply assemble into a set all those ordered pairs (x, y) for which the desired relation holds.

Definition. R is a *relation* on a set S if and only if $R \subseteq S \times S$. We say xRy (x is related to y) if and only if $(x, y) \in R$.

Referring back to the equivalence relation example on $\mathbb{Z}$ given before we see that the relation R is the set $\{(x, y) | x \in \mathbb{Z}$ and $y \in \mathbb{Z}$ and $x - y$ is even$\}$. It contains the pairs (5, 7), (12, 2), (8, 8), among others.

Definition. R is a *reflexive* relation on a set S if and only if $(x, x) \in R$ for all $x \in S$. R is a *symmetric* relation on a set S if and only if $(x, y) \in R$ implies $(y, x) \in R$. R is a *transitive* relation on a set S if and only if $(x, y) \in R$ and $(y, z) \in R$ imply $(x, z) \in R$. R is an *equivalence* relation on a set S if and only if R is a reflexive, symmetric, and transitive relation on S.

An order is another example of a relation. For example, let us take the set $\{0, 1, 2, 3\}$ and consider the elements related (ordered) as follows: $0 \leqslant 0$, $0 \leqslant 1$, $0 \leqslant 2$, $0 \leqslant 3$, $1 \leqslant 1$, $1 \leqslant 2$, $1 \leqslant 3$, $2 \leqslant 2$, $2 \leqslant 3$, $3 \leqslant 3$. We would then say that the order is the following set of pairs: $\{(0, 0), (0, 1), (0, 2), (0, 3), (1, 1), (1, 2), (1, 3), (2, 2), (2, 3), (3, 3)\}$.

Definition. R is an *antisymmetric* relation on a set S if and only if R is a relation and $(x, y) \in R$ and $(y, x) \in R$ imply $x = y$. A relation R on a set S is an *order* on S if and only if R is a reflexive, antisymmetric, and transitive relation.

An equivalent condition to $(x, y) \in R$ and $(y, x) \in R$ imply $x = y$ is the contrapositive statement $x \neq y$ implies not both (x, y) and (y, x) belong to R.

QUESTIONS

1. The relation $\{(0, 0), (1, 2), (2, 2)\}$ on the set $\{0, 1, 2\}$ is
- (A) reflexive
- (B) transitive
- (C) symmetric
- (D) antisymmetric.
- (E) The relation has none of the listed properties.

12. $A \times B$
    (A) always contains more members than A does
    (B) contains at least one ordered pair if B does
    (C) contains a pair (a, b) if $a \in B$ and $b \in A$
    (D) always has an empty intersection with $B \times A$.
    (E) None of the four possibilities completes a true sentence.

13. The Cartesian product of X and Y
    (A) contains all the members of X and all the members of Y
    (B) has the same number of members as $Y \times X$
    (C) is a subset of any relation on X if $X = Y$
    (D) is commutative.
    (E) None of the four possibilities completes a true sentence.

14. Which of these alternatives complete a false sentence? On any set X
    (A) $\varnothing$ is a symmetric relation
    (B) $\varnothing$ is a transitive relation
    (C) $\varnothing$ is an antisymmetric relation
    (D) $\varnothing$ is a reflexive relation.
    (E) All four statements are true.

15. We define on $\mathbb{Z}$ the following relation: xRy if and only if $x - y$ is divisible by 5. Which of the following are true?
    (A) R is an order on $\mathbb{Z}$.
    (B) The relation R fails to be symmetric.
    (C) $-7, -2, 2, 7, 12$ are all related to each other.
    (D) Although R fails to be an order it is reflexive and transitive.
    (E) None of the four statements is true.

Exercises

1. Prove this statement false: $X \times Y = Y \times X$ for all sets X, Y. Do this by giving an example of two sets A, B such that $A \times B \neq B \times A$.

2. Describe these subsets of the Cartesian plane $\mathbb{R} \times \mathbb{R}$:

   (a) $\mathbb{N} \times \mathbb{N}$ (b) $\mathbb{R} \times \mathbb{N}$ (c) $\mathbb{Z} \times \mathbb{Z}$.

3. Is it ever the case that $X \times Y = Y \times X$?

4. We define the disjoint union of sets X and Y to be $X \sqcup Y = (X \times \{1\}) \cup (Y \times \{2\})$. Construct both $X \sqcup Y$ and $X \cup Y$ for the sets $X = \{a, b, c\}$ and $Y = \{a, d\}$. Compare the results. What can you say in general about the sizes of the sets $X \sqcup Y$ and $X \cup Y$?

5. Let U be the set of all undergraduate students at Bucknell University. In each of the following examples decide whether or not the relation given is an equivalence relation. Interpretations may vary from reader to reader!
   (a) xRy if and only if x and y have family names beginning with the same letter.
   (b) xRy if and only if x is the same sex as y.
   (c) xRy if and only if x is a sibling of y.
   (d) xRy if and only if x and y are taking a course together.
   (e) xRy if and only if x and y are engaged to be married.

(f) xRy if and only if x and y are roommates.
(g) xRy if and only if x is not younger than y.

6. Give an example of
(a) a relation which is symmetric and transitive but not reflexive,
(b) a relation which is reflexive and symmetric but not transitive,
(c) a relation which is transitive and reflexive but not symmetric.

7. Construct all equivalence relations on $\{0, 1, 2\}$.

8. Match each one of these names (nonequality, less than or equal to, less than, equality, succession, parity) to one of the relations on $\{0, 1, 2\}$ following:
(a) $\{(0, 0), (1, 1), (2, 2)\}$,
(b) $\{(0, 0), (0, 1), (0, 2), (1, 1), (1, 2), (2, 2)\}$
(c) $\{(0, 1), (1, 2)\}$
(d) $\{(0, 1), (1, 0), (0, 2), (2, 0), (1, 2), (2, 1)\}$
(e) $\{(0, 0), (1, 1), (2, 2), (0, 2), (2, 0)\}$
(f) $\{(0, 1), (0, 2), (1, 2)\}$.

1.4 Quotient sets

In this section are defined quotient set and power set; the fundamental relation between equivalence relations and quotient sets is established.

A quotient set or a partition of a set is a division of the set into disjoint subsets. An example of a partition of the integers, $\mathbb{Z}$, is its separation into two sets, the even integers in one subset and the odd integers in the other. Another partition of the integers is to place each integer and its negative into a subset by themselves (see Figure 1.3).

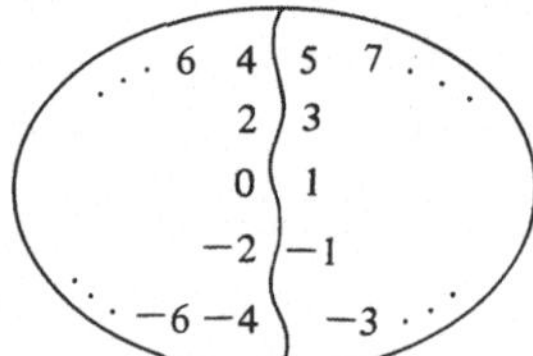

1 −1
2 −2
3 −3
0
5 4
−5 −4

Figure 1.3

We now give a formal definition of partition.

Definition. A *partition* or *quotient set* Q of a given set X is a collection of nonempty subsets of X such that every member of X is in some member of Q and the members of Q have no members in common.

The conditions of the formal definition are met by the above example of partitioning the integers into evens and odds. There is at least one even integer and at least one odd integer. Every integer is either even or odd. No integer is both even and odd.

There is a natural correspondence between the equivalence relations on a set X and the quotient sets of X.

Definition. Let R be a given equivalence relation on a set X. For any $x \in X$ define $x/R = \{y | y \in X \text{ and } yRx\}$. Define $X/R = \{x/R | x \in X\}$.

x/R is simply the subset of X of all members of X which are equivalent to the given element x. X/R is the collection of all such subsets which we will demonstrate to be a quotient set of X.

Theorem. *Let R be an equivalence relation on a set X. Then X/R is a quotient set of X.*

PROOF. We first show that two subsets of the form x/R and y/R are either disjoint or equal. If x/R and y/R fail to be disjoint then $x/R \cap y/R \neq \emptyset$. There exists $z \in X$ such that $z \in x/R$ and $z \in y/R$. zRx and zRy. xRz and zRy, xRy. We hold this result for a moment. We now propose to show $x/R = y/R$. Suppose r is any member of x/R. rRx. Now using xRy we get rRy. $r \in y/R$. $x/R \subseteq y/R$. In the same manner we show $y/R \subseteq x/R$. This gives $x/R = y/R$.

If $x \in X$ then we note the reflexivity of R implies $x \in x/R$. $x/R \in X/R$. For the same reason no subset x/R is empty. This completes proving X/R to be a quotient set. □

Corollary. *Given an equivalence relation R on a set X*

$$x/R = y/R \quad \textit{if and only if} \quad xRy.$$

PROOF. If xRy then $x \in y/R$. But $x \in x/R$. $x/R \cap y/R \neq \emptyset$. $x/R = y/R$. Conversely, if $x/R = y/R$ then $y \in x/R$ yielding yRx and xRy. □

It is convenient to have a name for the set of all subsets of a given set.

Definition. $\mathscr{P}X$, the *power set* of X, is defined to be the set of all subsets of X.

EXAMPLE. If $X = \{a, b\}$ then $\mathscr{P}X = \{\emptyset, \{a\}, \{b\}, \{a, b\}\}$. If $X = \emptyset$ then $\mathscr{P}X = \{\emptyset\}$.

Using the definition and notation of the power set we can restate the definition of a quotient set of X: Q is a quotient set of X if and only if

$Q \subseteq \mathscr{P}X$,
$A \in Q$ implies $A \neq \emptyset$,
$x \in X$ implies $x \in A$ for some $A \in Q$,
$A, B \in Q$ imply $A = B$ or $A \cap B = \emptyset$.

Questions

1. Given an equivalence relation R on X it follows that
(A) X/R has fewer members than X
(B) $x/R = y/R$ implies $x = y$
(C) $x = y$ implies $x/R = y/R$
(D) any member of X is a member of X/R.
(E) None of the possibilities is true.

2. Equality is an equivalence relation on $\mathbb{N}$. Which of the following is a member of $\mathbb{N}/=$?
(A) -2
(B) 5
(C) $\mathbb{N}$
(D) $\{6.02 \cdot 10^{23}\}$.
(E) None in the list is a member of $\mathbb{N}/=$.

3. For all $x, y \in \mathbb{Q}$ define xRy if and only if $x - y \in \mathbb{Z}$. R is an equivalence relation on $\mathbb{Q}$. Which of these statements are true?
(A) $\mathbb{Z} \in \mathbb{Q}/R$
(B) $(2/3)R(3/2)$
(C) $-1/3 \in (1/3)/R$
(D) $\{m/n | n \in \mathbb{N} \text{ and } n \neq 0 \text{ and } m \in \mathbb{N} \text{ and } m < n\} \in \mathbb{Q}/R$.
(E) None of the statements is true.

4. Which of these statements are true?
(A) $\mathscr{P}\varnothing = \varnothing$
(B) $X \subseteq \mathscr{P}Y$ implies $Y \in X$
(C) $\mathscr{P}(X \cup Y) \subseteq \mathscr{P}X \cup \mathscr{P}Y$
(D) $\mathbb{Q}^{\sim} \in \mathscr{P}\mathbb{R}$.
(E) None of the statements is true.

5. Let X be a finite, nonempty set and define on $\mathscr{P}X$ this relation: ARB if and only if $A = B$ or $A = B^{\sim}$. Which statements are true?
(A) R fails to be an equivalence relation on $\mathscr{P}X$.
(B) There are exactly two members of $(\mathscr{P}X)/R$.
(C) If X has n members then $(\mathscr{P}X)/R$ will have 2^{n-1} members.
(D) Members of $(\mathscr{P}X)/R$ are subsets of X.
(E) No statement given is true.

Exercises

1. Describe the quotient sets defined by each of the equivalence relations found in exercises 5 and 7 of Section 1.3.

2. Find $\mathscr{P}\{0, 1, 2\}$.

3. We define on $\mathbb{Z}$ the following relation: $x \sim y$ if and only if $x - y$ is divisible by 3. Is $\sim$ an equivalence relation? Show that $1/\sim\, = \{x | x = 3n + 1 \text{ for some } n \in \mathbb{Z}\}$.

4. We define T to be a relation on R as follows: xTy if and only if $x - y \in \mathbb{Q}$. Prove T is an equivalence relation. Prove $0/T = \mathbb{Q}$. Prove $\pi/T \neq \mathbb{Q}$.

1.5 Functions

In this section the ubiquitous mathematical concept of function is given a set meaning and special types of functions are considered.

Intuitively, a function is a rule that assigns to every member of a first set some member of a second set. For example, if we use as the first set $\mathbb{Z}$ and as the second set $\mathbb{N}$ we can assign to every member of $\mathbb{Z}$ its square in $\mathbb{N}$. To 3 is assigned 9, to 4 is assigned 16, to -2 is assigned 4, and so forth. If we plot these assignments on $\mathbb{Z} \times \mathbb{N}$ as ordered pairs (3, 9), (4, 16,) (-2, 4), etc., we have a graph of the function. It is this collection of ordered pairs that is the basis for a set theoretic definition of a function.

Definition. $f: X \to Y$ is a *function* from the set X to the set Y if and only if

0. $f \subseteq X \times Y$;
1. for each $x \in X$ there exists $y \in Y$ such that $(x, y) \in f$; and
2. if $(x, y_1) \in f$ and $(x, y_2) \in f$ then $y_1 = y_2$.

Condition 0 establishes the function as a set of ordered pairs selected from $X \times Y$. Condition 1 assures the existence of at least one ordered pair in the function containing any given member of the first set in the left side. Condition 2 assures that there can be at most one ordered pair with any given member of X in the left side. By abuse of language we frequently call f itself the function instead of $f: X \to Y$.

Definition. If $f: X \to Y$ is a function then X is called the *domain* of f and Y is called the *codomain* of f. If (x, y) belongs to the function f then we write $y = f(x)$ and call y the *value* of the function f at the *argument* x. We also, at times, express this in the symbolism $x \overset{f}{\rightsquigarrow} y$.

We distinguish between the different arrows. In $X \to Y$ the straight arrow goes from the domain to the codomain of f. In $x \rightsquigarrow y$ the curly arrow goes from an argument x to a value of the function. The set of all values of the function $f: X \to Y$ is the set $\{f(x) | x \in X\}$ and is called the *range* of the function. It is also written $f(X)$.

EXAMPLE. Let $X = \{a, b, c\}$ and $Y = \{d, e\}$. There are eight possible functions from X to Y in this example. Each function contains exactly three pairs.

$$
\begin{aligned}
f &= \{(a, d), (b, d), (c, d)\} \\
g &= \{(a, d), (b, d), (c, e)\} \\
h &= \{(a, d), (b, e), (c, d)\} \\
i &= \{(a, d), (b, e), (c, e)\} \\
j &= \{(a, e), (b, d), (c, d)\}
\end{aligned}
$$

$$k = \{(a, e), (b, d), (c, e)\}$$
$$l = \{(a, e), (b, e), (c, d)\}$$
$$m = \{(a, e), (b, e), (c, e)\}$$

All of the following statements are true about the above examples. $g: X \to Y$. Y is the codomain of k. X is not the codomain of j. $e = h(b)$. Range $m = \{e\}$. Range $h = Y$. $(b, e) \notin k$. $k \neq l$. $k(a) = l(a)$. $g(a) = g(b)$. $i \subseteq X \times Y$.

We now wish to single out certain special kinds of functions. First there are those functions which take all members of the codomain as values.

Definition. $f: X \to Y$ is a *surjection* if and only if

3. for each $y \in Y$ there is an $x \in X$ such that $(x, y) \in f$.

This is to say, for a surjection f we have for every $y \in Y$ some $x \in X$ such that $y = f(x)$. In other words, range f = codomain f. An older terminology still popular is to say that the function f is *onto* Y.

Second, there are those functions which do not duplicate values. This is to say, there are not two different ordered pairs (x_1, y) and (x_2, y) in the function.

Definition. $f: X \to Y$ is an *injection* if and only if

4. $(x_1, y) \in f$ and $(x_2, y) \in f$ imply $x_1 = x_2$.

Condition 4 may be alternately stated as $f(x_1) = f(x_2)$ implies $x_1 = x_2$. An injection is also called a one-to-one function.

Finally, a function with both the above delineated properties has its own name.

Definition. $f: X \to Y$ is a *bijection* if and only if f is both a surjection and an injection.

EXAMPLES. $f: \{a, b\} \to \{c, d\}$ such that $f(a) = c$ and $f(b) = d$ is a bijection. $g: \{a, b\} \to \{c, d\}$ such that $g(a) = c$ and $g(b) = c$ is not an injection because both (a, c) and (b, c) belong to g. g is also not a surjection because $d \notin$ range g.

Let $\mathbb{R}^+ = \{x | x \in \mathbb{R}$ and $x > 0\}$. Define $f: \mathbb{R}^+ \to \mathbb{R}^+$ such that $f(x) = x^2$. $f(x_1) = f(x_2)$ implies $x_1^2 = x_2^2$ which implies $x_1 = x_2$ when both x_1 and x_2 are positive. f is therefore an injection. Because every positive real number has a positive square root, f is a surjection.

Let $[0, \infty) = \{x | x \in \mathbb{R}$ and $x \geqslant 0\}$. Define $f: \mathbb{R} \to [0, \infty)$ such that $f(x) = x^2$. This function is a surjection but not an injection.

Definition. Let $f: X \to Y$ be a function. If $A \subseteq X$ then we define the *image* of A under f to be the subset of Y.

$$f(A) = \{f(x) | x \in A\}.$$

If $B \subseteq Y$ then we define the *inverse image* of B under f to be the subset of X,

$$f^{-1}(B) = \{x | f(x) \in B\}.$$

We note that the image of the domain of f, $f(X)$, is the range of f. We can also observe that the inverse image of the codomain Y, $f^{-1}(Y)$, is necessarily the domain X.

EXAMPLE. If $f: \mathbb{Z} \to \mathbb{Z}$ such that $f(x) = 9x + 2$ then $f(\{0, 1, 2\}) = \{2, 11, 20\}$ and $f^{-1}(\{1, 2, 3\}) = \{0\}$.

QUESTIONS

1. $X \times Y$
   (A) is always a function
   (B) is sometimes a function
   (C) never can be a function
   (D) is always nonempty.
(E) None of the alternatives completes a true sentence.

2. $f: X \to Y$ implies
   (A) Y is the range of f
   (B) f is a proper subset of $X \times Y$
   (C) if $(x_1, y) \in f$ and $(x_2, y) \in f$ then $x_1 = x_2$
   (D) $f \neq \varnothing$.
(E) None of the possibilities completes a true sentence.

3. Let f be a function with real arguments and real values given by the rule $f(x) = \sqrt{-x} + x^2$. Which of the following are true?
   (A) $f = \varnothing$
   (B) The largest possible domain is $\mathbb{R}^+$.
   (C) f is surjective with codomain $\mathbb{R}$.
   (D) $f(0) = 0$.
(E) None of the statements is true.

4. The function $f: \mathbb{R}^+ \to \mathbb{R}^+$ such that $f(x) = x^2$.
   (A) is not an injection
   (B) has a set of ordered pairs, f, which contains $(0, 0)$
   (C) has $f^{-1}(\mathbb{R}^+) = \mathbb{R}^+$
   (D) is not a surjection because the set of numbers of the form x^2 is properly smaller than the set of numbers of the form x.
(E) None of the possibilities completes a true sentence.

EXERCISES

1. Which of the following are injections? Which are surjections?
(a) $f: \mathbb{Z} \to \mathbb{Z}$ such that $f(x) = x + 2$
(b) $f: \mathbb{N} \to \mathbb{N}$ such that $f(x) = x + 2$
(c) $f: \mathbb{Z} \to \mathbb{Z}$ such that $f(x) = 2x$
(d) $f: \mathbb{R} \to \mathbb{R}$ such that $f(x) = 2x$

(e) $f:\mathbb{R} \to \mathbb{R}$ such that $f(x) = \sin x$
(f) $f:\mathbb{R} \to \mathbb{R}$ such that $f(x) = e^x$.

2. Consider the function $f:\mathbb{R} \to \mathbb{R}$ such that $f(x) = ax + b$. For what real numbers a, b is $f:\mathbb{R} \to \mathbb{R}$ a surjection? an injection?

3. Given $f:X \to Y$ prove $f(X) = Y$ if and only if f is a surjection.

4. Given $f:\mathbb{R} \to \mathbb{R}$ such that $f(x) = x^2$ verify all these results.

$$f(\mathbb{R}) = [0, \infty). \qquad f([0, 1]) = [0, 1].$$
$$f([2, 6]) = [4, 36]. \qquad f([-1, 1]) = [0, 1].$$
$$f^{-1}([-2, -1]) = \varnothing. \qquad f^{-1}([4, 36]) = [2, 6] \cup [-6, -2].$$
$$f^{-1}(\mathbb{R}) = \mathbb{R}. \qquad f^{-1}([0, 1]) = [-1, 1].$$
$$\mathbb{Z} \subseteq f^{-1}(\mathbb{N}). \qquad f(\mathbb{N}) \subseteq \mathbb{N}.$$
$$f^{-1}((-1, 1)) = (-1, 1).$$

5. Given $f:X \to Y$ prove $f^{-1}(B) = X$ if and only if range $f \subseteq B$.

6. Given $f:X \to Y$ prove $f(A_1 \cup A_2) = f(A_1) \cup f(A_2)$.

7. Given $f:X \to Y$ prove $f(A_1 \cap A_2) \subseteq f(A_1) \cap f(A_2)$.

8. Give an example of a function $f:X \to Y$ and two subsets A_1, A_2 of X so that $f(A_1) \cap f(A_2) \nsubseteq f(A_1 \cap A_2)$.

9. Given $f:X \to Y$ prove $f^{-1}(B_1 \cup B_2) = f^{-1}(B_1) \cup f^{-1}(B_2)$.

10. Given $f:X \to Y$ prove $f^{-1}(B_1 \cap B_2) = f^{-1}(B_1) \cap f^{-1}(B_2)$.

11. Show that $X \times (Y \times Z)$ is not equal to $(X \times Y) \times Z$ if X, Y and Z are non-empty. Show, however, there exists a bijection $\Phi:(X \times Y) \times Z \to X \times (Y \times Z)$.

1.6 Composition of functions

In this section the operation of composition of functions is defined; the identity function and results on functional inverses are established.

Definition. Given functions $f:X \to Y$ and $g:Y \to Z$ we define a function $g \circ f:X \to Z$ such that $(g \circ f)(x) = g(f(x))$ for all $x \in X$.

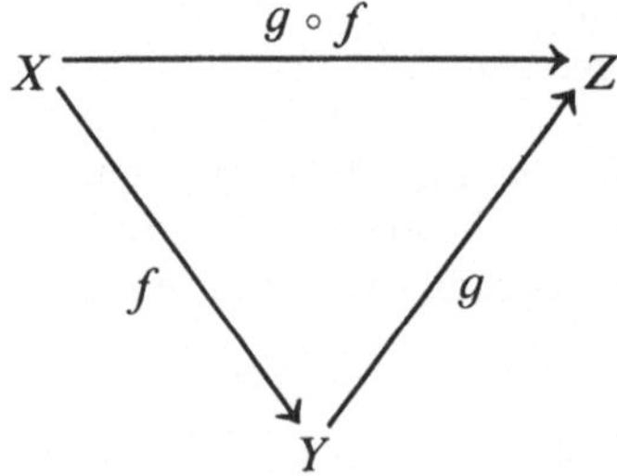

The diagram shows how the *composition* is a chaining together of the two given functions. It can be seen that the set of ordered pairs $g \circ f$ is $\{(x, z) | (x, y) \in f \text{ and } (y, z) \in g \text{ for some } y \in Y\}$. We note that in the composition $g \circ f$ it is the function f which is applied to the argument first.

EXAMPLE. If $f:\mathbb{Z} \to \mathbb{Z}$ such that $f(x) = 2x + 1$ and $g:\mathbb{Z} \to \mathbb{Z}$ such that $g(x) = -3y + 6$ then $g(f(x)) = -3(f(x)) + 6 = -3(2x + 1) + 6 = -6x + 3$. $g \circ f:\mathbb{Z} \to \mathbb{Z}$ such that $(g \circ f)(x) = -6x + 3$.

Composition preserves both injective and surjective properties of functions.

Theorem. *If $f:X \to Y$ and $g:Y \to Z$ are surjections then $g \circ f:X \to Z$ is a surjection. If $f:X \to Y$ and $g:Y \to Z$ are injections then $g \circ f:X \to Z$ is an injection.*

PROOF. Let $z \in Z$. There is an element $y \in Y$ such that $g(y) = z$. For $y \in Y$ there is an $x \in X$ such that $f(x) = y$. Therefore, given $z \in Z$ there exists an $x \in X$ such that $g(f(x)) = g(y) = z$. $g \circ f$ is a surjection.

Let $(g \circ f)(x_1) = (g \circ f)(x_2)$. $g(f(x_1)) = g(f(x_2))$. Because g is an injection we have $f(x_1) = f(x_2)$. Because f is an injection it follows that $x_1 = x_2$. This proves $g \circ f$ is an injection. □

Corollary. *If $f:X \to Y$ and $g:Y \to Z$ are bijections then $g \circ f:X \to Z$ is a bijection.*

PROOF. Combine the two results in the theorem. □

Theorem. *Composition of functions is associative.*

PROOF. Let $f:X \to Y$, $g:Y \to Z$, $h:Z \to W$ be given functions. $(h \circ (g \circ f)):X \to W$ and $((h \circ g) \circ f):X \to W$ are both functions defined by repeated composition and have the same domains and the same codomains. $(h \circ (g \circ f))(x) = h((g \circ f)(x)) = h(g(f(x))) = (h \circ g)(f(x)) = ((h \circ g) \circ f)(x)$ for all $x \in X$. □

We now explore the role of the identity function in the operation of composition of functions.

Definition. $I_X:X \to X$ such that $I_X(x) = x$ for all $x \in X$ is called the *identity function* on the set X.

Theorem. *If $f:X \to Y$ then $I_Y \circ f = f$ and $f \circ I_X = f$.*

PROOF. The following diagrams illustrate the functions involved.

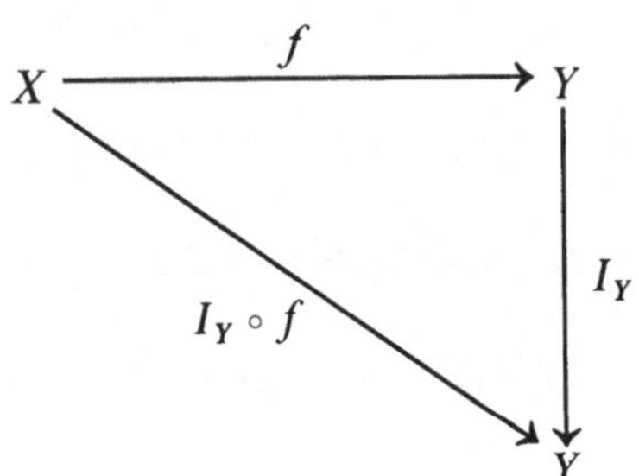

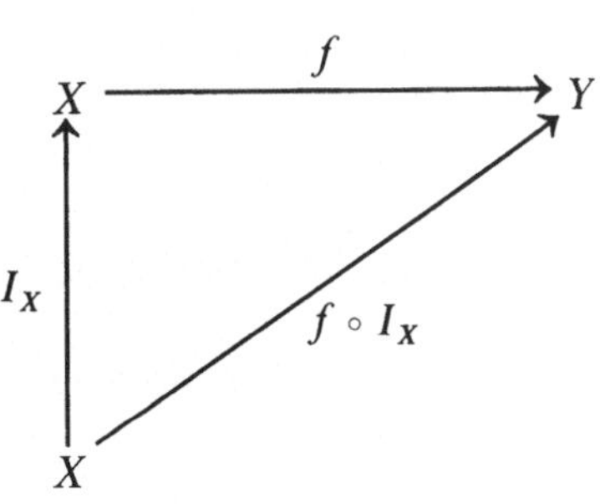

Clearly, $I_Y \circ f: X \to Y$ and $f \circ I_X: X \to Y$. It is easy to verify the equality of values. $(I_Y \circ f)(x) = I_Y(f(x)) = f(x)$ for all $x \in X$. $(f \circ I_X)(x) = f(I_X(x)) = f(x)$ for all $x \in X$. □

We consider finally in this section the more difficult problem of finding inverse functions. Given $f: X \to Y$ to find the inverse of f is to find a function $g: Y \to X$ so that the composition of f with g is the identity function.

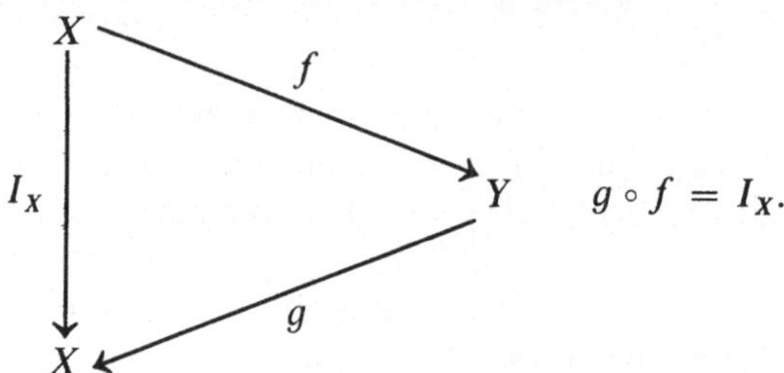

$$g \circ f = I_X.$$

Analogously, given $f: X \to Y$ we look for an $h: Y \to X$ to precede f so that $f \circ h = I_Y$.

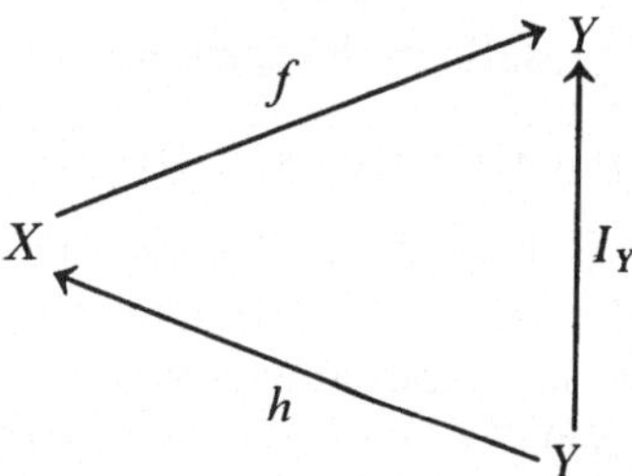

We will call g the *left inverse* of f and call h the *right inverse* of f. If, indeed, we can find some one function playing both roles ($g = h$) then we will call such a function an *inverse* of f.

EXAMPLE. Let $f: \mathbb{R} \to \mathbb{R}$ such that $f(x) = 2x - 3$. We define $g: \mathbb{R} \to \mathbb{R}$ such that $g(x) = \frac{1}{2}x + \frac{3}{2}$. Then $(g \circ f)(x) = g(f(x)) = \frac{1}{2}f(x) + \frac{3}{2} = \frac{1}{2}(2x - 3) + \frac{3}{2} = x$. $g \circ f = I_{\mathbb{R}}$. g is therefore a left inverse of f. The reader can verify that the given g is also a right inverse of f. We now proceed to the solution of the proposed problem.

Theorem. *Let $f: X \to Y$ and $X \neq \varnothing$. Then*

(a) *there exists $g: Y \to X$ such that $g \circ f = I_X$ if and only if f is an injection;*
(b) *there exists $h: Y \to X$ such that $f \circ h = I_Y$ if and only if f is a surjection; and*
(c) *there exists $k: Y \to X$ such that $k \circ f = I_X$ and $f \circ k = I_Y$ if and only if f is a bijection.*

PROOF. Suppose f is not an injection. Then there exist $x_1, x_2 \in X$ such that $x_1 \neq x_2$ and $f(x_1) = f(x_2)$. If there were a function $g: Y \to X$ such that $g \circ f = I_X$ then $(g \circ f)(x_1) = x_1$ and $(g \circ f)(x_2) = x_2$. This means $g(f(x_1)) \neq g(f(x_2))$. But $f(x_1) = f(x_2)$. This contradicts condition 2 in the definition of a function for g. There can be therefore no function $g: Y \to X$ such that $g \circ f = I_X$.

Next suppose f is not a surjection. There exists $y \in Y$ such that $f(x) \neq y$ for any $x \in X$. If there were a function $h: Y \to X$ such that $f \circ h = I_Y$ we would have $(f \circ h)(y) = y$. But then $h(y)$ is an element of X such that $f(h(y)) = y$. This contradicts the second sentence of this paragraph. There can be no function $h: Y \to X$ such that $f \circ h = I_Y$.

Suppose f is not a bijection. Then either f is not an injection or f is not a surjection. The statement that there exists $k: Y \to X$ such that $f \circ k = I_Y$ and $k \circ f = I_X$ is impossible.

We begin now the proofs of the three converses by assuming f to be an injection. We form the set $\{(y, x) | (x, y) \in f\}$ by reversing each of the pairs in f. Calling this set f^r we notice that if $(y, x_1) \in f^r$ and $(y, x_2) \in f^r$ then $x_1 = x_2$ because f is an injection. Furthermore, for each $y \in \text{range } f$ there exists $x \in X$ such that $(y, x) \in f^r$. We have f^r: range $f \to X$. f^r will not, in general, qualify as a function with domain Y because range f does not, in general, equal Y. We must therefore enlarge the set of ordered pairs f^r to include pairs with members of Y in the left side for every member of Y. For each $y \in Y - \text{range } f$ adjoin the pair (y, a) to the set f^r and call the result g, where a is some fixed member of X.

$$g = f^r \cup \{(y, a) | y \in (Y - \text{range } f)\}.$$

Now g is a function like f^r yet has domain Y and $g(f(x)) = x$ for all $x \in X$.

We now begin with a surjection $f: X \to Y$ and demonstrate the existence of a right inverse for f. We again form f^r from f by assembling all the reversed ordered pairs from f. For each $y \in Y$ there is at least one ordered pair in f^r with y in the left side because f is a surjection. There may be, however, several such ordered pairs for some $y \in Y$ so that f^r may not satisfy the condition 2 of uniqueness of value to be a function. We must then delete from f^r all pairs except for one pair containing a given y in the left side. We denote the resulting set of ordered pairs by h. h is a function from Y to X. Furthermore, $(f \circ h)(y) = f(h(y)) = y$ for all $y \in Y$.

Finally, we begin with a bijection $f: X \to Y$. Looking back to the part of this proof dealing with the injection we are dealing with the situation where $Y = \text{range } f$. $Y - \text{range } f = \varnothing$. The function g is therefore simply f^r which serves as the left inverse. Looking at the part of this proof dealing with the surjection again f^r is a function without any necessity of deleting ordered pairs to produce h. This is true, of course, because if f is an injection then $(y, x_1) \in f^r$ and $(y, x_2) \in f^r$ imply $x_1 = x_2$. Thus f^r is a right inverse of f. $f \circ f^r = I_Y$ and $f^r \circ f = I_X$. □

In the interest of simplicity we have avoided any detailed discussion in the previous proof of the mechanism of deleting pairs from the set f^r.

Questions

1. Which of the following alternatives complete a false sentence? If $f:X \to Y$ and $g:Y \to Z$ are functions then the composite function $g \circ f$
  (A) has domain X
  (B) has codomain Z
  (C) has range a subset of range g
  (D) has domain a subset of domain f.
(E) All of the alternatives make true sentences.

2. If $f:X \to Y$ and $g:Y \to Z$ are given then the composite function $g \circ f$
  (A) has domain X and codomain Z
  (B) contains the same number of ordered pairs as f
  (C) has range a subset of Z
  (D) can be empty.
(E) None of the alternatives completes a true sentence.

3. If $f:[0, \infty) \to \mathbb{R}$ such that $f(x) = (x)^{1/4}$ and $g:\mathbb{R} \to [0, \infty)$ such that $g(x) = x^4$ then
  (A) g is a left inverse of f but not a right inverse
  (B) g is a right inverse of f but not a left inverse
  (C) g is the inverse of f
  (D) $g \circ f$ exists but is not the identity.
(E) None of the alternatives completes a true sentence.

4. If $f(x) = x^2 - 3x$ and $g(x) = \sqrt{x}$ then the largest possible subset of the reals which can serve for domain f and have the composite $g \circ f$ defined is
  (A) $\mathbb{R}$
  (B) $[0, \infty)$
  (C) $(-\infty, 0] \cup [3, \infty)$
  (D) $\mathbb{R}^+$.
(E) No choice completes a true sentence.

5. g is the inverse of $f:X \to Y$
  (A) if and only if $g \circ f = I_X$
  (B) implies X and Y have the same number of elements
  (C) implies $g \circ f = \varnothing$
  (D) if and only if $g \circ f = X$.
(E) None of the choices completes a true sentence.

Exercises

1. Find two functions $f:\mathbb{R} \to \mathbb{R}$, $g:\mathbb{R} \to \mathbb{R}$ such that $f \circ g \neq g \circ f$.

2. Find two distinct functions $f:\mathbb{R} \to \mathbb{R}$, $g:\mathbb{R} \to \mathbb{R}$ such that $f \circ g = g \circ f$.

3. Let $f:\mathbb{R} \to \mathbb{R}$ such that $f(x) = 5x - 3$. Apply the theorem of this section to prove there exists $g:\mathbb{R} \to \mathbb{R}$ such that $g \circ f = I_{\mathbb{R}}$. Is it also true that $f \circ g = I_{\mathbb{R}}$? Find g.

4. Let $f:R \to R$ such that $f(x) = x^3$. Does there exist g or h such that $g \circ f = I_{\mathbb{R}}$ or $f \circ h = I_{\mathbb{R}}$? Which theorem applies? Find g or h if it exists.

5. Let $f:\mathbb{Z} \to \mathbb{Z}$ such that $f(n) = 2n$. Does there exist g or h such that $g \circ f = I_{\mathbb{Z}}$ or $f \circ h = I_{\mathbb{Z}}$? Which theorem applies? Find g or h if it exists.

6. Using the definition $g \circ f = \{(x, z) | (x, y) \in f \text{ and } (y, z) \in g \text{ for some } y \in Y\}$ for the composition of functions $f:X \to Y$ and $g:Y \to Z$ prove
   (a) $g \circ f \subseteq X \times Z$
   (b) for each $x \in X$ there exists $z \in Z$ such that $(x, z) \in g \circ f$
   (c) if $(x, z_1) \in g \circ f$ and $(x, z_2) \in g \circ f$ then $z_1 = z_2$.
   What do you conclude from these three conditions?

7. Using the definition of composition given in exercise 6 prove the two sets of ordered pairs $(h \circ g) \circ f$ and $h \circ (g \circ f)$ are equal for any given $f:X \to Y, g:Y \to Z, h:Z \to W$.

8. Prove this degenerate case of a theorem found in this section. Let $f:X \to Y$ and $X = \varnothing$. Prove there exists $k:Y \to X$ such that $k \circ f = I_X$ and $f \circ k = I_Y$ if and only if f is a bijection.

1.7 A factorization of a function

In this section we make some observations about relationships between injections and subsets and between surjections and quotient sets. We prove a factorization theorem for functions and entitle it the fundamental morphism theorem for sets.

Given any injection $q:A \to X$ there is defined a subset $q(A)$ of X which is the range of Q. This subset $q(A)$ of X and the set A are in one-to-one correspondence; this is to say, $q:A \to q(A)$ is a bijection (see Figure 1.4). Conversely, if we are given a set X and a subset A of X, then the identity function $I_A:A \to X$ is an injection from A to X.

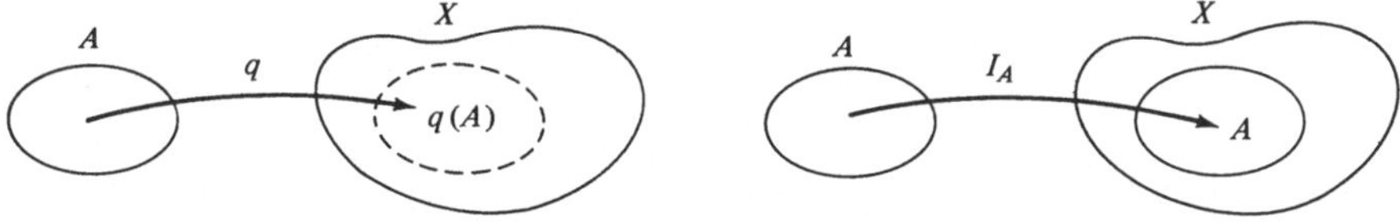

Figure 1.4

Definition. An injection $q:A \to X$ is sometimes called an *embedding* of A into X.

Given any surjection $\varphi:X \to B$ there is defined a quotient set X/R of X as follows:

1. ($x_1 R x_2$ if and only if $\varphi(x_1) = \varphi(x_2)$) defines an equivalence relation R on X
2. the equivalence relation R yields a quotient set X/R such that $x/R = \{z | \varphi(z) = \varphi(x)\}$.

In other words, the equivalence set x/R of the quotient set X/R consists of all those arguments in X which have the same value by the function φ.

Definition. By the equivalence relation and quotient set associated with a surjection $\varphi: X \to B$ we mean the relation $R = \{(x_1, x_2) | \varphi(x_1) = \varphi(x_2)\}$ and the quotient set X/R (see Figure 1.5).

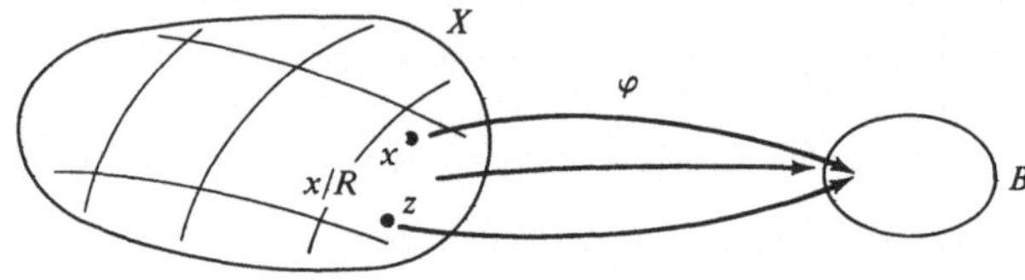

Figure 1.5

Conversely, associated with each quotient set there is a surjection. If a quotient set Q of a set X be given we can define a function $\varphi: X \to Q$ in which $\varphi(x)$ is the nonempty subset in Q which contains x. Since every x must belong to some subset, $\varphi(x)$ always exists. Since each subset is nonempty φ must be surjective, we call such a surjection a quotient map.

Definition. A surjection $\varphi: X \to Q$ such that Q is a quotient set of X and such that $x \in \varphi(x)$ for all $x \in X$ is called a *quotient map* from X to Q (see Figure 1.6).

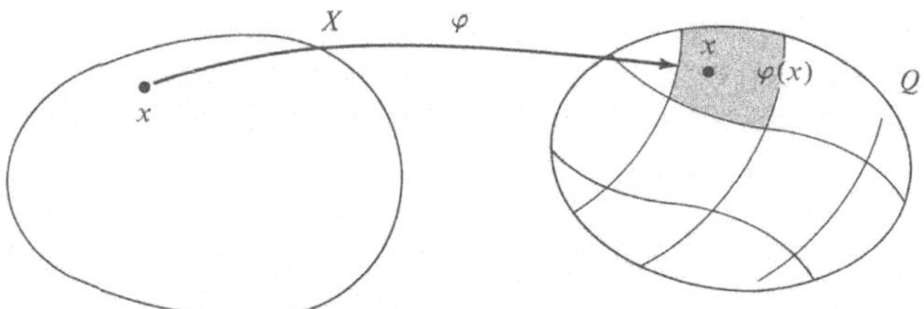

Figure 1.6

EXAMPLE. Let the set $\mathbb{Z}$ be given and Q be the partition of $\mathbb{Z}$ into two subsets, the odds and the evens. A quotient map $\varphi: \mathbb{Z} \to Q$ is then the surjection that takes any integer into the subset of all odds if it is odd and into the subset of all evens if it is even.

We now prove the theorem which is the main result of this section, that any function whatsoever can be factored into the composition of a surjection and an injection.

Theorem. *Given a function $f: X \to Y$ there exist*

an equivalence relation γ on X;
a surjection $\varphi: X \to X/\gamma$; and
an injection $f': X/\gamma \to Y$ such that $f = f' \circ \varphi$.

PROOF

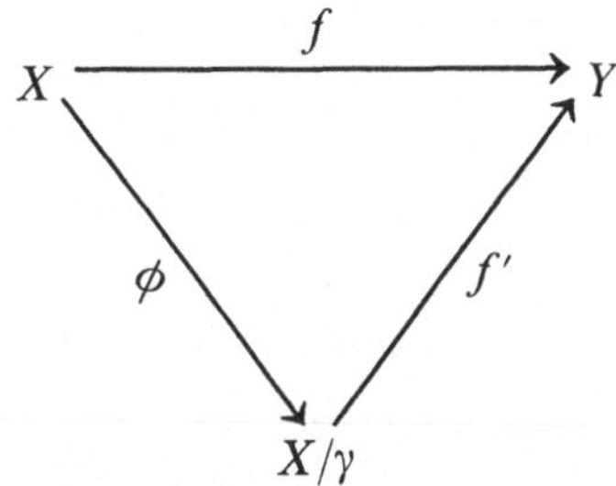

In terms of the function f we define a relation γ on $X: x_1 \gamma x_2$ if and only if $f(x_1) = f(x_2)$. As discussed earlier in this section, such a relation is an equivalence relation on X. Associated with the equivalence relation γ is the quotient set $X/\gamma = \{x/\gamma | x \in X\}$ in which $x/\gamma = \{z | z \in X \text{ and } z\gamma x\}$. We now define $\varphi: X \to X/\gamma$ such that $\varphi(x) = x/\gamma$. This function φ is a surjection because if $w \in X/\gamma$ then $w = x/\gamma$ for some $x \in X$. Then $\varphi(x) = x/\gamma = w$. The surjection φ is a quotient map.

We next define $f': X/\gamma \to Y$ such that $f'(x/\gamma) = f(x)$. f' is a well-defined function and an injection because $x_1/\gamma = x_2/\gamma$ if and only if $x_1 \gamma x_2$ if and only if $f(x_1) = f(x_2)$ if and only if $f'(x_1/\gamma) = f'(x_2/\gamma)$. Finally, we verify the equation $f = f' \circ \varphi$. $(f' \circ \varphi)(x) = f'(\varphi(x)) = f'(x/\gamma) = f(x)$ for all $x \in X$. □

Corollary. *Given a function $f: X \to Y$ there exists a bijection from a quotient set of X to the set $f(X)$.*

PROOF. The theorem asserts the existence of a quotient set of X and an injection $f': X/\gamma \to Y$. If the same set of ordered pairs f' is regarded as a function from X/γ to $f(X)$ then f' becomes a surjection as well as an injection. Thus we have a bijection $f': X/\gamma \to f(X)$. □

EXAMPLE. Suppose we consider the function $f: \mathbb{Z} \to \mathbb{Z}$ such that $f(x) =$ the remainder upon dividing x by 3 (see Figure 1.7). Then $x_1 \gamma x_2$ if and only if $f(x_1) = f(x_2)$ if and only if x_1 and x_2 have the same remainder upon dividing by 3 if and only if $x_1 - x_2$ is a multiple of 3. Any integer x will be γ-equivalent to precisely one of the three numbers 0, 1, and 2. This is to say $X/\gamma = \{0/\gamma, 1/\gamma, 2/\gamma\}$. $\varphi(x) = x/\gamma$. $f': X/\gamma \to \mathbb{Z}$ such that $f'(0/\gamma) = f(0) = 0$, $f'(1/\gamma) = f(1) = 1$, and $f'(2/\gamma) = f(2) = 2$. f' is an injection from $\{0/\gamma, 1/\gamma, 2/\gamma\}$ to $\mathbb{Z}$. According to the corollary, f' is a bijection from $\{0/\gamma, 1/\gamma, 2/\gamma\}$ to range $f = \{0, 1, 2\}$.

The analogue of the previous theorem will reappear later with various algebraic structures such as rings, groups, and vector spaces. When presented in an algebraic setting, the theorem—or more especially, the corollary—is sometimes called the fundamental theorem of (homo)morphisms or the law of (homo)morphism. We shall refer to the theorem of this section as the fundamental morphism theorem for sets.

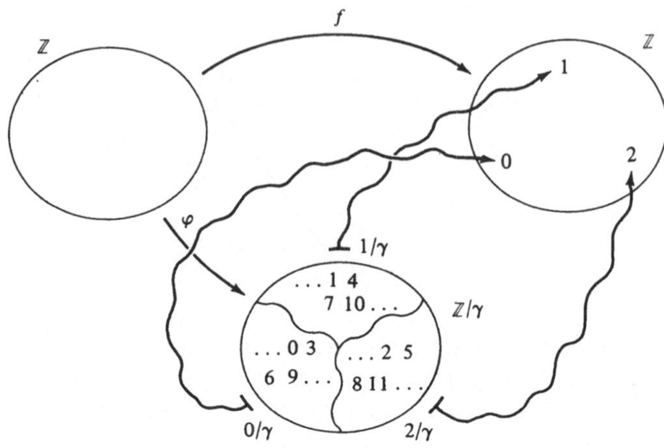

Figure 1.7

QUESTIONS

1. Which of these alternatives are false?
   A quotient map $\varphi: X \to Q$
   (A) has codomain Q a quotient set of X
   (B) has range φ = codomain φ
   (C) is not necessarily an injection
   (D) is necessarily a surjection.
   (E) None of the alternatives make a false sentence.

2. Given a function $f: X \to Y$
   (A) there exists a surjection $\varphi: X \to Y$
   (B) there exist an injection $f': X/\gamma \to Y$ and surjection $\varphi: X \to X/\gamma$ such that $f \circ \varphi = f'$ for some equivalence relation γ on X
   (C) there exists an injection $f': X/\gamma \to Y$ for some equivalence relation γ on X
   (D) there exists an injection $g: X \to Y$.
   (E) All of the alternatives complete a false sentence.

3. $I: \mathbb{N} \to \mathbb{Z}$ such that $I(x) = x$ for all $x \in \mathbb{N}$
   (A) is called an embedding
   (B) is called a quotient map
   (C) is not an injection
   (D) implies $I^{-1}(-12) = \{12\}$.
   (E) None of the alternative completes a true sentence.

4. Let B be the set of all Bucknell University undergraduate students and A be the set of all letters of the English alphabet. Let $f: B \to A$ such that $f(x)$ = the initial letter of x's last name. Furthermore, assume there are Bucknell undergraduate students named Leslie Jones and Dane Johnson. Which of the following statements are true?
   (A) f' (Leslie Jones) = f' (Dane Johnson)
   (B) f (Leslie Jones) $\in J$
   (C) Leslie Jones $\in \varphi^{-1}(J)$
   (D) Leslie Jones $\in f'^{-1}(J)$.
   (E) None of the statements is true.

1.8 The symmetric group

In this section we consider the collection of all functions on a set X and as a special case, the collection of all bijections on a set X, the symmetric group of X.

We begin with a class of functions.

Definition. If X and Y are sets we denote the collection of all functions from X to Y by Y^X.

$$Y^X = \{f | f:X \to Y\}.$$

We now collect some earlier results that apply to X^X, the set of all functions on the set X.

Theorem. *Composition is an associative operation on* X^X.

PROOF. The composition of two functions $f:X \to X$ and $g:X \to X$ is a function $g \circ f:X \to X$ which again is a member of X^X. The composition of two functions is associative. □

EXAMPLE. Composition is not necessarily commutative. $f:\mathbb{R} \to \mathbb{R}$ such that $f(x) = x + 1$ and $g:\mathbb{R} \to \mathbb{R}$ such that $g(x) = 2x$ are both members of $\mathbb{R}^\mathbb{R}$. $(g \circ f)(x) = 2x + 2$ whereas $(f \circ g)(x) = 2x + 1$. $g \circ f \neq f \circ g$.

Theorem. *I_X, the identity function on X, satisfies the equations $I_X \circ f = f \circ I_X = f$ for all functions $f \in X^X$. $f \in X^X$ is invertible if and only if f is a bijection.*

PROOF. These results are established in Section 1.6. □

We describe the property $I_X \circ f = f \circ I_X = f$ for all $f \in X^X$ by saying I_X is a neutral element for composition of functions. The result on invertibility of functions shows us that if we desire to have all functions have inverses we must limit ourselves to bijections. We now do just that. We use the symbol $\mathfrak{S}(X)$, an upper case German ess, for the set of all bijections of the set X.

Definition. $\mathfrak{S}(X) = \{f | f \in X^X$ and f is a bijection$\}$.

Because the composition of two bijections is a bijection, I_X is a bijection, and the inverse of a bijection is a bijection, we have this theorem:

Theorem. *Composition is an associative operation on $\mathfrak{S}(X)$, I_X is a neutral element for composition on $\mathfrak{S}(X)$, and every f in $\mathfrak{S}(X)$ is invertible.*

A set with the properties listed in the theorem is called a group.

Definition. A *group* is a set G together with an associative operation for which there exists a neutral element and every element of G is invertible.

$\mathfrak{S}(X)$ with the operation of composition is called the *symmetric group* of the set X.

The elements of the symmetric group are functions; in particular, they are bijections. Frequently, and especially if X is finite, bijections are called *permutations.* Not all groups are constructed with functions and composition as in the symmetric group. For example, the integers $\mathbb{Z}$ together with the operation of addition make a group with neutral element 0. We have been led to the concept of group at this time through our discussions of functions. We will later return to a study of groups in the broader sense.

EXAMPLE. We work out here the example of the symmetric group on the set $\{1, 2, 3\}$. We abbreviate $\mathfrak{S}(\{1, 2, 3\})$ with $\mathfrak{S}_3$. The set $\{1, 2, 3\}$ has the following bijections

$$\begin{array}{llll}
\sigma_1 \text{ such that } & \sigma_1(1) = 1, & \sigma_1(2) = 2, & \sigma_1(3) = 3, \\
\sigma_2 \text{ such that } & \sigma_2(1) = 1, & \sigma_2(2) = 3, & \sigma_2(3) = 2, \\
\sigma_3 \text{ such that } & \sigma_3(1) = 2, & \sigma_3(2) = 3, & \sigma_3(3) = 1, \\
\sigma_4 \text{ such that } & \sigma_4(1) = 3, & \sigma_4(2) = 2, & \sigma_4(3) = 1, \\
\sigma_5 \text{ such that } & \sigma_5(1) = 3, & \sigma_5(2) = 1, & \sigma_5(3) = 2, \\
\sigma_6 \text{ such that } & \sigma_6(1) = 2, & \sigma_6(2) = 1, & \sigma_6(3) = 3.
\end{array}$$

$\mathfrak{S}_3 = \{\sigma_1, \sigma_2, \sigma_3, \sigma_4, \sigma_5, \sigma_6\}$. The operation is composition. We verify, for example, that $\sigma_3 \circ \sigma_2 = \sigma_6$:

$$\begin{aligned}
(\sigma_3 \circ \sigma_2)(1) &= \sigma_3(\sigma_2(1)) = \sigma_3(1) = 2. \\
(\sigma_3 \circ \sigma_2)(2) &= \sigma_3(\sigma_2(2)) = \sigma_3(3) = 1. \\
(\sigma_3 \circ \sigma_2)(3) &= \sigma_3(\sigma_2(3)) = \sigma_3(2) = 3.
\end{aligned}$$

We tabulate all possible compositions in a composition table. The answers are worked out just as we found $\sigma_3 \circ \sigma_2 = \sigma_6$.

$\circ$	σ_1	σ_2	σ_3	σ_4	σ_5	σ_6
σ_1	σ_1	σ_2	σ_3	σ_4	σ_5	σ_6
σ_2	σ_2	σ_1	σ_4	σ_3	σ_6	σ_5
σ_3	σ_3	σ_6	σ_5	σ_2	σ_1	σ_4
σ_4	σ_4	σ_5	σ_6	σ_1	σ_2	σ_3
σ_5	σ_5	σ_4	σ_1	σ_6	σ_3	σ_2
σ_6	σ_6	σ_3	σ_2	σ_5	σ_4	σ_1.

Note that the neutral element for the set $\mathfrak{S}_3$ is σ_1, the identity function on the set $\{1, 2, 3\}$.

QUESTIONS

1. For a given set X we construct the power set $\mathscr{P}X$. Union ($\cup$) is an operation on $\mathscr{P}X$. Which of the following statements are true?
   (A) There is no neutral element in $\mathscr{P}X$.
   (B) Union fails to be associative.
   (C) Some elements of $\mathscr{P}X$ have no inverses in $\mathscr{P}X$.
   (D) X is a neutral element.
   (E) None of the four statements is correct.

2. Let L be the following set of functions: $\{f \mid f:\mathbb{R} \to \mathbb{R}$ and $f(x) = ax + b$ for some $a, b \in \mathbb{R}, a \neq 0\}$. Which of the following statements are correct?
   (A) L together with composition is a group with neutral element $I_{\mathbb{R}}$.
   (B) L together with composition is not a group because $g:\mathbb{R} \to \mathbb{R}$ such that $g(x) = x^3$ is a bijection and is not in L.
   (C) L together with composition is not a group because $h:\mathbb{R} \to \mathbb{R}$ such that $h(x) = 5$ is not a member of L.
   (D) L together with composition is not a group because there exist functions $f, g \in L$ such that $g \circ f \neq f \circ g$.
   (E) None of the statements is correct.

3. Which of the following statements about $\mathfrak{S}_3$ are correct?
   (A) $\sigma_4 \circ \sigma_3 = \sigma_6$.
   (B) σ_3 is the inverse of σ_5.
   (C) σ_4 is its own inverse.
   (D) σ_1 has no inverse.
   (E) None of the statements is correct.

4. $\mathbb{Z} - \{0\}$ together with multiplication is not a group because
   (A) 0 is absent
   (B) the product of odd numbers is always odd
   (C) the product of even numbers is not always even
   (D) inverses with respect to multiplication are not present.
   (E) $\mathbb{Z} - \{0\}$ is a group.

EXERCISES

1. In the symmetric group $\mathfrak{S}_3$ find the inverse of σ_3.

2. In the symmetric group $\mathfrak{S}_3$ solve these equations for τ:

$$\sigma_5 \circ \tau = \sigma_2, \quad \tau \circ \sigma_5 = \sigma_2, \quad \sigma_3 \circ \tau \circ \sigma_5 = \sigma_4 \circ \sigma_6, \quad \sigma_6 \circ \tau \circ \tau = \sigma_4, \quad \tau \circ \tau = \sigma_4.$$

3. Verify that each of the following subsets of the symmetric group $\mathfrak{S}_3$ are themselves groups (see that the composition of two elements in the subset is again in the subset, the subset contains inverses of each element in the subset, and the subset contains the neutral element):

$$\{\sigma_1, \sigma_3, \sigma_5\}, \quad \{\sigma_1, \sigma_6\},$$
$$\{\sigma_1, \sigma_4\}, \quad \{\sigma_1, \sigma_2\},$$
$$\{\sigma_1\}, \quad \{\sigma_1, \sigma_2, \sigma_3, \sigma_4, \sigma_5, \sigma_6\}.$$

4. Analyze the group $\mathfrak{S}_2$ as $\mathfrak{S}_3$ is done in the text.

5. If $X = \{1, 2, 3\}$ and $Y = \{1, 2\}$ construct X^Y.

6. Show that all functions $f: \mathbb{Q} \to \mathbb{Q}$ of the form $f(x) = ax + b$, a, $b \in \mathbb{Q}$, $a \neq 0$, make a group which is a subset of the group $\mathfrak{S}(\mathbb{Q})$. Prove there exists a member of $\mathfrak{S}(\mathbb{Q})$ which is not in the subset.

Rings: Basic theory 2

Chapters 2, 3, 4, and 5 compose the part of the book devoted to rings. In Chapter 2 the basic structure of the ring is explored using set theoretic concepts in such a manner that we can stress analogies when developing other algebraic structures. In Chapter 3 we develop the natural number system and the integers, and in Chapter 4 we use the natural number system and the integers in a further study of rings including a development of the rational numbers. Chapter 5 treats the ring of polynomials, some special rings, and factorization, field extensions, and complex numbers.

We begin this chapter with a discussion of operations, neutral elements, and other fundamental concepts of algebra. In this chapter we develop the concepts of ring, subring, quotient ring, and morphism based upon the corresponding set structures of set, subset, quotient set, and function. This organization holds throughout this book for the purpose of easing the learning for the student. Also included in this chapter are descriptions of special rings such as integral domains and fields.

A ring is defined in Section 2.1 as a listing $\langle R, +, \cdot, \theta\rangle$ in which R is a set, $+$ and $\cdot$ are binary operations, and θ is a neutral element for addition. Some students and practitioners of algebra may find our use of this listing too formal, too repetitious, or simply too much bother to write. Our advice in this regard is not to write the full listing $\langle R, +, \cdot, \theta\rangle$ but simply to write, "R is a ring," letting $+$ and $\cdot$ and θ be understood from context. It is an advantage for the author to be specific about the ring being a set together with the operations, and a necessity for the student to understand this, but certainly a person should not feel obligated to write out such an expression every time he wishes to discuss a ring. There are times in this text when it *is* important to be quite specific about the involved operations in an algebraic system; by having the notation and hopefully by having the reader prepared for such a point of view we hope to make our points better. The longer notation does make clearer to the learner the true nature of the ring.

2.1 Binary operations

We have, through the composition of functions, introduced in Section 1.8 the concept of a group. Rather than discuss the group more fully at this time, we move to the more familiar: the algebraic system with two binary operations. Our number systems, the natural numbers, the integers, the rational numbers, the real numbers, and the complex numbers are all examples of sets with two binary operations consistently called addition and multiplication. These operations enjoy various properties such as associativity and commutativity. There are other properties to be considered such as the existence of neutral elements and inverses. We will begin our study of algebraic systems by analyzing the familiar, gradually increasing our level of abstraction and sharpening our tools of analysis. It is through higher levels of abstraction that we find the most aesthetically satisfying organization of mathematics. A great strength of mathematics is its ability to cast aside the irrelevant and concentrate upon the very essence of any phenomenon. That the symmetries of art, the permutations of gamblers, and the behavior of quanta have identical aspects is one of the surprises of modern algebra.

We begin now with a study of the set of integers, $\mathbb{Z}$. We do this because of its present and historical importance, its central location in group and ring theory, and its relative simplicity for beginners.

We make some observations about the integers and their two familiar binary operations, addition and multiplication.

Addition. For any $x, y \in \mathbb{Z}$ there exists a unique $z \in \mathbb{Z}$ such that $x + y = z$.

Multiplication. For any $x, y \in \mathbb{Z}$ there exists a unique $z \in \mathbb{Z}$ such that $x \cdot y = z$.

Addition and multiplication are binary operations on the set $\mathbb{Z}$ and are therefore functions

$$\alpha : \mathbb{Z} \times \mathbb{Z} \to \mathbb{Z} \quad \text{such that} \quad \alpha(x, y) = x + y$$
$$\mu : \mathbb{Z} \times \mathbb{Z} \to \mathbb{Z} \quad \text{such that} \quad \mu(x, y) = x \cdot y \text{ (or } xy).$$

For example, $\alpha(3, 4) = 7$ and $\mu(3, 4) = 12$. The customary way of denoting a value of the addition function is to place the sign $+$ between the two arguments giving $x + y$. If we wish, however, to emphasize that we are dealing with a function and wish to disassociate ourselves from our prejudices concerning the behavior of $+$ we use α.

By way of recapitulation we state

Definition. A *binary operation* β on a set S is a function $\beta : S \times S \to S$. The value of β is written $x\beta y$ or $\beta(x, y)$.

Definition. A binary operation $\beta:S \times S \to S$ is *associative* if and only if $x\beta(y\beta z) = (x\beta y)\beta z$ for all $x, y, z \in S$.

Definition. A binary operation $\beta:S \times S \to S$ is *commutative* if and only if $x\beta y = y\beta x$ for all $x, y \in S$.

These last two conditions written as functions of two variables, as in calculus courses, are

$$\beta(x, \beta(y, z)) = \beta(\beta(x, y), z) \quad \text{for all } x, y, z \in S,$$
$$\beta(x, y) = \beta(y, x) \quad \text{for all } x, y \in S.$$

We proceed with other properties.

Definition. The binary operation $\beta:S \times S \to S$ has a *neutral element* v in S if and only if $x\beta v = v\beta x = x$ for all $x \in S$.

EXAMPLES. For $\mathbb{Z}$ both α and μ are associative and commutative. 0 is a neutral element for α on $\mathbb{Z}$. 1 is a neutral element for μ on $\mathbb{Z}$. In the set X^X, the collection of all functions from a given set X to itself, composition is associative, but not commutative, and I_X is the neutral element.

In some specific situations neutral elements will have special names such as zero, one, unity, identity.

Theorem. *There can be at most one neutral element for any binary operation on a set S.*

PROOF. Suppose v' and v'' are both neutral elements for some $\beta:S \times S \to S$. $v'\beta x = x$ for all $x \in S$ and therefore in particular $v'\beta v'' = v''$. $x\beta v'' = x$ for all $x \in S$ and in particular $v'\beta v'' = v'$. Comparing the two results we have $v' = v''$. □

A neutral element for any operation called addition is usually called a zero element. A neutral element for an operation called multiplication we shall call a unity. The word *identity* is used by many authors for this neutral element but we shall reserve the term identity for the identity function. Even though the usage of these terms varies considerably from author to author, it is usually evident from context what is meant.

Definition. Given a set S and a binary operation $\beta:S \times S \to S$ with neutral element v we say that an element x of S is *invertible* if and only if there exists an element y of S such that $x\beta y = y\beta x = v$. Moreover, if there is such an element y it is called the *inverse* of x.

EXAMPLES. For $\mathbb{Z}$, $+$, 0 we see -2 to be the inverse of 2. For $\mathbb{Q}$, $\cdot$, 1 we see $\frac{1}{2}$ to be the inverse of 2. For $\mathbb{Q}^{\mathbb{Q}}$, $\circ$, $I_{\mathbb{Q}}$ we see f such that $f(x) = 2x - 3$ to be the inverse of g such that $g(x) = \frac{1}{2}x + \frac{3}{2}$.

Theorem. *Let $\beta: S \times S \to S$ be an associative binary operation on a set S with neutral element v. If x is an invertible element of S then its inverse is unique.*

PROOF. We let both y' and y'' be inverses of x in S. $y' = y'\beta v = y'\beta(x\beta y'') = (y'\beta x)\beta y'' = v\beta y'' = y''$. □

The inverse of any element x with respect to the operation of addition is called its negative and denoted by $-x$. $-x$ means the negative of x. The symbol can be repeated; $-(-x)$ means the negative of the negative of x. The inverse of x with respect to multiplication we shall denote by x^-. We shall also use x^- to denote the inverse of x with respect to general operations such as β. The notation x^- will eventually yield to x^{-1} but this will await the introduction of exponents.

Theorem. *If x is any invertible element of a set S with a binary associative operation β with neutral element v then $(x^-)^- = x$.*

PROOF. x is invertible means there exists y in S such that $x\beta y = y\beta x = v$. It is clear from the symmetry of the equations that anytime y is the inverse of x then x is also the inverse of y. Since y is the inverse of x we have x is the inverse of the inverse of x. $x = (x^-)^-$. We have from an earlier theorem that the inverse of the inverse of x is unique. □

We observe that if the binary operation is addition and v is the neutral element 0 then the theorem states $-(-x) = x$.

Theorem. *If x and y are both invertible elements of a set S with an associative binary operation β having a neutral element v then $x\beta y$ is also invertible and its inverse is $y^-\beta x^-$.*

PROOF. Since it is given that both x and y are invertible the element $y^-\beta x^-$ belongs to S. $(x\beta y)\beta(y^-\beta x^-) = ((x\beta y)\beta y^-)\beta x^- = (x\beta(y\beta y^-))\beta x^- = (x\beta v)\beta x^- = x\beta x^- = v$. We can likewise show $(y^-\beta x^-)\beta(x\beta y) = v$. $y^-\beta x^-$ is therefore the unique inverse of $x\beta y$. □

We point out that the commutativity of β is not used in the previous theorem. In additive notation the previous theorem states $-(x + y) = (-y) + (-x)$ and in multiplicative notation the theorem reads $(xy)^- = y^-x^-$. In the conventional use of addition it is assumed to be a commutative operation which permits the result $-(x + y) = (-y) + (-x) = (-x) + (-y)$.

The existence of an inverse for an element x implies that element x is cancellable.

Theorem. *Let β be an associative operation on a set S with a neutral element v. If x is an invertible element of S then*

$$x\beta y = x\beta z \quad \textit{implies } y = z$$
$$y\beta x = z\beta x \quad \textit{implies } y = z$$

for any $y, z \in S$.

PROOF. We prove only one of the statements leaving the other to the reader. Suppose $x\beta y = x\beta z$. $x^{-}\beta(x\beta y) = x^{-}\beta(x\beta z)$. $(x^{-}\beta x)\beta y = (x^{-}\beta x)\beta z$. $v\beta y = v\beta z$. $y = z$. □

It is completely possible to have a cancellation theorem for a set and binary operation without the existence of inverses. For example, in $\mathbb{Z}$ and multiplication we have $2y = 2z$ implies $y = z$ yet 2 is not an invertible element of $\mathbb{Z}$.

In the integers the two operations of addition and multiplication are interrelated by means of a condition called *distributivity*:

$$x(y + z) = xy + xz \quad \text{for all } x, y, z \in Z$$
$$(y + z)x = yx + zx \quad \text{for all } x, y, z \in Z.$$

We actually call the first *left distributivity* and the second *right distributivity*. For a commutative multiplication each implies the other. We say in the case of the integers that multiplication is distributive with respect to addition. Note that the relation between addition and multiplication is not a symmetric one; addition fails to be distributive with respect to multiplication. $3 + (2 \cdot 1) = 5$ whereas $(3 + 2) \cdot (3 + 1) = 20$.

QUESTIONS

1. The fact that $x \cdot 1 = x$ and $1 \cdot x = x$ for all $x \in R$ means
   (A) x is invertible in R
   (B) $\cdot$ is a commutative operation
   (C) x is a binary operation on R
   (D) 1 is a neutral element of multiplication in R.
(E) None of the four answers is correct.

2. v' and v'' are both neutral elements of multiplication in some set R implies
   (A) $xv'x = v''xv''$ for all $x \in R$
   (B) $v' = v''$
   (C) $v'xv'' = xv'v''$ for all $x \in R$
   (D) $v' \neq 0$.
(E) None of the four answers is correct.

3. A binary operation on the natural numbers, $\mathbb{N}$, is given by $x\beta y = |x - y|$, the absolute value of the difference of x and y. Which of the following statements are true?
   (A) There is a unique neutral element for β.
   (B) β is associative.

(C) β is commutative.
(D) Every element of $\mathbb{N}$ has a unique inverse.
(E) All four statements are false.

4. We give a binary operation $*$ on the set $S = \{a, b, c, d\}$ by means of this table:

$*$	a	b	c	d
a	a	b	c	d
b	b	c	a	a
c	c	a	d	b
d	d	a	b	b.

Which of the following are true?
(A) $*$ is commutative.
(B) $*$ is associative.
(C) $*$ has a unique neutral element.
(D) Every element of S has an inverse.
(E) All invertible elements of S have unique inverses.

5. Let β be a binary operation on a set S. Then v is a neutral element of β if and only if
(A) $v\beta x = x$ for some $x \in S$
(B) $x\beta v = x$ for all $x \in S$
(C) $x\beta v - x = 0$ and $v\beta x - x = 0$ for all $x \in S$
(D) $f:S \to S$ such that $f(x) = x\beta v$ and $g:S \to S$ such that $g(x) = v\beta x$ are both the identity function I_S.
(E) None of the four alternatives completes a true sentence.

6. β is a binary operation on S implies
(A) $\beta \subseteq (S \times S) \times S$
(B) β is associative
(C) range $\beta = S$
(D) β has values 0 and 1 only.
(E) None of the four alternatives completes a true sentence.

7. Given a binary operation β on a set S with neutral element v, the element x in S is invertible if and only if
(A) v belongs to the range of the function $f_x:S \to S$ such that $f_x(y) = x\beta y$
(B) $x\beta y = v$ for all $y \in S$
(C) β is commutative and there exists $y \in S$ such that $x\beta y = v$
(D) x is the inverse of the inverse of x.
(E) None of the four alternatives completes a true sentence.

Exercises

1. Let the set be $\mathbb{Q}^{\mathbb{Q}}$, the binary operation be composition of functions, and the neutral element be the identity function, $I_{\mathbb{Q}}$.
Is $f:\mathbb{Q} \to \mathbb{Q}$ such that $f(x) = 27x - 7$ invertible?
Is $g:\mathbb{Q} \to \mathbb{Q}$ such that $g(x) = x^2 + 2$ invertible?
Is $h:\mathbb{Q} \to \mathbb{Q}$ such that $h(x) = x^3 - 6$ invertible?

2. Give an example of a nonassociative binary operation on a set.

3. Give three examples of binary operations with neutral elements.

4. Give an example of a binary operation which has no neutral element.

5. On $\mathbb{N}$, the set of natural numbers, we define a binary operation such that $x \wedge y = \min\{x, y\}$, the smaller of the two numbers. Show that the operation $\wedge$ is associative, commutative and has no neutral element.

6. On $\mathbb{N}$ we define the binary operation $\vee$ such that $x \vee y = \max\{x, y\}$, the larger of the two numbers. Show that $\vee$ is associative, commutative, has a neutral element, and only 0 is invertible.

7. v_1 is called a left neutral element for an operation $\beta: S \times S \to S$ if and only if $v_1 \beta x = x$ for all $x \in S$. v_r is a right neutral element for an operation $\beta: S \times S \to S$ if and only if $x \beta v_r = x$ for all $x \in S$. Prove that if a binary operation β on S has both a left and a right neutral element then β has a neutral element.

8. On the set $\mathbb{N}$ we define a binary operation $*$ such that $x * y = x$ for all $x, y \in \mathbb{N}$. Prove that $*$ is associative, noncommutative, has no left neutral element, but has an infinite number of right neutral elements.

9. Let X be a given set. On $\mathscr{P}X = \{S | S \subseteq X\}$, the set of all subsets of X, we define a binary operation $+$, called symmetric difference, such that $A + B = (A \cup B) - (A \cap B)$. Prove that $+$ is associative, commutative, has neutral element $\varnothing$, and every element of $\mathscr{P}X$ is invertible.

2.2 The ring

In this section we extend the concept of an operation, define a ring, and derive several elementary properties of a ring.

In Section 2.1 we used the concept of an operation as a function assigning a value in a set S to two given elements in S. This operation we called a binary operation. By increasing the number of arguments to three we can speak of ternary operations on a set. A ternary operation τ on a set S is a function $\tau: S \times S \times S \to S$. An example of such a ternary operation on the set $\mathbb{N}$ is to set $\tau(x, y, z) = \min\{x, y, z\}$, the minimum of the three numbers x, y, z. With examples of binary operations and ternary operations behind him the reader should be quite prepared to have an n-ary operation, for any non-negative integer n, defined for him. We prepare with Cartesian products of different sizes.

Definition. Let S be a set and n a positive integer. The nth *Cartesian product* is $S^n = \{(x_1, x_2, \ldots, x_n) | x_1, \ldots, x_n \in S\}$. We further define S^0 to be the set $\{0\}$.

In this definition $S^2 = \{(x_1, x_2) | x_1, x_2 \in S\}$ coincides with the earlier definition of the Cartesian product of two sets. $S^1 = \{(x_1) | x_1 \in S\}$ we will identify with S itself by ignoring the pair of parentheses. S^0 we have simply defined to be the set with the one element 0 (the natural number 0). Having defined Cartesian products of various sizes we can now define operations of various sizes.

Definition. Let n be a nonnegative integer and S a set. By an *n-ary operation* β on the set S we mean a function $\beta: S^n \to S$. An 0-ary operation is called a *nullary* operation. A 1-ary operation is called a *unary* operation. A 2-ary operation is called a *binary* operation.

EXAMPLES. $\beta(x, y) = x + y$ yields a binary operation β on $\mathbb{N}$. $\tau(x, y, z) = xy - z$ defines a ternary operation τ on $\mathbb{Z}$. $\nu(x) = -x$ defines a unary operation ν on $\mathbb{Z}$. An example of a nullary operation on $\mathbb{N}$ is $\nu: \mathbb{N}^0 \to \mathbb{N}$ so that $\nu(0) = 67$. Since $\mathbb{N}^0 = \{0\}$, a function that is a nullary operation on $\mathbb{N}$ has but one argument, namely 0. The value of ν for that one argument 0 must be in $\mathbb{N}$. When we know the one value, $\nu(0) = 67$, we know the entire function, the entire nullary operation.

We have seen in the previous example, following the consequences of the definition, that knowing a nullary operation $\nu: S^0 \to S$ is simply knowing one value, $\nu(0)$. A nullary operation, in effect, picks from S one element, the value of $\nu(0)$. Conversely, any choice of a single element from a set, or any designation of a single element from a set can be expressed in the form of a nullary operation on the set. We can, for example, designate the number 1 in the set of natural numbers, $\mathbb{N}$, by giving a nullary operation $\nu: \mathbb{N}^0 \to \mathbb{N}$ such that $\nu(0) = 1$. Or as another instance one might distinguish the neutral element of composition in $\mathfrak{S}(X)$ with a nullary operation $j: \mathfrak{S}(X)^0 \to \mathfrak{S}(X)$ such that $j(0) = I_X$. The logical advantage this use of nullary operations gives us is to allow us to give information about a set and a single element of that set in the form of an operation. This is an appealing esthetic consideration in this section; in Chapter 8 the use of nullary operations will be an integral and essential part of the study of general algebraic systems.

We now move toward a definition of the ring utilizing the material just outlined on operations. By analogy with the integers it is conventional to use the symbols $+$ and $\cdot$ for binary operations on many different sets even though the members of the set may not be numbers. In such cases, the operations must be clearly defined. By custom, $+$ is used only for commutative operations whereas $\cdot$ may or may not represent a commutative operation.

We now introduce the ring, of which the integers are the motivating example.

Definition. $\langle R, +, \cdot, \theta \rangle$ is a *ring* if and only if

R is a set;
$+$ is a binary operation on R;
$\cdot$ is a binary operation on R; and
θ is a nullary operation on R such that

$+$ is associative and commutative,
$\cdot$ is associative,
$\theta(0)$ is a neutral element for $+$,

$x \cdot (y + z) = x \cdot y + x \cdot z$ for all $x, y, z \in R$, and
$(y + z) \cdot x = y \cdot x + z \cdot x$ for all $x, y, z \in R$.

$\langle R, +, \cdot, \theta \rangle$ is a commutative ring if and only if $\langle R, +, \cdot, \theta \rangle$ is a ring, and $\cdot$ is commutative.

A certain amount of abuse of language is not uncommon. We frequently speak of the ring R, mentioning only the set R and leaving the operations to be understood by the reader. We also will use θ both as a symbol for the nullary operation and for the neutral element in its range. Both of these practices are abuses because a set, in and of itself, is not a ring and there is a difference between a function and the range of a function. Nevertheless, the first practice is universal. The second will save us from some cumbersome expressions.

Our principal example of a ring is $\langle \mathbb{Z}, +, \cdot, 0 \rangle$ which has inspired our definition. $\langle \mathbb{Z}, +, \cdot, 0 \rangle$ is, moreover, a commutative ring. We now prove some beginning results that apply to all rings, commutative or not.

Theorem. *Let $\langle R, +, \cdot, \theta \rangle$ be a ring. Then $\theta \cdot x = x \cdot \theta = \theta$ for all $x \in R$.*

PROOF. This result follows from distributivity and the neutral property of θ. $\theta \cdot x + \theta \cdot x = (\theta + \theta) \cdot x = \theta \cdot x = \theta \cdot x + \theta$. From both sides of $\theta \cdot x + \theta \cdot x = \theta \cdot x + \theta$ cancel $\theta \cdot x$ yielding $\theta \cdot x = \theta$. To prove $x \cdot \theta = \theta$ we repeat the proof from the right instead of the left. The cancellation is possible because every element of R is additively invertible; every element which is additively invertible is additively cancellable. □

Theorem. *Let $\langle R, +, \cdot, \theta \rangle$ be a ring. Then $x(-y) = (-x)y = -(xy)$ for all $x, y \in R$.*

PROOF. We have followed the usual custom of omitting the symbol for multiplication when no confusion can occur. $x(-y) + xy = x[(-y) + y] = x\theta = \theta$. Because $+$ is commutative $xy + x(-y) = \theta$ also. We conclude $x(-y)$ is the unique negative of xy which is written $-(xy)$. This is to say, $x(-y) = -(xy)$. In a symmetric manner we prove $(-x)y = -(xy)$. □

Theorem. *Let $\langle R, +, \cdot, \theta \rangle$ be a ring. Then $(-x)(-y) = xy$ for all $x, y \in R$.*

PROOF. Using the previous theorem twice: $(-x)(-y) = -[x(-y)] = -[-(xy)]$. But $-(-(xy)) = xy$ because the inverse of the inverse of any element is the element itself. □

The previous theorems demonstrate how some of the frequently performed manipulations of school algebra are valid in rings in general. We also can observe as we develop the theory of rings for what reasons our manipulations of school algebra are valid.

Theorem. *Let $\langle R, +, \cdot, \theta \rangle$ be a ring. Let the operation of multiplication have a neutral element v in R. Then $(-v)x = x(-v) = -x$ for all $x \in R$.*

PROOF. We notice that our ring in this theorem is not assumed to be a commutative ring so that $(-v)x = x(-v)$ does not follow from commutativity. However, $(-v)x = -(vx) = -x$ and $x(-v) = -(xv) = -x$ for all $x \in R$. □

In the case of $\langle \mathbb{Z}, +, \cdot, 0 \rangle$ (the integers have neutral element of multiplication 1) the theorem tells us $(-1)x = x(-1) = -x$ for all $x \in \mathbb{Z}$. In words, to find the negative of an integer x multiply that integer by the integer -1, the negative of 1. Our familiarity with these results should not be allowed to prevent our gaining a deeper understanding for why they are true.

QUESTIONS

1. Let $\langle R, +, \cdot, \theta \rangle$ be a ring.
$\theta x = x$ for all $x \in R$ implies
- (A) $R = \{\theta\}$
- (B) θ is the neutral element of multiplication for R
- (C) R is a commutative ring
- (D) there is an element of R which is invertible with respect to multiplication and is nonzero.
- (E) None of the possibilities completes a true sentence.

2. A ring $\langle R, +, \cdot, \theta \rangle$
- (A) must contain a neutral element of addition
- (B) must contain a neutral element of multiplication
- (C) must have its addition commutative
- (D) must have its multiplication commutative.
- (E) None of the four alternatives completes a true sentence.

3. In a ring $\langle R, +, \cdot, \theta \rangle$
- (A) every element must have an additive inverse in R
- (B) every element must have a multiplicative inverse in R
- (C) a negative of an element must itself have a negative in R
- (D) a multiplicative inverse of an element must itself have a multiplicative inverse in R.
- (E) None of the alternatives completes a true sentence.

4. Let $\langle R, +, \cdot, \theta \rangle$ be a ring with a neutral element of multiplication, v. Which of these statements are true?
- (A) $\langle R, +, \theta \rangle$ is a commutative group.
- (B) $\langle R, \cdot, v \rangle$ is a group.
- (C) $\langle R - \{\theta\}, \cdot, v \rangle$ is a group.
- (D) v is also a neutral element of addition.
- (E) None of the four statements is true.

EXERCISES

1. Which of the following examples are rings and which are not rings? You must, in each case, supply the understood operations.
- (a) The set of all even integers
- (b) The set of all odd integers

(c) The set of all nonnegative integers
(d) The set of all polynomials with real coefficients which have degree two or less: $\{Ax^2 + Bx + C | A, B, C \in \mathbb{R}\}$
(e) The set of all polynomials of degree two or less with even integer coefficients
(f) The set of all fractions which when reduced to lowest terms have even integer denominators
(g) The set of all expressions of the form $a + b\sqrt{2}$, $a, b \in \mathbb{Z}$.

2. We define on the set $\{a, b\}$ two operations $\star$, $\dagger$, by means of these operation tables:

$\star$	a	b
a	a	b
b	b	a

$\dagger$	a	b
a	a	a
b	a	b.

Verify that $\langle\{a, b\}, \star, \dagger, a\rangle$ is a ring, but $\langle\{a, b\}, \dagger, \star, b\rangle$ is not a ring.

3. In what ways does $\langle \mathscr{P}X, \cup, \cap, \varnothing\rangle$ fail to be a ring?

4. Using the definition of + given in Exercise 9, Section 2.1, show that $\langle \mathscr{P}X, +, \cap, \varnothing\rangle$ is a ring. Is there a neutral element of multiplication ($\cap$)? Which elements are $\cap$-invertible?

5. Show that left distributivity and commutativity of multiplication imply right distributivity.

6. Let $\langle R, +, \cdot, \theta\rangle$ be a ring with a neutral element of multiplication v. Prove $v = \theta$ if and only if $R = \{\theta\}$.

7. Let $\langle R, +, \cdot, \theta\rangle$ be a ring. The set of two by two matrices over R is the set $R^{2\times 2} = \left\{\begin{pmatrix} a & b \\ c & d \end{pmatrix} \middle| \, a, b, c, d \in R\right\}$. On this set we define + and $\cdot$ as follows:

$$\begin{pmatrix} a & b \\ c & d \end{pmatrix} + \begin{pmatrix} e & f \\ g & h \end{pmatrix} = \begin{pmatrix} a+e & b+f \\ c+g & d+h \end{pmatrix}, \quad \begin{pmatrix} a & b \\ c & d \end{pmatrix} \cdot \begin{pmatrix} e & f \\ g & h \end{pmatrix} = \begin{pmatrix} ae+bg & af+bh \\ ce+dg & cf+dh \end{pmatrix}.$$

Prove that $\left\langle R^{2\times 2}, +, \cdot, \begin{pmatrix} \theta & \theta \\ \theta & \theta \end{pmatrix}\right\rangle$ is a ring. Show that this is a noncommutative ring. Is there a neutral element of multiplication in this ring?

We will be studying matrices in considerable detail in later chapters, but we do wish to use this simple case at this time as a valuable example of a noncommutative ring.

8. Which of the following sets are rings?
(a) The set of all two by two matrices with values in $\mathbb{Z}$
(b) The set of all two by two matrices with values in $\mathbb{Z}$ but always with 0 in the lower right corner of the matrix
(c) The set of all two by two matrices with even integer values.

9. Let $\langle R, +, \cdot, \theta\rangle$ be a ring such that $xx = x$ for all $x \in R$. Prove $x + x = \theta$ for all $x \in R$. Prove the ring is commutative.

10. Give an example of a noncommutative ring obeying the condition $x + x = \theta$ for all x in the ring.

11. Let $\langle R, +, \cdot, \theta \rangle$ be a ring and S a nonempty set. Prove $\langle R^S, +, \cdot, z \rangle$ is a ring in which
R^S is the set of all functions with domain S and codomain R,
$(f + g): S \to R$ such that $(f + g)(x) = f(x) + g(x)$ for all $x \in S$,
$(f \cdot g): S \to R$ such that $(f \cdot g)(x) = f(x) \cdot g(x)$ for all $x \in S$, and
$z: S \to R$ such that $z(x) = \theta$ for all $x \in S$.

12. From calculus we borrow the definition of a continuous function. A function $f:[0, 1] \to \mathbb{R}$ is continuous on $[0, 1]$ if and only if for each a in the closed unit interval $[0, 1]$, $\lim_{x \to a} f(x) = f(a)$. Let $C^0[0, 1] = \{f \mid f \in \mathbb{R}^{[0, 1]}$ and f is continuous on $[0, 1]\}$. Show that $C^0[0, 1]$, the set of all functions continuous on $[0, 1]$, is a ring.

2.3 Special rings

In this section are defined some special rings: unitary ring, product ring, integral domain, division ring, and field.

A formalization of the existence of a neutral element for ring multiplication produces this definition:

Definition. $\langle R, +, \cdot, \theta, v \rangle$ is a *unitary ring* if and only if $\langle R, +, \cdot, \theta \rangle$ is a ring, v is a nullary operation on R such that $v(0)$ is a neutral element for $\cdot$. We shall call the neutral element of multiplication a *unity* of the ring.

We note that $\langle \mathbb{Z}, +, \cdot, 0, 1 \rangle$ is a unitary ring with unity 1.

Prior to the formal definition of the product ring we construct an example. This example makes a ring of the set $\mathbb{Z} \times \mathbb{Z}$ of all pairs of integers. $\mathbb{Z} \times \mathbb{Z} = \{(x, y) \mid x \in \mathbb{Z}$ and $y \in \mathbb{Z}\}$. We must define two binary operations on the set $\mathbb{Z} \times \mathbb{Z}$, calling the first addition and the second multiplication.

$$(s, t) + (u, v) = (s + u, t + v),$$
$$(s, t) \cdot (u, v) = (su, tv).$$

The operation $+$ is an associative and commutative operation with a neutral element (0, 0). We verify these assertions.

$$\begin{aligned}(s, t) + [(u, v) + (w, x)] &= (s, t) + (u + w, v + x) \\ &= (s + (u + w), t + (v + x)) \\ &= ((s + u) + w, (t + v) + x) \\ &= (s + u, t + v) + (w, x) \\ &= [(s, t) + (u, v)] + (w, x).\end{aligned}$$

$(s, t) + (u, v) = (s + u, t + v) = (u + s, v + t) = (u, v) + (s, t).$
$(0, 0) + (s, t) = (0 + s, 0 + t) = (s, t).$
$(s, t) + (0, 0) = (s + 0, t + 0) = (s, t).$

With respect to $+$, every element of $\mathbb{Z} \times \mathbb{Z}$ is invertible.

$$(s, t) + (-s, -t) = (0, 0); \qquad (-s, -t) + (s, t) = (0, 0).$$

In these proofs the properties of $+$ for $\mathbb{Z} \times \mathbb{Z}$ depend upon the underlying properties of $+$ in $\mathbb{Z}$. That $\cdot$ on $\mathbb{Z} \times \mathbb{Z}$ is also commutative and associative may be proved easily by the reader. $(1, 1)$ is furthermore a neutral element for multiplication. Left distributivity holds.

$$\begin{aligned}(s, t)[(u, v) + (w, x)] &= (s, t)(u + w, v + x) \\ &= (s(u + w), t(v + x)) \\ &= (su + sw, tv + tx) \\ &= (su, tv) + (sw, tx) \\ &= (s, t)(u, v) + (s, t)(w, x).\end{aligned}$$

The right distributive theorem is verified similarly. This completes the verification that $\langle \mathbb{Z} \times \mathbb{Z}, +, \cdot, (0, 0), (1, 1)\rangle$ is a unitary commutative ring.

The construction in the previous example motivates the following definition of a product ring.

Definition. Given rings $\langle R', +', \cdot', \theta'\rangle$ and $\langle R'', +'', \cdot'', \theta''\rangle$ we define the *product* of the two rings to be $\langle R' \times R'', +, \cdot, (\theta', \theta'')\rangle$ in which $+$ and $\cdot$ are defined by

$$\begin{aligned}(x', x'') + (y', y'') &= (x' +' y', x'' +'' y'') \\ (x', x'') \cdot (y', y'') &= (x' \cdot' y', x'' \cdot'' y'').\end{aligned}$$

It is not difficult to prove the product is a ring; the proof is like the example verification preceding the definition.

Returning to the integers we notice the integers enjoy the following property:

$$uv = 0 \text{ implies } u = 0 \text{ or } v = 0.$$

One says, to describe this property, that the integers have no *nontrivial divisors of zero.* This property is used to define a special kind of ring.

Definition. $\langle R, +, \cdot, \theta, v\rangle$ is an *integral domain* if and only if $\langle R, +, \cdot, \theta, v\rangle$ is a commutative unitary ring, $\theta \neq v$, and $xy = \theta$ implies $x = \theta$ or $y = \theta$ for all $x, y \in R$.

$\mathbb{Z}$, $\mathbb{Q}$, and $\mathbb{R}$ are all integral domains, but the product ring $\mathbb{Z} \times \mathbb{Z}$ is not an integral domain: $(1, 0)(0, 1) = (0, 0)$.

The condition $xy = \theta$ implies $x = \theta$ or $y = \theta$ is equivalent in a ring to cancellation, which we now prove.

Theorem. *A commutative unitary ring $\langle R, +, \cdot, \theta, v\rangle$ with $\theta \neq v$ is an integral domain if and only if nonzero multiplicative cancellation is always possible.*

PROOF. Suppose multiplicative cancellation is always possible for nonzero elements. Let $xy = \theta$. $xy = x\theta$ because $x\theta = \theta$. If $x \neq \theta$ then cancel x to get $y = \theta$. Therefore either $x = \theta$ or $y = \theta$.

For the converse, assume R is an integral domain and let $xy = xz$ with $x \neq \theta$. $xy + (-(xz)) = \theta$. $xy + x(-z) = \theta$. $x(y + (-z)) = \theta$. Since $x \neq \theta$ we must have $y + (-z) = \theta$. $y = z$. □

We finish this section with several more definitions.

Definition. $\langle R, +, \cdot, \theta, v\rangle$ is a *division ring* if and only if $\langle R, +, \cdot, \theta, v\rangle$ is a unitary ring, $\theta \neq v$, and every nonzero element of R is multiplicatively invertible.

Definition. $\langle R, +, \cdot, \theta, v\rangle$ is a *field* if and only if $\langle R, +, \cdot, \theta, v\rangle$ is a commutative division ring.

We observe that every field is an integral domain. $\mathbb{Q}$, $\mathbb{R}$, and $\mathbb{C}$ are all fields while $\mathbb{Z}$ is not a field.

Questions

1. Which of the following statements are true?
- (A) Some integral domains are fields.
- (B) Some division rings are fields.
- (C) Some fields are division rings.
- (D) Some integral domains are not fields.
- (E) Some division rings are not integral domains.

2. The set of even integers with the usual sum and product
- (A) is an integral domain
- (B) has no nontrivial divisors of zero
- (C) is a field
- (D) is a commutative ring
- (E) is a division ring.

3. The product ring $\mathbb{Z} \times \mathbb{Z}$
- (A) is a commutative ring
- (B) is a commutative unitary ring
- (C) is an integral domain
- (D) is a field.
- (E) None of the alternatives completes a true sentence.

4. The natural number system $\langle \mathbb{N}, +, \cdot, \theta\rangle$ is a
- (A) ring
- (B) unitary ring
- (C) integral domain
- (D) field.
- (E) None of the possibilities completes a satisfactory sentence.

5. In any unitary ring
- (A) some elements have inverses
- (B) there are multiplicatively cancellable elements

(C) θ never has a multiplicative inverse
(D) there are elements which commute with all other elements.
(E) None of the four alternatives completes a satisfactory sentence.

Exercises

1. Prove that the product of two rings is itself a ring.
2. Show that no product ring of nontrivial rings can be an integral domain.
3. Give an example of a ring without a unity.
4. Does the ring $\langle \mathscr{P}X, +, \cap, \varnothing \rangle$ have nontrivial divisors of zero? Is it an integral domain?
5. On the set $\mathbb{Z} \times \mathbb{Z}$ we define the following two operations:
$$(s, t) + (u, v) = (s + u, t + v)$$
$$(s, t) \boxdot (u, v) = (su + tv, sv + tu).$$
Show that $\langle \mathbb{Z} \times \mathbb{Z}, +, \boxdot, (0, 0) \rangle$ is a ring. Is the ring commutative? Does the ring have a unity? Is the ring an integral domain? Which elements of the ring are $\boxdot$-invertible?
6. Is $\mathbb{Q}^{2 \times 2}$ an integral domain? Which elements of the ring $\mathbb{Q}^{2 \times 2}$ are invertible?
7. Let $\langle R, +, \cdot, \theta \rangle$ be a ring with unity ν. Show that every nonzero element of R is multiplicatively invertible if and only if the equations
$$ax + b = \theta$$
$$xc + d = \theta \quad a, b, c, d \in R, a \neq \theta, b \neq \theta$$
always have unique solutions (for x) in R.
8. Show that the commutativity of addition is derivable from the other statements in the definition of a unitary ring.
9. Let $\langle R, +, \cdot, \theta, \nu \rangle$ be a commutative unitary ring. Prove that the set of all multiplicatively invertible elements of R is a group under multiplication.
10. Let S be a subset of a field $\langle K, +, \cdot, \theta, \nu \rangle$ closed under addition, multiplication, negatives, and reciprocals (of nonzero members). Prove S contains θ and ν if and only if S contains at least two members.
11. Let $\langle R, +, \cdot, \theta, \nu \rangle$ be an integral domain. Let S be a subset of R such that $\langle S, +, \cdot, \theta \rangle$ is itself a ring with a unity and such that S contains more than one element. Prove that the unity of S is ν.
12. Let $\langle R, +, \cdot, \theta \rangle$ be a ring without nontrivial divisors of zero. Suppose there exists an element $a \in R, a \neq \theta$, such that $aa = a$. Prove R has a unity. Beware of assuming the existence of a unity.
13. Prove that if a ring $\langle R, +, \cdot, \theta \rangle$ has a left unity which is unique then the ring has a unity.
14. Let $\langle R, +, \cdot, \theta, \nu \rangle$ be a unitary ring. Let x be an element of R which has at least one left multiplicative inverse and at least one right multiplicative inverse. Prove

that x has a unique multiplicative inverse in R. Prove furthermore that x has only one left multiplicative inverse and only one right multiplicative inverse.

15. Beginners in school algebra often write $(x + y)^2 = x^2 + y^2$. Give an example of a ring $\langle R, +, \cdot, \theta\rangle$ for which this statement is true: $(x + y)(x + y) = xx + yy$ for all $x, y \in R$. Give an example of a ring in which the statement is false.

16. Suppose $\langle R, +, \cdot, \theta\rangle$ is a ring such that $(x + y)(x + y) = xx + yy$ for all $x, y \in R$. Prove
(a) $xy = -yx$ for all $x, y \in R$
(b) $xx + xx$ for all $x \in R$
(c) $x + x = \theta$ for all $x \in R$ if R has a unity.

17. Give an example of a ring $\langle R, +, \cdot, \theta\rangle$ such that $xy = -yx$ for all $x, y \in R$ yet it is false that $x + x = \theta$ for all $x \in R$.

18. Give an example of a ring $\langle R, +, \cdot, \theta\rangle$ such that $x + x = \theta$ for all $x \in R$ yet it is false that $xx = x$ for all $x \in R$.

19. Let $\langle R, +, \cdot. \theta, v\rangle$ be a unitary ring. Let x be an element of R with a unique left multiplicative inverse. Prove x has an inverse in R.

20. An important example of a division ring which is not a field is given by the set of *quaternions*, $\{a + bi + cj + dk | a, b, c, d \in \mathbb{R}\}$. Addition of two quaternions is defined by the rule $(a_1 + b_1 i + c_1 j + d_1 k) + (a_2 + b_2 i + c_2 j + d_2 k) = (a_1 + a_2) + (b_1 + b_2)i + (c_1 + c_2)j + (d_1 + d_2)k$. This addition is commutative and associative, there is a neutral element $0 = 0 + 0i + 0j + 0k$ and every quaternion has a negative. Multiplication is defined using these reduction rules:

$$i^2 = j^2 = k^2 = -1, \qquad ij = -ji = k, \qquad jk = -kj = i, ki = -ik = j.$$

$$\begin{aligned}
&(a_1 + b_1 i + c_1 j + d_1 k)(a_2 + b_2 i + c_2 j + d_2 k)\\
&\quad = a_1a_2 + a_1b_2 i + a_1c_2 j + a_1d_2 k + b_1a_2 i + b_1b_2 i^2 + b_1c_2 ij + b_1d_2 ik + c_1a_2 j\\
&\qquad + c_1b_2 ji + c_1c_2 j^2 + c_1d_2 jk + d_1a_2 k + d_1b_2 ki + d_1c_2 kj + d_1d_2 k^2\\
&\quad = (a_1a_2 - b_1b_2 - c_1c_2 - d_1d_2) + (a_1b_2 + b_1a_2 + c_1d_2 - d_1c_2)i\\
&\qquad + (a_1c_2 + c_1a_2 + d_1b_2 - b_1d_2)j + (a_1d_2 + d_1a_2 + b_1c_2 - c_1b_2)k.
\end{aligned}$$

The multiplication is noncommutative since, for example, $ij \neq ji$. The multiplication is associative. Show that

$$(1/(a^2 + b^2 + c^2 + d^2))(a - bi - cj - dk)$$

is the inverse of $a + bi + cj + dk$ to prove that the quaternions are a division ring.

2.4 Subrings

In this section we define a subring, derive necessary and sufficient conditions for a subset to be a subring, and define the subring generated by a subset of a ring.

We begin with the subring.

Definition. S, a subset of the set R, is a *subring* of the ring $\langle R, +, \cdot, \theta\rangle$ if and only if $\langle S, +, \cdot, \theta\rangle$ is a ring.

It is to be understood in this definition that $+$ and $\cdot$, the binary operations on the subset S, are to have the same values on S that they have on the including set R. It is necessary, therefore, in order that $+$ and $\cdot$ be binary operations on S, that $x, y \in S$ imply $x + y$ and xy belongs to S. We speak then of $+$ and $\cdot$ as being *closed* on S.

EXAMPLE. We denote the even integers by $2\mathbb{Z} = \{2x | x \in \mathbb{Z}\}$. $\langle 2\mathbb{Z}, +, \cdot, 0\rangle$ is a subring of $\langle \mathbb{Z}, +, \cdot, 0\rangle$. Note that the subring must contain 0 to be a ring. In this example the subring $2\mathbb{Z}$ fails to contain a unity although the original ring $\mathbb{Z}$ does.

The odd integers, $2\mathbb{Z} + 1 = \{2x + 1 | x \in \mathbb{Z}\}$ do not form a subring. 0 fails to be a member. Furthermore, $3 + 7 = 10$, not an odd integer.

We now develop necessary and sufficient conditions for a given subset of a ring to be a subring.

Theorem. *S, a subset of R, is a subring of $\langle R, +, \cdot, \theta\rangle$ if and only if*

$$S \neq \varnothing \text{ and}$$
$$x, y \in S \quad \text{imply } x + y, xy, \text{ and } -x \text{ are in } S.$$

PROOF. First assume S is a subring. S is itself a ring. $S \neq \varnothing$ because S must contain θ, the neutral element of addition. Let x and $y \in S$. $x + y$ and xy must belong to S because the addition and multiplication are binary operations on S. $-x$ is in S because every element of S is $+$-invertible if S is a ring.

Secondly, to prove the converse, assume S is a subset of R such that $S \neq \varnothing$ and $x, y \in S$ imply $x + y$, xy, and $-x$ are in S. We must prove $\langle S, +, \cdot, \theta\rangle$ is a ring. The operations $+$ and $\cdot$ are binary operations on S. Given $S \neq \varnothing$ let $a \in S$. Then $-a \in S$. $(a) + (-a) \in S$. $\theta \in S$. Every $x \in S$ is $+$-invertible since $-x \in S$. Both $+$ and $\cdot$ are associative operations on R and therefore certainly on S. Addition is commutative on R and therefore also commutative on S. Thus $\langle S, +, \cdot, \theta\rangle$ is a ring, a subring of $\langle R, +, \cdot, \theta\rangle$. □

Every ring contains as subrings the trivial subring $\{\theta\}$ and the entire ring itself. Given an arbitrary subset A of a ring $\langle R, +, \cdot, \theta\rangle$ we ask the question whether or not there exist subrings of R which contain the given A as a subset. The answer is yes, there always exists at least one subring of R containing A, namely R itself. The following theorem proves a stronger result; there exists a smallest subring of R containing A.

Theorem. *Let A be any subset of a ring $\langle R, +, \cdot, \theta\rangle$. Then there exists a smallest subring of R which contains the set A as a subset.*

PROOF. We define a collection $\mathscr{C}$ of subrings of R as follows:

$$\mathscr{C} = \{S \mid A \subseteq S \text{ and } S \text{ is a subring of } R\}.$$

$R \in \mathscr{C}$ and therefore $\mathscr{C} \neq \varnothing$.

We now form the intersection of the collection $\mathscr{C}$. The intersection of the collection $\mathscr{C}$ is the set of all elements belonging to every subring in the collection $\mathscr{C}$.

$$\bigcap \mathscr{C} = \{x \mid x \in S \text{ for every } S \in \mathscr{C}\}.$$

We now demonstrate that this subset of R is actually a subring of R. Let x and y belong to $\bigcap \mathscr{C}$. By the definition of the intersection $x \in S$ for all $S \in \mathscr{C}$ and $y \in S$ for all $S \in \mathscr{C}$. Then $x + y$, xy, and $-x$ belong to S for all $S \in \mathscr{C}$. $x + y$, xy, and $-x$ belong to $\bigcap \mathscr{C}$. Furthermore, $\theta \in S$ for all $S \in \mathscr{C}$. $\theta \in \bigcap \mathscr{C}$. $\bigcap \mathscr{C}$ is a subring of R.

Let us now show $A \subseteq \bigcap \mathscr{C}$. Choose any $x \in A$. Then $x \in S$ for every $S \in \mathscr{C}$ because $A \subseteq S$ for every $S \in \mathscr{C}$. $x \in \bigcap \mathscr{C}$.

It remains to show that $\bigcap \mathscr{C}$ is the smallest subring of R containing A as a subset. We observe first that $\bigcap \mathscr{C} \subseteq S$ for every $S \in \mathscr{C}$. This is to say that the intersection of a collection of sets is a subset of every set in the collection. Thus $\bigcap \mathscr{C}$ is smaller than any other subring of R containing A. □

Definition. Given any subset A of a ring $\langle R, +, \cdot, \theta \rangle$ we define the subring of R generated by the set A to be the smallest subring of R containing A. We denote the subring of R generated by A by $[A]$.

It is to be observed that the theorem preceding makes the definition possible.

EXAMPLES. $\{0, 1\}$ generates the entire ring $\mathbb{Z}$. As a subset of $\mathbb{Z}$ we have $[2\mathbb{Z} + 1] = \mathbb{Z}$. In $\mathbb{Z}$, $[\{3, 6\}] = 3\mathbb{Z}$. As a subset of $\mathbb{Q}$, $[\mathbb{Z}] = \mathbb{Z}$. If we denote the positive fractions by $\mathbb{Q}^+$ then $[\mathbb{Q}^+] = \mathbb{Q}$.

For a finite set such as $\{a, b, c\}$ we frequently abbreviate $[\{a, b, c\}]$ with $[a, b, c]$ if there is no danger of error.

QUESTIONS

1. Let $\mathscr{C}$ be the collection of all subrings of a given ring $\langle R, +, \cdot, \theta \rangle$. Which of these statements are false?
   (A) $\{\theta\} \in \mathscr{C}$.
   (B) $\bigcap \mathscr{C} \in \mathscr{C}$.
   (C) $R \in \mathscr{C}$.
   (D) $\bigcup \mathscr{C} \in \mathscr{C}$.
   (E) All four statements are true.

2. Let $\mathscr{C}$ be the collection of all subrings containing some given subset A of the ring $\langle R, +, \cdot, \theta \rangle$. Which of these statements are true?

(A) $\{\theta\} \in \mathscr{C}$.
(B) $R \in \mathscr{C}$.
(C) $\bigcup\mathscr{C} \in \mathscr{C}$.
(D) $A \in \mathscr{C}$.
(E) None of the statements is true.

3. Given that $\mathscr{C}$ is a collection of subrings of a ring $\langle R, +, \cdot, \theta\rangle$ which of these statements are true?
(A) $\{\theta\} \in \mathscr{C}$.
(B) $R \in \mathscr{C}$.
(C) $\bigcup\mathscr{C} \in \mathscr{C}$.
(D) $\bigcap\mathscr{C} \in \mathscr{C}$.
(E) All four statements are false.

4. In the product ring $\mathbb{Z} \times \mathbb{Z}$ the subset $\{(x', x'')|x', x'' \in \mathbb{Z} \text{ and } x' = 2x''\}$
(A) is a subring of $\mathbb{Z} \times \mathbb{Z}$
(B) fails to be closed under negation
(C) fails to be closed under addition
(D) fails to be closed under multiplication.
(E) None of the four alternatives completes a true sentence.

5. Given the subset $S = \left\{\begin{pmatrix} a & b \\ c & d\end{pmatrix}\middle|\, ad - bc \geqslant 0\right\}$ of $\mathbb{Z}^{2\times 2}$ we can say that S is
(A) a subring of $\mathbb{Z}^{2\times 2}$
(B) not a subring of $\mathbb{Z}^{2\times 2}$ because $S = \varnothing$
(C) not a subring of $\mathbb{Z}^{2\times 2}$ because S is not closed under multiplication
(D) not a subring of $\mathbb{Z}^{2\times 2}$ because S is not closed under negation.
(E) None of the four choices satisfactorily completes the sentence.

Exercises

1. Given the ring $\langle \mathbb{Z}, +, \cdot, 0\rangle$ find $[0]$, $[1]$, $[2]$, $[7]$.

2. Given the ring $\langle \mathbb{R}, +, \cdot, 0\rangle$ find $[0]$, $[1]$, $[\pi]$, $[1, 2]$.

3. Given the ring $\left\langle \mathbb{Z}^{2\times 2}, +, \cdot, \begin{pmatrix} 0 & 0 \\ 0 & 0\end{pmatrix}\right\rangle$ find $\left[\begin{pmatrix} 1 & 0 \\ 0 & 0\end{pmatrix}\right]$, $\left[\begin{pmatrix} 2 & 0 \\ 0 & 0\end{pmatrix}\right]$, $\left[\begin{pmatrix} 0 & 2 \\ 2 & 0\end{pmatrix}\right]$.

4. Given the ring $\langle \mathbb{Q}, +, \cdot, 0\rangle$ find $[\frac{1}{2}]$, $[\{1/n | n \in \mathbb{Z} - \{0\}\}]$, $[1, \frac{1}{2}, \frac{1}{3}]$.

5. Find the collection of all subrings of $\mathbb{Z}$ containing the set $\{12\}$. What is the intersection of this collection?

6. Given the ring $\langle \mathscr{P}(\mathbb{N}), +, \cap, \varnothing\rangle$ find $[\{0, 1\}]$. Let $s(n) = \{x | x \in \mathbb{N} \text{ and } 0 \leqslant x < n\}$, the open segment of $\mathbb{N}$ determined by n. Find $[\{s(n) | n \in \mathbb{N}\}]$.

7. An expression such as $2 + 5X + 9X^2$ or $\frac{1}{2} + \frac{2}{3}X + \frac{19}{4}X^5$ is known as a polynomial. While we shall study polynomial rings in detail in Chapter 5 many aspects of polynomials are familiar and intuitive enough we can use them now as examples. The set of all polynomials of all degrees with rational coefficients, $\{a_0 + a_1X + \cdots + a_nX^n | a_0, a_1, \ldots, a_n \in \mathbb{Q}, n \in \mathbb{N}\}$, is a ring with the usual school algebra way of adding and multiplying polynomials. Find $[\frac{1}{2}]$, $[X]$, $[\mathbb{Q} \cup \{X\}]$.

2.5 Morphisms

In this section the structure-preserving functions, morphisms, are defined and investigated.

Let $\langle R, +, \cdot, \theta \rangle$ and $\langle R', +', \cdot', \theta' \rangle$ be two rings and $f: R \to R'$ a function. Consider the situation suggested by Figure 2.1. We might well obtain different results upon calculating $f(x + y)$ and calculating $f(x) +' f(y)$. Whether we first find the sum of x and y in R and second find the image in R' or whether we first find the images of x and y in R' and then add in R' may affect the outcome. If, however, the function f is of such a character that $f(x + y) = f(x) +' f(y)$ for all $x, y \in R$ we say that f *preserves* the first operation $+$ in the second $+'$. When f preserves all the operations of one mathematical system in the respective ones of the second then we call f a *morphism.*

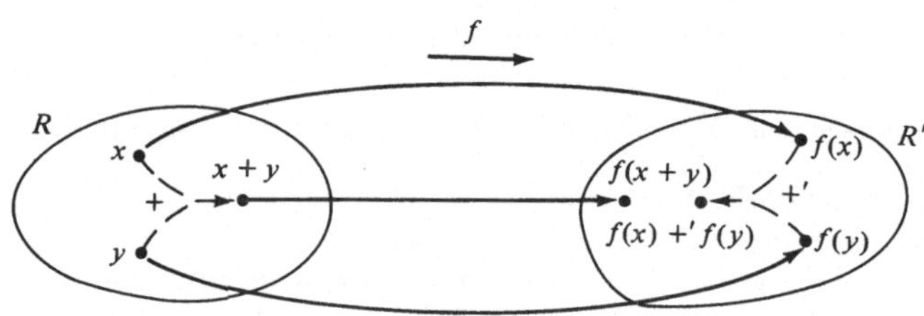

Figure 2.1

Definition. Given rings $\langle R, +, \cdot, \theta \rangle$, $\langle R', +', \cdot', \theta' \rangle$ and a function $f: R \to R'$ we say f is a ring *morphism* if and only if

$f(x + y) = f(x) +' f(y)$ for all $x, y \in R$;
$f(xy) = f(x) \cdot' f(y)$ for all $x, y \in R$; and
$f(\theta) = \theta'$.

We shall later consider morphisms of other algebraic systems such as groups and vector spaces. Only when there is a possibility of confusion need one say ring morphism rather than just morphism. An older term with the same meaning as morphism is *homomorphism.* We use the shorter term morphism and reserve the use of prefixes for special kinds of morphisms.

We now prove a theorem about morphisms which shows the condition $f(\theta) = \theta'$ to be superfluous for rings.

Theorem. *Given rings $\langle R, +, \cdot, \theta \rangle$, $\langle R', +', \cdot', \theta' \rangle$ and a function $f: R \to R'$, f is a morphism if and only if*

$$f(x + y) = f(x) +' f(y) \quad \textit{for all } x, y \in R, \textit{ and}$$
$$f(x \cdot y) = f(x) \cdot' f(y) \quad \textit{for all } x, y \in R.$$

PROOF. Obviously if f is a morphism then the two statements are true. It is the converse that requires the discussion. We must prove that if $f(x + y) = f(x) +' f(y)$ and $f(xy) = f(x)f(y)$ for all $x, y \in R$ then $f(\theta) = \theta'$.

$$f(\theta) = f(\theta + \theta) = f(\theta) +' f(\theta).$$
$$f(\theta) +' \theta' = f(\theta) +' f(\theta).$$

Because $f(\theta)$ is an element of R' and therefore has an additive inverse we can cancel $f(\theta)$ yielding $\theta' = f(\theta)$. □

Depending upon special properties of the function we have special names for morphisms which we now introduce.

Definition. Let $f:R \to R'$ be a morphism of the rings $\langle R, +, \cdot, \theta\rangle$ and $\langle R', +', \cdot', \theta'\rangle$.

If f is an injection then we call f a *monomorphism*.
If f is a surjection then we call f an *epimorphism*.
If f is a bijection then we call f an *isomorphism*.

EXAMPLES. Let S be a subring of the ring $\langle R, +, \cdot, \theta\rangle$. The identity injection $j:S \to R$ such that $j(x) = x$ is a monomorphism.

Let $\mathbb{Z} \times \mathbb{Z}$ be the product ring of $\mathbb{Z}$ with itself. The function $p_1:\mathbb{Z} \times \mathbb{Z} \to \mathbb{Z}$ such that $p_1(x, y) = x$ is an epimorphism but not a monomorphism.

The function $q_1:\mathbb{Z} \to \mathbb{Z} \times \mathbb{Z}$ such that $q_1(x) = (x, 0)$ is a monomorphism but not an epimorphism.

The range, $f(R)$, of a morphism $f:R \to R'$ has many of the properties of the domain R. We list some of these in a theorem.

Theorem. *Let $f:R \to R'$ be a morphism of the rings $\langle R, +, \cdot, \theta\rangle$ and $\langle R', +', \cdot', \theta'\rangle$. Then*

(a) $f(-x) = -f(x)$ *for all* $x \in R$
(b) $f(R)$ *is a subring of* R'
(c) $\langle R, +, \cdot, \theta\rangle$ *is a commutative ring implies* $\langle f(R), +', \cdot', \theta'\rangle$ *is a commutative ring*
(d) v *is a unity for* R *implies* $f(v)$ *is a unity for* $f(R)$
(e) *if both* R *and* R' *are unitary rings and* $f(v) = v'$ *and* x *is multiplicatively invertible in* R, *then* $f(x^-) = f(x)^-$.

PROOF

(a) $f(x) +' f(-x) = f(x + (-x)) = f(\theta) = \theta'$. Likewise, $f(-x) +' f(x) = \theta'$ for all $x \in R$. $f(-x) = -f(x)$ the unique negative of $f(x)$ in R'.

(b) $f(R) = \{f(x) | x \in R\}$. We must show $f(R)$ nonempty and closed under addition, multiplication, and negatives. θ' belongs to $f(R)$ because $f(\theta) = \theta'$. $f(R) \neq \varnothing$. Let $y_1, y_2 \in f(R)$. $y_1 = f(x_1)$ and $y_2 = f(x_2)$ for some $x_1, x_2 \in R$. $y_1 +' y_2 = f(x_1) +' f(x_2) = f(x_1 + x_2)$. $x_1 + x_2 \in R$ implies

$f(x_1 + x_2) \in f(R)$. $y_1 \cdot' y_2 = f(x_1) \cdot' f(x_2) = f(x_1x_2)$. $x_1x_2 \in R$ implies $f(x_1x_2) \in f(R)$. $-y_1 = -f(x_1) = f(-x_1)$. $-x_1 \in R$ implies $f(-x_1) \in f(R)$.

(d) We suppose R to have a unity v. Then $f(v) \in f(R)$. Any element $y \in f(R)$ is $f(x)$ for some $x \in R$. $y \cdot' f(v) = f(x) \cdot' f(v) = f(xv) = f(x) = y$. $f(v) \cdot' y = f(v) \cdot' f(x) = f(vx) = f(x) = y$. $f(v)$ is a neutral element for $\cdot'$ on $f(R)$, a unity for the subring $f(R)$.

(e) Let x^- represent the multiplicative inverse of x in R. $f(x^-)f(x) = f(x^-x) = f(v) = v'$ and $f(x)f(x^-) = f(xx^-) = f(v) = v'$. $f(x^-)$ is the unique inverse of $f(x)$; $f(x^-) = f(x)^-$. □

Several comments are in order at this point. In part (c) of the preceding theorem we have not asserted that the codomain R' is a commutative ring, only the range $f(R)$. In part (d), $f(v)$ is not necessarily a unity for R' even if R' has a unity; $f(v)$ is only a unity for $f(R)$.

Finally we prove that the inverse of an isomorphism is also an isomorphism.

Theorem. *If $f:R \to R'$ is an isomorphism of rings $\langle R, +, \cdot, \theta\rangle$ and $\langle R', +', \cdot', \theta'\rangle$ then $f^{-1}:R' \to R$ is an isomorphism.*

PROOF. Since f is a bijection $R \to R'$ then there exists a bijection $f^{-1}:R' \to R$ such that $f^{-1} \circ f = I_R$ and $f \circ f^{-1} = I_{R'}$. $y = f(x)$ if and only if $x = f^{-1}(y)$. $f^{-1}(y_1 +' y_2) = f^{-1}(f(x_1) +' f(x_2)) = f^{-1}(f(x_1 + x_2)) = x_1 + x_2 = f^{-1}(y_1) + f^{-1}(y_2)$. We use the fact that f is a surjection to find x_1 and x_2 in R so that $f(x_1) = y_1$ and $f(x_2) = y_2$. A parallel argument proves the formula for multiplication. f^{-1} is a morphism. □

QUESTIONS

1. Let $f:\mathbb{Z} \to \mathbb{Z}$ such that $f(x) = 2x$. Which of these statements are false?
   (A) f is a monomorphism.
   (B) $f(0) = 0$.
   (C) $\{x | f(x) = 0\} = \{0\}$.
   (D) $f(\mathbb{Z})$ is a subring of $\mathbb{Z}$.
   (E) All four statements are true.

2. The image $f(A)$ of a subring A under a ring morphism $f:R \to R'$ is
   (A) a subring of R'
   (B) a subring of R
   (C) a subset of R
   (D) nonempty.
   (E) None of the four possibilities completes a true sentence.

3. The preimage $f^{-1}(B)$ of a subring B by a ring morphism $f:R \to R'$ is
   (A) a subring of R'
   (B) a subring of R
   (C) a subset of R
   (D) nonempty.
   (E) None of the four alternatives completes a true sentence.

4. If $f:R \to R'$ is a ring morphism then
   (A) $f^{-1}(\theta)$ may be empty
   (B) $f^{-1}(\theta)$ is never empty
   (C) $f^{-1}(\theta)$ always contains exactly one member
   (D) $f^{-1}(\theta)$ may contain more than one member
   (E) $f^{-1}(\theta)$ always contains more than one member.

EXERCISES

1. Let $\langle R, +, \cdot, \theta\rangle$ and $\langle R', +', \cdot', \theta'\rangle$ be rings and $f:R \to R'$ be a morphism. Prove A is a subring of R implies $f(A)$ is a subring of R'. Prove B is a subring of R' implies $f^{-1}(B)$ is a subring of R.

2. Let $\langle R, +, \cdot, \theta\rangle$ and $\langle R', +', \cdot', \theta'\rangle$ be rings and $f:R \to R'$ be a morphism. Can R be commutative and R' be not commutative? Can R' be commutative and R be not commutative? Can R be commutative and $f(R)$ be not commutative? Can $f(R)$ be commutative and R be not commutative? In each case support your answer.

3. Define new operations on $\mathbb{Z}$ as follows:
$$x \circ y = x + y - xy, \qquad x * y = x + y - 1.$$
Using the two new binary operations and $\mathbb{Z}$ create a new ring. Show that the new ring is isomorphic to $\langle \mathbb{Z}, +, \cdot, 0\rangle$. *Warning*: it is not specified which of the two new operations plays the role of addition.

4. Let $\langle R, +, \cdot, \theta\rangle$ be a ring, R' be a set with binary operations $+'$ and $\cdot'$, and $f:R \to R'$ be a surjection preserving $+$ in $+'$ and preserving $\cdot$ in $\cdot'$. Prove there exists a neutral element θ' in R' such that $\langle R', +', \cdot', \theta'\rangle$ is a ring and $f:R \to R'$ is an epimorphism.

5. Let $\langle R, +, \cdot, \theta\rangle$ be a ring with neutral element v. For a, any multiplicatively invertible element of R, we define $\varphi_a:R \to R$ such that $\varphi_a(x) = a^{-}xa$ for all $x \in R$. Prove that $\varphi_a:R \to R$ is an isomorphism. φ_a is called an inner automorphism of R and the set of all inner automorphisms of R will be denoted by $\mathscr{I}(R)$.

6. Given the rings $\langle \mathbb{Q}, +, \cdot, 0\rangle$ and $\left\langle \mathbb{Q}^{2\times 2}, +, \cdot, \begin{pmatrix} 0 & 0 \\ 0 & 0 \end{pmatrix}\right\rangle$ we define $f:\mathbb{Q} \to \mathbb{Q}^{2\times 2}$ such that $f(x) = \begin{pmatrix} x & 0 \\ 0 & 0 \end{pmatrix}$. Prove (a) f is a monomorphism; (b) both rings have unities; (c) $f(1)$ is not the unity of $\mathbb{Q}^{2\times 2}$; (d) $f(x^{-}) \neq f(x)^{-}$ for all $x \in \mathbb{Q}$.

7. Given $S \neq \varnothing$ prove that any ring $\langle R, +, \cdot, \theta\rangle$ is isomorphic with a subring of the ring $\langle R^S, +, \cdot, z\rangle$ (cf. Exercise 11, Section 2.2).

8. Given rings $\langle R', +', \cdot', \theta'\rangle$ and $\langle R'', +'', \cdot'', \theta''\rangle$ prove that the projections
$$p_1:R' \times R'' \to R' \quad \text{such that } p_1(x_1, x_2) = x_1$$
$$p_2:R' \times R'' \to R'' \quad \text{such that } p_2(x_1, x_2) = x_2$$
of the product ring into R' and the product ring into R'' are epimorphisms.

9. Given rings $\langle R, +, \cdot, \theta\rangle$, $\langle R', +', \cdot', \theta'\rangle$, $\langle R'', +'', \cdot'', \theta''\rangle$ and morphisms $f:R \to R'$, $g:R \to R''$ prove there exists a morphism $\Phi:R \to R' \times R''$ such that $p_1 \circ \Phi = f$

and $p_2 \circ \Phi = g$ where p_1 and p_2 are the projections of the product $R' \times R''$ (cf. Exercise 8).

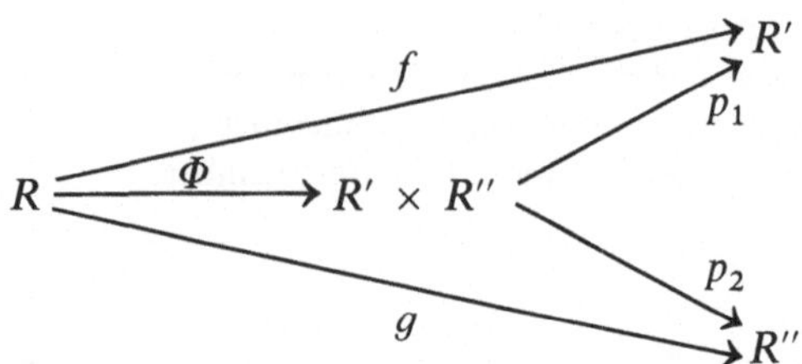

10. Let $\langle R, +, \cdot, \theta \rangle$ be a ring. Prove $\langle R^R, +, \circ, z \rangle$ is not a ring (the second binary operation is composition and z is the function with zero for all its values). Which distributive law fails?

11. Let $\langle R, +, \cdot, \theta \rangle$ and $\langle R', +', \cdot', \theta' \rangle$ be rings. We denote the set of all morphisms from R to R' by $\text{Mor}(R, R')$. Is $\text{Mor}(R, R')$ closed under the functional addition: $(f + g)(x) = f(x) + g(x)$? We call a morphism of a ring R into R, itself, an endomorphism and denote $\text{Mor}(R, R)$ with $\mathscr{E}(R)$. Show that $\langle \mathscr{E}(R), +, \circ, z \rangle$ is a ring.

12. If $\langle R, +, \cdot, \theta \rangle$ is a commutative ring then $\langle R^R, +, \cdot, z \rangle$ is also a commutative ring. Why? Is $\langle \mathscr{E}(R), +, \circ, z \rangle$ also a commutative ring? Again note that the second binary operation is now functional composition.

13. Given a unitary ring $\langle R, +, \cdot, \theta, v \rangle$ prove there exists a subring of $\mathscr{E}(R)$ isomorphic to R. [*Hint*: $f_a(x) = ax$.]

14. How many possible nonisomorphic rings are there with two elements? three elements? four elements? [*Hint*: Study all possible binary operation tables.]

15. In this example we construct $\mathbb{C}$, the complex numbers, from $\mathbb{R}$, the real numbers. The construction closely resembles the product ring construction but has a different multiplication.

(a) Prove $\langle \mathbb{R} \times \mathbb{R}, +, \odot, (0, 0), (1, 0) \rangle$ is a field in which $+$ is the product ring addition and $\odot$ is defined as follows: $(x_1, x_2) \odot (y_1, y_2) = (x_1 y_1 - x_2 y_2, x_1 y_2 + x_2 y_1)$. [*Hint*: The multiplicative inverse of (x_1, x_2) is $(x_1/(x_1^2 + x_2^2), -x_2/(x_1^2 + x_2^2))$.]

(b) Prove that if (x_1, x_2) is a member of $\mathbb{R} \times \mathbb{R}$ then $(x_1, x_2) = (x_1, 0) + (0, 1)(x_2, 0)$.

(c) Prove $\mathbb{R}$ and $\mathbb{R} \times \{0\}$ are isomorphic; i.e., show that $\Phi: \mathbb{R} \to \mathbb{R} \times \mathbb{R}$ such that $\Phi(x) = (x, 0)$ is a monomorphism with range $\mathbb{R} \times \{0\}$.

(d) We denote the number $(0, 1)$ by i and note $i^2 = (-1, 0)$.

(e) We use the isomorphism Φ to identify $\mathbb{R}$ with the subset $\mathbb{R} \times \{0\}$ of $\mathbb{R} \times \mathbb{R}$ so that we write $(x, 0)$ simply as the real x. Show that any (x_1, x_2) in $\mathbb{R} \times \mathbb{R}$ can thus be written as $x_1 + x_2 i$. We call the set $\{x_1 + x_2 i | x_1, x_2 \in \mathbb{R}\}$ the complex numbers and use the symbol $\mathbb{C}$ for the set.

2.6 Quotient rings

In this section we extend the concept of quotient set to quotient ring and thereby introduce ideals.

Preliminary to new material of substance we set out a useful notation.

Definition. If $+$ is a binary operation on a set R, and A and B are subsets of R and x is an element of R then

$$x + A = \{x + a | a \in A\}$$
$$A + B = \{a + b | a \in A \text{ and } b \in B\}.$$

Similarly, if $\cdot$ is a binary operation on a set R, A and B are subsets of R and x is an element of R then

$$xA = \{xa | a \in A\}$$
$$AB = \{ab | a \in A \text{ and } b \in B\}.$$

EXAMPLE. If $A = \{1, 3, 5\}$, $B = \{2, 5, 6\}$ are subsets of $\mathbb{Z}$ and $x = 7$ is an element of $\mathbb{Z}$ then $x + A = \{8, 10, 12\}$, $A + B = \{3, 5, 6, 7, 8, 9, 10, 11\}$, $xA = \{7, 21, 35\}$ and $AB = \{2, 5, 6, 10, 12, 15, 18, 25, 30\}$.

We recommend at this point a review of equivalence relations, quotient sets and quotient maps of Chapter 1. Forearmed, we begin the process of constructing the quotient ring.

Definition. If $\langle R, +, \cdot, \theta \rangle$ is a ring and A is a subset of R then we define $R/A = \{x + A | x \in R\}$.

Theorem. *If $\langle R, +, \cdot, \theta \rangle$ is a ring and A is a subring of R then R/A is a quotient set of R.*

PROOF. In terms of the given subring A we define the following relation on $R: x \sim y$ if and only if $x - y \in A$. This relation on R is an equivalence relation which we now verify:

Given any $x \in R$, $x - x = \theta \in A$ because A is a subring of R. Let $x \sim y$. $x - y \in A$. $-(x - y) \in A$. $y - x \in A$. $y \sim x$. Finally, let $x \sim y$ and $y \sim z$. $x - y \in A$ and $y - z \in A$. $x - z = (x - y) + (y - z) \in A$. $x \sim z$.

Associated with any equivalence relation is a quotient set, $R/\sim = \{x/\sim | x \in R\}$. Each equivalence set in the quotient set $R/\sim$ is in the form $x/\sim = \{z | z \in R \text{ and } z \sim x\}$. We wish now to demonstrate that $R/\sim = R/A$, or in other words, that any equivalence set $x/\sim$ in $R/\sim$ is equal to $x + A$. $x/\sim = \{z | z \sim x\} = \{z | z - x \in A\}$. But $z - x \in A$ if and only if $z - x = a$ for some $a \in A$ if and only if $z = x + a$ for some $a \in A$ if and only if $z \in x + A$. We conclude $x/\sim = x + A$. □

Members of the quotient set R/A are to be called cosets as well as equivalence classes. We wish now to prove that every coset in R/A has the same number of members an any other coset in R/A. This result is not true for quotient sets in general which may partition a set into subsets of varying sizes. We begin by introducing a term for sets of equal size.

Definition. Sets A and B are called *equipotent* if and only if there exists a bijection $f: A \to B$.

Theorem. *Let $\langle R, +, \cdot, \theta\rangle$ be a ring and A be a subring of R. Then any two cosets of the quotient set R/A are equipotent.*

PROOF. Let $b + A$ and $c + A$ be members of R/A. We define $f: b + A \to c + A$ such that $f(b + x) = c + x$. $b + x_1 = b + x_2$ if and only if $x_1 = x_2$ if and only if $c + x_1 = c + x_2$ using cancellation and the uniqueness of addition. This shows $b + x_1 = b + x_2$ if and only if $f(b + x_1) = f(b + x_2)$: f is a well-defined function and an injection. f is also a surjection because given any $c + x \in c + A$ there exists an $b + x \in b + A$ such that $f(b + x) = c + x$. □

EXAMPLE. $3\mathbb{Z}$, the set of all integral multiples of 3, is a subring of $\mathbb{Z}$. In the quotient set $\mathbb{Z}/3\mathbb{Z}$ two integers are equivalent if and only if their difference is a multiple of 3. $1 \sim 4$, $7 \sim 10$, $-2 \sim 4$, $0 \sim 6$, etc. $\mathbb{Z}/3\mathbb{Z} = \{3\mathbb{Z}, 1 + 3\mathbb{Z}, 2 + 3\mathbb{Z}\}$. We observe $1 + 3\mathbb{Z} = 4 + 3\mathbb{Z} = 1/\sim = 4/\sim$. $7 + 3\mathbb{Z} = 10 + 3\mathbb{Z} = 1 + 3\mathbb{Z}$. In this example each of the three cosets is an infinite set (see Figure 2.2).

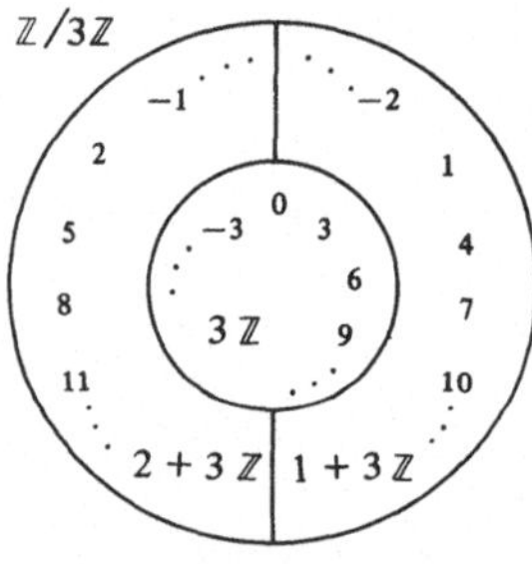

Figure 2.2

Having constructed a quotient set R/A from a given subring A we now move to make R/A into a ring by introducing operations on the set R/A. As we shall later show it is not sufficient for A to be a subring to accomplish this construction. We therefore introduce a new concept at this stage and use this new concept to construct the quotient ring R/A.

Definition. Let $\langle R, +, \cdot, \theta\rangle$ be a ring. A is an *ideal* or *normal subring* of R if and only if A is a subring of R and $r \in R$, $x \in A$ imply rx and xr belong to A.

EXAMPLES. The set $\{(r, 2s) | r \in \mathbb{Z} \text{ and } s \in \mathbb{Z}\}$ is an ideal of the product ring $\langle \mathbb{Z} \times \mathbb{Z}, +, \cdot, (0, 0)\rangle$. The set $3\mathbb{Z}$ is an ideal of the ring $\mathbb{Z}$. The set $\mathbb{Z}$ is a subring of the ring $\mathbb{Q}$ but not an ideal of $\mathbb{Q}$.

The condition $r \in R$ and $x \in A$ imply rx and $xr \in A$ can be equivalently stated as $RA \subseteq A$ and $AR \subseteq A$.

Theorem. *Let A be an ideal of the ring $\langle R, +, \cdot, \theta \rangle$. Then $\langle R/A, \mp, \bar{\cdot}, A \rangle$ is a ring in which $\mp$ and $\bar{\cdot}$ are defined by*

$$(x + A) \mp (y + A) = x + y + A$$
$$(x + A) \bar{\cdot} (y + A) = xy + A.$$

Proof. The definitions given for $\mp$ and $\bar{\cdot}$ are both definitions which depend upon particular representatives of the coset $x + A$ and $y + A$. Before we know the operations to be well defined we must show that the definitions are actually independent of the representatives x and y. We must show the following: if $x + A = x' + A$ and $y + A = y' + A$ then $x + y + A = x' + y' + A$ and $xy + A = x'y' + A$. $x + A = x' + A$ implies $x - x' \in A$. $y + A = y' + A$ implies $y - y' \in A$. $(x - x') + (y - y') \in A$. $(x + y) - (x' + y') \in A$. $x + y + A = x' + y' + A$. This proves the sum to be well defined. The product is more difficult and uses the ideal properties. $x - x' \in A$ and $y - y' \in A$. $(x - x')(y - y') \in A$ because A is a subring.

$$\begin{aligned} (x - x')(y - y') &= a \quad \text{for some } a \in A. \\ xy &= xy' + x'(y - y') + a \\ xy - x'y' &= xy' - x'y' + x'(y - y') + a \\ &= (x - x')y' + x'(y - y') + a. \end{aligned}$$

$(x - x')y'$ belongs to A because $AR \subseteq A$. $x'(y - y')$ belongs to A because $RA \subseteq A$. $a \in A$. The sum of all three terms belongs to A. $xy - x'y' \in A$. $xy + A = x'y' + A$. Multiplication on R/A is well defined.

We now verify that $\mp$ is associative and commutative.

$$(x + A) \mp (y + A) = x + y + A = y + x + A = (y + A) \mp (x + A).$$

$$\begin{aligned} (x + A) \mp [(y + A) \mp (z + A)] &= (x + A) \mp (y + z + A) \\ &= x + (y + z) + A \\ &= (x + y) + z + A \\ &= (x + y + A) \mp (z + A) \\ &= [(x + A) \mp (y + A)] \mp (z + A). \end{aligned}$$

We now show that $\bar{\cdot}$ is also associative. $(x + A) \bar{\cdot} [(y + A) \bar{\cdot} (z + A)] = (x + A) \bar{\cdot} (yz + A) = x(yz) + A = (xy)z + A = (xy + A) \bar{\cdot} (z + A) = [(x + A) \bar{\cdot} (y + A)] \bar{\cdot} (z + A)$. The distributive equations can be verified in a similar manner.

The neutral element for $\mp$ is $\theta + A$ which is equal to A. $A \mp (x + A) = (\theta + A) \mp (x + A) = \theta + x + A = x + A$. $(x + A) \mp A = x + A$ also. We next show every element of R/A is $\mp$-invertible. $(x + A) \mp (-x + A) = x + (-x) + A = \theta + A = A$. And because $\mp$ is commutative $(-x + A) \mp (-x + A) = A$ also. $-(x + A) = -x + A$. □

Corollary. *If $\langle R, +, \cdot, \theta \rangle$ is a ring with unity v and A is an ideal of R then $v + A$ is a unity for the quotient ring $\langle R/A, \mp, \bar{\cdot}, A \rangle$.*

PROOF. $(x + A)(v + A) = xv + A = x + A$. $(v + A)(x + A) = vx + A = x + A$. □

Corollary. *If $\langle R, +, \cdot, \theta\rangle$ is a ring which is commutative and A is an ideal of R then $\langle R/A, \mp, \bar{\cdot}, A\rangle$ is a commutative ring.*

PROOF. Left to the reader. □

Having constructed the quotient ring we now wish to make good our claim that it is necessary to take A to be an ideal for the construction.

Theorem. *Any quotient ring of a given ring $\langle R, +, \cdot, \theta\rangle$ in which the binary operations are defined by representatives, i.e.,*

$$x/\sim \mp y/\sim = (x + y)/\sim$$
$$x/\sim \bar{\cdot}\ y/\sim = (xy)/\sim$$

must be $\langle R/A, \mp, \bar{\cdot}, A\rangle$ for some ideal A of R.

PROOF. Let $R/\sim$ be a quotient set of R which is a ring with respect to the binary operations $\mp$, $\bar{\cdot}$ as defined in the hypothesis. Because the binary operations are well defined the following two statements hold:

$$x \sim x' \text{ and } y \sim y' \quad \text{imply } x + y \sim x' + y'$$
$$x \sim x' \text{ and } y \sim y' \quad \text{imply } xy \sim x'y' \quad \text{for all } x, y, x', y' \in R.$$

We will use the hypothesis in this form.

$\theta/\sim$ is clearly the neutral element of addition for the given quotient ring $R/\sim$. We proceed to show that the set $\theta/\sim$ is an ideal of R. Let $x, y \in \theta/\sim$. $x \sim \theta$ and $y \sim \theta$. $x + y \sim \theta + \theta = \theta$. $x + y \in \theta/\sim$. $\theta/\sim$ is closed under addition in R. Let $x \in \theta/\sim$. $x \sim \theta$. $-x \sim -x$. $x + (-x) \sim \theta + (-x)$. $\theta \sim (-x)$. $-x \in \theta/\sim$. $\theta/\sim$ is closed under negation. Next we consider product closure. Let $x, y \in \theta/\sim$. $x \sim \theta$ and $y \sim \theta$. $xy \sim \theta\theta = \theta$. $xy \in \theta/\sim$. This shows so far that $\theta/\sim$ is a subring of R. Now let $r \in R$ and $x \in \theta/\sim$. $x \sim \theta$ and $r \sim r$. $xr \sim \theta r$ and $rx \sim r\theta$. $xr \sim \theta$ and $rx \sim \theta$. $xr \in \theta/\sim$ and $rx \in \theta/\sim$. We have showed that $\theta/\sim$ is an ideal of R. Denote this ideal with the letter B.

$x \sim y$ if and only if $x + (-y) \sim y + (-y)$ if and only if $x - y \sim \theta$ if and only if $x - y \in \theta/\sim$ if and only if $x - y \in B$. The given equivalence relation $\sim$ defining the quotient set $R/\sim$ is exactly the same as the equivalence relation generated by the ideal B. It follows that the quotient set $R/\sim$ is identical with the quotient set R/B. Thus the quotient ring $\langle R/\sim, \mp, \bar{\cdot}, \theta/\sim\rangle = \langle R/B, \mp, \bar{\cdot}, B\rangle$. □

EXAMPLE. $n\mathbb{Z}$ is an ideal of $\langle \mathbb{Z}, +, \cdot, 0\rangle$ and therefore $\langle \mathbb{Z}/n\mathbb{Z}, \mp, \bar{\cdot}, n\mathbb{Z}\rangle$ is a quotient ring. We abbreviate $\mathbb{Z}/n\mathbb{Z}$ with $\mathbb{Z}_n$, which is general usage. The quotient ring $\mathbb{Z}_n$ has exactly n members. $\mathbb{Z}_n = \{n\mathbb{Z}, 1 + n\mathbb{Z}, 2 + n\mathbb{Z}, \ldots, n - 1 + n\mathbb{Z}\}$. In the theory of numbers $\mathbb{Z}_n$ is called the residue class ring of integers modulo n or simply the integers modulo n. The equivalence relation defining the quotient ring $\mathbb{Z}_n$ is called congruence and written $x \equiv y$ modulo n meaning $x - y \in n\mathbb{Z}$. We shall also denote $\mathbb{Z}_n$ by $\{\bar{0}, \bar{1}, \bar{2}, \ldots, \overline{n-1}\}$ to give a shorter notation for the cosets.

QUESTIONS

1. Let $\mathbb{Z}_6$ be denoted by $\{\bar{0}, \bar{1}, \bar{2}, \bar{3}, \bar{4}, \bar{5}\}$ in which $\bar{m} = m + 6\mathbb{Z}$. Which of these statements are true?
(A) The multiplicative inverse of $\bar{4}$ is $\bar{3}$.
(B) The multiplicative inverse of $\bar{5}$ is $\bar{1}$.
(C) The additive inverse of $\bar{2}$ is $\bar{3}$.
(D) The multiplicative inverse of $\bar{5}$ is $\bar{2} + \bar{3}$.
(E) None of the four statements is true.

2. Let $\langle R, +, \cdot, \theta \rangle$ be a ring. S is a subring of R implies
(A) $S \cup R = S$
(B) $\theta \in S \cap R$
(C) $SS \subseteq S$
(D) $S + S = S$.
(E) None of the four choices completes a true sentence.

3. Let S be a subring of the ring $\langle R, +, \cdot, \theta, v \rangle$, a unitary ring. If the multiplicative unity v belongs to S then
(A) $SS = S$
(B) $RS \subseteq S$
(C) $R + R \subseteq S$
(D) $R + S \subseteq S$.
(E) None of the four alternatives completes a true sentence.

4. In the ring $\mathbb{Z}_{12}$ the equation $\bar{4}x = \bar{3}$ has
(A) no solution
(B) one solution
(C) two solutions
(D) three solutions
(E) four solutions.

5. In the ring $\mathbb{Z}_{12}$ the equation $\bar{4}x = \bar{4}$ has
(A) no solution
(B) one solution
(C) two solutions
(D) three solutions
(E) four solutions.

EXERCISES

1. Give an example of a subring of a ring which fails to be an ideal.

2. In $\mathbb{Z}_6$ which elements have multiplicative inverses and which do not?

3. In $\mathbb{Z}_5$ which elements have multiplicative inverses and which do not?

4. Find all solutions to these equations in $\mathbb{Z}_6$:

$$(2 + 6\mathbb{Z})x = 4 + 6\mathbb{Z}$$
$$(2 + 6\mathbb{Z})x = 3 + 6\mathbb{Z}.$$

5. Find all solutions to these equations in $\mathbb{Z}_5$:

$$(2 + 5\mathbb{Z})x = 4 + 5\mathbb{Z}$$
$$(2 + 5\mathbb{Z})x = 3 + 5\mathbb{Z}.$$

6. Prove that a field has but two ideals. Does a division ring have only two ideals?

7. Give an example of a ring R and a quotient ring R/A and an element $r \in R$ such that r is a nontrivial divisor of zero in R yet $r + A$ is not a nontrivial divisor of zero in R/A.

8. Give an example of a ring R and a quotient ring R/A and an element $r \in R$ such that r is not a nontrivial divisor of zero in R yet $r + A$ is a nontrivial divisor of zero in R/A.

9. Let $\mathbb{Z}[X]$ stand for the set of all polynomials with integer coefficients (cf. Exercise 7 of Section 2.4). Show that the subset of all polynomials with even coefficients is a subring of the ring of all polynomials with integral coefficients.

2.7 Morphisms and quotient rings

In this section we prove the fundamental morphism theorem for rings and define kernel of a morphism.

We will, in this section, be extending a number of theorems we proved for sets to rings.

Theorem. *Let $\langle R, +, \cdot, \theta \rangle$ be a ring and A an ideal of R. Then the quotient map $\varphi: R \to R/A$ is an epimorphism.*

PROOF. The quotient map of a set into its quotient set is a function which sends each element of R into its containing coset. $\varphi: R \to R/A$ is a surjection such that $\varphi(x) = x + A$. $\varphi(x + y) = x + y + A = (x + A) \mp (y + A) = \varphi(x) \mp \varphi(y)$. $\varphi(xy) = xy + A = (x + A) \bar{\cdot} (y + A) = \varphi(x) \bar{\cdot} \varphi(y)$. □

Associated with every morphism $f: R \to R'$ are two distinguished sets, the kernel of f and the range of f. We have previously defined the range of f and proved it to be a subring of R'. We now define the kernel of f.

Definition. Let $f: R \to R'$ be a morphism of the rings $\langle R, +, \cdot, \theta \rangle$ and $\langle R', +', \cdot', \theta' \rangle$. By definition *kernel* $f = \{x | x \in R \text{ and } f(x) = \theta'\}$.

Theorem. *Let $f: R \to R'$ be a morphism of the rings $\langle R, +, \cdot, \theta \rangle$ and $\langle R', +', \cdot', \theta' \rangle$. Then kernel f is an ideal of R.*

PROOF. Let $x, y \in \ker f$. $f(x) = \theta'$ and $f(y) = \theta'$. $f(xy) = f(x)f(y) = \theta'\theta' = \theta'$. $xy \in \ker f$. $f(x + y) = f(x) +' f(y) = \theta' + \theta' = \theta'$. $x + y \in \ker f$. $f(-x) = -f(x) = -\theta' = \theta'$. $-x \in \ker f$. $f(\theta) = \theta'$. $\theta \in \ker f$. $\ker f \neq \varnothing$. Let $r \in R$

and $x \in \ker f$. $f(rx) = f(r)f(x) = f(r)\theta' = \theta'$. $rx \in \ker f$. $f(xr) = f(x)f(r) = \theta' f(r) = \theta'$. $xr \in \ker f$. □

EXAMPLE. The quotient map $\varphi: \mathbb{Z} \to \mathbb{Z}_7$ is $\varphi(x) = x + 7\mathbb{Z}$. The morphism $f: \mathbb{Z} \to \mathbb{Z}$ such that $f(x)$ is the remainder upon dividing x by 7 has as its kernel the ideal $7\mathbb{Z}$.

We now prove the fundamental morphism theorem for rings which extends the results on set theory of Section 1.7 to the ring operations.

Theorem. *Let $\langle R, +, \cdot, \theta\rangle$ and $\langle R', +', \cdot', \theta'\rangle$ be rings and $f: R \to R'$ a morphism. Then there exist*

an ideal A;
a quotient ring R/A;
an epimorphism $\varphi: R \to R/A$; and
a monomorphism $f': R/A \to R'$, such that $f' \circ \varphi = f$.

PROOF. From the fundamental morphism theorem for sets found in Section 1.7 we can assert the existence of an equivalence relation γ on R ($x\gamma y$ if and only if $f(x) = f(y)$), a quotient set R/γ, a surjection $\varphi: R \to R/\gamma$ such that $\varphi(x) = x/\gamma$, and an injection $f': R/\gamma \to R'$ such that $f'(x/\gamma) = f(x)$ and with $f' \circ \varphi = f$. What remains to prove or establish are the various algebraic or operational properties claimed in the conclusion of the theorem.

We define A to be $\{x | x \in R \text{ and } f(x) = \theta'\}$. A is therefore the kernel of f and an ideal of R. This ideal defines a quotient ring R/A. $x\gamma y$ if and only if $f(x) = f(y)$ if and only if $f(x) - f(y) = \theta'$ if and only if $f(x - y) = \theta'$ if and only if $x - y \in A$. Thus the quotient ring R/A is identical with the quotient set R/γ with $x/\gamma = x + A$ for all $x \in R$. Writing the defining equations for φ and f' in algebraic notation we have $\varphi: R \to R/A$ such that $\varphi(x) = x + A$ and $f': R/A \to R'$ such that $f'(x + A) = f(x)$. We have previously verified that the quotient map is a morphism and now we verify that f' is a morphism. $f'((x + A) \mp (y + A)) = f'(x + y + A) = f(x + y) = f(x) +' f(y) = f'(x + A) +' f'(y + A)$. $f'((x + A) \bar{\cdot} (y + A)) = f'(xy + A) = f(xy) = f(x) \cdot' f(y) = f'(x + A) \cdot' f'(y + A)$. □

Corollary. *Let $f: R \to R'$ be a morphism of the rings $\langle R, +, \cdot, \theta\rangle$ and $\langle R', +', \cdot', \theta'\rangle$. Then there exists an isomorphism $f': R/(\ker f) \to \text{range } f$.*

PROOF. By restricting the codomain of $f': R/A \to R'$ from R' to $f(R)$, the range of f, the monomorphism f' will become a surjection also, making it an isomorphism. The ideal A is the kernel of f. □

QUESTIONS

1. For a morphism $f: R \to R'$ of rings $\langle R, +, \cdot, \theta\rangle$ and $\langle R', +', \cdot', \theta'\rangle$ which of these statements are true?
   (A) $\ker f \subseteq \text{range } f$.
   (B) $R/\ker f = \{x + \ker f | x \in R\}$.

(C) $\{\theta\} \subseteq \ker f \subseteq R$.
(D) $f(\ker f) = \{\theta'\}$.
(E) $f^{-1}(R') = f^{-1}(\text{range } f)$.

2. The function $f:\mathbb{Z} \to \mathbb{Z}_n$ such that $f(x) = x + n\mathbb{Z}$
(A) is an epimorphism
(B) has kernel equal to $n\mathbb{Z}$
(C) is a quotient map
(D) is a monomorphism
(E) is identical with the function $g:\mathbb{Z} \to \mathbb{Z}_n$ such that $g(x) = nx + n\mathbb{Z}$.

3. Which of the following are subrings of the ring $2\mathbb{Z}$?
(A) $\mathbb{Z}_2$
(B) $\mathbb{Z}_4$
(C) $\mathbb{Z}_3$
(D) $6\mathbb{Z}$.
(E) None of the four is a subring of $2\mathbb{Z}$.

4. If R, R', R'' are rings and $f:R \to R'$, $g:R' \to R''$ then
(A) f and g monomorphic imply $g \circ f$ monomorphic
(B) f and g epimorphic imply $g \circ f$ epimorphic
(C) $g \circ f$ epimorphic implies g epimorphic
(D) $g \circ f$ monomorphic implies g monomorphic
(E) $g \circ f$ epimorphic implies f epimorphic
(F) $g \circ f$ monomorphic implies f monomorphic.

5. According to the fundamental morphism theorem if we have given a morphism $p_1:\mathbb{Z} \times \mathbb{Z} \to \mathbb{Z}$ such that $p_1(x_1, x_2) = x_1$ then there exist morphisms φ and p_1' such that
(A) $\ker p_1 = \{0\} \times \mathbb{Z}$
(B) $p_1':(\mathbb{Z} \times \mathbb{Z})/(\{0\} \times \mathbb{Z}) \to \mathbb{Z}$
(C) $p_1'((x_1, x_2) + \{0\} \times \mathbb{Z}) = x_1$
(D) $p_1' \circ \varphi = p_1$
(E) $\varphi((x_1, x_2) + \{0\} \times \mathbb{Z}) = (x_1, x_2)$.

Exercises

1. Let $f:R \to R'$ be a ring morphism. Prove f is a monomorphism if and only if $\ker f = \{\theta\}$.

2. Show that the only morphisms $\mathbb{Z} \to \mathbb{Z}_4$ are $f(x) = \bar{0}$ and $g(x) = \bar{x}$.

3. Show that the only morphisms $\mathbb{Z} \to \mathbb{Z}_6$ are $f(x) = \bar{0}$, $g(x) = \bar{x}$ and $h(x) = 3\bar{x}$.

4. Let m be different from 0 and 1 and belong to $\mathbb{N}$. Show that $f:\mathbb{Z} \to \mathbb{Z}_n$ such that $f(x) = mx$ is a morphism if and only if $n = m(m - 1)$.

5. Let $f:R \to R'$ and $g:R' \to R''$ be ring morphisms. Prove $\ker f \subseteq \ker(g \circ f)$ and $\text{range}(g \circ f) \subseteq \text{range } g$.

6. Given $f:R \to R'$ and $g:R \to R'$, both ring morphisms, we define $f \times g:R \times R \to R' \times R'$ such that $(f \times g)(x_1, x_2) = (f(x_1), g(x_2))$.
(a) Prove $f \times g$ is a morphism.

(b) Prove $\ker(f \times g) = \ker f \times \ker g$.
(c) Prove $\text{range}(f \times g) = \text{range } f \times \text{range } g$.
(d) Prove $f \times g$ is a monomorphism if and only if f and g are morphisms.
(e) Prove $f \times g$ is an epimorphism if and only if f and g are epimorphisms.
(f) Prove $(R \times R)/(\ker f \times \ker g)$ is isomorphic with $(R/\ker f) \times (R/\ker g)$. [*Hint*: Define a function $F: R \times R \to (R/\ker f) \times (R/\ker g)$ such that $F(x_1, x_2) = (x_1 + \ker f, x_2 + \ker g)$ and use the fundamental morphism theorem.

7. Show that $\mathbb{Z}_6$ and $\mathbb{Z}_2 \times \mathbb{Z}_3$ are isomorphic rings.

8. Show that $\mathbb{Z}_4$ and $\mathbb{Z}_2 \times \mathbb{Z}_2$ are not isomorphic rings.

9. Find all morphisms $\mathbb{Z} \to \mathbb{Z} \times \mathbb{Z}$.

10. Find all morphisms $\mathbb{Z} \times \mathbb{Z} \to \mathbb{Z}$.

2.8 Ideals

In this section we develop relationships between special kinds of ideals and special kinds of quotient rings.

We call any ideal which is not the entire ring itself a proper ideal of the ring and we call any ideal which is not the ideal consisting exactly of zero, $\{\theta\}$, a nontrivial ideal. Alternatively, we call R the improper ideal of R and call $\{\theta\}$ the trivial ideal of R.

We notice there is an order on any collection of ideals, the order of set inclusion that exists on any collection of sets.

Example. The following statements of order about the set of ideals of $\mathbb{Z}$ are all true. $4\mathbb{Z} \subseteq 2\mathbb{Z}$. $2\mathbb{Z} \nsubseteq 3\mathbb{Z}$. $3\mathbb{Z} \nsubseteq 2\mathbb{Z}$. One moral here is that not every pair of ideals are comparable with respect to inclusion; this order is often called partial for this reason.

Definition. An element M of an ordered set $\mathscr{C}$ is a maximal element of $\mathscr{C}$ if and only if no other element of $\mathscr{C}$ is strictly larger than M.

Example. Both $2\mathbb{Z}$ and $3\mathbb{Z}$ are maximal ideals in the set of all proper ideals of $\mathbb{Z}$. $2\mathbb{Z}$ is not called a maximum ideal because there are ideals, namely $3\mathbb{Z}$, which it does not surpass. The noncomparability of certain pairs permits this differentiation between the terms maximum and maximal.

Lemma. *An ideal A of a unitary commutative ring $\langle R, +, \cdot, \theta, v\rangle$ is proper if and only if $v \notin A$.*

Proof. Suppose $v \notin A$. Then clearly $A \neq R$ and A is proper. For the converse suppose $v \in A$. $A \subseteq R$. We now prove $A = R$ by proving $R \subseteq A$. Suppose $r \in R$. Then $rv \in A$. $r \in A$. □

Theorem. *Let A be an ideal of a unitary commutative ring $\langle R, +, \cdot, \theta, v\rangle$. Then A is a maximal proper ideal of R if and only if R/A is a field.*

Proof. First, we assume A is an ideal of R and that R/A is a field. We wish to demonstrate that A is a maximal proper ideal of R. Let B be any ideal of R strictly larger than A: $A \subset B \subseteq R$. There exists an element $b \in B$ such that $b \notin A$. There is an element $c \in R$ such that $(b + A)(c + A) = v + A$ because R/A is a field and $b + A \neq A$. $bc - v \in A$. $bc - v = a$ for some $a \in A$. $v = bc - a$. Since $bc \in B$ and $a \in B$ we have $v \in B$. $B = R$. Thus A is a maximal ideal of R. A is proper because R/A must have a unity $v + A$ different from A implying $v \notin A$.

For the converse, assume A is a maximal proper ideal of R. $A \neq R$ implies $v \notin A$ which implies $v + A \neq A$. The unity of the quotient ring R/A is different from the zero. We now demonstrate the existence of multiplicative inverses in R/A. Let $x + A \in R/A$ and $x + A \neq A$. $x \notin A$. Consider the set $\{rx + a | r \in R$ and $a \in A\}$. This set is an ideal of R containing A yet not equal to A. It is, therefore, R itself. Hence $v \in \{rx + a | r \in R$ and $a \in A\}$. $v = r'x + a'$ for some $r' \in R$, $a' \in A$. $r'x - v \in A$. $(r' + A)(x + A) = r'x + A = v + A$. This, with commutativity, proves $r' + A$ to be the multiplicative inverse of $x + A$ in the quotient ring R/A. R/A is a field. □

Examples. Both $\mathbb{Z}_2 = \mathbb{Z}/2\mathbb{Z}$ and $\mathbb{Z}_3 = \mathbb{Z}/3\mathbb{Z}$ are fields and $2\mathbb{Z}$ and $3\mathbb{Z}$ are maximal ideals of $\mathbb{Z}$. $\mathbb{Z}/4\mathbb{Z}$ is not a field and $4\mathbb{Z}$ is not a proper maximal ideal of $\mathbb{Z}$. $4\mathbb{Z} \subset 2\mathbb{Z} \subset \mathbb{Z}$.

We now introduce another special ideal.

Definition. An ideal A of a ring $\langle R, +, \cdot, \theta\rangle$ is *prime* if and only if $xy \in A$ implies $x \in A$ or $y \in A$.

Theorem. *Let $\langle R, +, \cdot, \theta, v\rangle$ be a unitary commutative ring and A be a proper ideal of R. Then A is prime if and only if R/A is an integral domain.*

Proof. Assume A is a proper prime ideal of R. We can furthermore take $v \neq \theta$ because $R = \{\theta\}$ has no proper ideals. Let $(x + A)(y + A) = A$. $xy + A = A$. $xy \in A$. $x \in A$ or $y \in A$. $x + A = A$ or $y + A = A$. Furthermore since A is proper, $v \notin A$ implying $v + A \neq A$. R/A is an integral domain.

To prove the converse assume R/A is an integral domain. Let $xy \in A$. $xy + A = A$. $(x + A)(y + A) = A$. $x + A = A$ or $y + A = A$. $x \in A$ or $y \in A$. A is a prime ideal. □

Theorem. *Let $\langle R, +, \cdot, \theta, v\rangle$ be a commutative ring with unity. Then every maximal proper ideal of R is a prime ideal.*

Proof. If A is a maximal proper ideal of R then R/A is a field. If R/A is a field then R/A is an integral domain. If R/A is an integral domain then A is a prime ideal. □

We will now define the ideal $\langle S\rangle$ generated by a set S in a way analogous to the way the subring $[S]$ generated by a set S was defined.

Theorem. *Let $\langle R, +, \cdot, \theta\rangle$ be a ring and S a subset of R. Then $\langle S\rangle = \bigcap\{A \mid A$ is an ideal of R and $S \subseteq A\}$ is an ideal of R and is the smallest ideal of R containing S.*

PROOF. The proof is left to the reader. □

Definition. Let $\langle R, +, \cdot, \theta\rangle$ be a ring and S a subset of R. We define $\langle S\rangle$ to be the ideal *generated* by the set S.

EXAMPLE. In $\mathbb{Z}$, $\langle 2\rangle = 2\mathbb{Z}$ and $\langle 2, 3\rangle = \mathbb{Z}$. In $\mathbb{Z}$, $\langle 2\mathbb{Z} \cup 3\mathbb{Z}\rangle = \mathbb{Z}$. In $\mathbb{Q}$, $\langle \mathbb{Z}\rangle = \mathbb{Q}$ whereas $[\mathbb{Z}] = \mathbb{Z}$.

QUESTIONS

1. $2\mathbb{Z}$ and $3\mathbb{Z}$ are noncomparable with respect to inclusion because
   (A) $2\mathbb{Z} \cap 3\mathbb{Z} = \varnothing$
   (B) $2\mathbb{Z} \cup 3\mathbb{Z} = \mathbb{Z}$
   (C) $21 \in 3\mathbb{Z}$ and $21 \notin 2\mathbb{Z}$ and $8 \notin 3\mathbb{Z}$ and $8 \in 2\mathbb{Z}$
   (D) both ideals are maximal proper ideals of $\mathbb{Z}$.
   (E) None of the four conditions are relevant.

2. $m\mathbb{Z} \subseteq n\mathbb{Z}$ if and only if
   (A) $m = kn$ for some $k \in \mathbb{Z}$
   (B) $n = km$ for some $k \in \mathbb{Z}$
   (C) $mn = 1$
   (D) $m\mathbb{Z} \cup n\mathbb{Z} = n\mathbb{Z}$.
   (E) None of the four possibilities completes a true sentence.

3. Which of the following imply $6\mathbb{Z}$ is a prime ideal of $\mathbb{Z}$?
   (A) $3 \cdot 2 \in 6\mathbb{Z}$ and $3 \notin 6\mathbb{Z}$ and $2 \notin 6\mathbb{Z}$.
   (B) $6\mathbb{Z} \subset 2\mathbb{Z} \subset \mathbb{Z}$.
   (C) $6\mathbb{Z} \subset 3\mathbb{Z} \subset \mathbb{Z}$.
   (D) $12\mathbb{Z} \subset 6\mathbb{Z} \subset \mathbb{Z}$.
   (E) None of the four imply $6\mathbb{Z}$ is a prime ideal of $\mathbb{Z}$.

4. Which of these statements are true?
   (A) $6\mathbb{Z}$ is a prime ideal of the ring $2\mathbb{Z}$.
   (B) $6\mathbb{Z}$ is a maximal ideal of the ring $2\mathbb{Z}$.
   (C) There are ideals strictly between ($\subset$) $6\mathbb{Z}$ and $2\mathbb{Z}$.
   (D) 2 is a unity for $2\mathbb{Z}$.
   (E) None of the four statements is true.

5. Which of the following statements are true?
   (A) A trivial ideal is improper.
   (B) An improper ideal is trivial.
   (C) A proper ideal is not trivial.
   (D) There exist nontrivial ideals which are proper.
   (E) None of the four statements is true.

Exercises

1. Give an example of a ring other than $\mathbb{Z}$ and a proper nontrivial ideal of that ring.

2. Show $\{m2 + n6 | m, n \in \mathbb{Z}\}$ to be an ideal of $\mathbb{Z}$.

3. Show that $3\mathbb{Z}$ is a maximal proper ideal of $\mathbb{Z}$.

4. If A and B are ideals of $\langle R, +, \cdot, \theta\rangle$ prove $A \cap B$ and $A + B$ are ideals of R. Prove $A + B$ is the smallest ideal containing both A and B. Prove $A \cap B$ is the largest ideal contained in both A and B.

5. If K and L are fields prove that the product ring $K \times L$ cannot be a field. Find all the ideals of $K \times L$.

6. Prove that if $\langle R, +, \cdot, \theta, v\rangle$ is a unitary commutative ring with exactly two ideals then R is a field.

7. Show that the ring $\left\langle \mathbb{Q}^{2\times 2}, +, \cdot, \begin{pmatrix} 0 & 0 \\ 0 & 0 \end{pmatrix} \right\rangle$ has exactly two ideals but is not a field nor a division ring.

8. Prove that if $\langle R, +, \cdot, \theta, v\rangle$ is a commutative unitary ring then $Ra = \{ra | r \in R\}$ is an ideal of R for all $a \in \mathrm{R}$; moreover, $\langle a\rangle = Ra$.

9. Give an example of a ring $\langle R, +, \cdot, \theta\rangle$ (noncommutative) with a set Ra which is not an ideal of R.

10. Give an example of a commutative ring R and an element $a \in R$ such that $\langle a\rangle \neq Ra$.

11. Let $a_1, a_2, \ldots, a_n$ be a finite number of elements in a commutative ring $\langle R, +, \cdot, \theta\rangle$. Show that $Ra_1 + Ra_2 + \cdots + Ra_n = \{r_1a_1 + r_2a_2 + \cdots + r_na_n | r_1, \ldots, r_n \in R\}$ is an ideal of R. Need a_1 belong to the ideal?

12. Find all the ideals of $\mathbb{Z} \times \mathbb{Z}$.

13. An ideal that is generated by a single element of a ring is called a principal ideal; i.e., A is principal if and only if $A = \langle a\rangle$ for some $a \in R$.
(a) Show that the image of any ideal under a morphism is an ideal.
(b) Show that the preimage of an ideal under a morphism is an ideal.
(c) Show that the image of a principal ideal under a morphism is a principal ideal.
(d) Show that the preimage of a principal ideal under a morphism is not necessarily principal.

14. Let $\langle R, +, \cdot, \theta\rangle$ be a ring. Prove that $C = \{x | x \in R \text{ and } xy = yx \text{ for all } y \in R\}$ is a subring of R. Prove that C is not necessarily an ideal by considering the example $R = \mathbb{Z}^{2\times 2}$.

Rings: Natural numbers and integers 3

The positive whole numbers are undoubtedly the oldest and most primitive objects of all mathematics. They formed, and still form, the basis from which all other mathematics sprang. Zero appeared upon the mathematical scene later. The natural numbers are so much the genesis of all mathematics that the 19th century mathematician Leopold Kronecker was led to say that God created the natural numbers and man created everything else in mathematics. The axiomization presented in this chapter of the natural numbers is credited to Giuseppe Peano, another of the great 19th century mathematicians who reworked and solidified the foundations of our number systems. The whole numbers as a key and insight to the nature of the universe were recognized by the Pythagoreans in the 6th century B.C. When Pythagoras said, "All is number," he meant the positive integers.

The Peano development of the natural numbers (we include zero with the positive whole numbers in our set $\mathbb{N}$) is a grand sequence of exercises in mathematical induction. Mathematical induction and the natural numbers are inseparable. A student of mathematics needs to master the technique of mathematical induction and feel perfectly comfortable with it. Mathematical induction to the student of mathematics should be as intuitively evident as the most obvious theorems of geometry or manipulations of algebra.

From the foundation of the natural number system we move in Section 3.5 to a construction of the integers. We adjoin the negative whole numbers to the natural numbers and build of the union the integers. Historically the negative numbers were nowhere so obvious to mankind as the positive ones. The name, negative, hardly indicates an affirmative attitude towards such numbers. Negative numbers as solutions to equations were still being rejected at the time of the Renaissance and even René Descartes in his analytical geometry did not see fit to accord negative numbers full status. By the time of Karl Friedrich Gauss, however, the integers as a number system including the negatives were fully established. Gauss called the study of the integers, or number theory as it is now known, the queen of mathematics. We construct in this chapter the integers from the natural numbers and then study the ring and order properties of the integers.

If the reader wishes to know the content of the chapter without engaging fully the details it is quite possible to read and understand the statements of the theorems; they are intuitively evident; they deal with familiar properties of familiar objects.

3.1 The Peano axioms

In this section are presented the Peano axioms for the natural numbers and definition by induction is discussed.

The set of natural numbers is the set we intuitively know as $\{0, 1, 2, 3, 4, \ldots\}$, the set of positive whole numbers and zero. Our aim is to analyze and to describe more precisely this set. We shall utilize the very important concept of mathematical induction.

Starting with a knowledge of set theory it is possible (with appropriate axioms) to construct a model of the natural numbers. This is to say, certain sets are generated which, for all intents and purposes, can be used as natural numbers after operations are appropriately defined for them. This construction of the natural numbers in set theory begins with the empty set $\varnothing$ as a model for the number zero and then proceeds to define 1 to be $\{\varnothing\} = \{0\}$. One must note that $\varnothing$ and $\{\varnothing\}$ are different sets because the first has no members and the second does have a member. The construction for 0, 1, 2, 3, 4 proceeds as follows: $0 = \varnothing, 1 = \{0\}, 2 = \{0, 1\}, 3 = \{0, 1, 2\}, 4 = \{0, 1, 2, 3\}$. Notice how these sets increase in size; 4 is not only fourth after zero but also contains 4 elements.

So far the construction has produced only 0, 1, 2, 3, and 4. If we continue in this manner we can produce quite a few more natural numbers. We will never, however, succeed in completing the project by this means. We must find a means of describing the totality of all natural numbers by finite means. The process of description must terminate even though the set being described be infinite.

We continue to analyze the construction. What is involved in this construction is a starting set $\varnothing$ and a mode of producing a successor which is repeated over and over. If we denote the *successor of n* by $s(n)$ we may discover that the operation is $s(n) = n \cup \{n\}$. A computational check for 1, 2, and 3 verifies the definition for those cases:

$$1 = s(0) = 0 \cup \{0\} = \varnothing \cup \{\varnothing\} = \{0\}.$$
$$2 = s(1) = 1 \cup \{1\} = \{0\} \cup \{1\} = \{0, 1\}.$$
$$3 = s(2) = 2 \cup \{2\} = \{0, 1\} \cup \{2\} = \{0, 1, 2\}.$$

The successor operation produces 1 from 0, 2 from 1, and 3 from 2. The successor of any number always contains some member that the number does not. The set of all the natural numbers, $\mathbb{N}$, must not only contain 0, 1, 2, and 3 but also the successor of 3, and the successor of the successor of 3, and so forth. $\mathbb{N}$ must contain the starting set 0 and all of its successors.

$$0 \in \mathbb{N}.$$
$$\text{If } n \in \mathbb{N} \text{ then } s(n) \in \mathbb{N}.$$

These two conditions are a way of ensuring that all natural numbers are in $\mathbb{N}$. There could be (and are), however, sets which satisfy both of these conditions, contain all natural numbers, yet are too big because they contain other things besides natural numbers. This is to say that they are too large, yet satisfy the two conditions stated. We seek the natural numbers as the smallest set satisfying the two conditions. We achieve this end by saying that any subset S of $\mathbb{N}$ which satisfies ($0 \in S$) and ($n \in S$ implies $s(n) \in S$) must be all of $\mathbb{N}$. This says that no set S smaller than $\mathbb{N}$ can satisfy the two conditions. This characterizes $\mathbb{N}$ as the smallest set containing 0 and all of its successors.

In a formal exposition of set theory the general order of procedure is to assume (along with previously assumed axioms) some axiom of infinity strong enough to produce a set containing 0 and all of its successors and then to prove the existence of $\mathbb{N}$. In summary, then, there exists a set $\mathbb{N}$ such that

$$0 \in \mathbb{N}$$
$$n \in \mathbb{N} \text{ implies } s(n) \in \mathbb{N}$$
$$\text{If } S \subseteq \mathbb{N} \text{ and } 0 \in S \text{ and } (n \in S \text{ implies } s(n) \in S) \text{ then } S = \mathbb{N}.$$

It is then not difficult to prove that $\mathbb{N}$ also has the properties:

$$s(m) = s(n) \quad \text{implies } m = n$$
$$s(n) \neq 0 \quad \text{for all } n \in \mathbb{N}.$$

The five conditions we have now listed are called Peano's axioms for the natural numbers. This completes our motivational sketch of how the natural numbers can be constructed within set theory.

We reword the previously given Peano axioms so that they do not depend upon the particular set constructions used above.

Axiom. *There exist a set* $\mathbb{N}$; *a member of* $\mathbb{N}$ *called* 0; *and an injection* $s: \mathbb{N} \to \mathbb{N}$ *such that* $0 \notin \text{range } s$ *and no proper subset* S *of* $\mathbb{N}$ *may have the properties* $0 \in S$ *and* ($n \in S$ *implies* $s(n) \in S$).

We comment on how the previously given set theoretic model of S satisfies this axiom. $\varnothing$ is, of course, 0. The operation of taking set successors is the injection s. No set successor is $\varnothing$ is equivalent to $0 \notin \text{range } s$. $s(m) = s(n)$ implies $m = n$ is the injective property of s. The statement that no proper subset S of $\mathbb{N}$ may have the properties $0 \in S$ and ($n \in S$ implies $s(n) \in S$) is known as the principle of mathematical induction.

Working now from our axiom we intend to construct addition and multiplication on $\mathbb{N}$ by means of definition by induction, sometimes called definition by recursion. Addition of natural numbers will be defined by this scheme:

$$m + 0 = m$$
$$m + s(k) = s(m + k).$$

This is a two-step definition; it involves first defining the result of adding zero to a natural number m and second defining the result of adding the successor of k in terms of the result of adding k to m. This procedure for definition strongly resembles the principle of mathematical induction in starting the definition with 0 and then moving the definition along from k to the successor $s(k)$. Some analysis of the situation reveals that we are in fact attempting to define a function from $\mathbb{N}$ to $\mathbb{N}$. Let us denote the result of adding n to m (finding $m + n$) by $\alpha_m(n)$. Then what we require is a function $\alpha_m: \mathbb{N} \to \mathbb{N}$ such that $\alpha_m(0) = m$ and $\alpha_m(s(k)) = s(\alpha_m(k))$. This statement in terms of α_m is merely a notational change, but it makes it much clearer that we are trying to find a function with certain properties when we are trying to define addition. Is it possible to define a function with domain $\mathbb{N}$ merely by defining what is the image of 0 and by defining the image of $s(k)$ in terms of the image of k? Moreover, is the result unique when possible? Within set theory it can be established that the answer is yes in both parts. The proof utilizes the principle of mathematical induction but is quite involved. It suits our purposes to take such a definition scheme as an axiom.

Axiom. *Let some set X be given as well as some element a of X and a function $f: X \to X$. Then there exists one and only one function $t: \mathbb{N} \to X$ such that $t(0) = a$ and $t(s(k)) = f(t(k))$.* (Figure 3.1 may help in picturing the situation.)

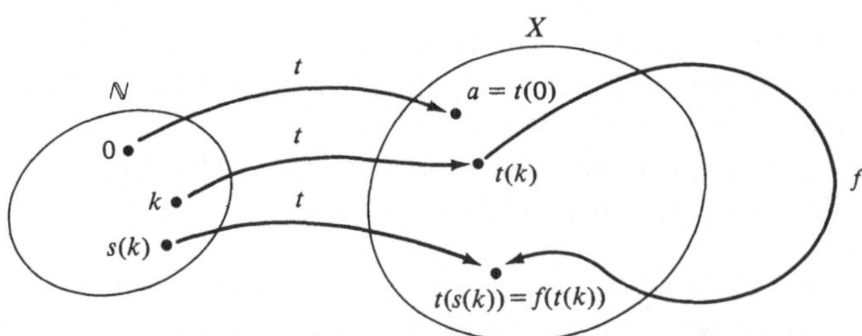

Figure 3.1

QUESTIONS

1. Considering the model of the natural numbers discussed in this section, $0 = \varnothing$, $1 = \{0\}$, $2 = \{0, 1\}$, $3 = \{0, 1, 2\}$, we have $2 \neq 1$ because
   (A) $1 \in 2$ and $1 \notin 1$
   (B) $1 \subseteq 2$
   (C) $2 \nsubseteq 1$
   (D) $s(2) = 1$.
   (E) None of the alternatives completes a true sentence.

2. Every natural number can be the successor of at most one natural number because
   (A) $s:\mathbb{N} \to \mathbb{N}$ is not a surjection
   (B) $s:\mathbb{N} \to \mathbb{N}$ is an injection
   (C) $\varnothing$ is a member of every natural number
   (D) natural numbers are sometimes unnatural.
(E) None of the choices completes a satisfactory sentence.

3. 0 is not the successor of any natural number because
   (A) $n \in \mathbb{N}$ implies $s(n) \in \mathbb{N}$
   (B) $0 \notin \text{range } s$
   (C) s is an injection
   (D) $s(\mathbb{N}) \cup \mathbb{N} = \mathbb{N}$.
(E) None of the choices completes a satisfactory sentence.

4. Let t be a function from $\mathbb{N}$ to $\mathbb{N}$ such that $t(0) = 4$ and $t(s(k)) = t(k) + 3$. Which of these statements are correct?
   (A) $t(1) = \varnothing$.
   (B) $t(2) = 5$.
   (C) $t(t(0)) = 16$.
   (D) $t(4) = 0$.
(E) None of the statements is correct.

3.2 Addition of natural numbers

This section treats addition of natural numbers and its properties.

This first theorem amounts to a definition of addition of natural numbers.

Theorem. *For each $m \in \mathbb{N}$ there exists a unique function $\alpha_m:\mathbb{N} \to \mathbb{N}$ such that $\alpha_m(0) = m$ and $\alpha_m(s(k)) = s(\alpha_m(k))$.*

PROOF. We apply the axiom for definition by induction of Section 3.1. The role of f in the general statement is played by the succession function $s:\mathbb{N} \to \mathbb{N}$. □

As defined for each $m \in \mathbb{N}$ the function $\alpha_m:\mathbb{N} \to \mathbb{N}$ is a unary operation on $\mathbb{N}$. We use all these unary operations, one for each $m \in \mathbb{N}$, to define one binary operation $\alpha:\mathbb{N} \to \mathbb{N}$.

Definition. We define *addition* on the natural numbers to be the binary operation $\alpha:\mathbb{N} \times \mathbb{N} \to \mathbb{N}$ such that $\alpha(m, n) = \alpha_m(n)$.

Once having defined α we now return to the conventional notation for addition; we denote $\alpha(m, n)$ by $m + n$ and $s(0)$ by 1. The statements in the following theorem will be a translation into conventional notation of the facts we have at hand.

Theorem. *$+$ is a binary operation on $\mathbb{N}$. $m + 0 = m$ for all $m \in \mathbb{N}$. $s(n) = n + 1$ for all $n \in \mathbb{N}$. $m + (n + 1) = (m + n) + 1$ for all $m, n \in \mathbb{N}$.*

PROOF. $\alpha_m(0) = m$ translates into $m + 0 = m$. $\alpha_m(s(0)) = s(\alpha_m(0))$ translates into $m + s(0) = s(m + 0)$. Using the symbol 1 for $s(0)$ and replacing $m + 0$ by m we have $m + 1 = s(m)$. $\alpha_m(s(n)) = s(\alpha_m(n))$ is true for all $m, n \in \mathbb{N}$. This translates into $m + (n + 1) = (m + n) + 1$. □

It should be observed that the use of $m + 1$ for $s(m)$ is only possible after the addition is defined.

We turn now to the usual theorems on binary operations.

Theorem. *Addition of natural numbers is associative.*

PROOF. We must demonstrate that $(m + n) + p = m + (n + p)$ for all $m, n, p \in \mathbb{N}$. We let S be the subset of all natural numbers p for which the equation $(m + n) + p = m + (n + p)$ for all $m, n \in \mathbb{N}$ is true. $S = \{p | p \in \mathbb{N}$ and $(m + n) + p = m + (n + p)$ for all $m, n \in \mathbb{N}\}$. Our procedure is to use the principle of mathematical induction to show that $S = \mathbb{N}$. We do this by showing that $0 \in S$ and $k \in S$ implies $s(k) \in S$. For $s(k)$ we can use $k + 1$.

$(m + n) + 0 = (m + n) = m + n = m + (n + 0)$ using the already established result that when 0 is added on the right to any natural number the number itself is the sum (0 is a right neutral element of addition). We conclude $0 \in S$.

It will be of use to us to refer to the equation

$$(m + n) + 1 = m + (n + 1) \quad \text{for all } m, n \in \mathbb{N} \qquad (*)$$

which we established earlier. We now assume $k \in S$, that is to say, $(m + n) + k = m + (n + k)$ for all $m, n \in \mathbb{N}$. We proceed to prove the appropriate equation for $k + 1$.

$$\begin{aligned}(m + n) + (k + 1) &= (m + n) + [k + 1] = [(m + n) + k] + 1 \\ &= [m + (n + k)] + 1 = m + [(n + k) + 1] \\ &= m + [n + (k + 1)] \quad \text{for all } m, n \in \mathbb{N}.\end{aligned}$$

The second, fourth, and fifth equality are because of equation (*). Thus $k + 1 \in S$. Upon an assumption of $k \in S$ we have proved $k + 1 \in S$. $k \in S$ implies $k + 1 \in S$. $S = \mathbb{N}$ and the theorem is proved. □

It is important to remember that in a proof by mathematical induction that it is implication $k \in S$ implies $k + 1 \in S$ that must be proved and not $k \in S$ or $k + 1 \in S$ separately.

In order to demonstrate the commutativity of addition it is efficient to prove first several lemmas.

Lemma. $0 + m = m$ *for all* $m \in \mathbb{N}$.

PROOF. Let $S = \{m | 0 + m = m\}$. $0 \in S$ because $0 + 0 = 0$. We assume $k \in S$; $0 + k = k$. Then $0 + (k + 1) = (0 + k) + 1 = k + 1$. $k + 1 \in S$. $S = \mathbb{N}$. □

The previous lemma together with the definition of addition establishes 0 as the neutral element of addition.

Lemma. $1 + m = m + 1$ *for all* $m \in \mathbb{N}$.

PROOF. Let $S = \{m|1 + m = m + 1\}$. $0 \in S$ because $1 + 0 = 1$ and $0 + 1 = 1$. Assume $k \in S$. $1 + k = k + 1$. $1 + (k + 1) = (1 + k) + 1 = (k + 1) + 1$. $k + 1 \in S$. $S = \mathbb{N}$. □

With these two preliminary results out of the way we can take on the general theorem on commutativity.

Theorem. $n + m = m + n$ *for all* $m, n \in \mathbb{N}$.

PROOF. Let $S = \{n|n + m = m + n$ for all $m \in \mathbb{N}\}$. $0 \in S$ because $0 + m = m + 0$ for all $m \in \mathbb{N}$. Assume $k \in S$. $k + m = m + k$ for all $m \in \mathbb{N}$. $(k + 1) + m = k + (1 + m) = k + (m + 1) = (k + m) + 1 = (m + k) + 1 = m + (k + 1)$ for all $m \in \mathbb{N}$. $k + 1 \in S$. $S = \mathbb{N}$. □

QUESTIONS

1. Which of these statements are correct?
   (A) $\alpha_m(s(k)) = s(\alpha_m(k))$.
   (B) $\alpha_m(0) = m$.
   (C) $\alpha(m, n) = \alpha_m(n)$.
   (D) $\alpha(0, n) = n$.
   (E) None of the statements is correct.

2. The statement $\alpha_m(s(n)) = s(\alpha_m(n))$ for all $n \in \mathbb{N}$ means
   (A) α is an injection
   (B) s is an injection
   (C) $m + (n + 1) = (m + n) + 1$ for all $n \in \mathbb{N}$
   (D) addition of natural numbers is commutative.
   (E) None of the alternatives completes a satisfactory sentence.

3. Addition of natural numbers is commutative is equivalent to
   (A) $\alpha(m, n) = \alpha(n, m)$ for all $m, n \in \mathbb{N}$
   (B) $\alpha_m(n) = \alpha_n(m)$ for all $m, n \in \mathbb{N}$
   (C) $\alpha_m(s(n)) = s(\alpha_m(n))$ for all $m, n \in \mathbb{N}$
   (D) $\alpha_m(\alpha_n(p)) = \alpha_m(\alpha_p(q))$ for all $p, q \in \mathbb{N}$.
   (E) None of the choices completes a true sentence.

4. In this list of results from Section 3.2, which result is out of the order of presentation? (One answer only.)
   (A) Definition of α_m.
   (B) Definition of α.
   (C) Commutativity of addition.
   (D) Associativity of addition.
   (E) 0 is a left neutral element of addition.

3.3 Multiplication of natural numbers

In this section the definition of multiplication is given and its important properties are proved.

Having defined addition for $\mathbb{N}$ and having proved it associative and commutative and having established the existence of a neutral element of addition, 0, we now turn our attention to multiplication. The guiding equations for the definition will be

$$m \cdot 0 = 0$$
$$m \cdot (n + 1) = m \cdot n + m.$$

These equations give sufficient information to use in the axiom for definition by mathematical induction. We use a function construction like the one used for addition. We let $\mu_m(n)$ represent the product (not yet defined) of m by n.

Theorem. *For each $m \in \mathbb{N}$ there exists a unique function $\mu_m: \mathbb{N} \to \mathbb{N}$ such that*

$$\mu_m(0) = 0$$
$$\mu_m(s(k)) = \mu_m(k) + m.$$

PROOF. This is an application of the axiom of definition by induction in which $X = \mathbb{N}$, $t = \mu_m$, and $f(x) = x + m$. □

Definition. We define multiplication of natural numbers to be the binary operation $\mu: \mathbb{N} \times \mathbb{N} \to \mathbb{N}$ such that $\mu(m, n) = \mu_m(n)$.

By discarding μ and replacing it with the conventional notation $m \cdot n$ or mn for product we get the following results.

Theorem. $m \cdot 0 = 0$ *for all* $m \in \mathbb{N}$. $m(n + 1) = mn + m$ for all $m, n \in \mathbb{N}$. $m \cdot 1 = m$ *for all* $m \in \mathbb{N}$.

PROOF. The equation $\mu_m(0) = 0$ translates into $m \cdot 0 = 0$. The equation $\mu_m(s(n)) = \mu_m(n) + m$ translates into $m(n + 1) = mn + m$. When $n = 0$ this reduces to $m \quad 1 = m$. □

The first theorem in developing the properties of multiplication is the left distributive law.

Theorem. $m(n + p) = mn + mp$ *for all* $m, n, p \in \mathbb{N}$.

PROOF. Let $S = \{p | m(n + p) = mn + mp$ for all $m, n \in \mathbb{N}\}$. $m(n + 0) = mn = mn + 0 = mn + m \cdot 0$ for all $m, n \in \mathbb{N}$ proves $0 \in S$. Now assume $k \in S$. $m(n + (k + 1)) = m((n + k) + 1) = m(n + k) + m = (mn + mk) + m = mn + (mk + m) = mn + m(k + 1)$. $k + 1 \in S$. $S = \mathbb{N}$. □

We continue with the associative law for multiplication.

Theorem. *$m(np) = (mn)p$ for all $m, n, p \in \mathbb{N}$.*

PROOF. Let $S = \{p|m(np) = (mn)p$ for all $m, n \in \mathbb{N}\}$. $m(n \cdot 0) = m \cdot 0 = 0 = (mn)0$ establishes $0 \in S$. Let $k \in S$. $(mn)(k + 1) = (mn)k + mn = m(nk) + mn = m(nk + n) = m(n(k + 1))$. $k + 1 \in S$. $S = \mathbb{N}$. □

We have thus far proved multiplication of natural numbers to be associative and that the left distributive relation holds. We now begin work on the right distributive law.

Lemma. *$0 \cdot m = 0$ for all $m \in \mathbb{N}$.*

PROOF. Let $S = \{m|0 \cdot m = 0\}$. $0 \in S$ because $0 \cdot 0 = 0$. Assume $k \in S$; that is, assume $0 \cdot k = 0$. $0(k + 1) = 0 \cdot k + 0 \cdot 1 = 0 + 0 = 0$. $k + 1 \in S$. $S = \mathbb{N}$. □

Lemma. *$1 \cdot m = m$ for all $m \in \mathbb{N}$.*

PROOF. Let $S = \{m|1 \cdot m = m\}$. $1 \cdot 0 = 0$ implies $0 \in S$. Assume $k \in S$. $1 \cdot (k + 1) = 1 \cdot k + 1 = k + 1$. $k + 1 \in S$. $S = \mathbb{N}$.

Theorem. *$(m + n)p = mp + np$ for all $m, n, p \in \mathbb{N}$.*

PROOF. Let $S = \{p|(m + n)p = mp + np$ for all $m, n \in N\}$. $(m + n)0 = 0 = 0 + 0 = m \cdot 0 + n \cdot 0$ implies $0 \in S$. Assume $k \in S$; that is, assume $(m + n)k = mk + nk$ for all $m, n \in \mathbb{N}$. $(m + n)(k + 1) = (m + n)k + (m + n) = (mk + nk) + (m + n) = (mk + m) + (nk + n) = m(k + 1) + n(k + 1)$ for all $m, n \in \mathbb{N}$. $k + 1 \in S$. $S = \mathbb{N}$. □

Finally we establish the commutativity of multiplication.

Theorem. *$mn = nm$ for all $m, n \in \mathbb{N}$.*

PROOF. Let $S = \{m|mn = nm$ for all $n \in \mathbb{N}\}$. $0 \cdot n = 0 = n \cdot 0$ yields $0 \in S$. Assume $kn = nk$ for all $n \in \mathbb{N}$, which is to say, $k \in S$. $n(k + 1) = nk + n = kn + 1 \cdot n = (k + 1)n$. $k + 1 \in S$. $S = \mathbb{N}$. □

In summary, we have defined $\mathbb{N}$, the set of natural numbers, two binary operations on $\mathbb{N}$ called addition and multiplication, proved both associative and commutative, proved that multiplication is distributive with respect to addition, and proved that 0 and 1 are respectively neutral elements of addition and multiplication.

QUESTIONS

1. Which of the following statements are correct?
   (A) $\mu_m(0) = 0$.
   (B) $\mu_m(s(k)) = \mu_m(k) + m$.

(C) $\mu(m, n) = \mu_m(n)$.
(D) $\mu_m(s(k)) = \alpha(\mu_m(k), m)$.
(E) None of the statements is correct.

2. $m(n + p) = mn + mp$ for all $m, n, p \in \mathbb{N}$
(A) is the left distributive law for natural numbers
(B) is the right distributive law for natural numbers
(C) has a proof which depends upon the commutative law of multiplication
(D) is the associative law for multiplication.
(E) None of the alternatives completes a true sentence.

3.4 Further properties of $\mathbb{N}$

In this section we complete our natural number constructions with some cancellation laws and related properties.

We begin by showing additive cancellation to be always possible in $\mathbb{N}$.

Theorem. *$m + p = n + p$ implies $m = n$ for all $m, n, p \in \mathbb{N}$.*

PROOF. Let $S = \{p | m + p = n + p$ for all $m, n \in \mathbb{N}\}$. $0 \in S$ because $m + 0 = n + 0$ implies $m = n$. $m + 1 = n + 1$ implies $m = n$ because the successor function is an injection. Now assume $k \in S$. $m + (k + 1) = n + (k + 1)$ implies $(m + k) + 1 = (n + k) + 1$ which implies $m + k = n + k$ which in turn implies $m = n$. $k + 1 \in S$. $S = \mathbb{N}$. □

Although additive cancellation is always possible most natural numbers do not have negatives (cf. Exercise 3 of this section).

Our axiom for $\mathbb{N}$ contains the information that 0 is the successor of no natural number. We now prove every nonzero natural number is the successor of some (other) natural number.

Theorem. *If $m \neq 0$ and $m \in N$ then $m = p + 1$ for some $p \in \mathbb{N}$.*

PROOF. In order to use induction a little twist is needed in the setting of S. Let $S = \{m | m = p + 1$ for some $p \in \mathbb{N}$ or $m = 0\}$. $0 \in S$ because S is defined in such a manner to contain 0. Now assume $k \in S$. $k = p + 1$ for some $p \in \mathbb{N}$ or $k = 0$. If $k = p + 1$ then $k + 1 = (p + 1) + 1$. Since $p + 1 \in \mathbb{N}$ we have written $k + 1$ as some natural number plus one. If, on the other hand, $k = 0$ then $k + 1 = 0 + 1$ which again is a natural number plus one. $k + 1 \in S$ in both cases. $S = \mathbb{N}$. Our conclusion is that $m = p + 1$ for some $p \in \mathbb{N}$ or $m = 0$ for all $m \in \mathbb{N}$. Since $m \neq 0$ is given in the hypothesis we conclude $m = p + 1$ for some $p \in \mathbb{N}$. □

This result allows us to handle the proof on zero divisors to come without using induction again.

Theorem. *$mn = 0$ implies $m = 0$ or $n = 0$ for all $m, n \in \mathbb{N}$.*

PROOF. The contrapositive of the statement to be proved is $m \neq 0$ and $n \neq 0$ imply $mn \neq 0$. We prove this instead. Since $m \neq 0$ and $n \neq 0$ there exist $p, q \in \mathbb{N}$ such that $m = p + 1$ and $n = q + 1$. Then $mn = (p + 1)(q + 1) = (pq + p + q) + 1$. mn is the successor of $pq + p + q$ and therefore cannot be zero. □

It was proved in Section 2.3 that in a ring multiplicative cancellation and no nontrivial zero divisors were equivalent conditions. That proof is inapplicable to $\mathbb{N}$ because $\mathbb{N}$ with its operations does not make a ring; the proof requires the existence of negatives. We must, therefore, produce a different proof for nonzero multiplicative cancellation for $\mathbb{N}$.

Theorem. *$mp = np$ and $p \neq 0$ imply $m = n$ for all $m, n, p \in \mathbb{N}$.*

PROOF. The setting of S requires the proper variable choice. Let $S = \{n | mp = np$ and $p \neq 0$ imply $m = n$ for all $m, p \in \mathbb{N}\}$. $0 \in S$ because $mp = 0 \cdot p$ implies $mp = 0$. By the previous theorem $m = 0$ or $p = 0$. But $p \neq 0$. Therefore $m = 0$. $0 \in S$.

Now let $k \in S$. $mp = kp$ and $p \neq 0$ imply $m = k$ for all $m, p \in \mathbb{N}$. Suppose $mp = (k + 1)p$ and $p \neq 0$. If m were 0 then $(k + 1)p = 0$ which is impossible because neither $k + 1$ nor p is 0. Thus m cannot be zero. $m = l + 1$ for some $l \in \mathbb{N}$. $(l + 1)p = (k + 1)p$. $lp + p = kp + p$. Additive cancellation gives us $lp = kp$. Since $k \in S$ we have $l = k$. But then $l + 1 = k + 1$. $m = k + 1$. This is the result we desired. $k + 1 \in S$. $S = \mathbb{N}$. □

QUESTIONS

1. Which of the following statements are true?
   (A) If a product of natural numbers is zero then one of the factors is zero.
   (B) Every natural number has a negative which is a natural number.
   (C) Multiplicative cancellation of nonzero natural numbers is valid.
   (D) Every natural number is the successor of some natural number.
   (E) None of the statements is true.

2. Which of the following statements are true?
   (A) For every $n \in \mathbb{N}$ there exists $m \in \mathbb{N}$ such that $n = m + 1$.
   (B) For every $n \in \mathbb{N}$ there exists $m \in \mathbb{N}$ such that $m = n + 1$.
   (C) For every $n \in \mathbb{N}$, $n \neq n + 1$.
   (D) ($u = s(p)$ for some $p \in \mathbb{N}$) and ($uv = 0$) imply $v = 0$.
   (E) $m + p = n + p$ and $p \neq 0$ imply $m = n$.

EXERCISES

1. Prove that $\mathbb{N}^+ = \mathbb{N} - \{0\}$ is closed under addition; i.e., $x, y \in \mathbb{N}^+$ imply $x + y \in \mathbb{N}^+$.

2. Prove $m + n = 0$ implies $m = 0$ and $n = 0$ for all $m, n \in \mathbb{N}$.

3. Prove that no natural number except 0 has a negative in $\mathbb{N}$.

4. Prove $\mathbb{N}^+$ is closed under multiplication.

5. Prove that every natural number excepting 0 and 1 is $p + 1$ for some $p \in \mathbb{N}^+$.

6. Prove that $mn = 1$ implies $m = 1$ and $n = 1$ for all $m, n \in \mathbb{N}$.

7. Prove that no natural number except 1 has a reciprocal (multiplicative inverse) in $\mathbb{N}$.

8. Find all functions $f: \mathbb{N} \to \mathbb{N}$ which preserve both addition and multiplication; i.e., $f(x + y) = f(x) + f(y)$ and $f(xy) = f(x)f(y)$ for all $x, y \in \mathbb{N}$.

9. A small study of additive functions on $\mathbb{N}$:
(a) Find all $f: \mathbb{N} \to \mathbb{N}$ such that $f(x + y) = f(x) + f(y)$.
(b) Let $E = \{f \mid f: \mathbb{N} \to \mathbb{N} \text{ and } f \text{ preserves } +\}$. Show that there exists a bijection $\varphi: \mathbb{N} \to E$.
(c) Define for E an addition and multiplication so that φ preserves $+$ and preserves $\cdot$.

3.5 Construction of the integers

In this section we construct the integers and prove they form an integral domain.

Our intuitive knowledge of the integers as natural numbers together with their negatives makes it seem we can somehow adjoin the negatives of natural numbers to the natural numbers and extend the operations of addition and multiplication from the natural numbers to all cases of "positives" and "negatives." Such a straightforward approach is possible but leads to awkward proofs because one must constantly refer back to definitions by cases. The approach we actually use is less direct but we are more than compensated by the resulting ease of proof. We also win, in the bargain, a new mathematical technique with applications elsewhere.

We begin with the set $\mathbb{N} \times \mathbb{N}$ with $+$ and $\boxdot$ defined as follows:

$$(r, s) + (u, v) = (r + u, s + v)$$
$$(r, s) \boxdot (u, v) = (ru + sv, rv + su).$$

This construction was previously discussed in Exercise 5 of Section 2.3. The two binary operations are associative and commutative and $\boxdot$ is distributive with respect to $+$. $(0, 0)$ is the neutral element of addition and $(1, 0)$ is the neutral element of multiplication. $\langle \mathbb{N} \times \mathbb{N}, +, \boxdot, (0, 0)\rangle$ fails to be a ring because only $(0, 0)$ has a negative in $\mathbb{N} \times \mathbb{N}$. We shall not go through the details of supporting these claims and refer the reader to Exercise 1.

On the set $\mathbb{N} \times \mathbb{N}$ we define a relation γ such that $(x, y)\gamma(u, v)$ if and only if $x + v = y + u$. That this relation γ is an equivalence relation follows without difficulty. We propose, by making appropriate definitions of binary operations, to make $\mathbb{N} \times \mathbb{N}/\gamma$ into a ring much as was done with the quotient ring in Section 2.6. This construction differs from the construction of R/γ from R in that this time $\mathbb{N} \times \mathbb{N}$ is not a ring.

Our tack now is to drop temporarily our specific problem with $\mathbb{N} \times \mathbb{N}$ and to take a more general view of procedure.

Definition. Given a set S and a binary operation β on S we say an equivalence relation ε on S is *compatible* with β if and only if $x\varepsilon x'$ and $y\varepsilon y'$ imply $(x\beta y)\varepsilon(x'\beta y')$ for all $x, y, x', y' \in S$.

Theorem. *Let S be a set with a binary operation β and an equivalence relation ε. ε is compatible with β if and only if $\bar{\beta}:(S/\varepsilon) \times (S/\varepsilon) \to S/\varepsilon$ such that $x/\varepsilon\bar{\beta}y/\varepsilon = (x\beta y)/\varepsilon$ is a binary operation on S/ε.*

PROOF. $x/\varepsilon\bar{\beta}y/\varepsilon = (x\beta y)/\varepsilon$ is well defined provided and only provided the definition is independent of the representatives chosen from the equivalence classes or cosets. In other words, $\bar{\beta}$ is well defined if and only if $x/\varepsilon = x'/\varepsilon$ and $y/\varepsilon = y'/\varepsilon$ imply $(x\beta y)/\varepsilon = (x'\beta y')/\varepsilon$. This condition is in turn equivalent to $x\varepsilon x'$ and $y\varepsilon y'$ imply $(x\beta y)\varepsilon(x'\beta y')$. □

EXAMPLE. In Section 2.6 the equivalence relation $\sim$ on R in the first theorem is compatible with both $+$ and $\cdot$.

We now apply the concept of compatible equivalence relation to our specific problem with γ and $\langle \mathbb{N} \times \mathbb{N}, +, \boxdot, (0, 0)\rangle$.

Theorem. *The equivalence relation γ on $\mathbb{N} \times \mathbb{N}$, $(x, y)\gamma(u, v)$ if and only if $x + v = y + u$, is compatible with the binary operations $+$ and $\boxdot$ on $\mathbb{N} \times \mathbb{N}$.*

PROOF. Suppose $(r, s)\gamma(r', s')$ and $(u, v)\gamma(u', v')$, all members of $\mathbb{N} \times \mathbb{N}$. Then $r + s' = s + r'$ and $u + v' = v + u'$. $(r + s') + (u + v') = (s + r') + (v + u')$. $(r + u) + (s' + v') = (s + v) + (r' + u')$. $(r + u, s + v)\gamma(r' + u', s' + v')$. $[(r, s) + (u, v)]\gamma[(r', s') + (u', v')]$. γ is compatible with $+$.

For multiplication we suggest proving $[(r, s) \boxdot (u, v)]\gamma[(r', s') \boxdot (u, v)]$ and $[(r', s') \boxdot (u, v)]\gamma[(r', s') \boxdot (u', v')]$ and using the transitivity of γ for the result. For one, $(r + s')u + (r' + s)v = (r' + s)u + (r + s')v$. $ru + sv + r'v + s'u = rv + su + r'u + s'v$. $(ru + sv, rv + su)\gamma(r'u + s'v, r'v + s'u)$. $[(r, s) \boxdot (u, v)]\gamma[(r', s') \boxdot (u, v)]$. □

Corollary. *$\mp$ and $\bar{\boxdot}$ are binary operations on $(\mathbb{N} \times \mathbb{N})/\gamma$ where $(r, s)/\gamma \mp (u, v)/\gamma = (r + u, s + v)/\gamma$ and $(r, s)/\gamma \bar{\boxdot} (u, v)/\gamma = (ru + sv, rv + su)/\gamma$.*

Having established our binary operations for $(\mathbb{N} \times \mathbb{N})/\gamma$ we move to prove this theorem:

Theorem. *$\langle(\mathbb{N} \times \mathbb{N})/\gamma, \mp, \bar{\boxdot}, (0, 0)/\gamma, (1, 0)/\gamma\rangle$ is a commutative unitary ring.*

PROOF. It remains to verify that every member of $(\mathbb{N} \times \mathbb{N})/\gamma$ has a negative in $(\mathbb{N} \times \mathbb{N})/\gamma$. Let $(x, y)/\gamma \in (\mathbb{N} \times \mathbb{N})/\gamma$. Then $(x, y)/\gamma \mp (y, x)/\gamma = (x + y, y + x)/\gamma = (0, 0)/\gamma$. □

Our next theorem utilizes a double induction.

Theorem. $\langle(\mathbb{N} \times \mathbb{N})/\gamma, \mp, \bar{\boxdot}, (0, 0)/\gamma, (1, 0)/\gamma\rangle$ *is an integral domain.*

PROOF. Let $(x, y)/\gamma \mathbin{\bar{\boxdot}} (u, v)/\gamma = (0, 0)/\gamma$. We must show $(x, y)/\gamma = (0, 0)/\gamma$ or $(u, v)/\gamma = (0, 0)/\gamma$. In other words we must show $xu + yv = xv + yu$ implies $x = y$ or $u = v$ for all $x, y, u, v \in \mathbb{N}$.

Let $S = \{y \mid xu + yv = xv + yu$ implies $x = y$ or $u = v$ for all $x, u, v \in \mathbb{N}\}$. $xu + yv = xv + yu$ implies $xu + 0 = xv + 0$ which implies $x = 0$ or $u = v$. $x = y$ or $u = v$. The statement is true for $y = 0$. $0 \in S$. We now assume $k \in S$. We intend to show that $k + 1 \in S$; that is, $xu + (k + 1)v = xv + (k + 1)u$ implies $x = k + 1$ or $u = v$ for all $u, v, x \in \mathbb{N}$. In order to do this let $T = \{x \mid xu + (k + 1)v = xv + (k + 1)u$ implies $x = k + 1$ or $u = v$ for all $u, v \in \mathbb{N}\}$. $0 \in T$ since $xu + (k + 1)v = xv + (k + 1)u$ implies $0 + (k + 1)v = (k + 1)u$ which implies $u = v$. Assume $l \in T$. Suppose $(l + 1)u + (k + 1)v = (l + 1)v + (k + 1)u$. $lu + u + kv + v = lv + v + ku + u$. $lu + kv = lv + ku$. Since we have assumed $k \in S$ we have $u = v$ or $k = l$. Thus $u = v$ or $k + 1 = l + 1$. $l + 1 \in T$. $T = \mathbb{N}$. $k + 1 \in S$. $S = \mathbb{N}$. □

This completes the construction of the integral domain which will be the integers.

QUESTIONS

1. In the system $\langle(\mathbb{N} \times \mathbb{N})/\gamma, \mp, \bar{\boxdot}, (0, 0)/\gamma, (1, 0)/\gamma\rangle$ constructed in this section
   (A) $(x, y)/\gamma \mp (u, v)/\gamma = (x + u, y + v)/\gamma$
   (B) $xu + yv = xv + yu$ implies $x = y$ or $u = v$ for all natural numbers x, y, u, v.
   (C) $((x, y)/\gamma)^2 = (0, 0)/\gamma$ implies $(x, y)/\gamma = (0, 0)/\gamma$
   (D) $(x, y)/\gamma = (u, v)/\gamma$ implies $x = u$ and $y = v$.
   (E) None of the alternatives completes a true sentence.

2. If $+$ is a binary operation on S and ρ is an equivalence relation compatible with $+$ then
   (A) $x/\rho + y/\rho = (x + y)/\rho$
   (B) $x + (-x) = \theta$ implies $x/\rho + -x/\rho = \theta/\rho$
   (C) $x/\rho = y/\rho$ if and only if $x\rho y$
   (D) there exists a y/ρ such that $x/\rho + y/\rho = x/\rho$.
   (E) None of the choices is satisfactory.

3. An equivalence relation ρ on a set S
   (A) is compatible with a binary operation $+$ if and only if $x_1\rho x_2$ and $y_1\rho y_2$ imply $(x_1 + y_1)\rho(x_2 + y_2)$
   (B) partitions a set S into a collection S/ρ of cosets

(C) is a subset of $S \times S$
(D) is compatible with a binary operation $\cdot$ if and only if $x_1 \rho x_2$ and $y_1 \rho y_2$ imply $(x_1 y_1) \rho (x_2 y_2)$.

(E) None of the alternatives completes a true sentence.

EXERCISES

1. Prove that for $\langle \mathbb{N} \times \mathbb{N}, +, \boxdot, (0, 0) \rangle$
   (a) $+$ and $\boxdot$ are associative and commutative;
   (b) $(0, 0)$ is a neutral element for $+$;
   (c) $(1, 0)$ is a neutral element for $\boxdot$; and
   (d) $\boxdot$ is distributive with respect to $+$.

2. Show that γ, the relation defined on $\mathbb{N} \times \mathbb{N}$ such that $(x, y)\gamma(u, v)$ if and only if $x + v = y + u$, is an equivalence relation.

3.6 Embedding $\mathbb{N}$ in the integers

In this section we show how we can regard $(\mathbb{N} \times \mathbb{N})/\gamma$ as an extension of $\mathbb{N}$.

Having constructed the integral domain $(\mathbb{N} \times \mathbb{N})/\gamma$ from $\mathbb{N}$ we proceed to show how we can regard $(\mathbb{N} \times \mathbb{N})/\gamma$ to be the integers. There are many ways of writing the same equivalence class in $(\mathbb{N} \times \mathbb{N})/\gamma$: $(0, 0)/\gamma = (1, 1)/\gamma = (2, 2)/\gamma = \cdots$ and $(1, 0)/\gamma = (2, 1)/\gamma = (3, 2)/\gamma = \cdots$. We now look for some unique representation of a member of $(\mathbb{N} \times \mathbb{N})/\gamma$.

Lemma. *For any pair $(m, n) \in \mathbb{N} \times \mathbb{N}$ there is a pair $(x, y) \in (m, n)/\gamma$ such that $x = 0$ or $y = 0$.*

PROOF. We use induction. Let $S = \{n \mid \text{for all } m \in \mathbb{N}, (m, n)/\gamma \text{ contains a pair } (x, y) \text{ in which } x = 0 \text{ or } y = 0\}$. $0 \in S$ because $(m, 0)/\gamma$ contains the pair $(m, 0)$. Suppose $k \in S$. Now consider $(m, k + 1)/\gamma$ for any $m \in \mathbb{N}$. If $m = 0$ then the pair $(0, k + 1) \in (m, k + 1)/\gamma$. If $m \neq 0$ then $m = p + 1$ for some $p \in \mathbb{N}$. We are then considering the equivalence class $(p + 1, k + 1)/\gamma$. This is the same set as $(p, k)/\gamma$ because $(p, k)\gamma(p + 1, k + 1)$. Because $k \in S$ we know $(p, k)/\gamma$ contains some pair (x, y) with $x = 0$ or $y = 0$. This same pair belongs to $(p + 1, k + 1)/\gamma = (m, k + 1)/\gamma$. $k + 1 \in S$. $S = \mathbb{N}$. □

Theorem. $(\mathbb{N} \times \mathbb{N})/\gamma = \{(m, 0)/\gamma \mid m \in \mathbb{N}\} \cup \{(0, n)/\gamma \mid n \in \mathbb{N}^+\}$.

PROOF. Let $(p, q)/\gamma \in (\mathbb{N} \times \mathbb{N})/\gamma$. $(p, q)\gamma(m, 0)$ for some $m \in \mathbb{N}$ or $(p, q)\gamma(0, n)$ for some $n \in \mathbb{N}^+$. Note $(p, q)\gamma(0, 0)$ is included in the first alternative. $(p, q)/\gamma = (m, 0)/\gamma$ for some $m \in \mathbb{N}$ or $(p, q)/\gamma = (0, n)/\gamma$ for some $n \in \mathbb{N}^+$. □

We now show that $(\mathbb{N} \times \mathbb{N})/\gamma$ consists exactly of the members $(0, 0)/\gamma$, $(1, 0)/\gamma$, $(0, 1)/\gamma$, $(2, 0)/\gamma$, $(0, 2)/\gamma$, $\ldots$.

Theorem. $(m, 0)/\gamma = (n, 0)/\gamma$ *if and only if* $m = n$. $(0, m)/\gamma = (0, n)/\gamma$ *if and only if* $m = n$. $(m, 0)/\gamma \neq (0, n)/\gamma$ *for any* $m \in \mathbb{N}$, $n \in \mathbb{N}^+$.

PROOF. $(m, 0)/\gamma = (n, 0)/\gamma$ if and only if $m + 0 = 0 + n$. Similarly for the second statement. $(m, 0)/\gamma = (0, n)/\gamma$ if and only if $m + n = 0 + 0$. But this cannot be for $n \neq 0$. □

We proceed to identify those members of the form $(m, 0)/\gamma$ with the natural numbers.

Theorem. *$j:\mathbb{N} \to (\mathbb{N} \times \mathbb{N})/\gamma$ such that $j(m) = (m, 0)/\gamma$ is an injection and preserves both binary operations of $\mathbb{N}$.*

PROOF. $j(m + n) = (m + n, 0)/\gamma = (m, 0)/\gamma \mathbin{\bar{+}} (n, 0)/\gamma = j(m) \mathbin{\bar{+}} j(n)$. $j(mn) = (mn, 0)/\gamma = (m, 0)/\gamma \mathbin{\bar{\boxdot}} (n, 0)/\gamma = j(m) \mathbin{\bar{\boxdot}} j(n)$. $j(m) = j(n)$ implies $(m, 0)/\gamma = (n, 0)/\gamma$ which implies $m = n$. □

Theorem. $(\mathbb{N} \times \mathbb{N})/\gamma = j(\mathbb{N}) \cup -j(\mathbb{N}^+)$.

PROOF. Those members of $(\mathbb{N} \times \mathbb{N})/\gamma$ of the form $(m, 0)/\gamma = j(m)$ form the set $j(\mathbb{N})$. By $-j(\mathbb{N}^+)$ we mean the set $\{-j(n) | n \in \mathbb{N}^+\}$ which is the set $\{-(n, 0)/\gamma | n \in \mathbb{N}^+\} = \{(0, n)/\gamma | n \in \mathbb{N}^+\}$. □

We now follow the practice of writing n for $j(n)$ and $-n$ for $-j(n)$. This is to say we no longer distinguish between the natural number n and the integer $j(n)$. We denote the resulting integral domain of integers with $\langle \mathbb{Z}, +, \cdot, 0, 1\rangle$.

QUESTIONS

1. Which of the following statements are true of $\langle(\mathbb{N} \times \mathbb{N})/\gamma, \bar{+}, \bar{\boxdot}, (0, 0)/\gamma, (1, 0)/\gamma\rangle$?
   (A) $(0, 3)/\gamma = (7, 10)/\gamma$.
   (B) $(0, 3) \in (4, 7)/\gamma$.
   (C) $(a + k, b + k)\gamma(a, b)$.
   (D) $(1, 4)/\gamma \cap (7, 10)/\gamma = (6, 6)/\gamma$.
   (E) None of the statements is true.
2. $j:\mathbb{N} \to (\mathbb{N} \times \mathbb{N})/\gamma$ such that $j(n) = (0, n)/\gamma$
   (A) preserves $+$ in $\bar{+}$
   (B) preserves $\cdot$ in $\bar{\boxdot}$
   (C) has $j(0) = (0, 0)/\gamma$
   (D) has $j(1) = (1, 0)/\gamma$.
   (E) None of the alternatives completes a true sentence.

EXERCISES

1. Prove $(m + k, n + k)/\gamma = (m, n)/\gamma$.
2. Prove $(m, n)/\gamma = (u, v)/\gamma$ if and only if $m + v - n + u$.
3. Solve this equation: $[(5, 3)/\gamma \mathbin{\bar{\boxdot}} (x, y)/\gamma] \mathbin{\bar{+}} (2, 8)/\gamma = (6, 4)/\gamma$.
4. Are there solutions to this equation? $(3, 6)/\gamma \mathbin{\bar{\boxdot}} (x, y)/\gamma = (2, 4)/\gamma$.

3.7 Ordered integral domains

In this section order is developed for integral domains.

Regarding the natural numbers $\mathbb{N}$ as a subset of the integers $\mathbb{Z}$ we call the set $\mathbb{N}^+$ of all nonzero natural numbers the positive subset of $\mathbb{Z}$ and denote it also by $\mathbb{Z}^+$. From our earlier results we can summarize our knowledge of $\mathbb{Z}^+$.

Theorem. $\mathbb{Z}^+ \subseteq \mathbb{Z}$. $0 \notin \mathbb{Z}^+$. *$x, y \in \mathbb{Z}^+$ imply $x + y$ and $xy \in \mathbb{Z}^+$. $x \neq 0$ implies $x \in \mathbb{Z}^+$ or $-x \in \mathbb{Z}^+$.*

Proof. We have noted in Section 3.4 that $\mathbb{N}^+$ is closed under addition and multiplication. Since $(\mathbb{N} \times \mathbb{N})/\gamma = j(\mathbb{N}) \cup -j(\mathbb{N}^+)$ we have the fourth property. □

Rather than just develop order for the integers we develop order for integral domains in general; it is no more difficult.

Definition. Let $\langle R, +, \cdot, \theta, v\rangle$ be a commutative unitary ring with $v \neq \theta$. A subset R^+ of R is called a *positive subset* of R if and only if

$\theta \notin R^+$,
$x, y \in R^+$ imply $x + y$ and $xy \in R^+$,
$x \in R$ and $x \neq \theta$ imply $x \in R^+$ or $-x \in R^+$.

The existence of a positive subset in a commutative unitary ring is enough to make the ring an integral domain.

Theorem. *If a commutative unitary ring $\langle R, +, \cdot, \theta, v\rangle$ with $\theta \neq v$ has a positive subset R^+ then R is an integral domain.*

Proof. Suppose that x and y are both elements of R and are both nonzero. We then have four cases: $x \in R^+$ and $y \in R^+$, $x \in R^+$ and $-y \in R^+$, $-x \in R^+$ and $y \in R^+$, $-x \in R^+$ and $-y \in R^+$. The four cases yield the following possibilities: $xy \in R^+$, $-xy \in R^+$, $-xy \in R^+$, $xy \in R^+$. In no case does $xy = \theta$. We have therefore proved $x \neq \theta$ and $y \neq \theta$ imply $xy \neq \theta$. This is the contrapositive of $xy = \theta$ implies $x = \theta$ or $y = \theta$. □

In view of this fact that every commutative unitary ring ($\theta \neq v$) with a positive subset is an integral domain we shall refer to such rings as integral domains with positive subsets.

Theorem. *If $\langle R, +, \cdot, \theta, v\rangle$ is an integral domain with a positive subset then $\{R^+, -R^+, \{\theta\}\}$ is a partition of R.*

Proof. We show first that $R^+ \cup -R^+ \cup \{\theta\} = R$. Let $x \in R$ and $x \neq \theta$. Then $x \in R^+$ or $-x \in R^+$. If $-x \in R^+$ then $-x = y$ for some $y \in R^+$. $x = -y$ for $y \in R^+$. $x \in -R^+$.

We next show that the three subsets are disjoint (have an empty intersection in pairs). $\theta \notin R^+$ and $\theta \notin -R^+$. Suppose $x \in R^+ \cap -R^+$. $x \in R^+$ and $x = -y$ for some $y \in R^+$. $-x = y \in R^+$. $x + (-x) \in R^+$. $\theta \in R^+$. This contradiction shows that no x can belong to both R^+ and $-R^+$. $R^+ \cap -R^+ = \varnothing$. □

We call the set $-R^+$ the set of negative elements of the integral domain. The previous theorem shows that all elements are either positive or negative or zero but never simultaneously more than one of these.

We now move to show how the existence of a positive subset of an integral domain allows the construction of an order on the integral domain. An order on a set S is called a *total order* if and only if every two elements of S are comparable: i.e., $x \in S$ and $y \in S$ imply $x \leqslant y$ or $y \leqslant x$. We note, for example, that inclusion on the set of all subsets of a given set (with at least two elements) is not a total order.

Definition. An integral domain $\langle R, +, \cdot, \theta, v\rangle$ is an *ordered integral domain* if and only if there exists a total order ($\leqslant$) on R such that

$$x \leqslant y \quad \text{implies } x + z \leqslant y + z$$
$$x \leqslant y \text{ and } z \geqslant \theta \quad \text{implies } xz \leqslant yz.$$

Theorem. *Let $\langle R, +, \cdot, \theta, v\rangle$ be an integral domain with a positive subset R^+. Then $\langle R, +, \cdot, \theta, v\rangle$ is an ordered integral domain with the order defined by ($x \leqslant y$ if and only if $y - x \in R^+$ or $y = x$).*

Proof. We must first show that the order in question is actually a total ordering of R. $x \leqslant x$ for all $x \in R$ because $x = x$. $x \leqslant y$ and $y \leqslant x$ imply ($y - x \in R^+$ or $y = x$) and ($x - y \in R^+$ or $x = y$). There are four cases here, three of which lead to the conclusion $x = y$. We show that the fourth case $y - x \in R^+$ and $x - y \in R^+$ is impossible. This case is impossible because the sum $(y - x) + (x - y)$ is θ which cannot belong to R^+. We conclude $x = y$ and have antisymmetry.

Now assume $x \leqslant y$ and $y \leqslant z$ which gives ($y - x \in R^+$ or $y = x$) and ($z - y \in R^+$ or $z = y$). We follow the four possibilities. $y - x \in R^+$ and $z - y \in R^+$ imply $z - x \in R^+$. $x \leqslant z$. $y - x \in R^+$ and $z = y$ imply $z - x \in R^+$. $x \leqslant z$. $y = x$ and $z - y \in R^+$ imply $z - x \in R^+$. $x \leqslant z$. $y = x$ and $z = y$ imply $z = x$. $x \leqslant z$. Now that we have proved $\leqslant$ to be reflexive, antisymmetric and transitive we know $\leqslant$ to be an order on R. We now prove the order to be total. Let $x, y \in R$. $x - y \in R$. $x - y \in R^+$ or $x - y \in -R^+$ or $x - y = \theta$ because $\{R^+, -R^+, \{\theta\}\}$ is a partition of R. $x - y \in R^+$ or $y - x \in R^+$ or $x = y$. $y \leqslant x$ or $x \leqslant y$.

We now prove the order to be compatible with the binary operations of the integral domain. If $x \leqslant y$ then $y - x \in R^+$ or $y = x$. If $x = y$ then $x + z = y + z$. On the other hand if $y - x \in R^+$ then $y + z - x - z \in R^+$. $(y + z) - (x + z) \in R^+$. $x + z \leqslant y + z$. For multiplication let $x \leqslant y$ and

$\theta \leqslant z$. ($y - x \in R^+$ or $x = y$) and ($z \in R^+$ or $z = \theta$). The four possibilities go as follows. *Case 1*: $y - x \in R^+$ and $z \in R^+$. $(y-x)z \in R^+$. $yz - xz \in R^+$. $xz < yz$. *Case 2*: $y - x \in R^+$ and $z = \theta$. $(y - x)z = \theta$. $xz = yz$. *Case 3*: $y - x = \theta$ and $z \in R^+$. $(y - x)z = \theta$. $xz = yz$. *Case 4*: $y - x = \theta$ and $z = \theta$. $(y - x)z = \theta$. $xz = yz$. Thus in all four cases $xz \leqslant yz$. □

We shall use $x \leqslant y$ and $y \geqslant x$ interchangeably. It is sometimes convenient to use the concept of strict order ($x < y$) instead of order ($x \leqslant y$). Either can be defined in terms of the other. Strict order can also be defined directly from the positive subset R^+: $x < y$ if and only if $y - x \in R^+$. See the exercises for details on this possibility.

We have showed how one can begin with a positive subset of an integral domain and then define an ordered integral domain. We complete a cycle and show that any ordered integral domain has a positive subset.

Theorem. *Let $\langle R, +, \cdot, \theta, v\rangle$ be an ordered integral domain with total order $\leqslant$. Then the subset of R, $\{x | x \geqslant \theta$ and $x \neq \theta\}$, is a positive subset of R.*

PROOF. Suppose x and y are both members of the subset $\{t | t \geqslant \theta$ and $t \neq \theta\}$. Then $x \geqslant \theta$, $y \geqslant \theta$, $x \neq \theta$, $y \neq \theta$. Using the compatibility of the order $x + y \geqslant x + \theta \geqslant \theta + \theta$. $x + y \geqslant \theta$. Suppose $x + y$ were zero. $x + y = \theta$. Since $x + y \geqslant x$ we would have $\theta \geqslant x$. $x \geqslant \theta$ and $\theta \geqslant x$ yields $x = \theta$. But $x \neq \theta$. $x + y \in \{t | t \geqslant \theta$ and $t \neq \theta\}$. For closure under multiplication we have $xy \geqslant x \cdot \theta$. $xy \geqslant \theta$. $xy \neq \theta$ because R is an integral domain. $xy \in \{t | t \geqslant \theta$ and $t \neq \theta\}$. We finally prove that if $x \in R$ and $x \neq \theta$ then $x \geqslant \theta$ or $-x \geqslant \theta$. Because any two elements of the totally ordered R are comparable either $x \geqslant \theta$ or $\theta \geqslant x$. To both sides of the second inequality we add $-x$ getting $-x \geqslant \theta$. □

We end this section with several brief results.

Theorem. *Let $\langle R, +, \cdot, \theta, v\rangle$ be an ordered integral domain. If $x \in R$ then $x^2 \geqslant \theta$.*

PROOF. By x^2 we mean, of course, xx. $x \in R$ means $x \geqslant \theta$ or $x \leqslant \theta$. If $x \geqslant \theta$ then $xx \geqslant \theta$. If $x \leqslant \theta$ then $-x \geqslant \theta$. $xx = (-x)(-x) \geqslant \theta$. □

Theorem. *Let $\langle R, +, \cdot, \theta, v\rangle$ be an ordered integral domain. Then $v > \theta$ and $-v < \theta$.*

PROOF. $v = vv \geqslant \theta$. $v \neq \theta$. $v \geqslant \theta$ implies $-v \leqslant \theta$. □

QUESTIONS

1. Which of the following are not part of a definition of positive subset R^+ of a commutative unitary ring R.
   (A) $\theta \notin R^+$.
   (B) $x \in R$ implies $x \in R^+$ or $-x \in R^+$.

(C) $x, y \in R^+$ imply $xy \in R^+$.
(D) $x, y \in R^+$ imply $x + y \in R^+$.
(E) None of the statements fails to be part of the definition.

2. Which of the following statements are correct for the ordered integral domain $\langle R, +, \cdot, \theta, v\rangle$?
(A) $x \neq \theta$ implies $-x \in -R^+$.
(B) $x, y \in -R^+$ imply $x + y \in -R^+$.
(C) $R^+ \cap -R^+ = \{\theta\}$.
(D) $(-R^+)(-R^+) \subseteq R^+$.
(E) None of the statements is correct.

3. Which of the following statements are true?
(A) Every integral domain has a positive subset.
(B) Every integral domain with a positive subset is a ordered integral domain.
(C) In an ordered integral domain every square is greater than or equal to zero.
(D) In an ordered integral domain $v \geqslant \theta$.
(E) None of the statements is true.

4. Let $\leqslant$ be an order on a set R. The order is total if and only if
(A) $x \leqslant x$ for all $x \in R$
(B) $x \leqslant y$ or $y \leqslant x$ for all $x, y \in R$
(C) $x \leqslant y$ and $y \leqslant z$ imply $x \leqslant z$ for all $x, y, z \in R$
(D) $x \leqslant y$ and $y \leqslant x$ imply $x = y$ for all $x, y \in R$.
(E) None of the alternatives completes a true sentence.

EXERCISES

1. Does the relation $x\kappa y$ if and only if $x - y \in \{3, 4, 5, \ldots\}$ make $\mathbb{Z}$ into an ordered integral domain?

2. Is $\langle \mathbb{Q}, +, \cdot, 0, 1\rangle$ with the usual ordering an ordered integral domain?

3. Show that the following statements are true in an ordered integral domain $\langle R, +, \cdot, \theta, v\rangle$
(a) $x^2 + y^2 \geqslant 2xy$
(b) $x \geqslant y \geqslant 0$ implies $x^2 \geqslant y^2$.

4. We remember that a binary operation β and an order $\leqslant$ are compatible if and only if $x_1 \leqslant x_2$ and $y_1 \leqslant y_2$ imply $x_1\beta y_1 \leqslant x_2\beta y_2$. Show that if $\langle R, +, \cdot, \theta, v\rangle$ is an ordered integral domain it is impossible for both $+$ and $\cdot$ to be compatible with the order.

3.8 A characterization of the integers

In this section we characterize the integers as an ordered integral domain with a well-ordered set of nonnegative elements. We also introduce a second form of mathematical induction.

We recall first that m is a *minimum* (or smallest) element of an ordered set if and only if $m \in S$ and $m \leqslant x$ for all $x \in S$.

Definition. An ordered set S is *well-ordered* if and only if every nonempty subset of S has a minimum element.

We now prove that the subset of nonnegative elements of $\mathbb{Z}$ is a well-ordered set (under the order of $\mathbb{Z}$).

Theorem. *$\mathbb{N}$ is a well-ordered set.*

PROOF. We must show that every nonempty subset of $\mathbb{N}$ has a minimum. Let S be any nonempty subset. Since $S \neq \varnothing$ we may choose an element from S, say n. We partition S into two subsets: $S = \{x | x \in S \text{ and } x \leqslant n\} \cup \{x | x \in S \text{ and } x > n\}$. No x can simultaneously be $\leqslant$ and $> n$. The two subsets are disjoint. Because the order on $\mathbb{Z}$ (and therefore on $\mathbb{N}$) is total every $x \in S$ must belong to at least one of the two subsets. The first subset must be nonempty because it must contain n. The second subset could possibly be empty and in that case there is but one subset. It is clear, however, that if we find an element m of the first set which is a minimum for the first set then it will be smaller than any member of the second set and therefore a minimum for all of S. We therefore content ourselves with the proving of this proposition. Any nonempty subset of $\mathbb{N}$ with elements not exceeding the natural number n has a minimum. We use induction on n. Let $n = 0$. The subset must then be $\{0\}$ which has a minimum 0. Assume the result true for k. Let A be any nonempty subset of $\mathbb{N}$ with elements not exceeding $k + 1$. If A consists exactly of $\{k + 1\}$ then $k + 1$ is the minimum. If not, then there are elements of A strictly smaller than $k + 1$. Let A' be the subset of A with members not exceeding k. This subset must have a minimum m by the inductive hypothesis. m is also a minimum for A because $m \leqslant k < k + 1$. □

We remark here that $\mathbb{Q}$ and $\mathbb{R}$, two other ordered integral domains, do not have well-ordered nonnegative elements. The set $\{(\frac{1}{2})^n | n \in \mathbb{N}\}$, for example, is a nonempty subset of both $\mathbb{Q}$ and $\mathbb{R}$ of nonnegative elements which has no minimum. Again, of course, we are relying upon the readers previous knowledge at this point because we have not yet formally constructed $\mathbb{Q}$ or $\mathbb{R}$.

The following theorem proves there is no integer between 0 and 1.

Theorem. *Let $\langle R, +, \cdot, \theta, v \rangle$ be an ordered integral domain with the set of nonnegative elements well ordered. Then there is no $y \in R$ such that $\theta < y < v$.*

PROOF. Suppose that $\{x | x \in R \text{ and } \theta < x < v\}$ is a nonempty set. It is a nonempty set of nonnegative elements of R, a nonempty subset of a well-ordered set and therefore has a minimum member, say, m. $m < v$ implies $mm < vm$. $mm < m$. On the other hand, $\theta < m$ implies $\theta < mm$. Thus mm is between θ and v and is properly smaller than the minimum such element, m.

This is a contradiction and this shows that $\{x|x \in R \text{ and } \theta < x < v\}$ is empty. □

We have demonstrated that $\mathbb{Z}$ is an ordered integral domain and that its set of nonnegative elements, $\mathbb{N}$, is well-ordered. Now let us begin with any ordered integral domain with well-ordered nonnegative elements and conversely prove that the nonnegative elements obey the principle of mathematical induction.

Theorem. *Let $\langle R, +, \cdot, \theta, v\rangle$ be an ordered integral domain such that $N = \{x|x \in R \text{ and } x \geqslant \theta\}$ is well ordered. If $S \subseteq N$ and $0 \in S$ and ($x \in S$ implies $x + v \in S$) then $S = N$.*

PROOF. Consider the set $N - S$, the relative complement of S in N. We wish to prove this set to be the empty set. Suppose $N - S \neq \varnothing$. Let r be the minimum of the set $N - S$. There is an element $r - v$ in R and certainly $r - v < r$. Since $r \neq \theta$ we know $r > \theta$. Since there are no elements of R between θ and v we know $r \geqslant v$. Thus we have $r - v \geqslant \theta$. $r - v \in N$. Since r is the smallest element of $N - S$ and $r - v$ is properly smaller than r we must have $r - v \notin N - S$. $r - v$ is therefore in S. But then $(r - v) + v$ also belongs to S. $r \in S$. This contradiction means $N - S = \varnothing$. $S = N$. □

This theorem establishing the principle of mathematical induction for any ordered integral domain with a subset of nonnegative well-ordered elements leads one to suspect that such an integral domain does not differ markedly from the integers. Such an integral domain is, in fact, isomorphic with the integers. We put this result into a theorem.

Theorem. *If $\langle R, +, \cdot, \theta, v\rangle$ is an ordered integral domain with $N = \{x|x \in R \text{ and } x \geqslant \theta\}$ well ordered then there exists an isomorphism $f:\mathbb{Z} \to R$. Moreover, $x \geqslant y$ implies $f(x) \geqslant f(y)$, the isomorphism preserves order.*

PROOF. We define $f':\mathbb{N} \to N$ such that $f'(0) = \theta$ and $f'(n + 1) = f'(n) + v$. Such a function exists by our axiom of definition by induction. We now prove f' to be a surjection. $\theta \in f'(\mathbb{N})$ because $f'(0) = \theta$. If $k \in f'(\mathbb{N})$ then $k = f'(a)$ for some $a \in \mathbb{N}$. $f'(a + 1) = f'(a) + v = k + v$. $k + v \in f'(\mathbb{N})$. $f'(\mathbb{N}) = N$. This has been an application of the previous theorem, induction on the nonnegative elements of R.

We now prove that the surjection $f':\mathbb{N} \to N$ is also an injection. Let $S = \{n|f'(m) = f'(n) \text{ implies } m = n \text{ for all } m \in \mathbb{N}\}$. Suppose $f'(m) = f'(0)$. $f'(0) = \theta$. $f'(m) = \theta$. m must be 0 for otherwise $m = p + 1$ for some $p \in \mathbb{N}$. We would then have $f'(m) = f'(p + 1) = f'(p) + v = \theta$. But this is impossible because θ is not the sum of v and any member of N. $0 \in S$. To complete this induction we must prove $k \in S$ implies $k + 1 \in S$. Suppose $k \in S$. $f'(m) = f'(k)$ implies $m = k$ for all $m \in \mathbb{N}$. Now if $f'(m) = f'(k + 1)$

we have $f'(m) = f'(k) + v$. m is not 0 because $f'(m) \neq \theta$. $m = q + 1$ for some $q \in \mathbb{N}$. $f'(m) = f'(q + 1)$. We have therefore $f'(q + 1) = f'(q) + v = f'(k) + v$. $f'(q) = f'(k)$. $q = k$. $q + 1 = k + 1$. $m = k + 1$. $k + 1 \in S$. The induction is complete proving f' to be a bijection.

We now show that $f': \mathbb{N} \to N$ preserves $+$ and $\cdot$. Let $S = \{n | f'(m + n) = f'(m) + f'(n) \text{ for all } m \in \mathbb{N}\}$. $0 \in S$ because $f'(m + 0) = f'(m) = f'(m) + \theta = f'(m) + f'(0)$. To prove $k \in S$ implies $k + 1 \in S$ let $f'(m + k) = f'(m) + f'(k)$ for all $m \in \mathbb{N}$. $f'(m + k + 1) = f'(m + k) + v = f'(m) + f'(k) + v = f'(m) + f'(k + 1)$. $S = \mathbb{N}$. This shows f' preserves $+$.

To demonstrate that f' preserves $\cdot$, let $S = \{n | f'(mn) = f'(m)f'(n) \text{ for all } m \in \mathbb{N}\}$. $0 \in S$ for $f'(m \cdot 0) = f'(0) = \theta = f'(m)\theta = f'(m)f'(0)$ for all $m \in \mathbb{N}$. To prove $k \in S$ implies $k + 1 \in S$ let $f'(mk) = f'(m)f'(k)$ for all $m \in \mathbb{N}$. Then $f'(m(k + 1)) = f'(mk + m) = f'(mk) + f'(m) = f'(m)f'(k) + f'(m) = f'(m)[f'(k) + v] = f'(m)f'(k + 1)$. $S = \mathbb{N}$.

The next part of the theorem proof is to extend $f': \mathbb{N} \to N$ to a function $f: \mathbb{Z} \to R$. The extension f must agree with f' on $\mathbb{N}$. We define $f: \mathbb{Z} \to R$ as follows: $f(x) = f'(x)$ for all $x \in \mathbb{N}$, $f(x) = -f'(-x)$ for $x \in -\mathbb{N}$. This function f is a bijection from $\mathbb{N}$ to N, a bijection from $-\mathbb{N}$ to $-N$ and therefore is a bijection from $\mathbb{Z}$ to R.

We now intend to demonstrate that f preserves $+$ and $\cdot$. *Case 1*: $x \in \mathbb{N}$ and $y \in \mathbb{N}$. $x + y \in \mathbb{N}$. $f(x) = f'(x)$. $f(y) = f'(y)$. $f(x + y) = f'(x + y) = f'(x) + f'(y) = f(x) + f(y)$. *Case 2*: $x \in \mathbb{N}$ and $y \in -\mathbb{N}$. $x + y \in -\mathbb{N}$. $f(x + y) = -f'(-(x + y)) = -f'(-x + (-y)) = -[f'(-x) + f'(-y)] = -f'(-x) - f'(-y) = f(x) + f(y)$. We remark that these two cases are not mutually exclusive but overlap at 0 just as the definition of f. *Case 3a*: $x \in \mathbb{N}$ and $y \in -\mathbb{N}$ and $x + y \in \mathbb{N}$. $x = (-y) + (x + y)$ with $-y \in \mathbb{N}$, $x + y \in \mathbb{N}$, $x \in \mathbb{N}$. $f'(x) = f'((-y) + (x + y)) = f'(-y) + f'(x + y)$. Solving for $f'(x + y)$ we have $f'(x + y) = f'(x) - f'(-y)$. $f(x + y) = f(x) + f(y)$. *Case 3b*: $x \in \mathbb{N}$ and $y \in -\mathbb{N}$ and $x + y \in -\mathbb{N}$. $-y = -(x + y) + x$ with $-y \in \mathbb{N}$, $-(x + y) \in \mathbb{N}$, and $x \in \mathbb{N}$. $f'(-y) = f'(-(x + y) + x) = f'(-(x + y)) + f'(x)$. Solving for $-f'(-(x + y))$ we have $-f'(-(x + y)) = f'(x) - f'(-y)$. $f(x + y) = f(x) + f(y)$. *Cases 4a*, ($x \in -\mathbb{N}$ and $y \in \mathbb{N}$ and $x + y \in \mathbb{N}$), and *4b*, ($x \in -\mathbb{N}$ and $y \in \mathbb{N}$ and $x + y \in -\mathbb{N}$), can be proved by interchanging the role of x and y in Cases 3a and 3b. This concludes the demonstration that f preserves $+$.

We now show how f preserves $\cdot$. *Case 1*: $x \in \mathbb{N}$ and $y \in \mathbb{N}$. Then $xy \in \mathbb{N}$. $f(xy) = f'(xy) = f'(x)f'(y) = f(x)f(y)$. *Case 2*: $x \in -\mathbb{N}$ and $y \in -\mathbb{N}$. Then $xy \in \mathbb{N}$. $f(xy) = f'(xy) = f'((-x)(-y)) = f'(-x)f'(-y) = (-f'(-x)) \cdot (-f'(-y)) = f(x)f(y)$. *Case 3*: $x \in \mathbb{N}$ and $y \in -\mathbb{N}$. Then $xy \in -\mathbb{N}$. $f(xy) = -f'(-xy) = -f'(x(-y)) = -f'(x)f'(-y) = f'(x)(-f'(-y)) = f(x)f(y)$. *Case 4*, ($x \in -\mathbb{N}$ and $y \in \mathbb{N}$), is similar to Case 3.

Finally in this lengthy proof we wish to prove $x \geqslant y$ if and only if $f(x) \geqslant f(y)$. This is equivalent to $x - y \in \mathbb{N}$ if and only if $f(x - y) \in N$. But this statement is clearly true from the construction of f. This completes the proof of the theorem. □

We complete this section with an alternate form of mathematical induction. This form is sometimes referred to as the second principle of mathematical induction. It is essentially a proof scheme based upon the well-ordering of $\mathbb{N}$.

Theorem. *Let S be a subset of $\mathbb{N}$ such that $\{k|k < n\} \subseteq S$ implies $n \in S$. Then $S = \mathbb{N}$.*

PROOF. In alternate words, we wish to show that if n belongs to S every time all elements smaller than n belong to S then S must be $\mathbb{N}$.

We must prove that the subset S mentioned in the hypothesis is all of $\mathbb{N}$. We do this by mathematical induction (the mathematical induction we have been consistently applying in this chapter). It is given that if $\{k|k < 0\} \subseteq S$ then $0 \in S$. But $\{k|k < 0\}$ is a subset of S because $\{k|k < 0\}$ is the empty set. Therefore $0 \in S$.

Suppose $m \in S$. We must show that $m + 1 \in S$. Assume there is some smallest number j, $0 < j < m$, which does not belong to S. j cannot be zero since $0 \in S$. Then all the numbers before j do belong to S. $\{x|x < j\} \subseteq S$. Using the hypothesis for the theorem $j \in S$. This contradicts $j \notin S$. There can be no integers smaller than m which fail to belong to S. $\{x|x \leqslant m\} \subseteq S$. $\{x|x < m + 1\} \subseteq S$. Again using the hypothesis we have $m + 1 \in S$. $S = \mathbb{N}$. □

Observe that the hypothesis of the theorem does not require proving $0 \in S$ separately. It should be clear from the theorem that once $\{k|k < n\} \subseteq S$ implies $n \in S$ is proved then $0 \in S$ is a consequence.

QUESTIONS

1. Which of the following statements are true?
   - (A) If $\langle R, +, \cdot, \theta, v\rangle$ is an ordered integral domain then there is no $y \in R$ such that $0 < y < v$.
   - (B) If $\langle R, +, \cdot, \theta, v\rangle$ is an ordered integral domain then $\theta < v$.
   - (C) An ordered integral domain with well-ordered nonnegative subset is order and ring isomorphic to the integers.
   - (D) If $S \subseteq \mathbb{N}$ and $n \in S$ implies $n + 1 \in S$ then $S = \mathbb{N}$.
   - (E) None of the statements is true.

2. Which of the following statements are true?
   - (A) 0 is a minimum element of $\{x|x \geqslant 0 \text{ and } x \in \mathbb{Q}\}$ shows that $\{x|x \geqslant 0 \text{ and } x \in \mathbb{Q}\}$ is well ordered.
   - (B) $\varnothing$ is a well-ordered subset of $\mathbb{Z}$.
   - (C) $\{1 - (\frac{1}{2})^n|n \in \mathbb{N}\}$ is a well ordered subset of $\mathbb{Q}$.
   - (D) $\{(\frac{1}{2})^n|n \in \mathbb{N}\}$ is a well ordered subset of $\mathbb{Q}$.
   - (E) None of the statements is true.

3. Let $f: R \to R'$ be a function from the ordered integral domain $\langle R, +, \cdot, \theta, v\rangle$ to the ordered integral domain $\langle R', +', \cdot', \theta', v'\rangle$. Which of the following conditions are not necessary for f to be an isomorphism and to preserve order?

(A) $x \leqslant y$ implies $f(x) \leqslant f(y)$ for all $x, y \in R$.
(B) $f(v) = v'$.
(C) $f(xy) = f(x)f(y)$ for all $x, y \in R$.
(D) $f(\theta) = \theta'$.
(E) All of the conditions are necessary.

4. Which of the following statements are correct?
(A) ($S \subseteq \mathbb{N}$ and $\{0, 1, \ldots, k - 1\} \subseteq S$ implies $k \in S$) imply $S = \mathbb{N}$.
(B) $-\mathbb{N}$ is well-ordered.
(C) Every subset of a well ordered set is also well ordered.
(D) Every nonempty subset of $\mathbb{N}$ has a maximum element.
(E) None of the statements is correct.

Exercises

1. Let $\langle R, +, \cdot, \theta, v\rangle$ be an ordered integral domain with well-ordered nonnegative subset. Let $f: R \to R$ be a morphism which also preserves order. Show that if $S \neq \varnothing$ and $S \subseteq \{x | x \geqslant 0 \text{ and } x \in R\}$ then $f(\min S) = \min f(S)$.

2. Show that if $\langle R, +, \cdot, \theta, v\rangle$ is an ordered integral domain with well-ordered nonnegative subset and $x, r, s \in R^+$ and $x = rs$ then $r \leqslant x$.

3. Let $\langle R, +, \cdot, \theta, v\rangle$ be an ordered integral domain. Show that if $x > v + v$ then $x^2 \geqslant x + v$.

4. Show that in any ordered integral domain there can be no maximum element.

5. Show that if S is a subset of the nonnegative integers and for all n, $\{x | x < n\} \subseteq S$ implies $n \in S$, then $0 \in S$.

4 Rings: Applications of the integers

In this chapter we assemble some results on rings which we obtain by using a specific knowledge of the natural numbers and the integers. We begin the chapter with some work refining our knowledge of finite and infinite sets. We then routinely study some theorems extending the associative, commutative, and distributive laws to any finite number of elements of a ring. We then extend to the integers the division algorithm earlier established for the natural numbers and discuss briefly prime numbers. After this we study the use in rings of the integers to indicate repeated additions and repeated multiplications: multiples and exponents. We consider in Section 4.5 the important result that every integral domain is included in some field. We show the existence of such a field and call it the field of fractions of the given integral domain. We specifically apply the theorem to the integers to construct the ordinary fractions or rational numbers. We finally, in Section 4.6, study the characteristic of a ring.

4.1 Finite sets

We have previously used the following criterion for two sets to be the same size: S and T are equipotent if and only if there exists a bijection $f:S \to T$. For example, $\{a, b, c\}$ and $\{0, 1, 2\}$ are the same size because $f = \{(a, 0), (b, 1), (c, 2)\}$ is one possible bijection between the two sets. A further example was contained in Section 2.6 where we defined equipotent and proved two cosets of R/A have the same number of members. We shall now use the term *cardinal number* of a set to mean the number of members of the set. We speak of two equipotent sets as having the same cardinal number. We abbreviate cardinal number of the set S with crd S.

Definition. crd S = crd T if and only if there exists a bijection $f:S \to T$.

One can verify that the following equivalence properties hold for the concept of equality of cardinal numbers.

Theorem. crd S = crd S. crd S = crd T *implies* crd T = crd S. crd S = crd T *and* crd T = crd U *imply* crd S = crd U.

It is possible to give crd S a specific identity as a set. The equality in the theorem and definition then becomes equality of sets. In particular when a set has the same cardinal number as some natural number then we define that natural number to be the cardinal number of the set.

Definition. Let $n \in \mathbb{N}$. crd $S = n$ if and only if there exists a bijection $f: n \to S$.

To realize this bijection we use the set model for the natural number as described in Section 3.1: $0 = \varnothing$, $1 = \{0\}$, $2 = \{0, 1\}$, $3 = \{0, 1, 2\}, \ldots,$ $n = \{0, 1, 2, \ldots, n - 1\}, \ldots$. For example, crd$\{a, b, c\} = 3$ by the bijection given in the first paragraph of this section. To legitimize this definition we now establish that a set can have at most one natural number as its cardinal number.

Theorem. *Given* $m, n \in \mathbb{N}$, $f: m \to n$ *is a bijection implies* $m = n$.

PROOF. We give a proof by induction on the first number m. If $m = 0$, the empty set, and $f: \varnothing \to n$ is a bijection then $y \in n$ implies there exists an $x \in \varnothing$ such that $(x, y) \in f$. But there is no $x \in \varnothing$ and so can be no y in n. $n = \varnothing$.

Assume the result is true for $m = k$. Let $f: k + 1 \to n$ be a bijection. $n \neq 0$ for otherwise f is not a function unless $k + 1 = 0$, which it is not. Because $n \neq 0$ it is $l + 1$ for some $l \in \mathbb{N}$. We now investigate the bijection $f: k + 1 \to l + 1$. If $f(k)$ happens to be l then $f - \{(k, l)\}: \{0, 1, 2, \ldots, k - 1\} \to \{0, 1, \ldots, l - 1\}$ is a bijection $k \to l$. By induction hypothesis $k = l$. Therefore $k + 1 = l + 1 = n$. If $f(k)$ happens not to be l then $f(k) = j$ for some $j \in \{0, 1, \ldots, l\}, j \neq l$. $f - \{(k, j)\}: \{0, 1, \ldots, k - 1\} \to \{0, 1, \ldots, j - 1, j + 1, \ldots, l\}$ is still a bijection. There exists an $i \in \{0, 1, \ldots, k - 1\}$ such that $f(i) = l$. $f - \{(k, j), (i, l)\}: \{0, 1, \ldots, i - 1, i + 1, \ldots, k - 1\} \to \{0, 1, \ldots, j - 1, j + 1, \ldots, l - 1\}$ is also a bijection. If we adjoin to this function the ordered pair (i, j) we have yet another bijection $(f - \{(k, j), (i, l)\}) \cup \{(i, j)\}: \{0, 1, \ldots, k - 1\} \to \{0, 1, \ldots, l - 1\}$. We have therefore a bijection from k to l. $k = l$. Therefore $k + 1 = l + 1 = n$. □

Corollary. *Given any set S there is at most one natural number n for which there exists a bijection $f: n \to S$.*

PROOF. Suppose $f: m \to S$ and $g: n \to S$ are both bijections of natural numbers into S. Then $g^{-1} \circ f: m \to n$ is a bijection. $m = n$. □

There are, of course, sets which do not have any natural number as their cardinal number. Two such sets are $\mathbb{N}$ and $\mathbb{R}$. Any such set not equipotent with some natural number will be called an infinite set. It can, incidently, be shown that $\mathbb{N}$ and $\mathbb{R}$ are not equipotent.

Definition. A set equipotent with some natural number is called a *finite set.* All other sets are called *infinite sets.*

It is possible to define certain sets to be cardinal numbers for infinite sets much as we have done for finite sets.

If S is a set with cardinal number n, a natural number, because of the existence of some bijection $f:n \to S$ it is possible and convenient to denote the members of S by a notation such as $x_0, x_1, x_2, \ldots, x_{n-1}$ where $x_i = f(i)$ for all $i \in \mathbb{N}$. An alternate equivalent notation preferred by some is $x_1, x_2, \ldots, x_n$. Using this indexed notation we develop a property peculiar to finite sets.

Theorem. *If* crd S = crd $T = n$ *for some* $n \in \mathbb{N}$ *and* $f:S \to T$ *is either an injection or a surjection then* $f:S \to T$ *is a bijection.*

Proof. The proof is by induction on n. If crd S = crd $T = 0$ then both S and T are equal to $\varnothing$ and any function $f:\varnothing \to \varnothing$ must be a bijection.

Assume the theorem to be true for the natural number k. Now let crd S = crd $T = k + 1$. *Case 1*: $f:\{x_0, x_1, \ldots, x_k\} \to \{y_0, y_1, \ldots, y_k\}$ is an injection. $f(x_k) = y_j$ for some j. $f - \{(x_k, y_j)\}:\{x_0, x_1, \ldots, x_{k-1}\} \to \{y_0, y_1, \ldots, y_{j-1}, y_{j+1}, \ldots, y_k\}$ is also an injection, but on sets which have k elements. By the induction hypothesis $f - \{(x_k, y_j)\}$ is a bijection. f is then also a bijection. *Case 2*: $f:\{x_0, x_1, \ldots, x_k\} \to \{y_0, y_1, \ldots, y_k\}$ is a surjection. There exists an x_i such that $f(x_i) = y_k$.

$$f - \{(x_i, y_k)\}:\{x_0, \ldots, x_{i-1}, x_{i+1}, \ldots, x_k\} \to \{y_0, y_1, \ldots, y_{k-1}\}$$

is also a surjection, but on sets with k elements. By the induction hypothesis $f - \{(x_i, y_k)\}$ is a bijection. Then f is also. □

We now prove a theorem about integral domains which shows some of the power of the previous theorem.

Theorem. *Any integral domain which is finite is a field.*

Proof. Let $\langle R, +, \cdot, \theta, v\rangle$ be an integral domain in which crd $R = n$, a natural number. We note that $n \geqslant 2$ because θ and v must be distinct elements in an integral domain. To show that R must be a field let $a \in R$ and $a \neq \theta$. Define $\varphi:R \to R$ such that $\varphi(x) = ax$. φ is an injection. $\varphi(x_1) = \varphi(x_2)$ implies $ax_1 = ax_2$ which implies $x_1 = x_2$. Since R is finite φ must be a bijection. Thus given $v \in R$ there exists $b \in R$ such that $\varphi(b) = v$. $ab = v$. By commutativity $ba = v$ also. b is the inverse of a. □

QUESTIONS

1. Set A has the same cardinal number as set B
   (A) implies there exists an injection from A to B
   (B) means A has the same size as B
   (C) only if there exists a bijection from B to A
   (D) implies there exists a surjection from A to B.
(E) None of the choices completes a true sentence.

2. Which of the following statements are true?
   (A) If there is a bijection from natural number m to natural number n then $m = n$.
   (B) If $f:m \to n$ and $g:n \to p$ are bijections and m, n, p are natural numbers then $m = p$.
   (C) If m and n are natural numbers and $m \subseteq n$ then there exists an injection from m into n.
   (D) A set is finite or infinite.
(E) None of the statements is true.

3. Which of the following statements are true?
   (A) If there is a bijection from A to B then there is also a bijection from B to A.
   (B) There exists at least one surjection from $\{0, 1, 2\}$ to $\{0, 1, 2\}$ which is not an injection.
   (C) Every finite field is an integral domain.
   (D) Integral domains which are finite have at least one element which is not multiplicatively invertible.
(E) None of the statements is true.

4. Which of the following statements are true?
   (A) $\text{crd}\{0, 2, 4, 6, \ldots\} = \text{crd}\{1, 3, 5, 7, \ldots\}$.
   (B) $\text{crd}\, \mathbb{N} \times \mathbb{N} = \text{crd}\, \mathbb{N}$.
   (C) $\text{crd}\{0, 2, 4, 6, \ldots\} = \text{crd}\, \mathbb{N}$.
   (D) $\text{crd}\{1, 2, 3, 4, \ldots\} = \text{crd}\, \mathbb{N}$.
(E) None of the statements is true.

5. The sum of two finite numbers is finite is another way of saying
   (A) the natural numbers are closed under addition
   (B) infinite numbers do not really exist
   (C) the range of a bijection is the same as the codomain
   (D) natural numbers, excepting zero, have no negatives.
(E) None of the listed possibilities is an equivalent statement.

6. $f:\mathbb{N} \to 2\mathbb{N}$ such that $f(x) = 2x$
   (A) is a bijection
   (B) implies there are the same number of even natural numbers as natural numbers
   (C) shows that infinite sets may have proper subsets with the same cardinal number as the entire set
   (D) shows that odd natural numbers cannot be in the range of any bijection.
(E) None of the alternatives completes a true sentence.

EXERCISES

1. Prove crd $\mathbb{N}$ = crd $2\mathbb{N}$.
2. Prove crd $\mathbb{N}$ = crd $\mathbb{Z}$.
3. Prove crd $\mathbb{Z}$ = crd $\mathbb{Q}$.
4. If $(0, 1)$ is the open unit interval of $\mathbb{R}$ prove crd$(0, 1)$ = crd $\mathbb{R}$.
5. Prove crd $\mathbb{R} \neq$ crd $\mathbb{N}$. [*Hint*: Use Exercise 4.] Suppose there is a function $f:\mathbb{N} \to (0, 1)$ such that $f(n) = 0.a_{n0}a_{n1}a_{n2}\ldots$ (decimal notation). Show that there exists a number $0.b_1b_2b_3\ldots$ between 0 and 1 and not in the range of f. Conclude f cannot be a surjection.
6. Show that if $n \in \mathbb{N}$ and $S \subseteq n$ then crd $S \in \mathbb{N}$.
7. Show that every subset of a finite set is finite.
8. Show that if $n \in \mathbb{N}$ and $S \subset n$ then crd $S \neq$ crd n.
9. Show that no finite set is equipotent with a proper subset of itself.
10. Show that $\mathbb{N}$ is not finite.
11. Show that $\mathbb{N}$ is equipotent with a proper subset of itself.
12. Show that every infinite set contains a subset equipotent with $\mathbb{N}$.
13. Show that every infinite set is equipotent with a proper subset of itself.
14. Show that every unitary ring (with $v \neq \theta$) with no divisors of zero (other than zero itself) is a division ring.
15. The order on cardinal numbers as natural numbers can be extended to cardinal numbers of infinite sets as follows: crd $S \leqslant$ crd T if and only if there exists an injection $f:S \to T$. Prove crd $S \leqslant$ crd S. Prove crd $S \leqslant$ crd T and crd $T \leqslant$ crd U imply crd $S \leqslant$ crd U. We have not listed antisymmetry because the proof is quite difficult. An intuitively appealing proof can be found on p. 340 of [1].
16. Prove that the order relation defined in Exercise 15 agrees with the order definition for natural numbers given in Section 3.4. This is to say, prove the following statement: there exists a natural number p such that $m + p = n$ if and only if there exists an injection $f:m \to n$.
17. Prove $S \subseteq T$ implies crd $S \leqslant$ crd T.
18. Prove crd $\mathbb{N} <$ crd $\mathbb{R}$.
19. Addition of cardinal numbers (infinite or finite) can be defined by the following equation: if $S \cap T = \varnothing$ then crd S + crd T = crd$(S \cup T)$. Show that this definition does not depend upon the particular sets S and T chosen.
20. Show that if S and T are both finite sets then the definition given in Exercise 19 agrees with the definition of addition of natural numbers given in Section 3.2. Use induction on crd T for the proof. We notice that this proves that the union of two finite sets is finite.
21. Prove crd S + crd T = crd$(S \cup T)$ + crd$(S \cap T)$.

4.2 Generalized associative, commutative, and distributive theorems

Theorems of associativity, commutativity, and distributivity, true for two and three variables, can be extended to any finite number of variables. The proofs are by induction on the number of terms, a natural number. We also review the sigma and pi notation for products and sums.

Let β be an associative binary operation on any set S. We know that $a_1\beta(a_2\beta a_3) = (a_1\beta a_2)\beta a_3$ for all $a_1, a_2, a_3 \in S$. This permits in practice the use of the symbol $a_1\beta a_2\beta a_3$ without parentheses because either way parentheses are inserted the two results are equal. We wish now to extend this principle to expressions of greater length.

Definition. We define $a_1\beta a_2\beta \cdots \beta a_n$ for all $n \in \mathbb{N}^+$ by defining (for any binary operation β on a set S)

$$a_1\beta a_2\beta \cdots \beta a_n = a_1 \quad \text{for } n = 1, \text{and}$$
$$a_1\beta a_2\beta \cdots \beta a_k\beta a_{k+1} = (a_1\beta a_2\beta \cdots \beta a_k)\beta a_{k+1}.$$

This is, of course, a definition by induction on the length of the expression. It yields, for example, for $n = 3$

$$a_1\beta a_2\beta a_3 = (a_1\beta a_2)\beta a_3$$

and for $n = 4$

$$a_1\beta a_2\beta a_3\beta a_4 = ((a_1\beta a_2)\beta a_3)\beta a_4.$$

In this definition we designate the expression without parentheses to be one of the possible expressions containing parentheses. We now prove a theorem which demonstrates that if the operation is associative all of the expressions formed by inserting parentheses in different ways are equal.

Theorem. *Let β be an associative binary operation on a set S. Let n and k be natural numbers, $n \geqslant 2$, $1 \leqslant k < n$. If $b = (a_1\beta a_2\beta \cdots \beta a_k)$ and $c = (a_{k+1}\beta \cdots \beta a_n)$ then $b\beta c = a_1\beta a_2\beta \cdots \beta a_k\beta a_{k+1}\beta \cdots \beta a_n$.*

PROOF. We are trying to prove

$$(a_1\beta a_2\beta \cdots \beta a_k)\beta(a_{k+1}\beta \cdots \beta a_n) = a_1\beta a_2\beta \cdots \beta a_n.$$

We give a proof by induction on the length n of the expression beginning our induction with $n = 2$. If $n = 2$ then $k = 1$ and we have as the only way the given expression can be split is $(a_1)\beta(a_2)$. $(a_1)\beta(a_2) = a_1\beta a_2$. We now assume the theorem to be true for expressions of length m and we wish to demonstrate that the result is true for expressions of length $m + 1$. Let $(a_1\beta a_2\beta \cdots \beta a_p)\beta(a_{p+1}\beta \cdots \beta a_m\beta a_{m+1})$ be any expression of length $m + 1$ which is split after the first p factors, $1 \leqslant p < m + 1$. Using associativity (expressions of length 3) and the inductive assumption we have

$$\begin{aligned}(a_1\beta a_2\beta\cdots\beta a_p)\beta(a_{p+1}\beta\cdots\beta a_m\beta a_{m+1}) \\ &= ((a_1\beta\cdots\beta a_p)\beta(a_{p+1}\beta\cdots\beta a_m))\beta a_{m+1} \\ &= (a_1\beta\cdots\beta a_m)\beta a_{m+1} \\ &= a_1\beta a_2\beta\cdots\beta a_m\beta a_{m+1}. \qquad\square\end{aligned}$$

We remark at this point that if the operation β has a neutral element v in S then one can define $a_1\beta a_2\beta\cdots\beta a_n = v$ for $n = 0$ (an empty product) and achieve consistency with the theorems.

We now consider a generalized commutativity theorem for an operation which is given associative.

Theorem. *Let S be a set with an associative, commutative binary operation β. Then if $(\sigma(1), \sigma(2), \ldots, \sigma(n))$ is any permutation of $(1, 2, \ldots, n)$ we have $a_{\sigma(1)}\beta a_{\sigma(2)}\beta\cdots\beta a_{\sigma(n)} = a_1\beta a_2\beta\cdots\beta a_n$ for any $n \in \mathbb{N}^+$.*

PROOF. We give a proof by induction on n, the length of the expression. There is only one permutation on $\{1\}$, namely, $\sigma(1) = 1$. Therefore, $a_{\sigma(1)} = a_1$.

Assuming the result true for the natural number k we demonstrate the result for $k + 1$. *Case 1*: $\sigma(k + 1) = k + 1$. $(a_{\sigma(1)}\beta a_{\sigma(2)}\beta\cdots\beta a_{\sigma(k)})\beta a_{\sigma(k+1)} = (a_{\sigma(1)}\beta\cdots\beta a_{\sigma(k)})\beta a_{k+1} = (a_1\beta a_2\beta\cdots\beta a_k)\beta a_{k+1}$ using first $a_{\sigma(k+1)} = a_{k+1}$ and then the inductive assumption. *Case 2*: $\sigma(k + 1) \neq k + 1$. Suppose $\sigma(k + 1) = j$ for some j, $1 \leqslant j < k + 1$.

$$\begin{aligned}(a_{\sigma(1)}\beta a_{\sigma(2)}\beta\cdots\beta a_{\sigma(k)})\beta a_{\sigma(k+1)} \\ &= (a_{\sigma(1)}\beta a_{\sigma(2)}\beta\cdots\beta a_{\sigma(k)})\beta a_j \\ &= (a_1\beta a_2\beta\cdots\beta a_{j-1}\beta a_{j+1}\beta\cdots\beta a_{k+1})\beta a_j \\ &= (a_1\beta a_2\beta\cdots\beta a_{j-1}\beta a_{j+1}\beta\cdots\beta a_k)\beta(a_{k+1}\beta a_j) \\ &= (a_1\beta a_2\beta\cdots\beta a_{j-1}\beta a_{j+1}\beta\cdots\beta a_k)\beta(a_j\beta a_{k+1}) \\ &= (a_1\beta a_2\beta\cdots\beta a_{j-1}\beta a_{j+1}\beta\cdots\beta a_k\beta a_j)\beta a_{k+1} \\ &= (a_1\beta a_2\beta\cdots\beta a_{j-1}\beta a_j\beta a_{j+1}\beta\cdots\beta a_k)\beta a_{k+1}. \qquad\square\end{aligned}$$

For notational simplicity we consider a generalized distributive theorem in a ring setting.

Theorem. *Let $\langle R, +, \cdot, \theta\rangle$ be a ring. Then $a(b_1 + \cdots + b_n) = ab_1 + \cdots + ab_n$ for all $a, b_1, \ldots, b_n \in R$, $n \in \mathbb{N}^+$.*

PROOF. The proof is by induction on n. $a(b_1) = ab_1$. Suppose

$$a(b_1 + \cdots + b_k) = ab_1 + \cdots + ab_k.$$

Then

$$\begin{aligned}a(b_1 + \cdots + b_k + b_{k+1}) &= a(b_1 + \cdots + b_k) + ab_{k+1} \\ &= ab_1 + \cdots + ab_k + ab_{k+1}\end{aligned}$$

using distributivity in the ring (case $n = 2$) and the inductive assumption. $\square$

For repeated additions and multiplications we remind the reader of the compact summation and product notation. This notation, at least the summation one, is usually encountered by calculus students.

$$a_1 + a_2 + \cdots + a_m = \sum_{i=1}^{m} a_i = \sum_{i \in \{1, 2, \ldots, m\}} a_i.$$

$$a_1 a_2 \cdots a_m = \prod_{i=1}^{m} a_i = \prod_{i \in \{1, 2, \ldots, m\}} a_i.$$

By $\sum_{i=p}^{q} a_i$, $1 < p \leqslant q$ we mean $\sum_{i=1}^{q} a_i - \sum_{i=1}^{p-1} a_i$. It is convenient to have a notation for the set $\{1, 2, \ldots, n\}$ just as n stands for the set $\{0, 1, \ldots, n-1\}$. We use $\hat{n}$ for $\{1, 2, \ldots, n\}$. $\sum_{i=1}^{m} a_i = \sum_{i \in \hat{m}} a_i$.

The summation notation is used in calculus for infinite series. We illustrate its use here as an example of definition by induction. Given a sequence $a: \mathbb{N}^+ \to \mathbb{R}$, i.e., $(a_1, a_2, \ldots)$, there is defined by induction a sequence of partial sums, $s: \mathbb{N}^+ \to \mathbb{R}$, such that $s_1 = a_1$, $s_2 = a_1 + a_2, \ldots, s_{k+1} = s_k + a_{k+1}, \ldots$. $(s_1, s_2, s_3, \ldots) = (a_1, a_1 + a_2, a_1 + a_2 + a_3, \ldots)$. Any particular partial sum s_n is written $\sum_{i=1}^{n} a_i$. If the sequence of partial sums has a limit (in the real numbers) then the series is convergent and the limit of the sequence of partial sums is called the sum of the series. That limit is often denoted by $\sum_{i=1}^{\infty} a_i$ which is not really an infinite sum but rather a limit of a sequence of finite sums.

Exercises

1. Prove $\sum_{i \in \hat{n}} i = \frac{1}{2}n(n+1)$ for all $n \in \mathbb{N}$.

2. Prove $\sum_{i \in \hat{n}} 1 = n$ for all $n \in \mathbb{N}$.

3. $n!$ is read *n factorial*. A mathematical induction definition of $n!$ is $0! = 1$, $(k+1)! = (k+1)k!$. Prove $n! > nn$ for all $n \geqslant 4$, $n \in \mathbb{N}$.

4. Let m, n be natural numbers and a_{ij} belong to a ring R for all $i \in \hat{m}$, $j \in \hat{n}$. Prove $\sum_{i \in \hat{m}} \sum_{j \in \hat{n}} a_{ij} = \sum_{j \in \hat{n}} \sum_{i \in \hat{m}} a_{ij}$.

5. Let m, n be natural numbers and a_i, b_j, c_{ij} be members of a commutative ring R for all $i \in \hat{m}$, $j \in \hat{n}$. Prove $\sum_{i \in \hat{m}} a_i \sum_{j \in \hat{n}} b_j c_{ij} = \sum_{j \in \hat{n}} b_j \sum_{i \in \hat{m}} a_i c_{ij}$.

4.3 The division algorithm for the integers

We develop in this section the division algorithm for natural numbers and integers and present some elementary facts on factorization.

Theorem. The division algorithm for natural numbers. *If $a, b \in \mathbb{N}$ and $a \neq 0$ then there exist unique natural numbers q, r with $a > r \geqslant 0$ such that that $b = qa + r$.*

Comments. In $a > r \geqslant 0$ the order being used is the order induced upon $\mathbb{N}$ by the order on $\mathbb{Z}$. As an example of the division algorithm, $13 = (4)(3) + 1$.

In words, 13 divided by 3 goes 4 times with a remainder of 1. In arithmetic b, q, a, r in the equation $b = qa + r$ are called, respectively, dividend, quotient, divisor, remainder. Further examples are $1 = 0 \cdot 3 + 1$; 1 divided by 3 goes 0 times with a remainder of 0. Later we shall extend the algorithm to $\mathbb{Z}$ giving such examples as $-11 = (-4)3 + 1$.

We now offer two existence proofs, one based upon the well-ordering of N and the other a direct induction proof.

PROOF 1. $S = \{b - xa | x \in \mathbb{Z} \text{ and } b - xa \geqslant 0\}$ is a subset of $\mathbb{N}$. S includes the integer $b - 0a = b \geqslant 0$ and is therefore nonempty. Let r be the first or smallest element of the set S. Since $r \in S$, $r = b - qa$ for some element $q \in \mathbb{Z}$. We now show $0 \leqslant r < a$. Suppose $r \geqslant a$. Then $r - a = r' \geqslant 0$. $r' = r - a = b - qa - a = b - (q + 1)a$. $r' \in S$ and $r' < r$ and $0 \leqslant r'$ contradicts the minimality of r. We have therefore $r < a$ and since $r \in \mathbb{N}$ we have $0 \leqslant r$.

We must now show $q \geqslant 0$. If $q \leqslant -1$ then $qa + r \leqslant (-1)a + r < 0$ which cannot be because b is given $\geqslant 0$. Therefore $q > -1$, which is to say, $q \geqslant 0$. $q \in \mathbb{N}$. □

PROOF 2. Let $a > 0$ be a given natural number. Let $S = \{n | \text{there exist } q, r \text{ such that } n = qa + r \text{ with } r, q \in \mathbb{N} \text{ and } 0 \leqslant r < a\}$. $0 \in S$ because $0 = 0a + 0$. Assume $k \in S$. $k = q_1 a + r_1$ for some $q_1, r_1 \in \mathbb{N}$ and $0 \leqslant r_1 < a$. Then $k + 1 = q_1 a + r_1 + 1$. If $r_1 + 1 < a$ then $k + 1 = q_1 a + (r_1 + 1)$ and the conclusion holds. If $r_1 + 1 = a$ then $k + 1 = (q_1 + 1)a + 0$ and again the conclusion holds. $k + 1 \in S$. □

We now give a proof of the uniqueness of q and r assuming their existence. Suppose we have $b = q''a + r''$ as well as $qa + r$. $0 = (q - q'')a + (r - r'')$. $r'' - r = (q - q'')a$. Use of the two inequalities $0 \leqslant r'' < a$ and $0 \leqslant r < a$ shows $-a < r - r'' < a$. This is impossible unless the integer $q - q''$ is zero. But then $r'' - r = 0$ also. The integers q and r are therefore unique. □

We extend the division algorithm for natural numbers to the integers, first defining *absolute value*.

Definition. If $a \in \mathbb{Z}$ and $a \geqslant 0$ then $|a| = a$. If $a \in \mathbb{Z}$ and $a < 0$ then $|a| = -a$.

EXAMPLE. $|-3| = -(-3) = 3$. $|3| = 3$.

Theorem. The division algorithm for integers. *If $a, b \in \mathbb{Z}$ and $a \neq 0$ then there exist unique integers $Q, R \in \mathbb{Z}$ such that $b = Qa + R$ and $0 \leqslant R < |a|$.*

PROOF

Case 1: $a > 0$ and $b \geqslant 0$. We may use the previous theorem for natural numbers yielding $b = qa + r$ with $0 \leqslant r < a$. We let $Q = q$ and $R = r$. Since $|a| = a$ the conclusion follows.

Case 2: $a > 0$ and $b < 0$. We apply the natural number result to a and $-b$, both natural numbers. This yields $q, r \in \mathbb{N}$ such that $-b = qa + r$ with $0 \leqslant r < a$. Again $|a| = a$. If $r = 0$ then we have $b = (-q)a + 0$ so that $Q = -q$ and $R = 0$. If $r \neq 0$ then we have $b = (-q)a + (-r)$, but $-r$ is not between 0 and $|a|$. We rewrite the equation as $b = (-q - 1)a + (a - r)$. We now let $Q = -q - 1$ and $R = a - r$. $0 < r < a$ yields $0 > -r > -a$ which in turn gives $a > a - r > 0$. Thus $|a| > R > 0$.

Case 3: $a < 0$ and $b < 0$. Using the natural number result on $-a$ and $-b$ we have $q, r \in \mathbb{N}$ so that $-b = q(-a) + r$ with $0 \leqslant r < -a$. Therefore $b = qa + (-r)$. If $r = 0$ set $Q = q$ and $R = 0$. If $r \neq 0$ then rewrite $b = (q + 1)a + (-a - r)$. $0 < r < -a$ implies $0 > -r > a$. $-a > -a - r > 0$. Set $Q = q + 1$, $R = -a - r$. Since $a < 0$, $|a| = -a$. $|a| > R > 0$.

Case 4: $a < 0$ and $b \geqslant 0$. There are $q, r \in \mathbb{N}$ such that $b = q(-a) + r$. $0 \leqslant r < -a$. $b = (-q)a + r$. Set $Q = -q$ and $R = r$. $0 \leqslant R < |a|$. □

In the case when the remainder is zero upon dividing b by a, a is said to divide b, to be a divisor of b. Thus the word divisor has several different meanings. We make a fresh definition to include the possibility of 0.

Definition. Given $a, b \in \mathbb{Z}$, a is a *factor* (*divisor*) of b if and only if $b = ca$ for some $c \in \mathbb{Z}$. b is a multiple of a if and only if $b = ca$ for some $c \in \mathbb{Z}$.

Examples. 3 is a factor of 21 and 21 is a multiple of 3. 0 is a factor of 0 and 0 is a factor of no other integer than 0. That the only factors of 1 and -1 are 1 and -1 follows from the next theorem.

Theorem. 1 and -1 are the only integers with multiplicative inverses.

Proof. Both 1 and -1 do have inverses because $(1)(1) = 1$ and $(-1)(-1) = 1$. We now prove that if $n > 1$ then n has no inverse in the integers. Exercise 6 of Section 3.4 with some additional argument gives the result. We offer here, however, a different proof using the order on $\mathbb{Z}$. Suppose n has an inverse u. Then $nu = un = 1$. Since the product is positive then $u > 0$ if $n > 1 > 0$. There are no positive integers between 0 and 1. $u \geqslant 1$. But $u \geqslant 1$ yields $nu \geqslant n > 1$ so that u cannot be the inverse of n. We conclude n cannot have an inverse if $n > 1$. On the other hand, if $-m$ is an integer < -1 and v is its inverse it is clear that $-v$ would be an inverse for $m > 1$. Therefore no integer strictly less than -1 can have an inverse either. 1 and -1 are the only integers with multiplicative inverses. □

Factorization will be studied in detail later for principal ideal domains, a generalization of the integers, rather than for the integers alone. We will now, however, give a few results for the integers which may assist with an understanding of the more general results later.

Definition. The multiplicatively invertible integers, 1 and -1, are called *units* of the ring $\mathbb{Z}$. Any integer p, not zero and not a unit, is called a *prime* of $\mathbb{Z}$ if and only if $p = qr$ implies q or r is a unit. Any integer, not zero, not a unit and not a prime is called a *composite*.

EXAMPLES. Some of the primes are 2, 3, 5, 7, 11, 13, 17, 19, 23. Some of the composites are 4, 6, 8, 9, 10, 12, 14, 15, 16. -2, -3, $-5, \ldots$ are also primes. -3 is a prime because the only factorizations of -3 are $(-1)(3)$, $(3)(-1)$, $(1)(-3)$, $(-3)(1)$, and in each case one factor is a unit.

Theorem. *Every integer not zero and not a unit must have at least one prime factor.*

PROOF. We use the second principle of induction. Assume all natural numbers strictly less than k and not zero have at least one prime factor. Consider the natural number k itself. If k is prime then k has at least one prime factor, namely, k. If k is a composite (not 0 or 1) then $k = rs$ for some r, s neither zero or a unit. $r < k$ or $s < k$ for otherwise $r \geqslant k$ and $s \geqslant k$ which would imply $rs \geqslant kk > k$. If $r < k$ then r has a prime factor by the inductive assumption and this prime factor of r is a prime factor of k. If $s < k$ then s has a prime factor by the inductive assumption and this prime factor of s is a prime factor of k. Finally if n is a negative integer then $-n$ is a natural number and a prime factor of $-n$ is a prime factor of n. □

Finally for this section we have this marvelous theorem dating from classical times.

Theorem. *There are an infinite number of primes in* $\mathbb{N}$.

PROOF. Let $\{p_1, p_2, \ldots, p_k\}$ be the set of the first k primes in $\mathbb{N}$. For example, if $k = 5$ then we are talking about the set $\{2, 3, 5, 7, 11\}$. Set $q = p_1 p_2 \cdots p_k + 1$. None of the primes $p_1, p_2, \ldots, p_k$ is a factor of q because the remainder upon dividing is always 1. But q must have a prime factor. There must exist then a prime other than $p_1, p_2, \ldots, p_k$. It is not possible that there exists only a finite number of primes. □

QUESTIONS

1. $\{14 - x3 | x \in \mathbb{Z} \text{ and } 14 - x3 \geqslant 0\}$
   (A) has minimum element 4
   (B) is a subset of $\mathbb{N}$
   (C) has maximum element 4
   (D) has minimum element 2.
   (E) None of the alternatives completes a true sentence.

2. Which of the following statements are true?
   (A) 0 is a divisor of 3.
   (B) 3 is a divisor of 0.

(C) 0 is a divisor of 0.
(D) 1 is a divisor of 1.
(E) None of the statements is true.

3. Which of the following statements are true?
(A) 1 is a prime integer.
(B) -3 is a prime integer.
(C) Every integer has at least one prime factor.
(D) 0 is a prime integer.
(E) None of the statements is true.

4. Which of the following statements are true?
(A) $|-x| = x$ if $x > 0$.
(B) $|-x| = -x$ implies $x \leqslant 0$.
(C) $|x| + x = 0$ implies $x \leqslant 0$.
(D) $|x - y| = |y - x|$ implies $x = y$.
(E) None of the statements is true.

EXERCISES

1. A common factor of a and b is also a factor of the remainder upon dividing b by a, $a \neq 0$. Prove.

2. Prove that 0 is a factor of no integer except 0.

3. Every integer is a factor of zero. Prove.

4. Prove: If $a, b \in \mathbb{Z}$, $a \neq b$, $a > 0$, $b > 0$ and a is a factor of b then $a < b$.

5. Prove that the only factors of 1 and -1 are 1 and -1.

6. Prove that if a is a factor of b and b is a factor of a then $a = b$ or $a = -b$. Be certain that your argument is complete.

4.4 Multiples and exponents in a ring

In this section we describe the use of integers as multiples and exponents to represent repeated sums and products in a ring.

If S is a set on which an associative binary operation is defined and represented by $+$ we wish to analyze the use of natural numbers to represent repetitive addition, such as $x = 1x$, $x + x = 2x$, and $x + x + x = 3x$. $3x$ is not to be thought of as a product of 3 and x in the usual sense of multiplication in a ring (3 may not be a member of the set S). $3x$ may be thought of as an abbreviation for $x + x + x$. We now define *multiples* inductively.

Definition. Let S be a set on which an associative binary operation is defined and represented by $+$. Let also $+$ have a neutral element θ in S. We then define for any $x \in S$

$$0x = \theta$$
$$(k + 1)x = kx + x.$$

In a perfectly analogous manner we may define *exponents* for an associative binary operation on a set S where the operation is represented multiplicatively. For example, $x = x^1, x \cdot x = x^2, x \cdot x \cdot x = x^3$.

Definition. Let S be a set on which an associative binary operation is defined and represented by $\cdot$. Let also $\cdot$ have a neutral element v. We then define for any x in S

$$x^0 = v$$
$$x^{k+1} = x^k \cdot x.$$

The assumption that the binary operation has a neutral element is not a necessary one for the use of multiples and exponents. The induction definitions can be begun with $1 \cdot x = x$ and $x^1 = x$. The exponential notation is also used for operations other than multiplicatively represented ones. $x\beta x\beta x = x^3$. $3x$ seems always to mean $x + x + x$.

Theorem. *Let $+$ be an associative binary operation on a set S with neutral element θ. Then if $m, n \in \mathbb{N}$ and $x \in S$*

$$(m + n)x = mx + nx.$$

Let $\cdot$ be an associative binary operation on a set S with neutral element v. Then if $m, n \in \mathbb{N}$ and $x \in S$

$$x^{m+n} = x^m x^n.$$

Proof. We first give the proof in additive notation. Let $x \in S$. Let $T = \{n | (m + n)x = mx + nx \text{ for all } m \in \mathbb{N}\}$. $0 \in T$ because $(m + 0)x = mx = mx + \theta = mx + 0x$. Suppose $k \in T$; $(m + k)x = mx + kx$. $(m + (k + 1))x = ((m + k) + 1)x = (m + k)x + x = (mx + kx) + x = mx + (kx + x) = mx + (k + 1)x$. $k + 1 \in T$. $T = \mathbb{N}$. □

In order to assist understanding we give the same proof in multiplicative notation. Let $x \in S$. Let $U = \{n | x^{m+n} = x^m x^n \text{ for all } m \in \mathbb{N}\}$. $0 \in U$ because $x^{m+0} = x^m = x^m v = x^m x^0$. Suppose $k \in U$; $x^{m+k} = x^m x^k$. $x^{m+(k+1)} = x^{(m+k)+1} = x^{m+k}x = x^m x^k x = x^m x^{k+1}$. $k + 1 \in U$. $U = \mathbb{N}$. □

It is clear both theorems are the same except for notation. We could express the theorem in such a manner that both statements are included in a general one, but we will not do this.

Continuing with the laws of multiples and exponents we have the next result.

Theorem. *Let $+$ be an associative binary operation on a set S with neutral element θ. Then if $m, n \in \mathbb{N}$ and $x \in S$*

$$n(mx) = (nm)x.$$

Let $\cdot$ be an associative binary operation on a set S with a neutral element. Then if $m, n \in \mathbb{N}$ and $x \in S$

$$(x^m)^n = x^{mn}.$$

PROOF. We prove only the first result. Let $T = \{n | n(mx) = (nm)x$ for all $m \in \mathbb{N}\}$. $0 \in T$ because $0(mx) = \theta = 0x = (0m)x$ for all $m \in \mathbb{N}$. Suppose $k(mx) = (km)x$ for all $m \in \mathbb{N}$; $k \in T$. $(k + 1)(mx) = k(mx) + mx = (km)x + (m)x = (km + m)x = ((k + 1)m)x$. $k + 1 \in T$. $T = \mathbb{N}$. □

Theorem. *If $+$ is an associative binary operation on a set S with neutral element θ and x and y are members of S which commute ($x + y = y + x$) then $n(x + y) = nx + ny$ for all $n \in \mathbb{N}$. If $\cdot$ is an associative binary operation on a set S with neutral element v and x and y are members of S which commute ($xy = yx$) then $(xy)^n = x^n y^n$ for all $n \in \mathbb{N}$.*

PROOF. We first prove x and ny commute. Let $T = \{n | x + ny = ny + x$ and $n \in \mathbb{N}\}$. $0 \in T$ because $x + 0y = x + \theta = \theta + x = 0y + x$. If $k \in T$ then $x + (k + 1)y = x + ky + y = ky + x + y = ky + y + x = (k + 1)y + x$ yielding $k + 1 \in T$. $T = \mathbb{N}$. x and ny commute for all $n \in \mathbb{N}$. Now let $U = \{n | n(x + y) = nx + ny\}$. $0(x + y) = \theta = \theta + \theta = 0x + 0y$ implies $0 \in U$. Assume $k \in U$. Then $(k + 1)(x + y) = k(x + y) + (x + y) = kx + ky + x + y = kx + x + ky + y = (k + 1)x + (k + 1)y$. $k + 1 \in U$. $U = \mathbb{N}$. □

Whenever a given element of a set with an associative binary operation with a neutral element has an inverse we can then define negative multiples or negative exponents.

Definition. Let S be a set with an associative binary operation $+$ and a neutral element θ. If the negative of x, the additive inverse, exists then we define $(-n)x = n(-x)$ for all $n \in \mathbb{N}$. Let S be a set with an associative binary operation $\cdot$ and a neutral element v. If the multiplicative inverse of any element x exists then we define $x^{-n} = (x^-)^n$ for all $n \in \mathbb{N}$.

We note in particular that $(-1)x = 1(-x) = -x$ and that $(x)^{-1} = (x^-)^1 = x^-$. This justifies the use of x^{-1} for the inverse of x.

We now have the project of proving the theorems which extend the laws of exponents and multiples from natural numbers to integers.

Theorem. *Let S be a set with an associative binary operation $+$ with neutral element θ. Let x be any element of S which has a negative in S. Then $(m + n)x = mx + nx$ for all $m, n \in \mathbb{Z}$. Let S be a set with an associative binary operation $\cdot$ with neutral element v. Let x be any element of S with multiplicative inverse. Then $x^{m+n} = x^m x^n$ for all $m, n \in \mathbb{Z}$.*

PROOF. We offer a proof only of the exponential result. The proof is by cases. *Case 1*: $m, n \in \mathbb{N}$. This is our earlier theorem for natural number exponents.

Case 2: $m < 0$ and $n < 0$. Let y be the inverse of x. $x^{m+n} = y^{-(m+n)} = y^{-m-n} = y^{-m}y^{-n} = x^m x^n$. *Case 3a*: $m \geqslant 0$ and $n < 0$ and $m + n \geqslant 0$. Let y be the inverse of x. $x^{m+n} = x^{n+n}v^{-n} = x^{m+n}(xy)^{-n} = x^{m+n}x^{-n}y^{-n} = x^m y^{-n} = x^m x^n$. *Case 3b*: $m \geqslant 0$ and $n < 0$ and $m + n < 0$. Again v is the neutral element and y is the inverse of x. $x^{m+n} = y^{-(m+n)} = y^{-m-n} = v^m y^{-m-n} = (xy)^m y^{-m-n} = x^m y^m y^{-m-n} = x^m y^{-n} = x^m x^n$. *Case a*: $m < 0$ and $n \geqslant 0$ and $m + n \geqslant 0$. *Case 4b*: $m < 0$ and $n \geqslant 0$ and $m + n < 0$. Cases 4a and 4b are similar to Cases 3a and 3b. In all cases the gist of the argument is to arrange to have nonnegative exponents so that the theorems already proved for $\mathbb{N}$ are applicable. □

Theorem. *Let S be a set on which $+$ is an associative binary operation with a neutral element θ. Let x be an element of S which has a negative. Then $n(mx) = (nm)x$ for all $m, n \in \mathbb{Z}$. Let S be a set on which $\cdot$ is an associative binary operation with a neutral element v. Let x be an element of S which has a multiplicative inverse. Then $(x^m)^n = x^{mn}$ for all $m, n \in \mathbb{Z}$.*

Proof. We prove only the exponential version. Let y be the inverse of x. *Case 1*: $m \geqslant 0$ and $n \geqslant 0$. This the theorem for natural numbers. *Case 2*: $m \geqslant 0$ and $n < 0$. y^m is the inverse of x^m because $x^m y^m = (xy)^m = v^m = v$ and $y^m x^m = (yx)^m = v^m = v$. We have $(x^m)^n = (y^m)^{-n} = y^{m(-n)} = y^{-mn} = x^{mn}$. *Case 3*: $m < 0$ and $n \geqslant 0$. $(x^m)^n = (y^{-m})^n = y^{(-m)n} = y^{-mn} = x^{mn}$. *Case 4*: $m < 0$ and $n < 0$. x^{-m} and y^{-m} are inverses because $x^{-m}y^{-n} = (xy)^{-m} = v^{-m} = v$. Also $y^{-m}x^{-m} = v$. $(x^m)^n = (y^{-m})^n = (x^{-m})^{-n} = x^{(-m)(-n)} = x^{mn}$. □

Theorem. *Let S be a set upon which $+$ is an associative binary operation with neutral element θ. Let x and y be elements of S which have negatives and which commute. Then $m(x + y) = mx + my$ for all $m \in \mathbb{Z}$. Let S be a set upon which $\cdot$ is an associative binary operation with neutral element v. Let x and y be elements of S which have multiplicative inverses and which commute. Then $(xy)^m = x^m y^m$ for all $m \in \mathbb{Z}$.*

Proof. We prove the exponential version. Let us represent the inverse of x by u and the inverse of y by v. We first prove that $xy = yx$ implies $uv = vu$. Suppose $xy = yx$. Then $vyxu = v$. $vxyu = v$. The inverse of $vxyu$, namely $xvuy$, is also equal to v. Multiplying $xvuy = v$ on the right by v and on the left by u we get $vu = uv$.

Knowing already that if two elements commute then all positive powers of the two elements commute we combine this with the result just proved and we then have proved that all integral powers commute. To demonstrate the conclusion of the theorem for $n < 0$, consider $(xy)^n$. $(xy)^n = (vu)^{-n} = v^{-n}u^{-n} = y^n x^n = x^n y^n$. We are, as we said, assuming the theorem proved for natural number exponents. □

Finally we have a theorem about multiples of elements in a ring and multiplication in the ring.

Theorem. *Let $\langle R, +, \cdot, \theta\rangle$ be a ring. Let $x, y \in R$. Then for any $n \in \mathbb{Z}$ we have $n(xy) = (nx)y = x(ny)$.*

PROOF. The proof is by induction for $n \geqslant 0$. $0(xy) = \theta = \theta y = (0x)y$. Assume the result true for k. $(k + 1)(xy) = k(xy) + xy = (kx)y + xy = (kx + x)y = ((k + 1)x)y$. The conclusion holds for all $n \in \mathbb{N}$. For $m \in \mathbb{Z}$ and $m < 0$ let $n = -m$. $n(xy) = (-m)(xy) = m(-(xy)) = m((-x)y) = (m(-x))y = ((-m)x)y = (nx)y$. The other equation is proved similarly. □

We have seen how the integers can be used to indicate multiples of ring elements. We have indicated this with juxtaposing the integer n and the ring element x:nx. In the case where the ring under consideration is $\mathbb{Z}$ itself we have a problem of ambiguity because we have sometimes indicated the product in this manner also. Suppose for the rest of this discussion we indicate the product in $\mathbb{Z}$ exclusively with the raised dot (never omitting it): $n \cdot m$ means the product in $\mathbb{Z}$ and nm is the nth multiple of the ring element m in $\mathbb{Z}$. We propose to show that the two are identical.

If we consider first n and m in $\mathbb{N}$ and compare the two definitions we find agreement:

$$0m = 0 \qquad\qquad 0 \cdot m = 0$$

$$(k + 1)m = km + m \qquad (k + 1) \cdot m = k \cdot m + m.$$

Because of the uniqueness provided by definition by mathematical induction we have $nm = n \cdot m$ for all $n, m \in \mathbb{N}$.

We secondly pass to $n \in \mathbb{N}$ and $m \in \mathbb{Z}$. In particular, we wish to compare nm and $n \cdot m$ where $n \in \mathbb{N}$ and $m < 0$. For multiples we had $0(-p) = 0$ and $(k + 1)(-p) = k(-p) + (-p)$ with $p > 0$. However, for multiplication in $\mathbb{Z}$ the procedure was different. The natural number system was extended by means of an equivalence class construction and multiplication was defined on the quotient set preserving the original multiplication on $\mathbb{N}$. We did prove, however, that in any ring, $(-x) \cdot y = x \cdot (-y) = -(x \cdot y)$. We have, therefore, for multiplication in $\mathbb{Z}$

$$\begin{aligned} 0 \cdot (-p) &= (-0) \cdot p = 0 \cdot p = 0 \\ (k + 1) \cdot (-p) &= -((k + 1) \cdot p) = -(k \cdot p + p) \\ &= -(k \cdot p) + (-p) = k \cdot (-p) + (-p). \end{aligned}$$

This agrees with the definition of multiples for all $n \in \mathbb{N}$, $m \in \mathbb{Z}$.

It remains to check agreement of nm and $n \cdot m$ for $n \in \mathbb{Z}$, $m \in \mathbb{Z}$, in particular for the case $n < 0$. Let $n = -q$. $nm = (-q)m = q(-m)$ according to the definition of negative multiples. On the other hand, for any ring, $n \cdot m = (-q) \cdot m = q \cdot (-m)$. But $q(-m) = q \cdot (-m)$ for $q \in \mathbb{N}$, $-m \in \mathbb{Z}$ by our previous result. We have now proved $nm = n \cdot m$ for all n, m in $\mathbb{Z}$. In view of this it is not necessary for us to distinguish between multiples and multiplication when we work entirely in $\mathbb{Z}$. This result also embodies the intuitive concept of multiplication as repeated addition in $\mathbb{Z}$.

Questions

1. Let $\langle R, +, \cdot, \theta, v\rangle$ be a unitary ring. Which of the following four are not the zero of R?
   (A) $(1)(\theta)$.
   (B) $(0)(v)$.
   (C) $(\theta)(v)$.
   (D) $(0)(1)$.
   (E) All four are θ.

2. Let $\langle R, +, \cdot, \theta\rangle$ be a ring and let $x \in R$ such that $3x = \theta$. We can conclude
   (A) $x = \theta$
   (B) $51x = \theta$
   (C) $3 \in R$
   (D) $2x = \theta$.
   (E) None of the four conclusions follow.

3. Let $\langle R, +, \cdot, \theta\rangle$ be a ring and let $x \in R$. We can conclude x^0 belongs to R provided
   (A) $x \neq \theta$
   (B) $x = \theta$
   (C) there exists a $y \in R$ such that $yz = zy = z$ for all $z \in R$
   (D) $x^2 = x$.
   (E) None of the four conditions is sufficient.

4. Given a ring $\langle R, +, \cdot, \theta, v\rangle$ with unity v and an invertible element $x \in R$, the function $f_x : \mathbb{Z} \to R$ such that $f_x(z) = x^z$
   (A) is a morphism of rings
   (B) preserves addition in multiplication
   (C) is a monomorphism
   (D) obeys the relation $f_x(z)f_x(-z) = v$.
   (E) None of the alternatives completes a true sentence.

5. Let $\langle R, +, \cdot, \theta\rangle$ be a ring and x an element of R. Then
   (A) $nx = \theta$ implies $(mn)x = \theta$
   (B) $0x = \theta$
   (C) $mx = \theta$ and $nx = \theta$ imply $(m - n)x = \theta$
   (D) $\{n | nx = \theta\}$ is an ideal of $\mathbb{Z}$
   (E) $\mathbb{Z}/\{n | nx = \theta\}$ is isomorphic with $[x]$.

Exercises

1. For a commutative ring $\langle R, +, \cdot, \theta\rangle$ prove $\langle a\rangle = Ra + \mathbb{Z}a$. [*Hint*: Review Exercises 8, 9, and 10 of Section 2.8.]

2. For a commutative ring $\langle R, +, \cdot, \theta\rangle$ prove $Ra = \langle a\rangle$ if and only if there exists an $r \in R$ such that $a = ra$.

3. Let $\langle R, +, \cdot, \theta\rangle$ be a ring. On the set $P = \mathbb{Z} \times R$ define operations:

$$(m, x) + (n, y) = (m + n, x + y)$$
$$(m, x) \cdot (n, y) = (mn, my + nx + xy).$$

Show that $\langle P, +, \cdot, (0, \theta)\rangle$ is a ring. Moreover, show that P has a unity and that R is isomorphic with an ideal of P.

4. For a ring $\langle R, +, \cdot, \theta\rangle$ prove that $(mx)(ny) = (mn)xy$ for any $x, y \in R$ and $m, n \in \mathbb{Z}$.

5. Let $\langle R, +, \cdot, \theta, v\rangle$ be a unitary ring. Prove $(nx)^m = n^m x^m$ for any $x \in R$ and all $n \in \mathbb{Z}, m \in \mathbb{N}$.

6. Let $\langle R, +, \cdot, \theta\rangle$ and $\langle R', +', \cdot', \theta'\rangle$ be rings and $f: R \to R'$ be a morphism. Prove $f(nx) = nf(x)$ for all $x \in R$ and $n \in \mathbb{Z}$.

7. Let $\langle R, +, \cdot, \theta, v\rangle$ and $\langle R', +', \cdot', \theta', v'\rangle$ be unitary rings and $f: R \to R'$ a unitary ring morphism ($f(v) = v'$ and f is a morphism). Prove $f(x^n) = (f(x))^n$ for all invertible x in R and all n in $\mathbb{Z}$. Prove that the hypothesis *f is a unitary ring morphism* can be replaced by *f is a ring epimorphism.*

8. By $\binom{n}{j}$ we mean $n!/(n-j)!j!$ for natural numbers n, j with $j \leqslant n$. Prove $\binom{n}{0} = \binom{n}{n} = 1$. Show that $\binom{n}{j}$ is always a natural number; the denominator is always a factor of the numerator. [*Hint*: Use induction on n; prove and use the formula $\binom{k}{j} + \binom{k}{j-1} = \binom{k+1}{j}$.]

9. Let $\langle R, +, \cdot, \theta, v\rangle$ be a commutative unitary ring. Prove $(x + y)^n = \sum_{j=0}^{n} \binom{n}{j} x^{n-j} y^j$ for all $n \in \mathbb{N}$. This is, of course, the binomial theorem from which one obtains the name binomial coefficients for the natural numbers $\binom{n}{j}$. Use induction and Exercise 8 to prove the result.

10. Study the three examples given below and then find a formula for n^k as a sum of lesser powers.

$$n = \sum_{j=1}^{n} (j - (j-1)) = \sum_{j=1}^{n} 1.$$

$$n^2 = \sum_{j=1}^{n} (j^2 - (j-1)^2) = \sum_{j=1}^{n} (2j - 1).$$

$$n^3 = \sum_{j=1}^{n} (j^3 - (j-1)^3) = \sum_{j=1}^{n} (3j^2 - 3j + 1).$$

11. Find $a, b \in \mathbb{N}$ such that

$$a\left(\sum_{j=1}^{n} (2j-1)\right) + b\left(\sum_{j=1}^{n} 1\right) = \sum_{j=1}^{n} j.$$

Use this principle to derive the formula in Exercise 1, Section 4.2.

12. Derive a formula for $\sum_{j=1}^{n} j^2$.

13. Derive a formula for $\sum_{j=1}^{n} j^k$ in terms of sums of lesser powers.

4.5 The field of fractions

In this section we embed a commutative unitary ring R in a larger ring $\bar{R}$ so that every cancellable element in R becomes an invertible element in $\bar{R}$; we embed an integral domain in a field and thereby construct the rational numbers from the integers.

If a ring does not contain multiplicative inverses we raise the question of whether or not it might be possible to adjoin elements to the commutative ring in some way so as to provide inverses. This is analogous to adjoining negatives to the natural numbers to construct the integers. In posing this problem we should remember we have proved that any element which is invertible in a ring is also multiplicatively cancellable. There is, therefore, no hope for adjoining inverses for those elements which do not cancel and are divisors of zero. If we deal with the special case of an integral domain so that no nonzero element can be a divisor of zero we might hope to adjoin inverses for every nonzero element in such a manner to produce a field. In fact, the hope is not in vain; the field exists and is called the field of fractions of the integral domain. The following theorem deals with the problem of adjoining inverses for cancellable elements in a ring not specialized to an integral domain. The enlarged ring will be called the ring of fractions.

Theorem. *Let $\langle R, +, \cdot, \theta, v\rangle$ be a commutative unitary ring. Let R^* denote the set of multiplicatively cancellable elements of R. Then there exists a ring $\langle \bar{R}, +, \cdot, \theta, v\rangle$ and a monomorphism $\varphi: R \to \bar{R}$ such that $\varphi(v) = v'$ and the image of every member of R^* is invertible in $\bar{R}$. Moreover, if $y \in \bar{R}$ then $y = \varphi(a)\varphi(b)^-$ for some $a \in R$, $b \in R^*$.*

Proof. We begin the proof by defining on $R \times R^*$ a relation $\sim$ such that $(x, y) \sim (u, v)$ if and only if $xv = yu$. We verify that $\sim$ is an equivalence relation. $(x, y) \sim (x, y)$ because $xy = yx$. $(x, y) \sim (u, v)$ implies $xv = yu$. Then $uy = vx$ which yields $(u, v) \sim (x, y)$ proving symmetry. For transitivity assume $(x, y) \sim (u, v)$ and $(u, v) \sim (r, s)$. $xv = yu$ and $us = vr$. $xvs = yus$; $yus = yvr$. $xvs = yvr$. $xsv = yrv$. $xs = yr$ because $v \in R^*$ and is a cancellable element of R. $(x, y) \sim (r, s)$.

On the quotient set $R \times R^*/\sim$ define two operations:

$$(x, y)/\sim + (u, v)/\sim = (xv + yu, yv)/\sim$$

and

$$(x, y)/\sim \cdot (u, v)/\sim = (xu, yv)/\sim .$$

Both pairs $(xv + yu, yv)$, (xu, yv) belong to $R \times R^*$ because the product yv of cancellable elements of R is itself cancellable. We comment that these operation definitions are simply the conventional rules for adding and multiplying fractions in somewhat disguised form. We now demonstrate the definition of addition to be independent of the representative chosen from the coset by proving $(u, v) \sim (u', v')$ and $(x, y) \sim (x', y')$ imply $(x, y)/\sim + (u, v)/\sim = (x, y)/\sim + (u', v')/\sim$ and $(x, y)/\sim + (u', v')/\sim = (x', y')/\sim + (u', v')/\sim$. As the two statements are quite similar we prove only the first. $(u, v) \sim (u', v')$ implies $uv' = vu'$. $xyvv' + yyu'v = xyvv' + yyuv'$. $(xv' + yu')yv = yv'(xv + yu)$. $(xv' + yu', yv') \sim (xv + yu, yv)$, which was to be proved.

We demonstrate that the definition of multiplication is also independent of the representative pair chosen by proving $(x, y)/\sim (u, v)/\sim = (x, y)/\sim (u', v')/\sim$ whenever $(u, v) \sim (u', v')$ and that $(x, y)/\sim (u', v')/\sim = (x', y')/\sim (u', v')/\sim$ whenever $(x, y) \sim (x', y')$. Again we prove only the first of the two assertions. Let $(u, v) \sim (u', v')$. $uv' = vu'$. $xuyv' = xu'yv$. $(xu, yv) \sim (xu', yv')$.

$\bar{R} = R \times R^*$ is a commutative ring with zero $(\theta, v)/\sim$. The commutative, associative and distributive properties can be routinely verified. To embed R into $\bar{R}$ define $\varphi: R \to \bar{R}$ such that $\varphi(x) = (x, v)/\sim$. $\bar{R}$ is a unitary ring with unity $\varphi(v) = (v, v)/\sim$. φ is easily seen to be a morphism. Let $\varphi(x) = \varphi(y)$. $(x, v)/\sim = (y, v)/\sim$. $xv = vy$. $x = y$. φ is a monomorphism.

Any element $x \in R^*$ has image $(x, v)/\sim$ under the monomorphism φ. $(v, x) \in R \times R^*$ because $x \in R^*$. $(v, x)/\sim$ is the inverse of $(x, v)/\sim$ in $\bar{R}$: $(v, x)/\sim (x, v)/\sim = (x, x)/\sim = (v, v)/\sim$, the unity of $\bar{R}$.

Finally let $y \in \bar{R}$. $y = (a, b)/\sim$ for some $a \in R$, $b \in R^*$. $(a, b)/\sim = (a, v)/\sim (v, b)/\sim = (a, v)/\sim [(b, v)/\sim]^- = \varphi(a)\varphi(b)^-$. □

It is interesting to note what the proof produces when an element x already has an inverse in R. Let $x \in R^*$ and $y \in R^*$ such that $xy = yx = v$. The images under φ remain inverses in $\bar{R}$. $(x, v)/\sim (y, v)/\sim = (xy, v)/\sim = (v, v)/\sim$. Furthermore the newly created inverse in $\bar{R}$ becomes identified with the image of the original inverse in R. $(v, x)/\sim = (y, v)/\sim$ and $(v, y)/\sim = (x, v)/\sim$.

Corollary. *Every integral domain is contained in a field.*

PROOF. As we have stated before if an integral domain is given then $R^* = R - \{\theta\}$ and the image of every nonzero element of R becomes invertible in $\bar{R}$. That $\bar{R}$ is a field is seen as follows. Let $(a, b)/\sim \in \bar{R}$ and $(a, b)/\sim \neq (\theta, v)/\sim$. $av \neq b\theta$. $a \neq \theta$. Because R is an integral domain, $a \in R^*$. $(b, a) \in R \times R^*$. $(b, a)/\sim \in \bar{R}$ and is the inverse of $(a, b)/\sim$. □

We will now discuss the construction of the rational numbers, $\mathbb{Q}$. We take the integral domain, $\mathbb{Z}$, of integers and apply the theorem to obtain $\bar{\mathbb{Z}}$, a field, and a monomorphism $\varphi: \mathbb{Z} \to \bar{\mathbb{Z}}$. Every member of $\bar{\mathbb{Z}}$ can be written as $\varphi(a)\varphi(b)^-$ for some $a, b \in \mathbb{Z}$, $b \neq \theta$, or as $(a, b)/\sim$. We write $\mathbb{Q}$ for $\bar{\mathbb{Z}}$. The conventional way to write the coset $(a, b)/\sim$ in $\mathbb{Q}$ is as the fraction a/b. We now summarize some of our results in the conventional fractional notation.

$x/y = u/v$ if and only if $xv = yu$, $\quad y, v \neq 0$.
$x/y + u/v = (xv + yu)/yv$, $\quad y, v \neq 0$.

$x/y \cdot u/v = (xu)/(yv), \quad y, v \neq 0.$
$\varphi(x) = x/1.$
0/1 is the zero of $\mathbb{Q}$.
1/1 is the unity of $\mathbb{Q}$.

We emphasize we have merely copied material from the theorem, changing notation to the conventional fractions. The fractions, $\frac{3}{4}$ and $\frac{6}{8}$, which stand for $(3, 4)/\sim$ and $(6, 8)/\sim$ actually are symbols which stand for the same equivalence class or coset in $\bar{\mathbb{Z}}$ ($=\mathbb{Q}$). By identifying the integer $x \in \mathbb{Z}$ with the fraction $x/1 = \varphi(x)$ in $\mathbb{Q}$ we embed the integers in the field $\mathbb{Q}$ of rational numbers.

Questions

1. Which of the following has $\mathbb{Q}$ as a field of fractions?
   (A) $\mathbb{R}$
   (B) $\mathbb{Z}$
   (C) $2\mathbb{Z}$
   (D) $\mathbb{Q}$
   (E) $\mathbb{N}$.

2. $\{m/2^n | m \in \mathbb{Z}, n \in \mathbb{N}\}$
   (A) is a subring of $\mathbb{Q}$
   (B) is an integral domain
   (C) is a field
   (D) has $\mathbb{Q}$ as its field of fractions.
   (E) None of the four alternatives completes a true sentence.

3. $S = \{m + n\sqrt{5} | m, n \in \mathbb{Z}\}$ is a subring of $\mathbb{R}$. The field of fractions of S
   (A) contains $1/\sqrt{5}$
   (B) is $\mathbb{Q} + \mathbb{Q}\sqrt{5}$
   (C) is $\mathbb{R}$
   (D) is $\mathbb{Q}$.
   (E) None of the four alternatives completes a true sentence.

4. $\mathbb{Z} + \mathbb{Z}\sqrt{5}$
   (A) is a field
   (B) is an integral domain
   (C) is its own ring of fractions
   (D) has as its ring of fractions, $\mathbb{Q} + \mathbb{Q}\sqrt{5}$.
   (E) None of the four alternatives is true.

Exercises

1. Show that any field containing a given integral domain also contains a subfield isomorphic to the field of fractions of the integral domain. Use the procedure outlined below in (a), (b), and (c).
   (a) Let $\langle R, +, \cdot, \theta, v\rangle$ be the given integral domain and K the enclosing field. Let $\varphi: R \to \bar{R}$ be the monomorphism of the integral domain onto the constructed

field of fractions of R described in the text of this section. Show that $R' = \{m/n | m, n \in R, n \neq \theta\}$ is a subfield of K.

(b) Show that the function $\Phi: R' \to \bar{R}$ such that $\Phi(m/n) = \varphi(m)\varphi(n)^{-}$ is well-defined and agrees with φ on R.

(c) Show that $\Phi: R' \to \bar{R}$ is an isomorphism.

2. Show that R' in Exercise 1 is the intersection of all subfields of K containing R and is therefore the smallest subfield of K containing R.

3. The subset $\mathbb{Z} + \mathbb{Z}i = \{m + ni | m, n \in \mathbb{Z}\}$ of the complex numbers is called the set of *Gaussian integers*. Show that $\mathbb{Z} + \mathbb{Z}i$ is a ring with field of fractions equal to $\mathbb{Q} + \mathbb{Q}i$.

4. Show if R is a finite ring then its ring of fractions is also finite. [*Hint*: If crd $R = n$ then crd $\bar{R} \leqslant n^2 - n$.

5. What is the field of fractions of the integral domain $\mathbb{Z} + \mathbb{Z}\sqrt{2}$?

6. What is the ring of fractions of the ring $\mathbb{Z} \times \mathbb{Z}$?

7. Fractional exponents in $\mathbb{R}$. For each positive $a \in \mathbb{R}$ and each positive integer n there exists a unique positive $x \in \mathbb{R}$ such that $x^n = a$. This x (the nth root of a) is denoted by $a^{1/n}$. By $a^{m/n}$ is meant $(a^{1/n})^m$ for all $m \in \mathbb{Z}$. Show that the following exponential laws are valid: $(a^{1/n})^n = a$; $a^{m/n}a^{p/q} = a^{m/n + p/q}$; $(a^{m/n})^{p/q} = a^{(m/n)(p/q)}$; $(ab)^{p/q} = a^{p/q}b^{p/q}$.

4.6 Characteristic of a ring

In this section we assign to each ring a natural number called the characteristic of the ring and explore its properties.

If we observe multiples of the elements of the ring $\mathbb{Z}_4$, denoted here by $\bar{0}, \bar{1}, \bar{2}, \bar{3}$, where $\bar{n}$ is the coset $n + 4\mathbb{Z}$, we see that the fourth multiple of every element of the ring is zero and no smaller multiple has that property.

$$\begin{array}{llll}
4(\bar{0}) = \bar{0} & 3(\bar{0}) = \bar{0} & 2(\bar{0}) = \bar{0} & 1(\bar{0}) = \bar{0} \\
4(\bar{1}) = \bar{0} & 3(\bar{1}) = \bar{3} & 2(\bar{1}) = \bar{2} & 1(\bar{1}) = \bar{1} \\
4(\bar{2}) = \bar{0} & 3(\bar{2}) = \bar{2} & 2(\bar{2}) = \bar{0} & 1(\bar{2}) = \bar{2} \\
4(\bar{3}) = \bar{0} & 3(\bar{3}) = \bar{1} & 2(\bar{3}) = \bar{2} & 1(\bar{3}) = \bar{3}.
\end{array}$$

We say that the natural number 4 is the characteristic of the ring $\mathbb{Z}_4$. If we look at the ring $\mathbb{Z}_2 \times \mathbb{Z}_2 = \{(\bar{0}, \bar{0}), (\bar{0}, \bar{1}), (\bar{1}, \bar{0}), (\bar{1}, \bar{1})\}$ which also has four elements we find the characteristic to be 2.

$$\begin{array}{ll}
2(\bar{0}, \bar{0}) = (\bar{0}, \bar{0}) & 1(\bar{0}, \bar{0}) = (\bar{0}, \bar{0}) \\
2(\bar{0}, \bar{1}) = (\bar{0}, \bar{0}) & 1(\bar{0}, \bar{1}) = (\bar{0}, \bar{1}) \\
2(\bar{1}, \bar{0}) = (\bar{0}, \bar{0}) & 1(\bar{1}, \bar{0}) = (\bar{1}, \bar{0}) \\
2(\bar{1}, \bar{1}) = (\bar{0}, \bar{0}) & 1(\bar{1}, \bar{1}) = (\bar{1}, \bar{1}).
\end{array}$$

The second multiple is always zero whereas the first is not. For convenience of language we shall say that the natural number n annihilates the element x when the nth multiple of x is zero.

Definition. Let $\langle R, +, \cdot, \theta \rangle$ be a ring. The natural number n *annihilates* x if and only if $nx = \theta$. The natural number n *annihilates* R if and only if $nx = \theta$ for all $x \in R$. The *characteristic* of a ring is the smallest positive natural number which annihilates R in the case that some positive natural number annihilates R and is the natural number 0 if no positive natural number annihilates R.

The rings $\mathbb{Z}$, $\mathbb{Q}$, and $\mathbb{R}$ all have characteristic 0 for there is no one positive multiple of every element which is zero. We remind the reader that n in the above definition is an integer and not a member of the ring R in general. The fact that $\mathbb{Z}$, $\mathbb{Q}$, and $\mathbb{R}$ all have characteristic zero shows that rings need not be isomorphic to share characteristic. We proceed then to explore what rings with the same characteristic have in common. The next theorem simplifies the procedure in finding the characteristic of a unitary ring.

Theorem. *Let $\langle R, +, \cdot, \theta, v \rangle$ be a unitary ring. If m is the smallest positive multiple of v which is θ then* $\operatorname{chr} R = m$. *If no positive multiple of v is θ then* $\operatorname{chr} R = 0$.

PROOF. If positive, $\operatorname{chr} R$ can be no smaller than m since $(\operatorname{chr} R)v = \theta$. Let x be any element of R. $mx = m(vx) = (mv)x = \theta x = \theta$. From this equation we conclude $\operatorname{chr} R$ is at least as small as m. Therefore $\operatorname{chr} R = m$. If $\operatorname{chr} R = 0$ then no positive multiple of v is θ provided v is distinct from θ because if $\theta = v$ we are dealing with the ring $\{\theta\}$ which is easily seen to have characteristic 1. □

A fruitful approach to these concepts is to consider all multiples of the ring which are zero rather than just the minimum one.

Definition. Let $\langle R, +, \cdot, \theta \rangle$ be a ring. The *annihilating ideal of an element* x in R is the set of all integers which annihilate x. The *annihilating ideal of* R is the set of all integers which annihilate R.

It must be verified that the set described in the definition is indeed an ideal of $\mathbb{Z}$.

Theorem. *Let $\langle R, +, \cdot, \theta \rangle$ be a ring. The annihilating ideal of an element x in R is an ideal of $\mathbb{Z}$ and the annihilating ideal of R is also an ideal of $\mathbb{Z}$.*

PROOF. If m, n annihilate x then $m + n$, $-m$, and km (for any $k \in \mathbb{Z}$) all annihilate x. □

We will need the following lemma about ideals in $\mathbb{Z}$.

Lemma. *If A is an ideal of $\langle \mathbb{Z}, +, \cdot, 0, 1 \rangle$ then A is generated by a single natural number μ: $A = \langle \mu \rangle = \mathbb{Z}\mu$.*

Proof. If $A = \{0\}$ then $A = \mathbb{Z}0 = \langle 0 \rangle$. If $A \neq \{0\}$ then A contains some positive integer (Why?). Let A^+ be the positive members of A. A^+, a nonempty subset of $\mathbb{N}$, contains a minimum member we call μ. Let a be any element of A. $a = q\mu + r$ with $0 \leqslant r < \mu$. $a, \mu \in A$. $a - q\mu = r \in A$. r must be 0 for otherwise r is a positive element of A smaller than μ. $a = q\mu$. $A \subseteq \mathbb{Z}\mu$. But $\mathbb{Z}\mu \subseteq A$. $A = \mathbb{Z}\mu$. □

Theorem. *Let $\langle R, +, \cdot, \theta \rangle$ be a ring. Then* chr $R = \mu$ *if and only if $\langle \mu \rangle$ is the annihilating ideal of R.*

Proof. If chr $R = \mu$ then $\langle \mu \rangle$ annihilates R. If $n \in \mathbb{Z}$ and annihilates R then by an argument similar to one in the lemma $n = q\mu$ for some $q \in \mathbb{Z}$. $\langle \mu \rangle$ is the annihilating ideal of R. Conversely, if A is the annihilating ideal of R then $A = \langle \mu \rangle$ for some μ in $\mathbb{N}$. If nonzero, μ will be the smallest positive integer annihilating R. If zero no positive integer will annihilate R. □

We continue with this structural theorem.

Theorem. *If $\langle R, +, \cdot, \theta, v \rangle$ is a unitary ring with characteristic μ then there exists a subring of R isomorphic to $\mathbb{Z}_\mu$ (including case $\mathbb{Z}_0 = \mathbb{Z}$).*

Proof. The morphism $f: \mathbb{Z} \to R$ such that $f(n) = nv$ has for its kernel $\mathbb{Z}\mu$, the annihilating ideal of R. By the fundamental morphism theorem $\mathbb{Z}/\ker f$ is isomorphic with $f(\mathbb{Z})$, a subring of R. $\mathbb{Z}/\mathbb{Z}\mu = \mathbb{Z}_\mu$ is isomorphic to a subring of R. □

The integers, rational numbers, real numbers, and complex numbers all contain a subring isomorphic with $\mathbb{Z}$ and are rings of characteristic zero. The product ring $\mathbb{Z} \times \mathbb{Z}_2$ is a ring of characteristic zero because there is no $m \in \mathbb{N}^+$ such that $m(1, \bar{1}) = (m, \bar{m}) = (0, \bar{0})$. The subring of $\mathbb{Z} \times \mathbb{Z}_2$ identified by the theorem which is isomorphic to $\mathbb{Z}$ is $f(\mathbb{Z}) = \{m(1, \bar{1}) | m \in \mathbb{Z}\}$. It is interesting to note that there are some elements of $\mathbb{Z} \times \mathbb{Z}_2$ such as $(0, \bar{1})$ which have finite multiples equal to zero: $2(0, \bar{1}) = (0, \bar{0})$.

The theorem just considered allows us to show that integral domains cannot have some integers as their characteristics.

Theorem. *If $\langle R, +, \cdot, \theta, v \rangle$ is an integral domain then* chr $R = 0$ *or* chr R *is a prime.*

Proof. R contains a subring which is isomorphic with $\mathbb{Z}_\mu$ where $\mu =$ chr R. Suppose $\mu = 1$. Then $1v = \theta$ yielding $v = \theta$. This is impossible in an integral domain. Therefore $\mu \neq 1$. Suppose now $\mu = \alpha\beta$ where α and β are not 1 and are positive natural numbers. $(\alpha v)(\beta v) = (\alpha\beta)v = \mu v = \theta$. αv and βv are therefore nontrivial divisors of zero in R, another contradiction. The only posibilities left for μ are that μ is prime or zero. □

Corollary. *If $\langle R, +, \cdot, \theta, v \rangle$ is a field then R has either prime or zero characteristic.*

Theorem. *Let $\langle R, +, \cdot, \theta, v\rangle$ be a field. If* chr $R = 0$ *then there is a subfield of R isomorphic to $\mathbb{Q}$; if* chr R *is a prime p then there is a subfield of R isomorphic to $\mathbb{Z}_p$.*

PROOF. If chr R is the prime p then there is a subfield of R isomorphic to $\mathbb{Z}_p$ which is a field as well as a ring. If chr R is zero then there is a subring isomorphic with $\mathbb{Z}$. Because R is a field every image of an element in $\mathbb{Z}$ must have an inverse in R. If $f:\mathbb{Z} \to R$ is the ring isomorphism such that $f(n) = nv$, then we define $\bar{f}:\mathbb{Q} \to R$ so that $\bar{f}(m/n) = f(m)f(n)^{-}$. $\bar{f}(\mathbb{Q})$ is a subfield of R isomorphic to $\mathbb{Q}$. This technique resembles the one described in Exercise 1, Section 4.5. □

QUESTIONS

1. The characteristic of $\mathbb{Z}_4 \times \mathbb{Z}_4$ is
   (A) 0
   (B) 1
   (C) 4
   (D) 8
   (E) 16.

2. Which of the following are possible?
   (A) A field with characteristic 4
   (B) An integral domain with characteristic 4
   (C) An infinite field with nonzero characteristic
   (D) A finite field with characteristic 0.
   (E) All four are impossible.

3. The characteristic of $\mathbb{Z}_6 \times \mathbb{Z}_4$ is
   (A) 0
   (B) 4
   (C) 6
   (D) 24.
   (E) None of the four numbers is the characteristic.

4. The characteristic of $\mathbb{Z}_8 \times \mathbb{Z}_4$ is
   (A) 0
   (B) 4
   (C) 8
   (D) 32.
   (E) None of the four numbers is the characteristic.

5. The multiples of the unity map ($f:\mathbb{Z} \to \mathbb{Z} \times \mathbb{Z}_2$ such that $f(n) = nv$) of $\mathbb{Z}$ into $\mathbb{Z} \times \mathbb{Z}_2$ has range
   (A) $\mathbb{Z} \times 2\mathbb{Z}$
   (B) $\{(2n, \bar{0}) | n \in \mathbb{Z}\} \cup \{(2n + 1, \bar{1}) | n \in \mathbb{Z}\}$
   (C) $\{n(1, \bar{1}) | n \in \mathbb{Z}\}$
   (D) $\mathbb{Z}$.
   (E) None of the four sets is the range.

6. If the characteristic of a ring R is zero then
(A) R is an integral domain
(B) $nx = \theta$ is impossible for nonzero x in R and positive n in $\mathbb{Z}$
(C) R has a unity
(D) R cannot be the zero ring, $\{\theta\}$.
(E) None of the alternatives makes a true sentence.

Exercises

1. Let $f:R \to R'$ be a ring epimorphism. Let anh R stand for the annihilating ideal of the ring R. Prove
(a) anh $R \subseteq$ anh R';
(b) chr R is a multiple of chr R';
(c) chr $R =$ chr R' if f is an isomorphism.

2. Find the characteristic of $\mathbb{Z}_2 \times \mathbb{Z}_3$. Prove $\mathbb{Z}_2 \times \mathbb{Z}_3$ is isomorphic to $\mathbb{Z}_6$.

3. Give an example of a ring with prime characteristic yet not an integral domain.

4. Let $\langle R, +, \cdot, \theta, v\rangle$ be a unitary commutative ring. If chr R is the prime p prove $(x + y)^p = x^p + y^p$ for all $x, y \in R$.

5. Find a noncommutative unitary ring R and a prime p such that chr $R = p$ and the equation $(x + y)^p = x^p + y^p$ fails for some x and y.

6. Find all morphisms of $\mathbb{Z}$ into an integral domain R.

7. Find a ring R and a morphism $f:\mathbb{Z} \to R$ which is not the zero morphism nor of the type $f(n) = nv$.

8. Show that the only morphism of the ring $\mathbb{Q}$ into the ring $\mathbb{Z}$ is the zero morphism.

9. We define the additive order of an element x in a ring R to be $\min\{n|nx = \theta$ and $n \in \mathbb{N}^+\}$ if the set is nonempty and 0 otherwise.
(a) Show that the order of an element is the natural number generator of the annihilating ideal of x.
(b) If A is the annihilating ideal of R and A_x is the annihilating ideal of the element x in R show that $A = \bigcap\{A_x|x \in R\}$.
(c) Let R be a ring. Show chr $R = 0$ if any element of R has order 0. Show chr $R = \operatorname{lcm}\{n_x|x \in R\}$ if every $x \in R$ has order $n_x \in \mathbb{N}^+$. The notation lcm means least common multiple.
(d) Are there elements of $\mathbb{Z}_2 \times \mathbb{Z}_3$ with positive order smaller than the characteristic?

10. Let V be the set of all sequences $(a_2, a_3, a_4, \ldots)$ where $a_i \in \mathbb{Z}_i$ and all but a finite number of the a_i equal $\bar{0}$. Verify that V is a commutative ring without unity under the operations $(a_i) + (b_i) = (a_i + b_i)$, $(a_i)(b_i) = (a_ib_i)$. Show that every element of V has finite order yet chr $V = 0$.

11. Show that the field of fractions of an integral domain has the same characteristic as the integral domain.

5 Rings: Polynomials and factorization

We begin with the properties of polynomials found in school algebra and then move in Section 5.2 to a relatively formal construction of the polynomial ring. Particularly interesting to students ought to be the altered properties of polynomials when the ring of coefficients is not a field. Section 5.3 deals with polynomial functions arising from polynomials emphasizing the distinction between these two concepts. Although this is a small point to a professional mathematician, an understanding of such distinctions helps mature the student of mathematics. The matters at stake in the factor theorem are then clearer. We are then led to the important results counting the number of roots of a polynomial and the concept of multiplicity of roots.

We now make a change of subject and discuss rings enjoying the division algorithm. We study factorization of these rings having both the polynomials and integers as examples. A distinction is made between irreducible element and prime element which becomes active in the more general examples. We prove the fundamental theorem of arithmetic for principal ideal domains. We then introduce greatest common divisors. Greatest common divisor as we use it is not unique; any unit multiple of a greatest common divisor is also a greatest common divisor. We have chosen not to formalize the equivalence of all greatest common divisors of a pair of elements. We do connect the greatest common divisor study with the use of partial fractions in finding antiderivatives in elementary calculus. Unique factorization is studied for its own sake with unique factorization seen as the distinction between irreducible and prime elements. We use again polynomials this time to study field extensions and in turn we apply field extensions to construct the complex numbers.

5.1 The ring of polynomials

In this section we review formal properties of polynomials and their operations. We assume some previous familiarity with polynomials.

By a *polynomial* in X with coefficients in a ring R we shall mean any expression of the form $a_nX^n + a_{n-1}X^{n-1} + \cdots + a_2X^2 + a_1X + a_0$ in which $a_0, a_1, \ldots, a_n \in R$ and $a_n \neq \theta$ or the expression θ which is the zero element of R. $a_n, a_{n-1}, \ldots, a_2, a_1, a_0$ are called the *coefficients* of the polynomial. θ is called the *zero polynomial*. The natural number n is called the *degree* of the polynomial and a_n is called the *leading coefficient*. The zero polynomial has no leading coefficient and no degree.

Examples. $3X^2 + X + 1$ (or $1 + X + 3X^2$) has leading coefficient 3 and degree 2. $X^{64} + 12$ is a polynomial with leading coefficient 1 and degree 64. All coefficients $a_1, a_2, \ldots, a_{63}$ are zero. 5 is a polynomial with leading coefficient 5 and degree 0.

The definition of polynomial is provisional; it is somewhat unsatisfactory until the meaning of the symbol X is resolved. A more abstract formulation will come up later, but we have chosen to begin with the familiar.

It is often convenient to regard a nonzero polynomial as having an infinite number of coefficients, all zero beyond a certain one. In this vein $a_0 + a_1X + a_2X^2 + \cdots + a_nX^n$ with $a_n \neq \theta$ (the leading coefficient) may be regarded as $a_0 + a_1X + a_2X^2 + \cdots + a_nX^n + \theta X^{n+1} + \theta X^{n+2} + \cdots$. In summation notation $\sum_{i=0}^{n} a_iX^i = \sum_{i=0}^{\infty} a_iX^i$ in which $a_i = \theta$ for $i > n$. We also can regard $\theta = \theta + \theta X + \theta X^2 + \cdots$ as having all coefficients zero.

We denote the set of all polynomials with coefficients in the ring $\langle R, +, \cdot, \theta\rangle$ by the symbol $R[X]$ and define operations for the set as follows.

Addition

$\sum_{i=0}^{m} a_iX^i + \sum_{i=0}^{n} b_iX^i = \sum_{i=0}^{p} (a_i + b_i)X^i$. m is the degree of the first polynomial and n is the degree of the second polynomial; $a_m \neq \theta$ and $b_n \neq \theta$. $a_i = \theta$ for $i > m$ and $b_i = \theta$ for $i > n$. p denotes the maximum of the two natural numbers, m and n.

Examples. $(6X^3 + 4X^2 + 5) + (X^2 + 4X) = (6X^3 + 4X^2 + 0X + 5) + (0X^3 + 1X^2 + 4X + 0) = 6X^3 + 5X^2 + 4X + 5$. $(X^2 + 2X + 5) + (-X^2 + X + 2) = 3X + 7$. This second example shows that p is not necessarily the degree of the sum.

Multiplication

$(\sum_{i=0}^{m} a_iX^i)(\sum_{j=0}^{n} b_jX^j) = \sum_{k=0}^{m+n} (\sum_{i+j=k} a_ib_j)X^k$. m is the degree of the first polynomial and n is the degree of the second polynomial; $a_m \neq \theta$ and $b_n \neq \theta$. $a_i = \theta$ for $i > m$ and $b_i = \theta$ for $i > n$.

Examples. $(5X^4 + 6X^3 - 4X^2 + 3X + 2)(2X^3 - 3X^2 + 6X - 7) = (5 \cdot 2)X^7 + (5(-3) + 6 \cdot 2)X^6 + (5 \cdot 6 + 6(-3) + (-4)2)X^5 + (5(-7) + 6(6) + (-4)(-3) + 3(2))X^4 + (6(-7) + (-4)6 + (-3)3 + 2(2))X^3 + ((-4)(-7) + 3(6) + 2(-3))X^2 + (3(-7) + 2(6))X + 2(-7) = 10X^7 + (-3)X^6 + 4X^5 + 19X^4 + (-71)X^3 + 40X^2 + (-9)X + (-14)$. The degree of the product is not necessarily $m + n$ in general.

Utilizing these definitions of addition and multiplication of polynomials with coefficients in a ring we state a theorem.

Theorem. *If $\langle R, +, \cdot, \theta \rangle$ is a ring then $\langle R[X], +, \cdot, \theta \rangle$ is a ring. If $\langle R, +, \cdot, \theta, v \rangle$ is a unitary ring then $\langle R[X], +, \cdot, \theta, v \rangle$ is a unitary ring. If $\langle R, +, \cdot, \theta \rangle$ is a commutative ring then $\langle R[X], +, \cdot, \theta \rangle$ is also a commutative ring.*

We do not prove these theorems but rather leave proofs to the reader. The proofs are routine. Relevant to the last statement in the theorem we point out that even if R is not a commutative ring the definition of multiplication for $R[X]$ implies that the symbol X must at least commute with every member of R; $Xa = (vX + \theta)(a) = (va)X + \theta = aX$. If there is no unity in R we can still be assured that $(cX)(a) = (cX + \theta)(a) = (ca)X + \theta = (ca)X$ for all $a, c \in R$.

We give one more theorem of the same type and do prove this one.

Theorem. *$\langle R, +, \cdot, \theta, v \rangle$ is an integral domain if and only if $\langle R[X], +, \cdot, \theta, v \rangle$ is an integral domain.*

Proof. Let R be an integral domain and let $p(X)q(X)$ be a product of polynomials in $R[X]$ which is zero. Then all of the coefficients of the product are zero including that formed from the leading coefficients of $p(X)$ and $q(X)$, namely $a_m b_n$. But if $a_m b_n = \theta$ and R is an integral domain we have $a_m = \theta$ or $b_n = \theta$, which is a contradiction. The only alternative to the contradiction is that either $p(X)$ or $q(X)$ is the zero polynomial.

To prove the converse assume $R[X]$ is an integral domain. Let ab be a product in R which is zero. Consider the polynomial a of degree zero with leading coefficient a and the polynomial b of degree zero with leading coefficient b assuming both a and b are different from zero. The product of the two polynomials is the polynomial ab also of degree zero. But this polynomial is the zero polynomial. Therefore either a or b must be the zero polynomial. This contradicts both a and b being different from zero. Either $a = \theta$ or $b = \theta$. □

We summarize in a theorem some facts about degree that are hinted in the previous theorem. We will leave the proofs to the reader.

Theorem. *If* $\langle R, +, \cdot, \theta, v\rangle$ *is an integral domain and* $p(X)$, $q(X)$ *are nonzero polynomials of* $R[X]$ *then* $\deg(p(X) + q(x)) \leqslant \max\{\deg p(X), \deg q(X)\}$, $\deg(p(X)q(X)) = \deg p(X) + \deg q(X)$, $\deg(p(X)q(X)) \geqslant \deg q(X)$.

We now establish a division algorithm for polynomials with coefficients in a field.

Theorem. *Let* $\langle K, +, \cdot, \theta, v\rangle$ *be a field. If* $a(X)$ *and* $b(X)$ *belong to* $K[X]$ *with* $a(X) \neq \theta$ *then there exist* $q(X)$, $r(X)$ *in* $K[X]$ *such that* $b(X) = q(X)a(X) + r(X)$ *and* $0 \leqslant \deg r(X) < \deg a(X)$ or $r(X) = \theta$.

PROOF. The proof is by induction on the degree of $b(X)$. Assume the theorem to be true for all polynomials of degree $<k$. Let $b(X)$ be of degree k.

Case 1: $\deg a(X) > \deg b(X) = k$. Then $b(X) = \theta a(X) + b(X)$. This satisfies the conclusion of the theorem with $q(X) = \theta$ and $r(X) = b(X)$.

Case 2: $\deg a(X) \leqslant \deg b(X) = k$. $b_0 + b_1X + \cdots + b_kX^k - (b_k/a_m)\cdot(a_mX^m + \cdots + a_1X + a_0)X^{k-m} = c_{k-1}X^{k-1} + \cdots + c_1X + c_0$, some polynomial of degree smaller than k or possibly zero. The values of the $c_0, c_1, \ldots, c_{k-1}$ are determined by comparing coefficients. We have, however, no interest in their actual values in terms of the a_i's and b_j's because it is sufficient for our purposes to know $\deg c(X) \leqslant k - 1$ or $c(X) = \theta$. Applying the induction hypothesis to $c(X)$ we have $c_{k-1}X^{k-1} + c_{k-2}X^{k-2} + \cdots + c_1X + c_0 = q^*(X)a(X) + r(X)$ for some $q^*(X)$, $r(X)$ in $K[X]$ with $\deg r(X) < \deg a(X)$ or $r(X) = \theta$. Substituting this back for $c(X)$ we have $b(X) - (b_k/a_m)a(X)X^{k-m} = q^*(X)a(X) + r(X)$. Rearrangement yields $b(X) = (q^*(X) + (b_k/a_m)X^{k-m})a(X) + r(X)$ with $\deg r(X) < \deg a(X)$ or $r(X) = \theta$. The second case is now proved and the induction is complete. The conclusion is true for polynomials $b(X)$ of all degrees. The theorem is also true for polynomials of no degree, namely, the zero polynomial, because if $b(X) = \theta$ then $\theta a(X) + \theta = \theta$. □

EXAMPLE. $2X^5 + \frac{9}{2}X^3 + X^2 + \frac{7}{4}X + 5 = (2X^3 + \frac{1}{2}X + 1)(X^2 + 2) + \frac{3}{4}X + 3$.

QUESTIONS

1. In the ring $\mathbb{Z}_4[X]$ the polynomial $(2X^2 + 3X + 2)(2X^3 + 3)$ has degree
   (A) 6
   (B) 5
   (C) 4
   (D) 3.
   (E) None of the listed numbers is the degree.

2. With dividend $X^3 + X + 1$ and divisor $2X + 1$ the division algorithm in $\mathbb{Z}_3[X]$ yields a remainder
   (A) 0
   (B) 1

(C) 2
(D) X.
(E) None of the listed alternatives is the remainder.

3. The degree of $f(X) + g(X)$, the sum of two polynomials, is
(A) $\leqslant \deg f(X) + \deg g(X)$
(B) $= \deg f(X) + \deg g(X)$
(C) $\leqslant \max\{\deg f(X), \deg g(X)\}$
(D) $\leqslant \min\{\deg f(X), \deg g(X)\}$.
(E) None of the alternatives completes a true sentence.

4. How many different polynomials of degree $\leqslant 1000$ are there in $\mathbb{Z}_4[X]$?
(A) $3 \cdot 4^{1000}$
(B) 4000
(C) 1004
(D) $1000!/4!$
(E) None of the above numbers is correct.

5. The product $(2X + 2)(2X + 1)$ in $\mathbb{Z}_4[X]$ is
(A) a second degree polynomial
(B) not the only possible factorization of $2X + 2$
(C) equal to $2(X + 1)$
(D) equal to $(2X + 1)(2X + 3)$.
(E) None of the four alternatives is acceptable.

6. The product of two even degree polynomials with integral coefficients
(A) always has even degree also
(B) may in some cases have odd degree
(C) can be zero
(D) is sometimes not defined.
(E) None of the alternatives makes a true sentence.

Exercises

1. We define a polynomial to be monic if and only if its leading coefficient is unity. Let $\langle R, +, \cdot, \theta, v\rangle$ be a commutative unitary ring. Given $a(X)$, $b(X)$ *in* $R[X]$ with $a(X)$ a monic polynomial, prove there exist polynomials $q(X)$, $r(X)$ such that $b(X) = q(X)a(X) + r(X)$ with $\deg r(X) < \deg a(X)$ or $r(X) = \theta$. Moreover, $q(X)$ and $r(X)$ are unique.

2. Let $\langle R, +, \cdot, \theta, v\rangle$ be a unitary ring, not necessarily commutative. Given $a(X)$, $b(X)$ in $R[X]$ with $a(X)$ a monic polynomial (leading coefficient unity), prove there exist $q_1(X)$, $r_1(X)$, $q_2(X)$, $r_2(X)$ such that $b(X) = q_1(X)a(X) + r_1(X)$ and $b(X) = a(X)q_2(X) + r_2(X)$ with $\deg r_1(X) < \deg a(X)$ or $r_1(X) = \theta$, $\deg r_2(X) < \deg a(X)$ or $r_2(X) = \theta$.

3. Is the ideal of $\mathbb{Z}[X]$ generated by 2 and X a principal ideal? [*Hint*: cf. Exercise 13, Section 2.8.]

4. Show that if the elements of $\mathbb{Q}[X]$ are ordered in a manner extending the ordering on $\mathbb{Q}$, i.e., $p(x) < q(x)$ if and only if $q(x) - p(x)$ has positive leading coefficient, then the result is an ordered integral domain (cf. Section 3.7).

5. Is $\mathbb{Q}[X]$ an ideal of $\mathbb{R}[X]$?

6. If A is an ideal of a ring $\langle R, +, \cdot, \theta, v\rangle$ which is commutative and has a unity, is $A[X]$ an ideal of $R[X]$? Are all ideals of $R[X]$ of this form?

7. Let $\langle K, +, \cdot, \theta, v\rangle$ be a field. Show that all ideals of $\langle K[X], +, \cdot, \theta, v\rangle$ are simply generated (principal) ideals. [*Hint*: Reread the lemma proof in Section 4.6 which shows that all ideals of $\mathbb{Z}$ are principal and build a similar proof for $K[X]$.]

8. Prove the following proposition or find a counterexample. No polynomial of odd degree can be the square of another polynomial.

5.2 A formal definition of a polynomial ring

In this section we reconstruct the polynomial ring. We wish to give a definition which dispenses with the convenient, but somewhat mysterious symbol X. The construction is more abstract; the original conventional notation just used should be kept in mind as guidance.

Our aim is to take an approach utilizing the coefficients, but deleting the X. It is clear enough that in a polynomial it is the coefficients which convey the information; the powers of X serve as placeholders for the coefficients. One can, for instance, write $9X^3 + 2X + 3$ as $(9, 0, 2, 3)$ and convey the same information. The sum of $9X^3 + 2X + 3$ and $2X^2 + 7X + 1$ can be computed as $(9, 0, 2, 3) + (0, 2, 7, 1) = (9, 2, 9, 4)$. As polynomials may be of arbitrarily high degree we cannot be satisfied with finite n-ples of one given length. Infinite sequences of coefficients are therefore to be used. We recall the notation $R^{\mathbb{N}}$, meaning the set of all infinite sequences with values in R, or, equivalently, the set of all functions with domain $\mathbb{N}$ and with values in R. Because polynomials all have leading coefficients (unless the polynomial is zero) we introduce the next definition.

Definition. Let $\langle R, +, \cdot, \theta\rangle$ be a ring. We define $(R^{\mathbb{N}})^w$ to be the set $\{p \mid p: \mathbb{N} \to R$ and there exists $m \in \mathbb{N}$ such that $p(n) = \theta$ for all $n > m\}$.

We shall refer to $(R^{\mathbb{N}})^w$ as the *weak power* to distinguish it from the regular power $R^{\mathbb{N}}$. The set, $(R^{\mathbb{N}})^w$, is the set of all functions defined on the natural numbers with values in the ring R which, except for a finite number of them, are all zero. $(R^{\mathbb{N}})^w$ is the set of all infinite sequences with values in R in which all but a finite number of the values are zero. On this set of functions we define the operations $+$ and $\cdot$ as follows:

$$p + q: \mathbb{N} \to R \quad \text{such that} \quad (p + q)(n) = p(n) + q(n),$$
$$p \cdot q: \mathbb{N} \to R \quad \text{such that} \quad (p \cdot q)(n) = \textstyle\sum_{j+k=n} p(j)q(k).$$

The set and addition here are those found in Exercise 11 of Section 2.2. The multiplication is different from that found in Exercise 11; several examples should be tried to verify for oneself that this definition is the same as that used for polynomials in Section 5.1.

We identify $z:\mathbb{N} \to R$ as the mapping $z(n) = \theta$ for all $n \in \mathbb{N}$.

Theorem. *If $\langle R, +, \cdot, \theta\rangle$ is a ring then $\langle (R^{\mathbb{N}})^w, +, \cdot, z\rangle$ is a ring.*

We leave the proof as an exercise.

Definition. $\langle (R^{\mathbb{N}})^w, +, \cdot, z\rangle$ is defined to be the *polynomial ring* over the given ring $\langle R, +, \cdot, \theta\rangle$.

Having constructed the polynomial ring by the use of sequences of coefficients we now move to connect the new formulation with the conventional notation for polynomials reviewed in Section 5.1. We do this by identifying each member of $(R^{\mathbb{N}})^w$ with a member of $R[X]$. The sequence $(a_0, a_1, a_2, \ldots, a_m, \theta, \theta, \theta, \ldots)$, $a_m \neq \theta$, is identified with the expression $a_0 + a_1X + a_2X^2 + \cdots + a_mX^m$. In other words a function $p:\mathbb{N} \to R$ is identified with the polynomial $p(0) + p(1)X + p(2)X^2 + \cdots + p(m)X^m$ where $p(n) = \theta$ for all $n > m$. Let us call this identification G.

Definition. *$G:(R^{\mathbb{N}})^w \to R[X]$ such that $G(p) = \sum_{i=0}^{m} p(i)X^i$ where m is an integer such that $p(n) = \theta$ for all $n > m$.*

Theorem. *$G:(R^{\mathbb{N}})^w \to R[X]$ is an isomorphism for all rings R.*

Proof. Given any $a_0 + a_1X + \cdots + a_mX^m$ in $R[X]$, $a_m \neq \theta$, choose p in $(R^{\mathbb{N}})^w$ such that $p(n) = a_n$ for all $n \leqslant m$ and $p(n) = \theta$ for all $n > m$. Then $G(p) = a_0 + a_1X + \cdots + a_mX^m$. G is a surjection. To verify that G is a bijection let $G(p) = G(q)$. Since the two polynomials are equal then all their coefficients are equal in respective pairs. $p(n) = q(n)$ for all $n \in \mathbb{N}$. Then $p = q$.

We now check the morphism properties.

$$G(p) + G(q) = \sum_{i=0}^{m} p_iX^i + \sum_{i=0}^{n} q_iX^i = \sum_{i=0}^{\max(m,n)} (p_i + q_i)X^i = G(p + q).$$

$$G(p)G(q) = \left(\sum_{i=0}^{m} p_iX^i\right)\left(\sum_{j=0}^{n} q_jX^j\right) = \sum_{k=0}^{m+n}\left(\sum_{i+j=k} p_iq_j\right)X^k = G(pq). \qquad \square$$

These previous arguments have established $(R^{\mathbb{N}})^w$ to be a ring isomorphic with $R[X]$. As the symbol X does not occur in the construction of $(R^{\mathbb{N}})^w$ we have therefore succeeded in our task. Before seeing where the X has gone we notice that the constants occur in $(R^{\mathbb{N}})^w$ as a subring.

Theorem. *If $\langle R, +, \cdot, \theta\rangle$ is a ring then there is a monomorphism $\psi:R \to (R^{\mathbb{N}})^w$ such that $\psi(x) = (x, \theta, \theta, \ldots)$.*

Proof. $\psi(x + y) = (x + y, \theta, \theta, \ldots) = (x, \theta, \theta, \ldots) + (y, \theta, \theta, \ldots) = \psi(x) + \psi(y)$. $\psi(xy) = (xy, \theta, \theta, \ldots) = (x, \theta, \theta, \ldots)(y, \theta, \theta, \ldots) = \psi(x)\psi(y)$.

$\psi(x) = \psi(y)$ implies $(x, \theta, \theta, \ldots) = (y, \theta, \theta, \ldots)$, which in turn implies $x = y$. □

We have showed that the ring R, called the constants, is isomorphic with the subring of $(R^{\mathbb{N}})^w$ denoted by $\{(a_0, \theta, \theta, \ldots) | a_0 \in R\}$. These sequences are simply the ones corresponding by G to the polynomials of $R[X]$ of degree zero and the zero polynomial.

The sequence $(\theta, \theta, \ldots, \theta, a_j, \theta, \ldots)$ of $(R^{\mathbb{N}})^w$ corresponds under G to the polynomial $a_j X^j$. If j is, in particular, 1, then $(\theta, a_1, \theta, \theta, \ldots)$ is identified with $a_1 X$. If there is a unity in the ring R we see by letting $a_1 = v$ that $(\theta, v, \theta, \theta, \ldots)$ corresponds to vX. If we consider vX and X to be one and the same then we have identified the sequence $(\theta, v, \theta, \theta, \ldots)$ with X. Thus X is the function from $\mathbb{N}$ to R which is θ for all $n \in \mathbb{N}$ save 1 where it has the value v.

Having demonstrated that it is possible to construct the polynomial ring free of the symbol X we return from here on to the conventional notation for the polynomial.

Questions

1. Which of the following are true?
(A) $(R^{\mathbb{N}})^w \subseteq R^{\mathbb{N}}$ for all rings R.
(B) $R^{\mathbb{N}} = (R^{\mathbb{N}})^w$ for all rings R.
(C) $R^{\mathbb{N}} \subseteq (R^{\mathbb{N}})^w$ for all rings R.
(D) $R^{\mathbb{N}} \neq (R^{\mathbb{N}})^w$ for all rings R.
(E) None of the four is true.

2. Under the correspondence G between the ring of functions, $(R^{\mathbb{N}})^w$, and the polynomial ring, $R[X]$, the polynomial $3X^2 + 2$ is the image of
(A) (3, 0, 2)
(B) (2, 3)
(C) (2, 0, 3, 0, 0, . . .)
(D) $3X^2 + 0X + 2$.
(E) None of the four answers is the preimage.

3. The existence of the function (1, 1, 1, . . .) in $\mathbb{Z}^{\mathbb{N}}$ shows that
(A) $\mathbb{Z}^{\mathbb{N}}$ is not isomorphic to $\mathbb{Z}[X]$
(B) $\mathbb{Z}^{\mathbb{N}} \neq (\mathbb{Z}^{\mathbb{N}})^w$
(C) the square of a polynomial is not necessarily a polynomial
(D) $\mathbb{Q}^{\mathbb{N}} = \mathbb{Z}^{\mathbb{N}}$.
(E) None of the four answers follows.

4. In the ring $(\mathbb{Z}^{\mathbb{N}})^w$ which functions $(a_0, a_1, a_2, \ldots)$ satisfy the equation $a^2 = a$?
(A) all functions with a square root in $(\mathbb{Z}^{\mathbb{N}})^w$
(B) (1, 0, 0, . . .)
(C) (0, 0, 0, . . .)
(D) (0, 1, 0, . . .).
(E) No functions in $(\mathbb{Z}^{\mathbb{N}})^w$ satisfy the equation.

Exercises

1. In proving $(R^{\mathbb{N}})^w$ to be a ring it must be demonstrated that if p and q both belong to $(R^{\mathbb{N}})^w$ then the product pq is also in $(R^{\mathbb{N}})^w$. Since $(pq)(k) = \sum_{i+j=k} p(i)q(j)$ it is clear enough that pq is defined for all $k \in \mathbb{N}$. It is not so clear that there exists $m \in \mathbb{N}$ such that for all $n > m$ we have $(pq)(n) = \theta$. Prove this.
2. Prove chr $R[X]$ = chr R for any ring R.
3. Prove there exist rings which are infinite yet have characteristic not equal to zero.
4. Is $(R^{\mathbb{N}})^w$ an ideal of $R^{\mathbb{N}}$ as well as a subring? Are the constants an ideal of $(R^{\mathbb{N}})^w$?
5. Find the fields of fractions of the rings $\mathbb{Z}[X]$ and $\mathbb{Q}[X]$.

5.3 Polynomial functions

In this section we see how a polynomial defines a function in a natural way and we investigate the correspondence between polynomials and polynomial functions. Finally we use polynomials to analyze field extensions and the complex numbers in particular.

Replacement of the symbol X in a polynomial of $R[X]$, $a_mX^m + a_{m-1}X^{m-1} + \cdots + a_2X^2 + a_1X + a_0$, $a_m \neq \theta$, by any element x of R yields a member $a_mx^m + a_{m-1}x^{m-1} + \cdots + a_2x^2 + a_1x + a_0$ of R. For example, if we replace X in $X^2 - 2X + 4$ by 3 in $\mathbb{Z}$ we obtain $3^2 - 2(3) + 4$ or 7, a member of $\mathbb{Z}$. We wish now to discuss this substitution phenomenon at some length. We will denote a polynomial in $R[X]$ by $p(X)$ and denote the function from R to R with $p(x)$ as its value with the letter p. This function is called a polynomial function and is a different function from that discussed in Section 5.2. Because the operation of multiplication requires that the coefficients of the polynomials commute with the symbol X and therefore with the substituted x of R we limit our discussion to commutative rings.

Definition. Let $\langle R, +, \cdot, \theta\rangle$ be a commutative ring. If $p(X)$ belongs to $R[X]$ and we denote $p(X)$ by $a_mX^m + \cdots + a_1X + a_0$, $a_m \neq \theta$, then the function $p: R \to R$ such that $p(x) = a_mx^m + \cdots + a_1x + a_0$ is called the *polynomial function* corresponding to the polynomial $p(X)$. The zero function $R \to R$ will correspond to the zero polynomial. We denote the correspondence taking polynomials into polynomial functions by $\Phi: R[X] \to R^R$ such that $\Phi(p(X)) = p$. (R^R is a previously used notation meaning the set of all functions with domain R and codomain R.) Range Φ is the set of all polynomial functions from R to R and we denote this set by $\mathscr{p}(R, R)$.

Theorem. *Let $\langle R, +, \cdot, \theta\rangle$ be a commutative ring. Then $\Phi: R[X] \to R^R$ such that $\Phi(p(X)) = p$ is a morphism.*

Proof. The values of Φ are uniquely determined; therefore Φ is a function. To demonstrate the morphism properties, $\Phi(p(X) + q(X)) = \Phi(p(X)) + \Phi(q(X))$ and $\Phi(p(X)q(X)) = \Phi(p(X))\Phi(q(X))$, we show the following equations hold for all $x \in R$:

$$\Phi(p(X) + q(X))(x) = \Phi(p(X))(x) + \Phi(q(X))(x),$$
$$\Phi(p(X)q(X))(x) = \Phi(p(X))(x) \cdot \Phi(q(X))(x).$$

$$\begin{aligned}\Phi(p(X)+q(X))(x) &= \Phi(a_mX^m+\cdots+a_1X+a_0+b_nX^n+\cdots+b_1X+b_0)(x)\\ &= \Phi((a_\mu+b_\mu)X^\mu+\cdots+(a_1+b_1)X+(a_0+b_0))(x)\\ &= (a_\mu+b_\mu)x^\mu+\cdots+(a_1+b_1)x+(a_0+b_0)\\ &= a_mx^m+\cdots+a_1x+a_0+b_nx^n+\cdots+b_1x+b_0\\ &= \Phi(a_mX^m+\cdots+a_1X+a_0)(x)\\ &\quad+\Phi(b_nX^n+\cdots+b_1X+b_0)(x)\\ &= \Phi(p(X))(x)+\Phi(q(X))(x).\end{aligned}$$

In the equation $\mu = \max\{m, n\}$ and we include the possibility of all coefficients zero to avoid handling the zero case separately.

$$\begin{aligned}\Phi(p(X)q(X))(x) &= \Phi((a_mX^m+\cdots+a_1X+a_0)(b_nX^n+\cdots+b_1X+b_0))(x)\\ &= \Phi\left(\sum_{k=0}^{m+n}\left(\sum_{i+j=k} a_ib_j\right)X^k\right)(x)\\ &= \sum_{k=0}^{m+n}\left(\sum_{i+j=k} a_ib_j\right)x^k\\ &= (a_mx^m+\cdots+a_1x+a_0)(b_nx^n+\cdots+b_1x+b_0)\\ &= \Phi(a_mX^m+\cdots+a_1X+a_0)(x)\Phi(b_nX^n+\cdots+b_1X+b_0)(x)\\ &= \Phi(p(X))(x)\Phi(q(X))(x).\end{aligned}$$

□

Example. The polynomial $p(X) = X^2 + 2X - 3$ belongs to $\mathbb{Z}[X]$. $p(x) = x^2 + 2x - 3$ is the value of the function $p:\mathbb{Z} \to \mathbb{Z}$ for any $x \in \mathbb{Z}$. $p(0) = -3$. $p(4) = 21$.

Example. The polynomials $p(X) = X + \bar{1}$ and $q(X) = \bar{2}X^3 + \bar{2}X + \bar{1}$ are distinct polynomials in $\mathbb{Z}_3[X]$. The coefficients $\bar{2}$ and $\bar{1}$ really stand, of course, for equivalence classes in $\mathbb{Z}_3$ with representatives 2 and 1 respectively. The two polynomial functions, $p:\mathbb{Z}_3 \to \mathbb{Z}_3$ and $q:\mathbb{Z}_3 \to \mathbb{Z}_3$, are, however, identical. Check this by noting that $p(\bar{0}) = q(\bar{0})$, $p(\bar{1}) = q(\bar{1})$ and $p(\bar{2}) = q(\bar{2})$.

In general the polynomials $R[X]$ and the polynomial functions $\not p(R, R)$ are not in one-to-one correspondence; Φ is not a monomorphism. An instance of this is seen in the previous example. For another example, consider in $\mathbb{Z}_2[X]$ all of the following distinct polynomials: $\bar{0}$, $\bar{1}$, X, $X + \bar{1}$, X^2, $X^2 + X + \bar{1}$, $X^2 + X$, $X^2 + \bar{1}$. There are eight polynomials in $\mathbb{Z}_2[X]$

of degree two or less (and the polynomial $\bar{0}$). There are, however, only four possible functions $\mathbb{Z}_2 \to \mathbb{Z}_2$ which we denote by a, b, c and d.

x	$a(x)$	$b(x)$	$c(x)$	$d(x)$
$\bar{0}$	$\bar{0}$	$\bar{1}$	$\bar{0}$	$\bar{1}$
$\bar{1}$	$\bar{0}$	$\bar{1}$	$\bar{1}$	$\bar{0}$.

We can represent a, b, c, and d by the following formulas: $a(x) = \bar{0}$, $b(x) = \bar{1}$, $c(x) = x$, $d(x) = x + \bar{1}$. We note

$$\begin{aligned} &\Phi(\bar{0}) = a, && \Phi(\bar{1}) = b, && \Phi(X) = c, && \Phi(X + \bar{1}) = d, \\ &\Phi(X^2 + X) = a, && \Phi(X^2 + X + \bar{1}) = b, && \Phi(X^2) = c, && \Phi(X^2 + \bar{1}) = d. \end{aligned}$$

Furthermore, if we continue into third- and higher-degree polynomials they will all correspond or give rise to one or more of the four functions a, b, c, or d. The number of polynomials in $\mathbb{Z}_2[X]$ is infinite yet the number of functions in $\mathscr{p}(R, R)$ is 4.

We now establish some results important for themselves and then eventually use them to cast some light upon the relative sizes of $R[X]$ and $\mathscr{p}(R, R)$. We begin with a theorem usually seen by elementary algebra students called the factor theorem.

Definition. Let $\langle R, +, \cdot, \theta\rangle$ be a ring. An element r in R is a *root* of a polynomial $p(X)$ in $R[X]$ if and only if $p(r) = \theta$.

Example. 2 is a root of the polynomial $X^2 - 4$ in the ring $\mathbb{Z}$. The polynomial $4X^2 - 1$ has no roots in the ring $\mathbb{Z}$. Notice how r is a root of the polynomial $p(X)$ is really a statement about the polynomial function p; $p: R \to R$ is zero at the argument r.

Theorem. *Let $\langle R, +, \cdot, \theta, v\rangle$ be a unitary commutative ring. $X - r$ is a factor of $p(X)$ in $R[X]$ if and only if r is a root of $p(X)$ in R.*

Proof. Suppose $X - r$ is a factor of $p(X)$. Then in $R[X]$ the polynomial $p(X)$ has a factorization

$$\begin{aligned} p(X) &= q(X)(X - r) \quad \text{for some } q(X) \text{ in } R[X]. \\ \Phi(p(X)) &= \Phi(q(X)(X - r)). \\ \Phi(p(X)) &= \Phi(q(X))\Phi(X - r). \text{ Why?} \\ p(x) &= q(x)(x - r) \quad \text{for all } x \in R. \text{ Why?} \\ p(r) &= q(r)(r - r). \\ p(r) &= \theta. \end{aligned}$$

$$r \text{ is a root of } p(X).$$

Now in order to prove the converse assume r to be a root of $p(X)$. Use the division algorithm in $R[X]$ for monic polynomials (cf. Exercise 1 of Section

5.1) to obtain

$$p(X) = q(X)(X - r) + s(X) \quad \text{with } \deg s(X) = 0 \text{ or } s(X) = \theta.$$

Applying the morphism Φ which takes polynomials into polynomial functions we get

$$\Phi(p(X)) = \Phi(q(X)(X - r) + s(X)).$$
$$\Phi(p(X)) = \Phi(q(X))\Phi(X - r) + \Phi(s(X)).$$
$$p(x) = q(x)(x - r) + s(x) \quad \text{for every } x \in R.$$
$$p(r) = q(r)(r - r) + s(r).$$

But since r is a root of $p(X)$ we know $p(r) = \theta$.

$$\theta = q(r)\theta + s(r).$$
$$\theta = s(r).$$

$s(X)$ is a polynomial of degree less than 1 which is zero at r or $s(X)$ is the zero polynomial. But polynomials of degree less than 1, namely degree 0, are constants and a nonzero constant cannot be zero at r. Ruling out the first alternative we therefore know $s(X)$ to be the zero polynomial. $p(X) = q(X)(X - r)$. This completes the proof of the factor theorem. □

We now use the factor theorem to place an upper bound on the number of roots of a polynomial.

Theorem. *Let $\langle R, +, \cdot, \theta, v\rangle$ be an integral domain. Then any polynomial in $R[X]$ of degree n can have at most n distinct roots.*

PROOF. Suppose $p(X) \in R[X]$ and $\deg p(X) = n$. We shall assume $p(X)$ has more than n distinct roots and obtain a contradiction. Let $r_1, r_2, \ldots, r_n, r_{n+1}$ be $n + 1$ of the distinct roots. $p(r_1) = \theta$ implies $X - r_1$ is a factor of $p(X)$. $p(X) = q_1(X)(X - r_1)$ for some $q_1(X) \in R[X]$. Evaluating at r_2 gives $p(r_2) = q_1(r_2)(r_2 - r_1)$. The left side is zero and therefore the product on the right is zero. $(r_2 - r_1)$ is not zero and therefore $q_1(r_2)$ is zero because of the integral domain hypothesis. r_2 is a root of $q_1(X)$. $q_1(X) = q_2(X)(X - r_2)$ for some $q_2(X)$ in $R[X]$. We have then $p(X) = q_2(X)(X - r_2)(X - r_1)$ with degree $q_2(X) = n - 2$. Using induction we arrive at $p(X) = c(X - r_n)\cdots(X - r_2)(X - r_1)$ for some $c \in R[X]$ and $\deg c = 0$. Evaluating at r_{n+1} we obtain a contradiction. □

EXAMPLE. $X^3 - 6X^2 + 11X - 6$ has roots 1, 2, and 3 and factored form $(X - 1)(X - 2)(X - 3)$. The polynomial is a member of $\mathbb{Z}[X]$ and $\mathbb{Z}$ is an integral domain. If we choose a ring which fails to be an integral domain such as $\mathbb{Z}_{12}$ then a polynomial such as $X^2 - \overline{4}$ turns out to have roots $\overline{2}, -\overline{2}, \overline{4}, -\overline{4}, \overline{8}, -\overline{8}, \overline{10}, -\overline{10}$ with the number of roots greater than two. Actually $\overline{2}$ and $-\overline{10}$ represent the same equivalence class in $\mathbb{Z}_{12}$ so that a complete list of distinct roots is $\overline{2}, \overline{4}, \overline{8}, \overline{10}$. The number of roots is larger than two. The possible factorizations of $X^2 - \overline{4}$ are $(X - \overline{2})(X - \overline{10})$ and $(X - \overline{4})(X - \overline{8})$.

Corollary. *Let $\langle R, +, \cdot, \theta, v\rangle$ be an integral domain. Let $p(X)$ and $q(X)$ be polynomials both of degree strictly smaller than n and let p and q agree for n distinct values of R: $p(x_i) = q(x_i)$ for $x_1, x_2, \ldots, x_n$ in R and distinct. Then $p(X) = q(X)$.*

PROOF. Show that the polynomial $p(X) - q(X)$ will have n roots and use the theorem. □

We return now to the correspondence Φ between polynomials and polynomial functions to show a case in which every polynomial must give rise to a distinct polynomial function.

Theorem. *Let $\langle R, +, \cdot, \theta, v\rangle$ be a nonfinite integral domain. Then $\Phi: R[X] \to \not{p}(R, R)$ is an isomorphism.*

PROOF. As $\not{p}(R, R)$ is by definition the range of Φ we only need to show Φ to be a monomorphism. Φ is an injection if and only if kernel $\Phi = \{\theta\}$. We now show the kernel to the zero polynomial alone. Suppose $p(X)$ belongs to kernel Φ. Then $p = \Phi(p(X))$ must be the constantly zero function of R^R. $p(x) = \theta$ for all $x \in R$. Suppose the polynomial $p(X)$ has some degree, say n. Since R is not finite there are in R more than n distinct values, each of which makes p be zero. This is to say the polynomial $p(X)$ of degree n has strictly more than n roots in R. This contradicts our theorem on the number of distinct roots and means $p(X)$ can have no degree. $p(X)$ is therefore the zero polynomial. Kernel $\Phi = \{\theta\}$. □

We can push the techniques of this theorem somewhat further to give some information even for finite integral domains.

Theorem. *Let $\langle R, +, \cdot, \theta, v\rangle$ be a finite integral domain. Then the ring of all polynomial functions $\not{p}(R, R)$ is isomorphic with $R[X]/\langle\varphi(X)\rangle$ where $\langle\varphi(X)\rangle$ is an ideal generated by some polynomial in $R[X]$.*

PROOF. $\Phi: R[X] \to \not{p}(R, R)$ is an epimorphism. There exists by the theory of Section 2.7 an isomorphism $f': R[X]/\ker \Phi \to \text{range } \Phi$ such that $f'(\ker \Phi + p(X)) = p$. Kernel Φ is an ideal of $R[X]$, a principal ideal, and therefore is generated by some polynomial $\varphi(X)$. (cf. Exercise 7 of Section 5.1). Remember that all finite integral domains are fields. □

EXAMPLE. Since $\mathbb{Z}$ is an infinite integral domain the polynomial ring $\mathbb{Z}[X]$ and the set of polynomial functions $\not{p}(\mathbb{Z}, \mathbb{Z})$ are isomorphic; each polynomial gives rise to a distinct polynomial function. We remark that not all functions in $\mathbb{Z}^{\mathbb{Z}}$ are polynomial functions; for example, the function $f: \mathbb{Z} \to \mathbb{Z}$ such that $f(2n) = 0$ and $f(2n + 1) = 1$ for all $n \in \mathbb{Z}$ cannot be a polynomial function because the number of roots would be infinite.

If we look now at the example $\mathbb{Z}_2[X]$ we are dealing with the case of a finite integral domain $\mathbb{Z}_2$. $X^2 + X$ gives rise to the zero polynomial function and is the polynomial of smallest degree with that property. $\mathbb{Z}_2[X]/\langle X^2+X\rangle$ is isomorphic with $\wp(\mathbb{Z}_2, \mathbb{Z}_2)$.

Questions

1. The polynomial function $p:\mathbb{Z} \to \mathbb{Z}$ defined by the polynomial $p(X) = X^2 - X$ is
 (A) an injection
 (B) a surjection
 (C) a bijection
 (D) a different function than that defined by the polynomial $q(X) = X^3 - X$.
(E) None of the alternatives is correct.

2. The polynomial function $p:\mathbb{Z}_3 \to \mathbb{Z}_3$ defined by the polynomial $p(X) = X^2 + \bar{2}X$ has a range with
 (A) 0 members
 (B) 1 member
 (C) 2 members
 (D) 3 members.
(E) None of the numbers is correct.

3. The number of members in $\wp(\mathbb{Z}_3, \mathbb{Z}_3)$ is
 (A) 0
 (B) 3
 (C) 6
 (D) 9.
(E) None of the alternatives is correct.

4. The number of roots of $X^2 + \bar{1}$ in $\mathbb{Z}_4$ is
 (A) 0
 (B) 1
 (C) 2
 (D) 3
 (E) 4.

5. The number of real polynomials and the number of polynomial functions in $\wp(\mathbb{R}, \mathbb{R})$ is
 (A) the same
 (B) different.

6. Which of the following statements are true?
 (A) $X^2 + \bar{2}X$ has two roots in $\mathbb{Z}_3$.
 (B) $X^3 + \bar{2}X = X(X - \bar{1})(X - \bar{2})$ in $\mathbb{Z}_3[X]$.
 (C) $X^3 + \bar{2}X$ corresponds to the zero function in $\wp(\mathbb{Z}_3, \mathbb{Z}_3)$.
 (D) $\wp(\mathbb{Z}_3, \mathbb{Z}_3) = \mathbb{Z}_3^{\mathbb{Z}_3}$.
(E) None of the statements is true.

EXERCISES

1. The following theorem occurs in college algebra texts along with the factor theorem and is called the remainder theorem. If $\langle R, +, \cdot, \theta, v\rangle$ is an integral domain and $p(X)$ belongs to $R[X]$ then $p(c) = a$ if and only if $p(X) = q(X)(X - c) + a$ for some $q(X)$ in $R[X]$. In other words, the value of a polynomial function p at c is the remainder upon dividing $p(X)$ by $X - c$. In the special case the remainder is zero then one has the factor theorem. Prove the remainder theorem.

2. Prove that if $\langle R, +, \cdot, \theta, v\rangle$ is an integral domain then the only invertible elements of $R[X]$ are those of R.

3. Give an example of a polynomial ring with invertible elements of positive degree.

4. What is the field of fractions of $K[X]$ where K is a given field?

5. From calculus we recall that the derivative of a function $p(x) = a_n x^n + a_{n-1}x^{n-1} + \cdots + a_2x^2 + a_1x + a_0$ is the function $p'(x) = na_n x^{n-1} + (n-1)a_{n-1}x^{n-2} + \cdots + 2a_2x + a_1$. If we deal only with polynomials we can forget all about limits and make the previous equation be our definition of $p'(x)$. Given a polynomial $p(X) = a_nX^n + \cdots + a_1X + a_0$ we *define* $p'(X) = na_nX^{n-1} + (n-1)a_{n-1}X^{n-2} + \cdots + 2a_2X + a_1$. With this definition prove the following result. Let $\langle R, +, \cdot, \theta, v\rangle$ be an integral domain. Then $p(r)$ and $p'(r)$ are both zero if and only if $(X - r)^2$ is a factor of $p(X)$.

6. Prove that the sine function is not a polynomial function (not in $\mathscr{p}(\mathbb{R}, \mathbb{R})$).

7. *Lagrange interpolation formula.* Show that

$$f(X) = \sum_{j=1}^{n} \frac{(X - a_1)\cdots(X - a_{j-1})(X - a_{j+1})\cdots(X - a_n)}{(a_j - a_1)\cdots(a_j - a_{j-1})(a_j - a_{j+1})\cdots(a_j - a_n)} b_j$$

is a polynomial of degree $\leqslant n$ such that $f(a_i) = b_i$, $i = 1, 2, \ldots, n$.

5.4 Euclidean and principal ideal domains

As a prelude to the study of factorization the division algorithm property is abstracted from the integers and the polynomial ring in order to define a special ring: the Euclidean ring.

Definition. $\langle E, +, \cdot, \theta, v\rangle$ is a *Euclidean domain* if and only if $\langle E, +, \cdot, \theta, v\rangle$ is an integral domain and there exists a *gauge* $g: E - \{\theta\} \to \mathbb{N}$ such that $g(ab) \geqslant g(a)$ for all $a, b \in E - \{\theta\}$ and $a, b \in E$, $a \neq \theta$ imply there exist $q, r \in E$ such that $b = qa + r$ with $0 \leqslant g(r) < g(a)$ or $r = \theta$.

EXAMPLES. Since we are generalizing another property of the integers we note that $\mathbb{Z}$ with $g(x) = |x|$ is a Euclidean domain. Note that even though $|0|$ exists g is only defined for nonzero integers. If K is any field then $K[X]$ with $g(p(X)) = \deg p(X)$ is a Euclidean domain.

We now wish to expand on a notion introduced in Exercise 7 of Section 5.1.

Definition. Let $\langle R, +, \cdot, \theta \rangle$ be a ring. An ideal of R is called *principal* if and only if the ideal is generated by a single element. A *principal ideal ring* is a ring in which every ideal is principal. We shorten "principal ideal integral domain" to "principal ideal domain" or "principal domain."

We next compare principal ideal rings with Euclidean rings.

Theorem. *If $\langle R, +, \cdot, \theta, v \rangle$ is a Euclidean domain then R is a principal ideal domain. Moreover, if A is some ideal of R then $A = Ra$ for some $a \in R$.*

Proof. Let A be any ideal of R. If $A = \{\theta\}$ then A is principal for it is the smallest ideal containing θ, its generator. Furthermore, $\{\theta\} = R\theta$, the set of multiples of θ. If $A \neq \{\theta\}$ then let $a \neq \theta$ be a member of A which has minimum g value: $g(a) \leqslant g(x)$ for all $x \in A$, $x \neq \theta$. Such an a exists because $\{g(x) | x \in A, x \neq \theta\}$ is a subset of $\mathbb{N}$, a well-ordered set. Now let b be any member whatsoever of A. $b = qa + r$ for some $q, r \in R$ with $0 \leqslant g(r) < g(a)$ or $r = \theta$. $b, a \in A$ imply $r = b - qa$ is in A also. Since it is impossible for r to be in A and $g(r) < g(a)$ we must have $r = \theta$. $b = qa$. Thus the ideal Ra is A itself and A is the ideal simply generated by the single element a.

We now classify the elements of an integral domain according to factorization properties.

Definition. Let $\langle R, +, \cdot, \theta, v \rangle$ be an integral domain. $x \in R$ is a *unit* if and only if x is multiplicatively invertible. $x \in R$ is an *irreducible element* if and only if x is not zero and not a unit and $x = ab$ implies a or b is a unit. $x \in R$ is a *reducible element* if and only if x is not zero and not a unit and x is not an irreducible element.

Some further terminology well known to students of elementary algebra: If $x = ab$ then a and b are called *factors* or *divisors* of x. On the other hand x is called a *multiple* of a or b. If $x = ab$ and a and b are not units then a and b are called *proper divisors* or *proper factors* of x. We also say a divides x.

Examples. 1 and -1 are units of $\mathbb{Z}$ and the only units of $\mathbb{Z}$. 2 is an irreducible element of $\mathbb{Z}$ while 6 is a reducible element. 2 and 3 are proper divisors of 6 in $\mathbb{Z}$. -1 is a divisor of 6, but not a proper divisor of 6 in $\mathbb{Z}$. $X^2 + 1$ is an irreducible element of $\mathbb{Q}[X]$. 2 and $X - 1$ are proper divisors of $2X^2 - 2$ in $\mathbb{Z}[X]$. $X - 1$ is a proper divisor of $2X^2 - 2$ but 2 is a unit in $\mathbb{Q}[X]$.

We now investigate some relationships between ideals and elements of an integral domain. The definitions of prime and maximal ideals were introduced in Section 2.8. We recall now that an ideal A is prime if and only if $ab \in A$ implies $a \in A$ or $b \in A$.

Definition. Let $\langle R, +, \cdot, \theta, v\rangle$ be an integral domain. A nonzero, nonunit element p of R is called a *prime* if and only if p is a divisor of ab implies p is a divisor of a or p is a divisor of b.

We now match the definition of prime element to prime ideal.

Theorem. *Let $\langle R, +, \cdot, \theta, v\rangle$ be an integral domain. A nonzero, nonunit element p of R is prime if and only if the ideal $\langle p\rangle$ is prime.*

PROOF. p is prime if and only if (p is a divisor of ab implies p is a divisor of a or p is a divisor of b). Equivalent to this is: ab is a multiple of p implies (a is a multiple of p or b is a multiple of p). This is true if and only if $ab \in \langle p\rangle$ implies ($a \in \langle p\rangle$ or $b \in \langle p\rangle$). But this is the definition of prime ideal. That $\langle p\rangle = Rp$ is assured by Exercise 8 of Section 2.8. □

The property of primeness has been introduced primarily for its use in proving the unique factorization theorem in the next section. Of the two properties, primeness and irreducibility, primeness is the stronger in an arbitrary domain. The two concepts coincide in the integers.

Theorem. *Let $\langle R, +, \cdot, \theta, v\rangle$ be an integral domain. Any prime element is an irreducible element.*

PROOF. Suppose p is a prime element and $p = ab$. Since $p = ab$, p is, a fortiori, a divisor of ab. It follows that p is a divisor of a or p is a divisor of b. If p is a divisor of a then $a = rp$ for some $r \in R$. Substituting into equation $p = ab$ we have $p = rpb$, yielding $v = rb$ because R is an integral domain. b is then a unit. On the other hand, if p is a divisor of b then by symmetric reasoning a is a unit. We have showed that either a or b is a unit proving p is irreducible. □

We pause before the next theorem in order to give an example of an integral domain which has irreducible elements which fail to be prime. It is also an integral domain in which factorization is not unique. The integral domain we have in mind is the set $\{a + b\sqrt{-5} | a, b \in \mathbb{Z}\}$ with addition and multiplication defined in the obvious way:

$$(a + b\sqrt{-5}) + (c + d\sqrt{-5}) = (a + c) + (b + d)\sqrt{-5},$$
$$(a + b\sqrt{-5})(c + d\sqrt{-5}) = ac + ad\sqrt{-5} + bc\sqrt{-5} + bd(-5)$$
$$= (ac - 5bd) + (ad + bc)\sqrt{-5}.$$

The reader can assure himself that the set is a ring. If one realizes the set to be a subset of the field of complex numbers then it follows there can be no divisors of zero. We will, however, verify this property anew in a different manner. It is convenient to define a norm of each element of the domain as follows: $N(a + b\sqrt{-5}) = a^2 + 5b^2$. This is the square of the modulus of the complex number $a + b\sqrt{-5}$, if the reader happens to be familiar with the modulus of a complex number. It is easily seen that

$$N(a + b\sqrt{-5}) = 0 \quad \text{if and only if} \quad a = b = 0,$$

and

$$N((a + b\sqrt{-5})(c + d\sqrt{-5})) = N(a + b\sqrt{-5})N(c + d\sqrt{-5}).$$

Now suppose

$$(a + b\sqrt{-5})(c + d\sqrt{-5}) = 0.$$
$$N((a + b\sqrt{-5})(c + d\sqrt{-5})) = 0.$$
$$N(a + b\sqrt{-5})N(c + d\sqrt{-5}) = 0.$$
$$N(a + b\sqrt{-5}) = 0 \quad \text{or} \quad N(c + d\sqrt{-5}) = 0.$$
$$a + b\sqrt{-5} = 0 \quad \text{or} \quad c + d\sqrt{-5} = 0.$$

This shows that $\mathbb{Z} + \mathbb{Z}\sqrt{-5}$ has no divisors of zero (other than zero itself).

We now use the norm to find the units of the integral domain $\mathbb{Z} + \mathbb{Z}\sqrt{-5}$. Clearly 1 and -1 are both in $\mathbb{Z} + \mathbb{Z}\sqrt{-5}$ and both are units. We now show that they are the only units. Suppose $(a + b\sqrt{-5})(c + d\sqrt{-5}) = 1$. Then $N(a + b\sqrt{-5})N(c + d\sqrt{-5}) = 1$. The only nonnegative integers which can have product 1 are 1 and 1. Thus any unit of $\mathbb{Z} + \mathbb{Z}\sqrt{-5}$ must have norm 1. If $N(a + b\sqrt{-5}) = 1$ we have $a^2 + 5b^2 = 1$. If $|b| \geqslant 1$ then $a^2 + 5b^2 \geqslant 5$. b must be 0. $a^2 = 1$ has two solutions $a = 1$ and $a = -1$. The only units of $\mathbb{Z} + \mathbb{Z}\sqrt{-5}$ are $1 + 0\sqrt{-5} = 1$ and $-1 + 0\sqrt{-5} = -1$. Furthermore, $N(a + b\sqrt{-5}) = 1$ if and only if $a + b\sqrt{-5}$ is a unit.

We continue the exploration of $\mathbb{Z} + \mathbb{Z}\sqrt{-5}$ by showing 3 is an irreducible element. If 3 has some factors,

$$3 = (a + b\sqrt{-5})(c + d\sqrt{-5}),$$

then $9 = N(3) = N(a + b\sqrt{-5})N(c + d\sqrt{-5})$. The positive integer $N(a+b\sqrt{-5})$ must be a factor of 9, other than 1 or 9, if the factor $(a+b\sqrt{-5})$ is to be proper. We have, therefore, $N(a + b\sqrt{-5}) = 3$. $a^2 + 5b^2 = 3$. b must be zero for if $|b| \geqslant 1$ then $a^2 + 5b^2 \geqslant 5$. Setting $b = 0$ we have $a^2 = 3$. However, the equation $a^2 = 3$ has no solution in integers. We conclude there can be no proper factors of 3 and it is irreducible in $\mathbb{Z} + \mathbb{Z}\sqrt{-5}$. By similar arguments the reader can show that $2 + \sqrt{-5}$ and $2 - \sqrt{-5}$ are also irreducible.

An examination of the following equations will convince one that 3 is not a prime.

$$9 = 3 \cdot 3. \quad 9 = (2 + \sqrt{-5})(2 - \sqrt{-5}).$$

3 is a factor of 9 and therefore of the product $(2 + \sqrt{-5})(2 - \sqrt{-5})$. 3 is not a factor of either $2 + \sqrt{-5}$ or $2 - \sqrt{-5}$. In other words, 3 does not satisfy the definition of a prime element: 3 is a factor of ab implies 3 is a factor of a or 3 is a factor of b. Finally, we note the two factorizations given of 9 are quite different factorizations of 9. $\mathbb{Z} + \mathbb{Z}\sqrt{-5}$ is not an integral domain in which we can expect unique factorization.

We now show that in a principal ideal domain the concept of primeness and the concept of irreducibility coincide just as in $\mathbb{Z}$.

Theorem. *If $\langle R, +, \cdot, \theta, v\rangle$ is a principal ideal domain then every irreducible element of R is prime.*

PROOF. Let p be an irreducible element of R. We proceed to show the ideal $\langle p\rangle$ is a maximal ideal of R. Let A be any ideal of R such that $\langle p\rangle \subset A \subseteq R$. All ideals of R are principal and we can therefore write $\langle p\rangle \subset \langle a\rangle \subseteq R$ for some $a \in R$. $p \in \langle a\rangle$ giving $p = ra$ for some $r \in R$. r cannot be a unit because if it were we would have $a = r^{-1}p \in \langle p\rangle$ and then $\langle a\rangle = \langle p\rangle$. If r is not a unit then a is because p is irreducible. But then $v = a^{-1}a \in \langle a\rangle$ and $\langle a\rangle = R$. This proves the ideal $\langle p\rangle$ to be maximal. $\langle p\rangle$ is therefore also a prime ideal (cf. Section 2.8). $\langle p\rangle$ is generated by a nonzero, nonunit element $p \cdot p$ is therefore a prime element. □

QUESTIONS

1. Which of these statements are true?
   (A) Every Euclidean domain is a principal ideal domain.
   (B) Every prime element in an integral domain is an irreducible element.
   (C) Every irreducible element in an integral domain is a prime element.
   (D) Every unit in an integral domain is irreducible.
   (E) None of the four statements is true.

2. Which of these statements are true?
   (A) $\mathbb{Z}[X]$ is an integral domain.
   (B) $\mathbb{Z}[X]$ is a subring of $\mathbb{R}[X]$.
   (C) $\mathbb{Z}[X]$ contains an irreducible element $2 + 2X$.
   (D) $\mathbb{R}[X]$ contains an irreducible element $2 + 2X$.
   (E) None of the four statements is true.

3. If S is a subdomain of the integral domain R then
   (A) an irreducible element of S is an irreducible element of R
   (B) an irreducible element of R is an irreducible element of S
   (C) a unit of S is a unit of R
   (D) a unit of R is a unit of S.
   (E) None of the four alternatives completes a true sentence.

4. In an integral domain R, element a is a factor of element b implies
   (A) $\langle b\rangle \subseteq \langle a\rangle$
   (B) $\langle a, b\rangle = \langle a\rangle$
   (C) $\langle a\rangle \neq R$
   (D) $a \neq \theta$.
   (E) None of the four alternatives completes a true sentence.

5. In an integral domain R
   (A) no element except v can generate the ideal R
   (B) an ideal A equals R if and only if $v \in A$
   (C) an ideal containing no prime is $\{\theta\}$

(D) an ideal containing a composite cannot contain a prime.
(E) None of the four alternatives completes a true sentence.

6. The number 2 is
(A) an irreducible element of $\mathbb{Z}$
(B) a unit of $\mathbb{Z}$
(C) a proper divisor of $X^2 - 16$ in $\mathbb{Z}[X]$
(D) an irreducible element of $\mathbb{Z}_5$.
(E) None of the four alternatives completes a true sentence.

7. $\mathbb{Z}_5[X]$
(A) is an integral domain
(B) is a Euclidean domain
(C) is a principal ideal domain
(D) contains an irreducible element $X^2 + 1$.
(E) None of the alternatives completes a true sentence.

8. $X^2 - 2$
(A) is irreducible in $\mathbb{Q}[X]$
(B) is irreducible in $\mathbb{R}[X]$
(C) is irreducible in $\mathbb{Z}[X]$
(D) is irreducible in $\mathbb{Z}_7[X]$.
(E) None of the alternatives completes a true sentence.

9. Factors of $X^2 + 4$ in $\mathbb{Z}_7[X]$ are
(A) $(X + \bar{2})(X - \bar{2})$
(B) $(X + \bar{5})(X + \bar{2})$
(C) $(X + \bar{2})(X + \bar{5})$
(D) $(X + \bar{5})(X + \bar{5})$.
(E) None of the alternatives completes a true sentence.

10. Factors of $X^2 + 4$ in $\mathbb{Z}_5[X]$ are
(A) $(X + \bar{4})(X + \bar{1})$
(B) $(X + \bar{2})(X + \bar{2})$
(C) $(X - \bar{4})(X - \bar{1})$
(D) $(X - \bar{2})(X + \bar{2})$.
(E) None of the four possibilities is satisfactory.

Exercises

1. Show that $\mathbb{Z}_6$ with $g(x) = x$ is not a Euclidean domain.
2. Show that if x is a unit in a Euclidean domain with gauge g then $g(x) = g(v)$.
3. Show that if a Euclidean domain is finite then the gauge g must be a constant function.
4. If a and b are nonzero members of a Euclidean domain and b is a proper divisor of a then $g(b) < g(a)$.
5. Give a direct proof by induction that every nonzero, nonunit element of a Euclidean domain is a product of irreducible elements. [*Hint*: the induction should be on the value of $g(x)$.]

6. Show that the existence of a unity follows from the others statements in the definition of a Euclidean domain.

7. Find two distinct irreducible factorizations of 9 in $\mathbb{Z} + \mathbb{Z}\sqrt{-2}$.

8. Make a list of all polynomials of $\mathbb{Z}_2[X]$ of degree $\leqslant 3$. Mark each as unit, zero, irreducible, or reducible element. Give a factorization into irreducible elements of each reducible element.

9. Do reducible elements in integral domains generate nonprime, nonmaximal ideals?

10. Prove every field is a principal ideal domain.

11. Show that $\mathbb{Z}[X]$ is not a principal ideal domain.

12. Let $\langle R, +, \cdot, \theta \rangle$ be a given ring. Let $R[X, Y]$ be the ring of all polynomials in two indeterminants X and Y. $\mathbb{Z}[X, Y]$ contains for example, $X + Y$ and $X^2 + 2XY + Y^3$. $R[X, Y]$ is simply $R[X][Y]$. Show that even if R is a field $R[X, Y]$ is not a principal ideal domain.

13. Show that $\langle X \rangle$, $\langle X, Y \rangle$, $\langle X, Y, 2 \rangle$ are all prime ideals in $\mathbb{Z}[X, Y]$ but only $\langle X, Y, 2 \rangle$ is maximal.

14. Show by example that there exist principal ideal domains with subdomains which are not principal ideal domains.

15. If every ideal of an integral domain is finitely generated then every nonzero, nonunit element is the product of irreducible elements. Show this to be true.

5.5 Factorization in principal ideal domains

In this section we establish the existence and uniqueness of an irreducible factorization for elements of a principal ideal domain which are nonzero and not invertible. The result applies also to Euclidean domains and to the integers in particular, since they are both principal ideal domains. The result for the integers is called the fundamental theorem of arithmetic.

We begin with a lemma showing the nonexistence of an infinite sequence of properly increasing ideals.

Lemma. *Let $\langle R, +, \cdot, \theta, v \rangle$ be a principal ideal domain. If $A_1, A_2, A_3, \ldots$ is an increasing sequence of ideals of R, each ideal included in the next, then there exists a $k \in \mathbb{N}^+$ such that $A_k = A_{k+1} = \cdots$.*

PROOF. $\bigcup\{A_i | i \in \mathbb{N}^+\}$ is an ideal. It is $\langle a \rangle$ for some $a \in R$. $a \in A_k$ for some $k \in \mathbb{N}^+$. Hence, $\bigcup\{A_i | i \in \mathbb{N}^+\} \subseteq A_k$. But $A_k \subseteq \bigcup\{A_i | i \in \mathbb{N}^+\}$. Therefore

$$A_k = \bigcup\{A_i | i \in \mathbb{N}^+\}.$$

$$A_k \subseteq A_{k+1} \subseteq A_{k+2} \subseteq \cdots \subseteq \bigcup\{A_i | i \in \mathbb{N}^+\}.$$

$$A_k = A_{k+1} = \cdots. \qquad \square$$

This lemma will be used to prove the next theorem.

Before reading the next theorem we wish the reader to observe the following diagrammatic examples of step by step factorization of some integers.

The factorization proceeds from one step to the next by writing each reducible integer as the product of a pair of nonunits of $\mathbb{Z}$. The procedure is not unique.

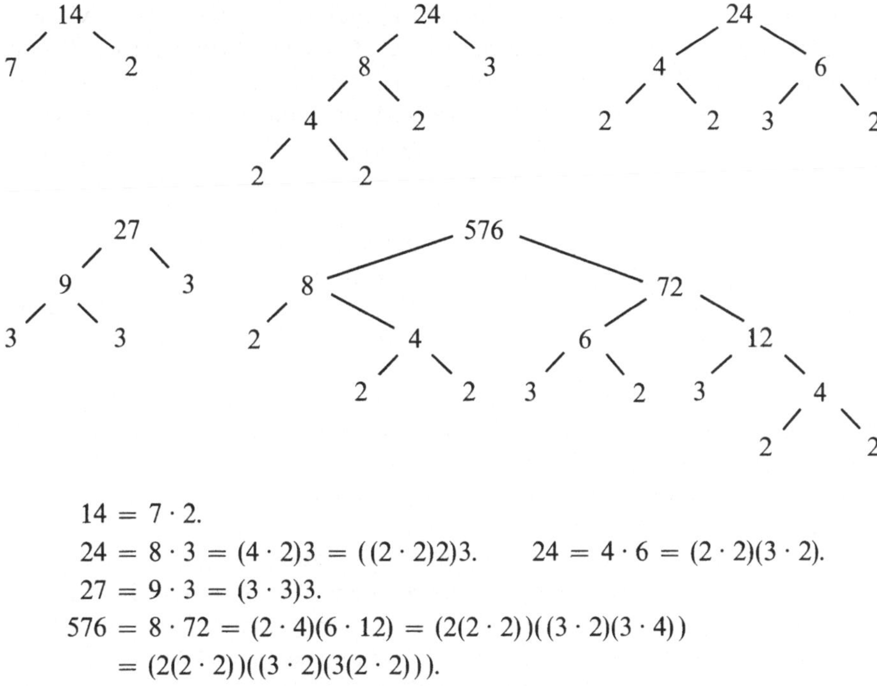

$$\begin{aligned}
14 &= 7 \cdot 2. \\
24 &= 8 \cdot 3 = (4 \cdot 2)3 = ((2 \cdot 2)2)3. \qquad 24 = 4 \cdot 6 = (2 \cdot 2)(3 \cdot 2). \\
27 &= 9 \cdot 3 = (3 \cdot 3)3. \\
576 &= 8 \cdot 72 = (2 \cdot 4)(6 \cdot 12) = (2(2 \cdot 2))((3 \cdot 2)(3 \cdot 4)) \\
&= (2(2 \cdot 2))((3 \cdot 2)(3(2 \cdot 2))).
\end{aligned}$$

We now prove the existence of an irreducible factorization.

Theorem. *Let $\langle R, +, \cdot, \theta, v\rangle$ be a principal ideal domain. If x is a nonzero, nonunit member of R then x is a product of irreducible elements of R.*

PROOF. If x is irreducible the theorem is proven. If x belongs to R and is reducible then x can be written as a product of a pair of nonunit members of R. Each reducible component of the pair producing the product can then be written as a product of a pair of nonunit members of R. If any factor is irreducible then the factoring process ceases for that factor. After k steps each factor which is reducible can be written as the product of a pair of nonunit elements of R. There are possibly as many as 2^k factors which have product x after k steps. There may be fewer than 2^k factors if any irreducible elements are produced before the kth step. If, after some finite number of steps, all factors are irreducible then x is the product of irreducible elements. If not, then it is always possible to find still more nonunit factors after any finite number of steps. In this case there would exist a sequence of nonunits of R, $x, x_1, x_2, x_3, \ldots$, such that each term of the sequence is a proper factor of the preceding element. The principal ideals generated by the sequence terms are then properly included as follows: $\langle x\rangle \subset \langle x_1\rangle \subset \langle x_2\rangle \subset \langle x_3\rangle \subset \cdots$. This result contradicts the lemma. We must then conclude that no such nonterminating factorization sequence exists. After some finite number of steps x must be the product of irreducible elements. □

We now argue the uniqueness of the irreducible factorization. It will not be claimed that all factorizations are exactly the same. For example, 24 can be written several ways as the product of irreducibles: $24 = 3 \cdot 2 \cdot 2 \cdot 2 = 2 \cdot 2 \cdot 3 \cdot 2 = (-2)(-2) \cdot 3 \cdot 2$. What we wish to claim is that any irreducible factorization is merely a rearrangement of unit multiples of any other factorization.

Theorem. *Let $\langle R, +, \cdot, \theta, v\rangle$ be a principal ideal domain. If $p_1p_2 \cdots p_m$ and $q_1q_2 \cdots q_n$ are any two irreducible factorizations of the same element of R then $m = n$ and there exist units $e_1, e_2, \ldots, e_n$ and a permutation σ of $1, 2, \ldots, n$ such that $q_1 = e_1p_{\sigma(1)}, q_2 = e_2p_{\sigma(2)}, \ldots, q_n = e_np_{\sigma(n)}$.*

Proof. We prove the theorem by induction. Suppose the result is true for all $n < k$ where k is the number of irreducible factors $q_1, q_2, \ldots, q_k$. We note that in a principal ideal domain primes and irreducible elements are the same. $p_1p_2 \cdots p_m = q_1q_2 \cdots q_k$. q_1 is a factor of $p_1p_2 \cdots p_m$ and therefore is a factor of one of the p_i's. $q_1 = e_1p_{\sigma(1)}$ for some unit e_1 and some number $\sigma(1)$ between 1 and m. By cancellation $p_1p_2 \cdots \widehat{p_{\sigma(1)}} \cdots p_m = e_1q_2q_3 \cdots q_k$ where the circumflex indicates the deleted prime. We note that the product e_1q_2 of a unit and a prime is itself a prime. We can therefore apply the induction hypothesis and conclude $m - 1 = k - 1$ and there exist units $e_2', e_3, \ldots, e_k$ and a permutation of the numbers $1, 2, \ldots, \widehat{\sigma(1)}, \ldots, k$ which we denote by $\sigma(2), \ldots, \sigma(k)$ such that $(e_1q_2) = e_2'p_{\sigma(2)}, q_3 = e_3p_{\sigma(3)}, \ldots, q_k = e_kp_{\sigma(k)}$. We need only to let $e_2 = e_1^{-1}e_2'$ to complete the proof. □

For future convenience we introduce a name for a concept found in the previous theorem.

Definition. Two elements of an integral domain $\langle R, +, \cdot, \theta, v\rangle$ which are unit multiples of each other are called *associates*.

For example, in $\mathbb{Z}$ the elements -2 and 2 are associates. In $\mathbb{Q}[X]$, $2X + 9$ and $X + \frac{9}{2}$ are associates.

Questions

1. $X^2 + 1$
   (A) is an irreducible polynomial in $\mathbb{R}[X]$
   (B) is an irreducible polynomial in $\mathbb{Q}[X]$
   (C) factors uniquely in $\mathbb{C}[X]$ (up to unit multiples)
   (D) is irreducible in $\mathbb{Z}_{10}[X]$.
   (E) None of the above completes a true sentence.

2. Which of the following statements are true?
   (A) In a principal ideal domain any two irreducible factorizations of an element must have the same factors although possibly permuted in order.
   (B) $\mathbb{Z} + \mathbb{Z}i$ cannot be a principal ideal domain because $7 - i$ has two distinct factorizations $(2 - i)(3 + i)$ and $(1 - 3i)(1 + 2i)$.
   (C) Every principal ideal domain is a Euclidean domain.
   (D) In a principal ideal domain all ideals are generated by a single element and are therefore maximal ideals.
   (E) None of the alternatives is true.

3. The ideal $\langle 21, 56, 147\rangle$ in $\mathbb{Z}$
   (A) contains the ideal $\langle 14\rangle$
   (B) is contained in the ideal $\langle 28\rangle$
   (C) is equal to the ideal $\langle 49, 14\rangle$
   (D) is an improper ideal.
   (E) None of the alternatives completes a true sentence.

4. Which of the following sentences are true?
   (A) The existence of the infinite sequence of ideals $\langle 2\rangle \subset \langle 4\rangle \subset \langle 8\rangle \subset \cdots$ shows that $\mathbb{Z}$ is not a principal ideal domain.
   (B) $\mathbb{Z}$ is not a prime ideal of $\mathbb{Z}$.
   (C) The ideal $\langle 2, 4, 8, \ldots\rangle$ is not principal.
   (D) In the sequence $\langle 24\rangle \subset \langle 8\rangle \subset \langle 2\rangle \subset \mathbb{Z}$ more ideals of $\mathbb{Z}$ than the two given ones cannot be properly inserted between $\langle 24\rangle$ and $\mathbb{Z}$.
   (E) None of the statements is true.

EXERCISES

1. Prove that $\mathbb{Z}^{2\times 2}$, the noncommutative unitary ring of two by two matrices with integral entries, is a principal ideal ring. Prove that every ideal of $\mathbb{Z}^{2\times 2}$ consists of left multiples of its generator or of right multiples of its generator (cf. Exercise 7 of Section 2.2).

2. Since $\mathbb{Z} + \mathbb{Z}\sqrt{-5}$ is not a unique factorization domain it cannot be a principal ideal domain. Give an example of an ideal which fails to be simply generated in $\mathbb{Z} + \mathbb{Z}\sqrt{-5}$.

3. Prove that every element x in R, a principal ideal domain, can be expressed in the form $x = p_1^{\alpha_1}p_2^{\alpha_2}\cdots p_s^{\alpha_s}$ in which $p_1, p_2, \ldots, p_s$ are nonassociated primes and $\alpha_1, \alpha_2, \ldots, \alpha_s$ are positive integers. Further, if $x = q_1^{\beta_1}q_2^{\beta_2}\cdots q_t^{\beta_t}$ then $t = s$ and there exists a permutation σ of $\{1, 2, \ldots, s\}$ such that $q_1 = u_1p_{\sigma(1)}, q_2 = u_2p_{\sigma(2)}, \ldots, q_s = u_sp_{\sigma(s)}$ and $\beta_1 = \alpha_{\sigma(1)}, \beta_2 = \alpha_{\sigma(2)}, \ldots, \beta_s = \alpha_{\sigma(s)}$ for some units $u_1, u_2, \ldots, u_s$.

5.6 Greatest common divisor

In this section we define a greatest common divisor of two elements, find a representation of a greatest common divisor as a linear combination of the two elements, and establish an algorithm for finding the greatest common divisor. We then study partial fractions.

We begin this section with a definition of greatest common divisor.

Definition. Let $\langle R, +, \cdot, \theta, v\rangle$ be an integral domain. Given $a, b \in R$ we define d to be a *greatest common divisor* of a and b if and only if

1. d is a divisor of a and b, and
2. If e is any divisor of a and b then e is a divisor of d.

Also, a and b are said to be *relatively prime* if and only if v is a greatest common divisor of a and b.

We will abbreviate greatest common divisor of a and b with $\gcd(a, b)$. We now have a theorem on the linear representation of the greatest common divisor.

Theorem. *Let $\langle R, +, \cdot, \theta, v\rangle$ be a principal ideal domain and let d be a divisor of a and b. Then d is a greatest common divisor of a and b if and only if there exist $x, y \in R$ such that $xa + yb = d$.*

Proof. If there exist an x and y such that $xa + yb = d$ and e is a divisor of both a and b then e is a divisor of d. d is therefore greater than every divisor of a and b. To prove the converse assume d is a greatest common divisor of a and b. Let $A = \{\alpha a + \beta b | \alpha, \beta \in R\}$. This set is easily seen to be an ideal of R. By hypothesis every ideal of R is principal. Let A be generated by $c \in R$; $A = \langle c\rangle$. a and b both belong to the ideal A because $a = va + \theta b$ and $b = \theta a + vb$. As members of the ideal, $\langle c\rangle$, both a and b are multiples of c. As a common divisor of a and b, the element c must be a divisor of d, the given greatest common divisor. Because $c \in A$ there exist $x', y' \in R$ such that $c = x'a + y'b$. d is a divisor of c. c and d are associates and $\langle d\rangle = \langle c\rangle = A$. There exist $x, y \in R$ such that $xa + yb = d$. □

The next theorem, which presents a procedure for finding greatest common divisors, is called the Euclidean algorithm.

Theorem. *Let $\langle R, +, \cdot, \theta, v\rangle$ be a Euclidean domain with gauge g. Let a and b be two nonzero members of R. Then either there exists $q \in R$ such that $b = qa$ and $\gcd(a, b) = a$ or there exist two (finite) sequences $q_0, q_1, \ldots, q_{k+1}$ of R and $r_0, r_1, \ldots, r_k$, nonzero elements of R, $0 \leqslant k \leqslant g(a)$ such that*

$$\begin{aligned}
b &= q_0 a + r_0, & 0 \leqslant g(r_0) < g(a);\\
a &= q_1 r_0 + r_1, & 0 \leqslant g(r_1) < g(r_0);\\
r_0 &= q_2 r_1 + r_2, & 0 \leqslant g(r_2) < g(r_1);\\
r_1 &= q_3 r_2 + r_3, & 0 \leqslant g(r_3) < g(r_2);\\
&\cdots\\
r_{k-2} &= q_k r_{k-1} + r_k, & 0 \leqslant g(r_k) < g(r_{k-1});\\
r_{k-1} &= q_{k+1} r_k + \theta;
\end{aligned}$$

and $\gcd(a, b) = r_k$.

PROOF. Since each division lowers the gauge value of the remainder by at least one there can be only a finite number of repeated divisions as described; the process must terminate after a finite number of steps not exceeding $g(a) + 1$. The process actually terminates when a remainder of θ is obtained. As r_k is a divisor of r_{k-1} (from the last equation) and also is a divisor of r_{k-2} (from the next to last equation) one can prove inductively that r_k is a divisor of both a and b. On the other hand, any divisor of both a and b is a divisor of r_0 (using the first equation) and a divisor of r_1 (using the second equation). Continuing we conclude that any divisor of both a and b is a divisor or r_k. r_k is a gcd(a, b). □

The theorem demonstrating the existence of x, y in R such that $xa + yb =$ gcd(a, b) can be constructively affirmed by the Euclidean algorithm. By using the procedure of the theorem, solving the first equation for r_0, substituting in the second, solving the second for r_1, substituting in the third, etc., we can obtain r_k, gcd(a, b), as a linear combination of a and b.

EXAMPLE. In $\mathbb{Z}$, the integers, a greatest common divisor of 576 and 243 is 9.

$$\begin{aligned}
576 &= (2)(243) + 90.\\
243 &= (2)(90) + 63.\\
90 &= (1)(63) + 27.\\
63 &= (2)(27) + 9.\\
27 &= (3)(9) + 0.
\end{aligned}$$

To represent the greatest common divisor 9 as a linear combination of the two integers 576 and 243 we begin with the first equation and substitute in the second, and so forth.

$$\begin{aligned}
90 &= 576 - (2)(243).\\
63 &= 243 - (2)(576 - (2)(243))\\
&= (5)(243) - (2)(576).\\
27 &= 90 - (1)(63) = (576 - (2)(243)) - ((5)(243) - (2)(576))\\
&= (-7)(243) + (3)(576).\\
9 &= 63 - (2)(27)\\
&= (5)(243) - (2)(576) - (2)((3)(576) + (-7)(243))\\
&= (19)(243) + (-8)(576).
\end{aligned}$$

EXAMPLE. Find a greatest common divisor of the polynomials $X^3 + \bar{2}X^2 + \bar{3}X + \bar{2}$ and $X^2 + \bar{4}$ in the Euclidean domain $\mathbb{Z}_5[X]$.

$$\begin{aligned}
X^3 + \bar{2}X^2 + \bar{3}X + \bar{2} &= (X + \bar{2})(X^2 + \bar{4}) + (\bar{4}X + \bar{4}).\\
X^2 + \bar{4} &= (\bar{4}X + \bar{1})(\bar{4}X + \bar{4}) + \bar{0}.
\end{aligned}$$

A greatest common divisor is $\bar{4}X + \bar{4}$. One can notice that $\bar{2}$ is a unit in $\mathbb{Z}_5[X]$ and therefore $\bar{2}(\bar{4}X + \bar{4}) = \bar{3}X + \bar{3}$ is another greatest common divisor of the two given polynomials.

In calculus courses a technique for finding antiderivatives of rational functions involves finding partial fractions. The method generally used for resolving a fraction into its component additive parts involves comparing undetermined coefficients and solving linear equations. The Euclidean algorithm is perhaps a simpler method.

Given polynomials $a(X)$, $b(X)$ in $\mathbb{R}[X]$ which are relatively prime there exist polynomials $x(X)$, $y(X)$ such that $x(X)b(X) + y(X)a(X) = 1$. In fractional form this becomes $x(X)/a(X) + y(X)/b(X) = 1/(a(X)b(X))$. The polynomials $x(X)$ and $y(X)$ can be found using the Euclidean algorithm.

EXAMPLES. Resolve $(X^2 + 2X + 3)/(X^2 + 1)(X - 1)$ into partial fractions. By the division algorithm $X^2 + 1 = (X + 1)(X - 1) + 2$. Thus

$$-(X + 1)(X - 1) + 1(X^2 + 1) = 2$$

or

$$\frac{-\frac{1}{2}(X + 1)}{X^2 + 1} + \frac{\frac{1}{2}}{X - 1} = \frac{1}{(X^2 + 1)(X - 1)}.$$

$$\frac{-\frac{1}{2}(X + 1)(X^2 + 2X + 3)}{X^2 + 1} + \frac{\frac{1}{2}(X^2 + 2X + 3)}{X - 1} = \frac{X^2 + 2X + 3}{(X^2 + 1)(X - 1)}.$$

$$-\tfrac{1}{2}\left\{(X + 3) + \frac{4X}{X^2 + 1}\right\} + \tfrac{1}{2}\left\{(X + 3) + \frac{6}{(X - 1)}\right\} = \frac{2X}{X^2 + 1} + \frac{3}{X - 1}.$$

And as we learn in calculus, an antiderivative of the sum is $\ln|X^2 + 1| + 3\ln|X - 1|$.

Resolve into partial fractions the rational fraction $X^2/(X - 2)^2(X + 3)$. $X^2 - 4X + 4 = (X - 7)(X + 3) + 25$ using the division algorithm.

$$\frac{-(X - 7)X^2/25}{X^2 - 4X + 4} + \frac{X^2/25}{X + 3} = \frac{X^2}{(X - 2)^2(X + 3)}.$$

$$\tfrac{1}{25}\left\{(-X + 3) + \frac{16X - 12}{(X - 2)^2}\right\} + \tfrac{1}{25}\left\{(X - 3) + \frac{-9}{X + 3}\right\}$$

$$= \tfrac{1}{25}\left\{\frac{1}{X - 2}\left[16 + \frac{20}{X - 2}\right]\right\} + \tfrac{1}{25}\frac{-9}{X + 3}$$

$$= \frac{\frac{16}{25}}{X - 2} + \frac{\frac{4}{5}}{(X - 2)^2} + \frac{-\frac{9}{25}}{X + 3}$$

This function has antiderivative

$$\tfrac{16}{25}\ln|X - 2| - \tfrac{4}{5}(X - 2)^{-1} - \tfrac{9}{25}\ln|X + 3|.$$

QUESTIONS

1. Which of the following are greatest common divisors of $X^2 - 3X + 2$ and $X^2 - X - 2$ in $\mathbb{Q}[X]$?
   (A) X^2.
   (B) $X + 1$.

(C) $10 - 2X$.
(D) $(X - 2)/3$.
(E) None of the four is a greatest common divisor.

2. The ideal D generated by the greatest common divisor d of a and b in a principal ideal domain R is related to the ideal A generated by a and the ideal B generated by b by which of the following?
(A) $A \cap B = D$.
(B) $B \subseteq D$.
(C) $A \subseteq D$.
(D) $\langle A \cup B \rangle = D$.
(E) None of the alternatives is correct.

3. A least common multiple of a and b is m if and only if (a is a divisor of m and b is a divisor of m) and (a is a divisor of n, b is a divisor of n imply m is a divisor of n). Denoting the ideals generated by m, a, and b by M, A, and B, respectively, which of the following are correct?
(A) $A \cap B = M$.
(B) $B \subseteq M$.
(C) $A \subseteq M$.
(D) $\langle A \cup B \rangle = M$.
(E) None of the alternatives is correct.

4. The equation $2x + 4y = 1$
(A) has an infinite number of solutions in integers
(B) has unique integral solution
(C) cannot be solved because there are two unknowns and only one equation
(D) has no solution in integers because 1 is not a multiple of gcd(2, 4).
(E) None of the alternatives is correct.

Exercises

1. Find a greatest common divisor of 54 and 102 using the Euclidean algorithm in $\mathbb{Z}$.

2. Find a generator of the ideal $\langle 576, 48, 27, 54, 60 \rangle$ in $\mathbb{Z}$.

3. Find a greatest common divisor of 1617 and 1683 using the Euclidean algorithm in $\mathbb{Z}$. Find the greatest common divisor as a linear combination of 1617 and 1683; that is, find x, y in $\mathbb{Z}$ such that $x(1617) + y(1683) = \text{gcd}$.

4. Resolve $X/(X - 2)(X - 3)$ into partial fractions in $\mathbb{Q}[X]$.

5. Resolve each of the following into partial fractions in $\mathbb{Q}[X]$:
(a) $(X - 1)/X(X + 3)$ (b) $X/(X^2 + X + 1)(X - 3)^2$
(c) $1/(X^3 + 1)$ (d) $(X^3 + 1)/(X - 1)(X - 2)(X^2 + X + 2)$.

6. The equation $mx + ny = p$ has solution in integers if and only if p is a multiple of a greatest common divisor of m and n. Euler used an interesting method for solving this equation. We demonstrate Euler's method through an example: $216x + 270y = 108$.

$$x = \frac{108 - 270y}{216} = \frac{108 - 54y}{216} - y.$$

In order that there be a solution for x and y in integers one must have

$$\frac{108 - 54y}{216} = w \quad \text{for some integer } w.$$

Solving for y:

$$y = \frac{108 - 216w}{54} = 2 - 4w$$

Substituting back for x:

$$x = \frac{108 - 270y}{216} = \frac{108 - 270(2 - 4w)}{216} = -2 + 5w.$$

The solution in parametric form is

$$x = -2 + 5w \qquad y = 2 - 4w.$$

Solve this equation using Euler's method: $222x + 246y = 24$. [*Hint*: The process must be continued one step further than in the example.]

5.7 Unique factorization domains

We have previously established the existence and uniqueness of factorizations in Euclidean and principal ideal domains. We now look at some results about factorization with weaker hypotheses.

Definition. $\langle R, +, \cdot, \theta, v\rangle$, an integral domain, is a *unique factorization domain* or *Gaussian domain* if and only if every nonzero, nonunit element of R is the product of irreducible elements of R and given any two irreducible factorizations of a nonzero, nonunit x of R, $x = p_1p_2 \cdots p_m$ and $x = q_1q_2 \cdots q_n$, it follows that $n = m$, there exist a permutation σ of $\{1, 2, \ldots, n\}$ and units $u_1, u_2, \ldots, u_m$ of R such that $p_1 = u_1q_{\sigma(1)}, p_2 = u_2q_{\sigma(2)}, \ldots, p_m = u_mq_{\sigma(m)}$.

This definition is, of course, the conclusion of the principal theorem of Section 5.5, the fundamental theorem of arithmetic. Our first result of this section will show that when factorization exists then uniqueness of factorization will eliminate the distinction between prime and irreducible elements in an integral domain. We remember that prime elements are always irreducible, but not conversely, in an arbitrary integral domain.

Theorem. *If $\langle R, +, \cdot, \theta, v\rangle$ is an integral domain such that every nonunit, nonzero element has an irreducible factorization then the irreducible factorization is unique if and only if every irreducible element is prime.*

PROOF. First let unique factorization be given and suppose x to be an irreducible element of R. Let x divide a product ab. $ab = xy$ for some y in R. y can be expressed as the product of irreducible elements of R yielding the factorization $xp_1p_2 \cdots p_k$ of ab. On the other hand, both a and b each have irreducible factorizations giving a factorization $q_1q_2 \cdots q_m r_1 r_2 \cdots r_n$ of ab with $q_1, q_2, \ldots, q_m, r_1, r_2, \ldots, r_n$ all irreducible elements of R. By the assumption of unique factorization x is a unit multiple of some q_i or some r_j and therefore divides either a or b.

To prove the converse assume every irreducible element to be prime and consider two irreducible factorizations of an element x. The proof is exactly that given in the last theorem of Section 5.5. □

We saw in Section 5.5 that an integral domain with ideals generated by a single element is a unique factorization domain. An integral domain with ideals generated by a finite number of elements does not necessarily have unique factorization. It does, however, have irreducible factorizations.

We now proceed towards the basic theorem that the polynomial ring of a unique factorization domain is itself a unique factorization domain. This result will prove $\mathbb{Z}[X]$ to be a unique factorization domain. We recall that $\mathbb{Z}[X]$ is not a principal ideal domain.

Lemma. *Let $\langle R, +, \cdot, \theta, v\rangle$ be a unique factorization domain. An irreducible element p of R divides the product of polynomials $f(X)g(X)$ if and only if p divides $f(X)$ or p divides $g(X)$.*

PROOF. We prove p divides all the coefficients of $f(X)g(X)$ if and only if p divides all the coefficients of $f(X)$ or p divides all the coefficients of $g(X)$. If each coefficient a_i of the polynomial $f(X) = a_0 + a_1X + \cdots + a_mX^m$ is a multiple of p then clearly each coefficient $c_k = \sum_{i+j=k} a_ib_j$ of the polynomial $f(X)g(X) = \sum_{k=1}^{m+n} c_kX^k$ is also a multiple of p where $g(X)$ is denoted by $b_0 + b_1X + \cdots + b_nX^n$. By the same reasoning if each coefficient of $g(X)$ is a multiple of p then each coefficient of $f(X)g(X)$ is also.

The converse requires somewhat more care. Suppose the irreducible element p does not divide $f(X)$ and does not divide $g(X)$. Let a_r be the first (smallest power of X) coefficient of $f(X)$ which p does not divide and let b_s be the first coefficient of $g(X)$ which p does not divide. The coefficient of the power $r + s$ of X in $f(X)g(X)$ is $a_0b_{r+s} + a_1b_{r+s-1} + \cdots + a_rb_s + \cdots + a_{r+s}b_0$ with the understanding some of the coefficients may be zero. Or in summation notation the coefficient of X^{r+s} in $f(X)g(X)$ is $\sum_{i+j=r+s} a_ib_j$ with $0 \leqslant i \leqslant m$, $0 \leqslant j \leqslant n$. In this sum a_rb_s fails to be divisible by p. Any terms $a_0b_{r+s}, a_1b_{r+s-1}, \ldots, a_{r-1}b_{s+1}$ will be divisible by p because $a_0, a_1, \ldots, a_{r-1}$ are all divisible by p. Likewise, any of the terms $a_{r+1}b_{s-1}, a_{r+2}b_{s-2}, \ldots, a_{r+s}b_0$ will all be divisible by p because $b_0, b_1, \ldots, b_{s-1}$ are all divisible by p. The sum $\sum_{i+j=r+s} a_ib_j$ fails to be a multiple of p. p does not divide the coefficient of X^{r+s} in $f(X)g(X)$. □

Definition. Let $\langle R, +, \cdot, \theta, v\rangle$ be a unique factorization domain. A polynomial $f(X)$ in $R[X]$ is called *primitive* if and only if the greatest common divisors of the coefficients of $f(X)$ are units. Denote any greatest common divisor of the coefficients of $f(X)$ by $c(f(X))$.

Lemma. *Let $\langle R, +, \cdot, \theta, v\rangle$ be a unique factorization domain. $f(X)g(X)$ is primitive if and only if $f(X)$ and $g(X)$ are primitive. $c(f(X)g(X)) = uc(f(X))c(g(X))$ for some unit $u \in R$.*

PROOF. The proofs here are consequences of the previous lemma. □

We now relate factorization in $R[X]$ to factorization in the polynomial ring of the field of fractions of R.

Lemma. *Let $\langle R, +, \cdot, \theta, v\rangle$ be a unique factorization domain and R' its field of fractions. Then a nonconstant polynomial $f(X)$ in $R[X]$ reducible in $R'[X]$ is also reducible in $R[X]$.*

PROOF. Every polynomial $g(X)$ in $R'[X]$ can be expressed uniquely as $rg_1(X)$ where $r \in R'$ and $g_1(X)$ is primitive in $R[X]$. We express $f(X)$ as $af_1(X)$ where a is a greatest common divisor of the coefficients of $f(X)$ in R and $f_1(X)$ is a primitive polynomial in $R[X]$. $f(X)$ in $R[X]$ and of degree one or higher is reducible in $R'[X]$ if and only if $f_1(X)$ is also. Now suppose $f_1(X) = g(X)h(X)$, each polynomial $g(X)$ and $h(X)$ in $R'[X]$. $f_1(X) = sg_1(X)th_1(X)$ with $g_1(X)$, $h_1(X)$ primitive in $R[X]$, $s, t \in R'$. Let $s = s_1/s_2$, $t = t_1/t_2$, $\gcd(s_1, s_2) = \gcd(t_1, t_2) = v$, $s_1, s_2, t_1, t_2 \in R$. We have $s_2t_2f_1(X) = s_1t_1g_1(X)h_1(X)$. Then s_2t_2 divides s_1t_1; s_1t_1 divides s_2t_2. $s_1t_1 = us_2t_2$. $f_1(X) = ug_1(X)h_1(X)$. $f_1(X)$ is reducible. Likewise, $f(X)$ is reducible. □

Theorem. *If $\langle R, +, \cdot, \theta, v\rangle$ is a unique factorization domain then $\langle R[X], +, \cdot, \theta, v\rangle$ is also.*

PROOF. A polynomial of degree zero is reducible in $R[X]$ if and only if it is reducible in R. Its factorization into irreducibles in R and in $R[X]$ are identical.

Denote again the field of fractions of R by R'. A polynomial $f(X)$ of degree one or greater can be factored uniquely in $R'[X]$ into irreducibles because $R'[X]$ is a unique factorization domain. But an irreducible factorization of an element of $R[X]$ in $R'[X]$ implies a factorization in $R[X]$. Each factor in $R[X]$ can be factored into a primitive polynomial and an element of R. Thus every polynomial in $R[X]$ is the product of irreducibles in R and primitive polynomials which are also irreducible in $R[X]$. Any other irreducible factorization in $R[X]$ must be an irreducible factorization in $R'[X]$. Each factor is a rational multiple of some factor in the original factorization. But the only rational multiple of a primitive polynomial in $R[X]$ is a unit multiple. The zero degree irreducibles must be unit multiples because irreducible factorizations in R are unique. □

QUESTIONS

1. In the ring of Gaussian integers, $\mathbb{Z} + \mathbb{Z}i$, the equation $5 = (1 + 2i)(1 - 2i)$
   (A) implies $\mathbb{Z} + \mathbb{Z}i$ is not a unique factorization domain
   (B) implies 5 is reducible

(C) implies 5 is a unit
(D) implies 5 has a square root in $\mathbb{Z} + \mathbb{Z}i$
(E) None of the four alternatives completes a true sentence.

2. Let S be an integral subdomain of the integral domain R. If an element x is irreducible in S then
(A) x is irreducible in R
(B) x is a nonzero element of R
(C) x is not a unit of R
(D) x is not reducible in R.
(E) None of the four sentence completions is correct.

3. Let S be an integral subdomain of the integral domain R. A unit element of S is
(A) a unit element of R
(B) never irreducible in R
(C) possible reducible in R
(D) never zero in R.
(E) None of the four alternatives is satisfactory.

4. $\mathbb{Z}$ is a subdomain of $\mathbb{Z}[\frac{1}{2}]$, also an integral domain. Which of these statements are true?
(A) 6 is irreducible in $\mathbb{Z}[\frac{1}{2}]$.
(B) 3 is irreducible in $\mathbb{Z}[\frac{1}{2}]$.
(C) $6 \cdot 1 = 3 \cdot 2$ implies irreducible factorizations are not necessarily unique in $\mathbb{Z}[\frac{1}{2}]$.
(D) $3 = [\frac{1}{2}] \cdot 6$ indicates 3 is reducible in $\mathbb{Z}[\frac{1}{2}]$.
(E) None of the four statements is true.

Exercises

1. Let R be a commutative, unitary ring in which every ideal is finitely generated. Prove that it is not possible to have a strictly increasing sequence (infinite) of ideals from R. [*Hint*: Examine the proof for principal ideal rings given in Section 5.5.]

2. Let R be an integral domain in which every ideal is finitely generated. Prove that every nonzero, nonunit is the product of irreducible elements.

3. The following result is known as Eisenstein's irreducibility criterion. Let $f(X) = a_nX^n + \cdots + a_1X + a_0$ be a polynomial such that a_n is not a multiple of some prime p in $\mathbb{Z}$; $a_{n-1}, a_{n-1}, \ldots, a_1, a_0$ are all in $\mathbb{Z}$ and are multiples of p, and a_0 is not a multiple of p^2. Then prove $f(X)$ is irreducible in $\mathbb{Q}[X]$ (and $\mathbb{Z}[X]$). [*Hint*: Suppose $f(X)$ reducible and use a coefficient argument to get a contradiction.]

4. Prove $X^{p-1} + X^{p-2} + \cdots + X + 1$ irreducible in $\mathbb{Q}[X]$. [*Hint*: $X^{p-1} + \cdots + X + 1 = (X^p - 1)/(X - 1) = [(Y + 1)^{p-1} - 1]/[(Y + 1) - 1] = Y^{p-1} + pY^{p-2} + \cdots + \binom{p}{j} Y + \cdots + p$.

5. $X^2 + 1$ is irreducible in $\mathbb{Z}_3[X]$. Prove.

6. Let P be the set of all infinite series with coefficients in $\mathbb{Q}$. $P = \{a_0 + a_1X + a_2X^2 + \cdots | a_0, a_1, a_2, \ldots \in \mathbb{Q}\}$. It is not assumed that the series converge in any sense. Prove that P is an integral domain with sums and products defined in the usual way (as with polynomials). Show that $a_0 + a_1X + \cdots$ is invertible if and only if $a_0 \neq 0$. Show that every element of P can be written as $u(X)X^n$ for some $n \in \mathbb{N}$ where $u(X)$ is a unit in P. Show that X is the only irreducible element of P (up to unit multiples). Show that P is a unique factorization domain. Show that the only ideals of P are P, $\langle X \rangle$, $\langle X^2 \rangle$, $\langle X^3 \rangle, \ldots, \langle 0 \rangle$ and therefore P is a principal ideal domain. Show that P is a Euclidean domain (in a rather trivial way). Show that $\langle X \rangle$ is a maximal ideal, the only maximal ideal of P, and that $P/\langle X \rangle$ is isomorphic to $\mathbb{Q}$.

7. Which of the following polynomials are irreducible in $\mathbb{Q}[X]$?
(a) $2X^3 + 3X^2 + 27X + 3$
(b) $X^9 + 121X + 55$
(c) $X^9 + 10!X^8 + 8!X^6 + 6!X^4 + 4!X^2 + 2!$
(d) $2X^3 + 4X^2 + 20X + 4$
(e) $(X + 3)^5 - 3^5$.

5.8 Field extensions and complex numbers

We now use our knowledge of polynomials to analyze field extensions and in particular to construct again the complex numbers from the real numbers.

We begin by classifying elements outside a given field K into two types according to whether or not they satisfy a polynomial equation with coefficients in K.

Definition. Let K be a field included in a commutative ring L (K is a subring of L and K is a field). An element α in L is called *algebraic* over K if and only if α is the root of some nonzero polynomial in $K[X]$. If α is not algebraic over K then α is called *transcendental* over K.

EXAMPLES. $\sqrt{2}$ is a root of $X^2 - 2$ in $\mathbb{Q}[X]$ and is therefore algebraic over $\mathbb{Q}$. The base for the natural logarithms, e, is not a root of any polynomial except zero and is therefore transcendental over $\mathbb{Q}$. (This fact is nontrivial; one must consult works in analysis for the proof.) The square root of negative one, i, is a root of $X^2 + 1$, a polynomial in $\mathbb{Q}[X]$, and is therefore algebraic over $\mathbb{Q}$.

It may be intuitively useful to think of algebraic elements as closer to the original field K than transcendental elements. Whereas an algebraic element is not itself necessarily a member of K, some sum of multiples of powers of the algebraic element does lie in K.

Definition. If K is a field included in a commutative ring L and $\alpha \in L$ we define $K[\alpha]$ to be the smallest subring of L which includes $K \cup \{\alpha\}$. If L is more especially a field and $\alpha \in L$ we further define $K(\alpha)$ to be the smallest subfield of L including $K \cup \{\alpha\}$.

We notice that the possibly ambiguous notation $K[\alpha]$ could be construed to mean the set of all polynomials in α with coefficients in K. This interpretation is consistent with the new definition, because any ring containing K and α must also contain any polynomial in α. Furthermore, the set of all polynomials in α clearly is a subring of L.

We now wish to discuss the structure of the ring $K[\alpha]$. The discussion falls naturally into the two cases: algebraic and transcendental. The substitution morphism introduced in Section 5.3 is now utilized to analyze the structure of $K[\alpha]$. The morphism $\Phi_\alpha: K[X] \to L$ replaces each polynomial $a_nX^n + a_{n-1}X^{n-1} + \cdots + a_1X + a_0$ by the value $a_n\alpha^n + a_{n-1}\alpha^{n-1} + \cdots + a_1\alpha + a_0$ in L obtained by substituting α for X. $\Phi_\alpha(p(X)) = p(\alpha)$. For convenience we shall diminish what we regard as the codomain of Φ_α from L to coincide with the range $K[\alpha]$. The mapping $\Phi_\alpha: K[X] \to K[\alpha]$ is then a surjection, an epimorphism of rings.

Theorem. *If K is a field and $\alpha \in L$, a ring including K, then there exists a principal ideal $\langle f(X)\rangle$ of $K[X]$ such that $K[\alpha]$ is isomorphic with $K[X]/\langle f(X)\rangle$, a quotient ring of $K[X]$.*

Proof. $\Phi_\alpha: K[X] \to K[\alpha]$ such that $\Phi_\alpha(p(X)) = p(\alpha)$ is a ring morphism. The kernel of the epimorphism Φ_α is an ideal of the principal ideal domain $K[X]$. Let $f(X)$ be a generator of the principal ideal, kernel Φ_α. Kernel $\Phi_\alpha = \langle f(X)\rangle$. By the fundamental morphism theorem for rings (Section 2.7) there exists an isomorphism $\Phi'_\alpha: K[X]/\langle f(X)\rangle \to K[\alpha]$. This proves the theorem. □

We make a few remarks about the theorem and its proof. The ideal, kernel $\Phi_\alpha(=\langle f(X)\rangle)$, consists precisely of those polynomials in $K[X]$ which have α as a root. There are two cases: $\langle f(X)\rangle = \{\theta\}$ and $\langle f(X)\rangle \neq \{\theta\}$. If $\langle f(X)\rangle = \{\theta\}$ then kernel Φ_α is trivial, Φ_α is already an isomorphism, and the only polynomial with α as a root is the zero polynomial. α is transcendental. If $\langle f(X)\rangle \neq \{\theta\}$ then the kernel is not trivial, the generator $f(X)$ is nonzero, and α is an algebraic element.

Examples. For the transcendental element e over $\mathbb{Q}$ the subring $\mathbb{Q}[e]$ of $\mathbb{R}$ is simply the set of all polynomials in e with rational coefficients. This ring of polynomials in e is isomorphic to $\mathbb{Q}[X]$. If, alternatively, we consider the algebraic element $\sqrt{2}$ over $\mathbb{Q}$, a generator of the kernel $\Phi_{\sqrt{2}}$ is $X^2 - 2$. The ring $\mathbb{Q}[\sqrt{2}]$ is isomorphic to $\mathbb{Q}[X]/\langle X^2 - 2\rangle$.

For a further example we let K be the field $\mathbb{Z}_3$. We let $L = \{m + n\alpha | m, n \in \mathbb{Z}_3\}$ with operations defined by

$$(m + n\alpha) + (p + q\alpha) = (m + p) + (n + q)\alpha$$
$$(m + n\alpha)\cdot(p + q\alpha) = mp + (mq + np)\alpha.$$

K is a subfield of the ring L. Since $\alpha^2 = 0$ we notice this ring L is not an integral domain. We consider $\mathbb{Z}_3[\alpha]$ by considering $\Phi_\alpha: \mathbb{Z}_3[X] \to Z_3[\alpha]$.

Kernel Φ_α is generated by some polynomial $f(X)$ in $\mathbb{Z}_3[X]$ which cannot be of degree zero or one because $m + n\alpha = 0$ implies $m = n = 0$. X^2 is in kernel Φ_α and must therefore be a generator. $\Phi'_\alpha:\mathbb{Z}_3[X]/\langle X^2\rangle \to \mathbb{Z}_3[\alpha]$ is the isomorphism.

We now consider the more important case in which L is assumed to be a field. The previously developed machinery is adequate to the purpose.

Theorem. *If K is a field included in a field L and $\alpha \in L$ then $K(\alpha)$ is isomorphic to $K[X]/\langle f(X)\rangle$ for some irreducible $f(X)$ in $K[X]$ and α is algebraic over K or $K(\alpha)$ is isomorphic to $K(X)$ and α is transcendental over K.*

PROOF. If α is algebraic over K the isomorphism $\Phi'_\alpha: K[X]/\langle f(X)\rangle \to K[\alpha]$, derivable from the substitution morphism Φ_α, has range $K[\alpha]$, a subring of the field L. $K[\alpha]$ must therefore be an integral domain. Its isomorphic preimage $K[X]/\langle f(X)\rangle$ is also an integral domain. The ideal $\langle f(X)\rangle$ is proper and prime. In a principal ideal domain prime ideals are maximal ideals. $\langle f(X)\rangle$ is then proper and maximal. $K[X]/\langle f(X)\rangle$ is therefore a field. Its isomorphic image $K[\alpha]$ is also a field and may be written $K(\alpha)$. $\langle f(X)\rangle$'s being a maximal ideal also implies $f(X)$ is irreducible.

Alternately, if α is transcendental over K then $\Phi_\alpha: K[X] \to K[\alpha]$ is an isomorphism of two integral domains which are not fields. Each integral domain can be embedded in its field of fractions and these fields of fractions are isomorphic also. This is to say, the isomorphism Φ_α can be extended to an isomorphism $\Psi_\alpha: K(X) \to K(\alpha)$ such that $\Psi_\alpha(p(X)/q(X)) = p(\alpha)/q(\alpha)$. □

EXAMPLES. The transcendental element e over $\mathbb{Q}$ gives rise to the field extension $\mathbb{Q}(e)$ inside $\mathbb{R}$. $\mathbb{Q}(e)$ is the field of all rational functions of e with coefficients in $\mathbb{Q}$. $\mathbb{Q}(e) = \{p(e)/q(e) \mid p(X)$ and $q(X)$ are polynomials with rational coefficients and $q(X)$ is not the zero polynomial$\}$. The algebraic element $\sqrt{2}$ is a root of $X^2 - 2$, a polynomial with rational coefficients. $\mathbb{Q}(\sqrt{2})$ is a field isomorphic to $\mathbb{Q}[X]/\langle X^2 - 2\rangle$ and is the set $\{m + n\sqrt{2} \mid m, n \in \mathbb{Q}\}$ with operations $(m + n\sqrt{2}) + (p + q\sqrt{2}) = m + p + (n + q)\sqrt{2}$ and $(m + n\sqrt{2})(p + q\sqrt{2}) = mp + 2nq + (mq + np)\sqrt{2}$.

We offer one more example in detail, that of extending the real number system, $\mathbb{R}$, to the complex number system, $\mathbb{C}$. This is one more step in the chain constructing the number systems: $\mathbb{N}$, $\mathbb{Z}$, $\mathbb{Q}$, $\mathbb{R}$, and $\mathbb{C}$. This construction is an alternative to the brief one offered in Exercise 15, Section 2.5 and offers a different kind of insight. We are acquainted with the fact that $X^2 + 1 = 0$ has no solution in the field $\mathbb{R}$ of real numbers because the square of every real number is nonnegative ($\mathbb{R}$ is an ordered field). $X^2 + 1$ has no root in $\mathbb{R}$ and can therefore have no linear factor. If $X^2 + 1$ can have no first degree factors then it is irreducible for it is only of second degree. A root to $X^2 + 1$ contained in $\mathbb{C}$ and represented by i is an algebraic element over $\mathbb{R}$. The field extension $\mathbb{R}(i)$ is a subfield of the field $\mathbb{C}$. $\Phi'_i: \mathbb{R}[X]/\langle X^2 + 1\rangle \to \mathbb{R}(i)$ is the derived isomorphism from the substitution morphism Φ_i. A complete set of representatives for $\mathbb{R}[X]/\langle X^2 + 1\rangle$ will be the set of all real polynomials of degree zero and one and the zero polynomial.

$$\mathbb{R}[X]/\langle X^2 + 1\rangle = \{a_0 + a_1X + \langle X^2 + 1\rangle | a_0, a_1 \in \mathbb{R}\}.$$

Under the isomorphism Φ_i', $a_0 + a_1X + \langle X^2 + 1\rangle$ corresponds to $a_0 + a_1 i$ in $\mathbb{R}(i)$. We can determine the operations in the field $R(i)$ by noting the corresponding operations in $\mathbb{R}[X]/\langle X^2 + 1\rangle$.

$$a_0 + a_1X + \langle X^2 + 1\rangle + b_0 + b_1X + \langle X^2 + 1\rangle$$
$$= a_0 + b_0 + (a_1 + b_1)X + \langle X^2 + 1\rangle.$$
$$(a_0 + a_1X + \langle X^2 + 1\rangle)(b_0 + b_1X + \langle X^2 + 1\rangle)$$
$$= a_0b_0 + (a_0b_1 + a_1b_0)X + a_1b_1X^2 + \langle X^2 + 1\rangle$$
$$= (a_0b_0 - a_1b_1) + (a_0b_1 + a_1b_0)X + \langle X^2 + 1\rangle.$$

Thus in $\mathbb{R}(i)$ we have

$$a_0 + a_1i + b_0 + b_1i = a_0 + b_0 + (a_1 + b_1)i \quad \text{and}$$
$$(a_0 + a_1i)(b_0 + b_1i) = (a_0b_0 - a_1b_1) + (a_0b_1 + a_1b_0)i.$$

The reciprocal of any nonzero element $b_0 + b_1i$ is easily found by equating a product to 1. $(a_0b_0 - a_1b_1) + (a_0b_1 + a_1b_0)i = 1$.

$$a_0b_0 - a_1b_1 = 1 \quad \text{and} \quad a_0b_1 + a_1b_0 = 0.$$

We solve the two equations for a_0 and a_1.

$$a_0b_0b_1 = (1 + a_1b_1)b_1 \quad \text{and} \quad a_0b_0b_1 = -a_1b_0^2.$$
$$b_1 + a_1b_1^2 = -a_1b_0^2.$$
$$a_1 = (-b_1)/(b_0^2 + b_1^2). \qquad a_0 = b_0/(b_0^2 + b_1^2).$$
$$(b_0 + b_1i)^{-1} = b_0/(b_0^2 + b_1^2) + [-b_1/(b_0^2 + b_1^2)]i.$$

We delineate a few more properties of complex numbers in the exercises, but a complete discussion of complex numbers and real numbers belongs in the field of analysis. There the concepts of limit are developed. The reader who wishes to see a construction of the real numbers using an algebraic setting should consult reference [10, p. 234].

Some more remarks concerning the existence of the field $\mathbb{R}(i)$ are in order. The theorem we have applied begins with assuming the existence of an extension L of $\mathbb{R}$ containing a root i of the polynomial $X^2 + 1$ and concludes with the isomorphism of $\mathbb{R}(i)$ and $\mathbb{R}[X]/\langle X^2 + 1\rangle$. Whether or not the extension L of $\mathbb{R}$ containing i actually exists can be handled as follows. $\mathbb{R}[X]/\langle X^2 + 1\rangle$ is itself a model field extension of $\mathbb{R}$ containing $X + \langle X^2 + 1\rangle$ which is a root of $X^2 + 1$. At least one such extension L of $\mathbb{R}$ containing a root of $X^2 + 1$ does therefore exist. Furthermore, our theorem tells us that any extension $\mathbb{R}(i)$ containing a root of $X^2 + 1$ will be isomorphic to $\mathbb{R}[X]/\langle X^2 + 1\rangle$.

That our simple extension $\mathbb{R}(i)$ of $\mathbb{R}$ is actually all of $\mathbb{C}$, the complex numbers, is in fact the case. We cannot, however, prove such an assertion without first having some definition of $\mathbb{C}$ for comparison. But a full description of $\mathbb{C}$ would presumably involve its analytical properties. We shall simply take the position that we have defined $\mathbb{C}$ and leave the verification of its analytical properties to texts on analysis. Of course then we are unable to prove the outstanding theorems about complex numbers which are consequences of limit properties. Such an important theorem is the one perhaps inappropriately called the fundamental theorem of algebra: *Given any polynomial $f(X)$ in $\mathbb{C}[X]$ there exists an r in $\mathbb{C}$ such that* $f(r) = 0$. This theorem, requiring deep analytic properties of complex numbers to prove, was first correctly proved by C. F. Gauss in his doctoral thesis for the University of Helmstedt. Gauss himself used the name "fundamental theorem of algebra" (or more precisely, he used the German equivalent).

QUESTIONS

1. The product of the two complex numbers $2 + 3i$ and $7 - i$ is
   (A) $11 - 19i$
   (B) $17 + 19i$
   (C) $9 + 2i$
   (D) $14 - 3i$.
(E) None of the suggested answers is the correct product.

2. The equation $(2 + 3i)x = 4 - 3i$ has solution
   (A) $(4 - 3i)/(2 + 3i)$
   (B) 0
   (C) $-\frac{1}{13} - \frac{18}{13}i$
   (D) $(2 - 3i)/(4 + 3i)$.
(E) None of the four previous answers is a solution.

3. The equation $x^2 + (-6 + i)x + 14 - 8i = 0$ has solution
   (A) $2 - 3i$
   (B) $4 + 2i$
   (C) $3 + \frac{1}{2}(-i + \sqrt{-21 + 20i})$
   (D) $3 + \frac{1}{2}(-i - \sqrt{-21 + 20i})$.
(E) None of the four answers is a solution.

4. $\sqrt{2} + \sqrt{3}$ is
   (A) algebraic over $\mathbb{Q}$
   (B) algebraic over $\mathbb{Q}(\sqrt{2})$
   (C) transcendental over $\mathbb{Q}$
   (D) a root of an irreducible polynomial in $\mathbb{Q}[X]$ of degree 6
   (E) algebraic over $\mathbb{R}$.

5. $\mathbb{Q}(\sqrt[3]{2} + \sqrt{3})$ is isomorphic to
   (A) $\mathbb{Q}[X]/\langle X^3 + 6X + 2\rangle$
   (B) $\mathbb{Q}[X]/\langle X^6 - 9X^4 - 4X^3 + 27X^2 - 36X - 23\rangle$
   (C) $\mathbb{Q}[X]/\langle X^3 + X^2\rangle$
   (D) $\mathbb{Q}(X)$.
(E) None of the answers completes a true sentence.

6. $2 = \sqrt{4} = \sqrt{(-2)(-2)} = \sqrt{(-2)}\sqrt{(-2)} = -2$.
The statement given
   (A) is correct because 2 and -2 belong to the same coset of the quotient field defined by the ideal $\langle X^2 + 1\rangle$
   (B) is correct because there are two square roots of 4
   (C) is correct but neither reason offered in (A) and (B) is relevant
   (D) is simply incorrect
   (E) illustrates the unreliability of mathematics.

7. The mapping $f:\mathbb{Z} \to \mathbb{C}$ such that $f(m) = (a + bi)^m$ with $a + bi \neq 0$
   (A) is a morphism of the additive group $\mathbb{Z}$ to the multiplicative group $\mathbb{C} - \{0\}$
   (B) is an injection
   (C) is a surjection
   (D) is a ring morphism.
(E) None of the choices completes a true sentence.

Exercises

1. Argue the case that $\{a + b\sqrt{2} | a, b \in \mathbb{Q}\}$ is a field. What is the inverse of a nonzero $a + b\sqrt{2}$?

2. For $a_0 + a_1 i$ in $\mathbb{C}$ the nonnegative real number $\sqrt{(a_0^2 + a_1^2)}$ is called the modulus or absolute value and written $|a_0 + a_1 i|$. Show the following to be true: $|z| = 0$ if and only if $z = 0$; $|yz| = |y|\,|z|$, $|y + z| \leqslant |y| + |z|$.

3. The conjugate of a complex number $a_0 + a_1 i$ is defined to be $a_0 - a_1 i$ and is written $(a_0 + a_1 i)^*$. Prove these results:

$$(y + z)^* = y^* + z^*, \qquad (yz)^* = y^* z^*,$$
$$y = y^* \text{ if and only if } y \in \mathbb{R},$$
$$yy^* = |y|^2, \qquad y^{-1} = y^*/|y|^2,\ y \neq 0.$$

4. Show that $k:\mathbb{C} \to \mathbb{C}$ such that $k(z) = z^*$ is a (ring) automorphism of $\mathbb{C}$.

5. Let $p(X)$ be in $\mathbb{R}[X]$. Show that if z is a root of $p(X)$ then z^* is also a root.

6. Prove that every cubic real polynomial has either one or three real roots, never exactly two. To get the correct number each root's multiplicity must be counted; for example, X^3 has roots 0, 0, 0.

7. Show that every real polynomial can be factored into real polynomials of degree two or less. Assume the fundamental theorem of algebra.

8. Is the ideal $\langle X^4 + X^2 + 1\rangle$ maximal or prime in $\mathbb{Q}[X]$?

9. Find a smallest field extension of $\mathbb{Q}$ in which $1 + \sqrt{17}$ is a member; write as $\mathbb{Q}[X]/\langle f(X)\rangle$ for some $f(X)$.

10. Find a smallest field extension of $\mathbb{Q}$ such that $X^2 + 3X + 2$ is reducible.

11. For any field of characteristic p show that $f:K \to K$ such that $f(x) = x^p$ is a monomorphism. [*Hint*: Use the binomial theorem and the kernel.] Further prove that if K is finite then f is an automorphism. Demonstrate by example that f need not be an epimorphism if K is not finite.

12. Show that if R is a principal ideal domain then every irreducible element of R generates a nontrivial, maximal proper ideal.

13. Show that if a simply generated ideal is a nontrivial, proper maximal ideal of a commutative unitary ring R then the generator of the ideal is an irreducible element of R.

14. What has $\langle X\rangle \subset \langle X, 2\rangle \subset \mathbb{Z}[X]$ to say about the converse of Exercise 13?

15. $X^2 + \bar{1}$ is irreducible in $\mathbb{Z}_3[X]$. Obtain a field extension $\mathbb{Z}_3(i)$ by adjoining i, a root of $X^2 + \bar{1}$. Are there any second degree irreducible polynomials in $\mathbb{Z}_3(i)[X]$?

16. Let T be an extension of $\mathbb{Q}$ containing all the rational powers of 2. Show that T is an algebraic extension of $\mathbb{Q}$; i.e., show T contains only elements algebraic over $\mathbb{Q}$. Show that T is not $\mathbb{Q}[X]/\langle f(X)\rangle$ for any $f(X)$.

17. Solve the equation $x^2 = c + di$, $c, d \in \mathbb{R}$, for x and put the solution in the form $a + bi$ with $a, b \in \mathbb{R}$.

18. Prove that the set of all complex numbers which are algebraic over $\mathbb{Q}$ is a field. [*Hint*: If $\alpha, \beta \in \mathbb{A}$ work with $\mathbb{Q} \subset \mathbb{Q}(\alpha, \beta) \subset \mathbb{A}$ where $\mathbb{A}$ is the subset of $\mathbb{C}$ algebraic over $\mathbb{Q}$.]

19. Prove that the set of all real numbers algebraic over $\mathbb{Q}$ is a field.

20. Prove that the field $\mathbb{A}$, of Exercise 18, is an algebraically closed field; that is, $\mathbb{A}$ contains any complex number algebraic over $\mathbb{A}$ (The fundamental theorem of algebra states that $\mathbb{C}$ is algebraically closed). [*Hint*: Let α be a root of $a_nX^n + \cdots + a_1X + a_0$ and consider $\mathbb{Q}(a_0, a_1, \ldots, a_n)$.]

Linear algebra: Modules 6

Chapters 6, 7, and 10 are devoted to material mathematicians call linear algebra. Here are developed the elementary properties of modules over rings and modules over fields (vector spaces). This includes a study of matrices as they arise in linear algebra. We use at the beginning for examples and motivation spaces of functions. We feel it is a natural intuitive way of approaching the vector space concept for beginners and ties in neatly with coordinate ideas. We have not treated modules over division rings which are not commutative but the experience gained here by the reader should allow him to handle that variation should he meet it. We have interpolated in Section 6.2 and later in the chapter discussions relating vector spaces and modules on the abstract level with vector spaces as directed line segments as used on the more intuitive level in engineering and physics classes. Hopefully this will assist the beginning student in seeing that the subjects are the same.

Following Section 6.2 is Appendix 6A placed out of logical sequence. The purpose of the appendix is to give the reader a *modus operandi* so that he can solve the exercises efficiently and in the proper form. The results of the appendix are not used in the logical development of the text until justified in Section 7.6. Our apology, if one is necessary, for the use in exercises of material not yet developed formally is a plea that people do not always learn best by taking everything is strict logical order. Linear equations are studied in Appendix 6A, Section 7.6, and again in Section 10.3. Again with the goal of connecting the beginner's intuition with the abstract, we connect ordinary analytical geometry, lines and planes, with the concepts of subspaces and quotient spaces.

The concept of independence dominates linear algebra and leads to the study of bases. We have used the device of *coordinate morphism* to tie together the concepts of linear combinations, bases, coordinates, and the isomorphism between spaces of the same dimension. The coordinate morphism will be found throughout our treatment of linear algebra both in theorems and diagrams. Along with this device we employ family notation (Section 6.6) for the dual purpose of handling matrices and linear combinations. We wish to avoid the often erroneous formulation of linear dependence (e.g., if two vectors in a set $\{x_1, x_2, \ldots, x_n\}$ are equal then the set is linearly dependent) and have therefore used linear combinations of families. We hasten to add that there are equally viable alternatives to the one we chose. Family notation of functions is used throughout mathematics and it is important that a student realize that family and function are the same.

In Section 6.10 we treat and carry through some more special properties and techniques available for vector spaces and not for modules in general. A basis for a vector subspaces can always be extended to a basis for the entire vector space and bases for vector spaces always exist.

Appendices 6B and 6C contain material that is above the general level of the text but that is included for completeness. They form a kind of a justification for a number of assumed statements in the text proper.

6.1 Function spaces, modules, and vector spaces

In this section we define module and vector space, prove a few basic theorems, and explore examples.

We shall introduce the module and vector space through some examples. These examples are in all cases sets of functions. We begin by noticing that all of the following objects are functions:

(a) an ordered pair (a_1, a_2)
(b) an ordered n-ple $(a_1, a_2, \ldots, a_n)$
(c) a two-rowed, two-columned matrix $\begin{pmatrix} a_{11} & a_{12} \\ a_{21} & a_{22} \end{pmatrix}$
(d) an m-rowed, n-columned matrix $\begin{pmatrix} a_{11} & a_{12} & \cdots & a_{1n} \\ a_{21} & a_{22} & \cdots & a_{2n} \\ \cdots & & & \\ a_{m1} & a_{m2} & \cdots & a_{mn} \end{pmatrix}$
(e) a sequence $(a_1, a_2, a_3, \ldots)$
(f) a real-valued function defined on the real unit interval $[0, 1]$.

We notice this in the following ways.

(a) An ordered pair is a function $f:\{1, 2\} \to E$ where E is some set containing the values $f(1) = a_1$ and $f(2) = a_2$.
(b) Assuming E is some set containing all the values $a_1, a_2, a_3, \ldots, a_n$, the n-ple is a function $f:\{1, 2, \ldots, n\} \to E$.

(c) Again assume E is some set large enough to contain all the values of our function. We note $\{1, 2\} \times \{1, 2\} = \{(1, 1), (1, 2), (2, 1), (2, 2)\}$. The 2 by 2 matrix is a function $f:\{1, 2\} \times \{1, 2\} \to E$ such that $f(1, 1) = a_{11}$, $f(1, 2) = a_{12}$, $f(2, 1) = a_{21}$, $f(2, 2) = a_{22}$.
(d) The m-rowed, n-columned matrix is a function $f:\{1, 2, \ldots, m\} \times \{1, 2, \ldots, n\} \to E$ such that $f(i, j) = a_{ij}$ for all $i = 1, 2, \ldots, m$; $j = 1, 2, \ldots, n$ and E is a set containing the values.
(e) The sequence $(a_1, a_2, a_3, \ldots)$ is a function $f:\mathbb{N} \to E$ or $f:\mathbb{N}^+ \to E$ such that $f(i) = a_i$ and E is a set containing the values of f.
(f) $f:[0, 1] \to \mathbb{R}$ has domain the real unit interval and codomain the set of real numbers.

We recall from Section 1.8 that if E and S are sets then the notation E^S stands for the set of all functions from domain S to codomain E

$$E^S = \{f \mid f:S \to E\}.$$

Again in parallel with the earlier given examples of functions we list now the encompassing collections.

(a) $E^{\{1, 2\}}$ is the set of all ordered pairs with values in E. We usually abbreviate this with E^2.

2 can be used to stand for the set $\{0, 1\}$ consistent with the discussion of Section 3.1. Ordered pairs are then numbered (a_0, a_1) instead of (a_1, a_2). Infinite sequences can begin with index 0 instead of $1:(a_0, a_1, a_2, \ldots)$. All this is acceptable notation. The consistent beginning with 0 becomes not totally desirable when we must begin all of our matrices with row number 0 instead of row number 1. An alternative which we employ is to define also sets $\hat{2} = \{1, 2\}$, $\hat{3} = \{1, 2, 3\}$, etc. The set of ordered pairs indexed with 1 and 2 is then written $E^{\hat{2}}$ instead of E^2. However, we pull the usual gambit of simply not writing symbols which prove too tedious. On must often find out from context when E^2 means $E^{\hat{2}}$ or E^2. Of course, if the distinction is important we will use the extra notation.

(b) E^n is the set of all n-ples with values in E.
(c) $E^{2 \times 2}$ is the set of all 2-rowed, 2-columned matrices with entries in E.
(d) $E^{m \times n}$ is the set of all m-rowed, n-columned matrices with entries in E.
(e) $E^{\mathbb{N}}$ is the set of all sequences with values in E.
(f) $\mathbb{R}^{[0, 1]}$ is the set of all functions with domain $[0, 1]$ and codomain $\mathbb{R}$.

If addition is possible in E then there is a natural way to add functions with values in E.

$$(a_1, a_2) + (b_1, b_2) = (a_1 + b_1, a_2 + b_2).$$

$$\begin{pmatrix} a_{11} & a_{12} \\ a_{21} & a_{22} \end{pmatrix} + \begin{pmatrix} b_{11} & b_{12} \\ b_{21} & b_{22} \end{pmatrix} = \begin{pmatrix} a_{11} + b_{11} & a_{12} + b_{12} \\ a_{21} + b_{21} & a_{22} + b_{22} \end{pmatrix}.$$

$$(a_1, a_2, \ldots) + (b_1, b_2, \ldots) = (a_1 + b_1, a_2 + b_2, \ldots).$$

If $f(x) = 3x$ and $g(x) = x^2$ are the values of $f:[0, 1] \to \mathbb{R}$ and $g:[0, 1] \to \mathbb{R}$ then $(f + g):[0, 1] \to \mathbb{R}$ has the value $(f + g)(x) = 3x + x^2$. These are examples of pointwise or placewise addition of functions.

We now formalize in a definition what has just been demonstrated in examples.

Definition. Let S be a set and $\langle E, +\rangle$ a set with an associative binary operation. Given $f:S \to E$ and $g:S \to E$ we define $(f + g):S \to E$ such that $(f + g)(x) = f(x) + g(x)$ for all x in S.

It follows that the $+$ just defined is a binary operation on E^S. We have used the same symbol $+$ both for an operation on E and the new operation on E^S; even though the second arises from the first they are different operations.

Theorem. *If S is a set and $\langle E, +\rangle$ is a set together with an associative binary operation then $\langle E^S, +\rangle$ is a set together with an associative binary operation. If θ is a neutral element for $\langle E, +\rangle$ then $z:S \to E$ such that $z(x) = \theta$ for all x in S is a neutral element for $\langle E^S, +\rangle$. If $\langle E, +, \theta\rangle$ is a commutative group then $\langle E^S, +, z\rangle$ is also a commutative group.*

PROOF. To demonstrate that $+$ is an associative operation on E^S let f, g, h belong to E^S. Both $(f + g) + h$ and $f + (g + h)$ are functions with domain S and codomain E. Furthermore, $((f + g) + h)(x) = (f + g)(x) + h(x) = (f(x) + g(x)) + h(x) = f(x) + (g(x) + h(x)) = f(x) + (g + h)(x) = (f + (g + h))(x)$. In a similar fashion we can prove commutativity. Let $z:S \to E$ such that $z(x) = \theta$ for all x in S. $(f + z)(x) = f(x) + z(x) = f(x) + \theta = f(x)$. Similarly, $z + f = f$. To complete the demonstration of the group properties let f be in E^s. Define $g:S \to E$ such that $g(x) = -f(x)$ for all x in S. $(f + g)(x) = f(x) + g(x) = f(x) - f(x) = \theta. f + g = z$. Hence, $g = -f$. □

This theorem applies to all the given examples; it is only the set S that changes the character of the examples.

If the set E has a second binary operation besides the original $+$ we have discussed we can define multiplication of a function in E^S by a member of E. For example,

$$a(a_1, a_2) = (aa_1, aa_2).$$

$$a\begin{pmatrix} a_{11} & a_{12} \\ a_{21} & a_{22} \end{pmatrix} = \begin{pmatrix} aa_{11} & aa_{12} \\ aa_{21} & aa_{22} \end{pmatrix}.$$

$$a(a_1, a_2, a_3, \ldots) = (aa_1, aa_2, aa_3, \ldots).$$

The kind of multiplication illustrated in these examples is usually called multiplication by a number or scalar. Because the operation multiplies a member of E^S by a member of E outside of E^S it is a function $E \times E^S \to E^S$. We shall refer to this type of operation as an exterior multiplication on E^S by E or more briefly as an E-exterior multiplication on E^S.

Definition. If S is a set and $\langle R, +, \cdot, \theta \rangle$ is a ring then we define *R-exterior multiplication* on the set R^S to be the binary operation $\boxdot: R \times R^S \to R^S$ such that $(a \boxdot f)(x) = a \cdot f(x)$ for all x in S.

The box about the $\cdot$ is temporary. It, as well as the dot itself, are most frequently omitted as with multiplication. Note that $\boxdot$ is not a binary operation on R nor on R^S alone but is a "mixed" binary operation defined on $R \times R^S$.

We now summarize in a theorem results with the new operation together with the previous results.

Theorem. *If S is a set and $\langle R, +, \cdot, \theta, v \rangle$ is a unitary ring then*

$\langle R^S, +, z \rangle$ is a commutative group
$\boxdot: E \times E^S \to E^S$ is an E-exterior multiplication on E^S such that

$$(a \cdot b) \boxdot f = a \boxdot (b \boxdot f)$$
$$(a + b) \boxdot f = (a \boxdot f) + (b \boxdot f)$$
$$a \boxdot (f + g) = (a \boxdot f) + (a \boxdot g)$$
$$v \boxdot f = f$$

for all $a, b \in E$; $f, g \in E^S$.

PROOF. The temporary symbolism with $\boxdot$ allows us to see how the new operation is involved with the old:
$((a \cdot b) \boxdot f)(x) = (a \cdot b) \cdot f(x) = a \cdot (b \cdot f(x)) = a \cdot ((b \boxdot f)(x)) = (a \boxdot (b \boxdot f))(x)$. $((a + b) \boxdot f)(x) = (a + b) \cdot f(x) = (a \cdot f(x)) + (b \cdot f(x)) = (a \boxdot f)(x) + (b \boxdot f)(x) = (a \boxdot f + b \boxdot f)(x)$. $(a \boxdot (f + g))(x) = a \cdot (f + g)(x) = a \cdot (f(x) + g(x)) = a \cdot f(x) + a \cdot g(x) = (a \boxdot f)(x) + (a \boxdot g)(x) = (a \boxdot f + a \boxdot g)(x)$. $(v \boxdot f)(x) = v \cdot f(x) = f(x)$. □

The properties listed in the previous theorem form the basis for a definition of a module. Once we have made this definition it will immediately follow that R^S is a module.

We now make the principal definition after our long preliminaries.

Definition. $\langle M, +, \zeta; \boxdot \rangle$ is a *module* over the ring $\langle R, +, \cdot, \theta, v \rangle$ if and only if

$\langle M, +, \zeta \rangle$ is a commutative group
$\langle R, +, \cdot, \theta, v \rangle$ is a unitary ring
$\boxdot: R \times M \to M$ is an R-exterior multiplication on M such that

$$(r + s) \boxdot x = (r \boxdot x) + (s \boxdot x)$$
$$r \boxdot (x + y) = r \boxdot x + r \boxdot y$$
$$r \boxdot (s \boxdot x) = (r \cdot s) \boxdot x$$
$$v \boxdot x = x$$

for all $r, s \in R$, $x, y \in M$.

For brevity we shall most often simply say M is an R-module.

The elements of the ring R are called *scalars* or numbers while the elements of the module M are called *vectors* or module elements. When we follow the usual practice of omitting product symbols the four properties above read

$$(r + s)x = rx + sx$$
$$r(x + y) = rx + ry$$
$$r(sx) = (rs)x$$
$$vx = x.$$

In the case when R is a division ring or field we use the name vector space instead of module.

Definition. An R-module is called a *vector space* if R is a division ring or field.

The theorem previous to the definition of module demonstrated that if R is a ring and S is a set then R^S is an R-module. If K is a field then K^S is a vector space.

A very simple example of a module is the $\mathbb{Z}$-module given by any commutative group $\langle G, +, \theta\rangle$. For any n in $\mathbb{Z}$ and x in G we define nx to be the nth multiple of x. The taking of multiples can be envisioned as a $\mathbb{Z}$-exterior operation $\cdot : \mathbb{Z} \times G \to G$ with the value nx. It is furthermore true that

$$m(x + y) = mx + my$$
$$m(nx) = (mn)x$$
$$(m + n)x = mx + nx$$
$$1x = x \quad \text{for all } m, n \in \mathbb{Z} \text{ and } x, y \in G.$$

Thus we may take any commutative group, regard multiples as $\mathbb{Z}$-exterior multiplication, and have a $\mathbb{Z}$-module. The theory of commutative groups is thereby subsumed within the theory of modules.

Another somewhat trivial but important example is to notice that any unitary ring $\langle R, +, \cdot, \theta, v\rangle$ is an R-module. The R-exterior multiplication is simply the ring multiplication. Any ring R is a module over itself.

An alternative that should be mentioned is the possibility of using exterior right multiplication in place of left multiplication:

$$\cdot : M \times R \to M \quad \text{with } xr \in M \text{ for each } x \in M, r \in R.$$

In our elementary work we will avoid this second possibility.

To complete this introductory section we prove some elementary results about modules.

Theorem. *Let M be an R-module. Then $\theta x = \zeta$ and $(-v)x = -x$ for all x in M.*

Proof. $\theta x = (\theta + \theta)x = \theta x + \theta x$. $\theta x + \zeta = \theta x + \theta x$. It follows by cancellation (possible because of the existence of additive inverses) that $\theta x = \zeta$.

Secondly, $x + (-\nu)x = \nu x + (-\nu)x = (\nu + (-\nu))x = \theta x = \zeta$. With an associative binary operation there can be at most one inverse. $(-\nu)x = -x$. □

QUESTIONS

1. $\mathbb{Z}^{m\times n}$ is the module of all m-rowed, n-columned matrices with integer entries. Which of these statements are true?
(A) $\mathbb{Z}^{m\times n}$ is a vector space.
(B) Each x in $\mathbb{Z}^{m\times n}$ is a function with mn arguments.
(C) z, the zero function or zero vector of $\mathbb{Z}^{m\times n}$, is the m by n matrix of all zero entries.
(D) The negative of a module element x in $\mathbb{Z}^{m\times n}$ is not necessarily an integer.
(E) None of the alternatives is true.

2. $\mathbb{Q}^{\mathbb{N}}$ is a vector space over the field $\mathbb{Q}$ with
(A) (0, 0, 0 . . .) as zero vector
(B) no negative vector for (0, 0, 0, . . .)
(C) more than a finite number of vectors
(D) no zero vector.
(E) None of the alternatives completes a true sentence.

3. $\mathbb{R}[X]$, the ring of polynomials with coefficients in $\mathbb{R}$, is
(A) a vector space over the field $\mathbb{Q}$
(B) a vector space over the field $\mathbb{R}$
(C) a module over the ring $\mathbb{Z}$
(D) a module over the ring $\mathbb{N}$.
(E) None of the possibilities completes a satisfactory sentence.

4. $\mathbb{Q}^{m\times n}$ with appropriate operations is
(A) a vector space over the field $\mathbb{Q}$
(B) a module over the ring $\mathbb{Q}$
(C) a group
(D) a commutative group.
(E) None of the possibilities completes a true sentence.

5. Which of the following sentences are true?
(A) $\mathbb{Z}$ is an $\mathbb{R}$-module.
(B) $\mathbb{R}$ is a $\mathbb{Z}$-module.
(C) $\mathbb{R}$ is an $\mathbb{R}$-module.
(D) $\mathbb{Z}$ is a $\mathbb{Q}$-module.
(E) $\mathbb{Q}$ is a $\mathbb{Z}$-module.

6. In the $\mathbb{Z}$-module $\mathbb{Z}_4$
(A) $mx = \bar{0}$ implies $m = 0$ or $x = \bar{0}$
(B) $\bar{2} + \bar{2} = \bar{0}$
(C) $8x = \bar{0}$ for all $x \in \mathbb{Z}_4$
(D) $\{n \mid n\bar{2} = \bar{0}\} = 2\mathbb{Z}$.
(E) None of the alternatives completes a true sentence.

7. Which of the following statements are not true for a vector space M over a field K?
(A) $\nu x = x$ for all $x \in M$.
(B) $ax = \zeta$ implies $a = \theta$ for all $a \in K$, $x \in M$.

(C) $(-a)(-x) = ax$ for all $a \in K$, $x \in M$.
(D) $ax = y$ and $x \neq \zeta$ imply $a = y/x$ for all $a \in K$, $x, y \in M$.
(E) All of the statements are true.

8. Which of the following statements do not appear among the axioms (definition) for a vector space M over a field K?
(A) $r(x + y) = rx + ry$ for all $r \in K$, $x, y \in M$.
(B) $\theta x = \zeta$ for all $x \in M$.
(C) $r(xy) = (rx)y$ for all $r \in K$, $x, y \in M$.
(D) $(r + s)x = rx + sx$ for all $r, s \in K$, $x \in M$.
(E) All of the statements appear in the definition of a vector space.

Exercises

1. What is the neutral element z in the commutative group $\langle \mathbb{Z}^{2\times 2}, +, z\rangle$? Solve the equation $X + \begin{pmatrix} 6 & -2 \\ 4 & 1 \end{pmatrix} = \begin{pmatrix} -1 & 3 \\ 4 & -1 \end{pmatrix}$ in $\mathbb{Z}^{2\times 2}$.

2. Let M be an R-module. Prove $r\zeta = \zeta$ for all $r \in R$. Prove $sx = \zeta$ for all $s \in R$ implies $x = \zeta$.

3. Give an example of an R-module M, a nonzero $r \in R$, a nonzero x in M so that $rx = \zeta$.

4. Show that if M is an R-vector space then $rx = \zeta$ and $r \neq \theta$ imply $x = \zeta$.

5. Let $\langle R, +, \cdot, \theta, v\rangle$ be a unitary ring. Prove that any ideal A of R is an R-module. Give an example to show that it is not sufficient for A to be a subring of R in order that A be an R-module.

6. $\left\langle \mathbb{Z}^{2\times 2}, +, \begin{pmatrix} 0 & 0 \\ 0 & 0 \end{pmatrix}\right\rangle$ is a $\mathbb{Z}$-module, one of our earlier examples. Show that $\mathbb{Z}^{2\times 2} \times \mathbb{Z}^{2\times 2}$, the set of all ordered pairs of 2-rowed, 2-columned matrices, is also a $\mathbb{Z}$-module.

7. $\langle \mathbb{R}^{\mathbb{N}^+}, +, z\rangle$, as we have seen earlier, is an $\mathbb{R}$-vector space. Relying on our calculus background we let $\mathscr{L}$ be the subset of $\mathbb{R}^{\mathbb{N}^+}$ of all those sequences which have real finite limits. For example, $(1, \frac{1}{2}, \frac{1}{3}, \ldots) \in \mathscr{L}$ yet $(1, 2, 3, \ldots) \notin \mathscr{L}$. Show that $\mathscr{L}$ is a submodule of $\mathbb{R}^{\mathbb{N}^+}$; that is, show that $\mathscr{L}$ is itself an $\mathbb{R}$-module.

8. A module which contains only one element is called a trivial module. What must the one module element be? Show that $\langle R^{\varnothing}, +, z\rangle$ is a trivial R-module regardless of the nature of the ring R.

6.2 Submodules

In this section we define submodules, develop necessary and sufficient conditions for a subset to be a submodule and discuss the submodule generated by a given subset.

Definition. Let $\langle M, +, \zeta\rangle$ be a module over the unitary ring $\langle R, +, \cdot, \theta, v\rangle$. A subset N of M is a *submodule* if and only if the operations of the module M when restricted to N make N be an R-module.

EXAMPLES. The subset $\mathbb{R} \times \{0\} = \{(x, 0) | x \in \mathbb{R}\}$ of $\mathbb{R} \times \mathbb{R}$ is a submodule. It is easily seen that the sum of two members of the subset is again in the subset and that any scalar multiple of a member of the subset is again in the subset. Furthermore, the entire list of properties of a module are trivially satisfied. A different submodule of the same module $\mathbb{R} \times \mathbb{R}$ is the subset $\Delta = \{(s, s) | s \in \mathbb{R}\}$. It is easy to visualize these two submodules as collections of points in $\mathbb{R} \times \mathbb{R}$ (see Figure 6.1).

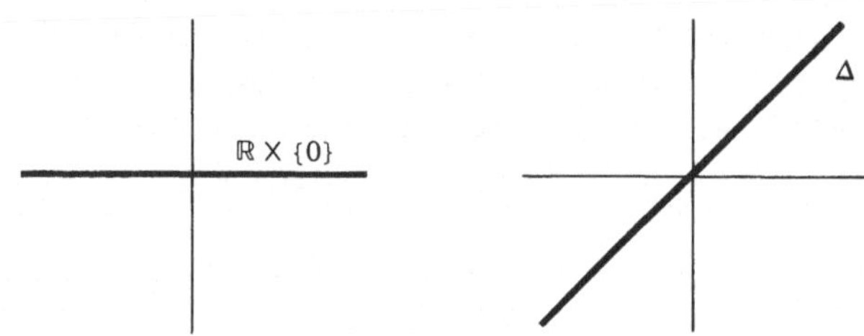

Figure 6.1

We now develop a simpler criterion for a subset to be a submodule.

Theorem. *Let $\langle M, +, \zeta\rangle$ be an R-module. A subset N of M is a submodule if and only if $N \neq \varnothing$; $x, y \in N$ imply $x + y \in N$; and $r \in R$, $x \in N$ imply $rx \in N$.*

PROOF. If N is a submodule then N contains ζ, the zero vector, and N is therefore nonempty. N is also closed under addition and R-exterior multiplication. Thus the conditions hold if N is a submodule.

For the converse, if N is nonempty then there is some vector x in N. Since N is closed under multiplication $-x = (-\nu)x$ is in N. The sum $x + (-x) = \zeta$ must be in N. This shows that N contains the zero vector. The argument $-x = (-\nu)x \in N$ whenever x does belong to N shows that N is closed under negatives. The associativity and commutativity of addition in N follows immediately from the associativity and commutativity in M. N is therefore a commutative group with respect to addition. The closure of R-exterior multiplication is given. The distributive laws and other properties are true in N because they are true in M. N is a submodule. □

EXAMPLE. To verify that $\Delta = \{(s, s) | s \in \mathbb{R}\}$ is a submodule of $\mathbb{R} \times \mathbb{R}$ it is enough to verify that $(1, 1) \in \Delta$, $(s, s) \in \Delta$ and $(t, t) \in \Delta$ imply $(s + t, s + t) \in \Delta$, $(s, s) \in \Delta$ implies $r(s, s) = (rs, rs)$ for all $r \in \mathbb{R}$.

For a subset of M which fails to be a submodule one can construct a smallest possible submodule of the given module which contains the given subset.

Theorem. *Let $\langle M, +, \zeta\rangle$ be an R-module and S any subset of M. Then there exists a submodule $[S]$ of M such that $S \subseteq [S]$ and if N is any submodule of M including S then $[S] \subseteq N$.*

Proof. Alternatively put, the theorem states that there is a smallest submodule of M which includes S. We begin the proof by letting $\mathscr{S}$ be the set of all submodules of M which have S as a subset.

$$\mathscr{S} = \{N | S \subseteq N \text{ and } N \text{ is a submodule of } M\}.$$

Note that each element of $\mathscr{S}$ is a submodule of M; $\mathscr{S}$ is a collection of submodules of M and is not itself a subset of M. $\mathscr{S}$ is nonempty because it contains at least M itself. The intersection of the collection is then well defined.

$$\bigcap \mathscr{S} = \{x | x \in N \text{ for all } N \in \mathscr{S}\}.$$

By its definition $\bigcap \mathscr{S}$ is a subset of M. It is furthermore a superset of S, for let $x \in S$. Then $x \in N$ for all $N \in \mathscr{S}$. $x \in \bigcap \mathscr{S}$. One may readily prove $\bigcap \mathscr{S}$ is a submodule of M. Finally, if N is any submodule of M which has S as a subset then $N \in \mathscr{S}$ showing $\bigcap \mathscr{S} \subseteq N$. □

Definition. Let $\langle M, +, \zeta\rangle$ be an R-module and S any subset of M. We define $[S]$ to be the submodule of M generated by the subset S.

Examples. The set $\{(2, 0), (3, 0)\}$ generates the submodule $\mathbb{Z} \times \{0\}$ of the $\mathbb{Z}$-module $\mathbb{Z} \times \mathbb{Z}$. The set $\{(1, 0), (0, 1)\}$ generates the entire $\mathbb{Z}$-module $\mathbb{Z} \times \mathbb{Z}$. The set $\{(2, 2)\}$ generates the submodule $\{n(2, 2) | n \in \mathbb{Z}\}$ of the $\mathbb{Z}$-module $\mathbb{Z} \times \mathbb{Z}$. The set $\{a\}$ generates the entire $\mathbb{Q}$-module $\mathbb{Q}$ if $a \neq 0$. The set $\varnothing$ generates the space $\{\zeta\}$, the trivial submodule.

Example. This example we call $\mathscr{D}_0$, the vector space of directed line segments in $\mathbb{R}^3$. This set provides a good intuitive example of a vector space and is often used in calculus and physics books. We define $\mathscr{D}_0$ to be the set of all directed line segments emanating from the origin in $\mathbb{R}^3$. Corresponding to each point $q = (q_1, q_2, q_3)$ of $\mathbb{R}^3$ there exists a directed line segment (or arrow, if you will) from $(0, 0, 0)$ to (q_1, q_2, q_3). This directed line segment we call $\overrightarrow{0q}$ (see Figure 6.2). There is a one-to-one correspondence between the terminal

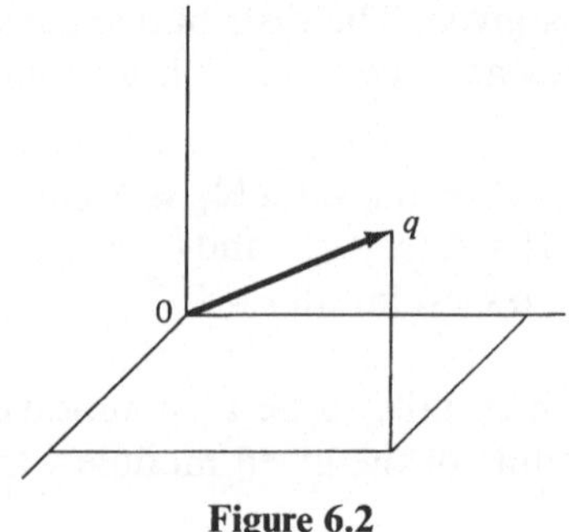

Figure 6.2

points of the line segments and the set of all points in $\mathbb{R}^3$. We call this bijection $\mathscr{D}_0 \to \mathbb{R}^3$ by the name F and set $F(\overrightarrow{0q}) = q$.

Corresponding to the operations of $\mathbb{R}^3$ there are operations in $\mathscr{D}_0$. The sum of (p_1, p_2, p_3) and (q_1, q_2, q_3) in $\mathbb{R}^3$ yields $(p_1 + q_1, p_2 + q_2, p_3 + q_3)$. By taking differences in coordinates we see that the line through (p_1, p_2, p_3) and $(p_1 + q_1, p_2 + q_2, p_3 + q_3)$ is parallel to the line passing through $(0, 0, 0)$ and (q_1, q_2, q_3). So also is the line passing through (q_1, q_2, q_3) and $(p_1 + q_1, p_2 + q_2, p_3 + q_3)$ parallel to the line passing through $(0, 0, 0)$ and (p_1, p_2, p_3). The plane containing the four lines contains the parallelogram determined by the four points. The sum of the arrows $\overrightarrow{0p}$ and $\overrightarrow{0q}$ is the arrow $\overrightarrow{0(p + q)}$ (see Figure 6.3).

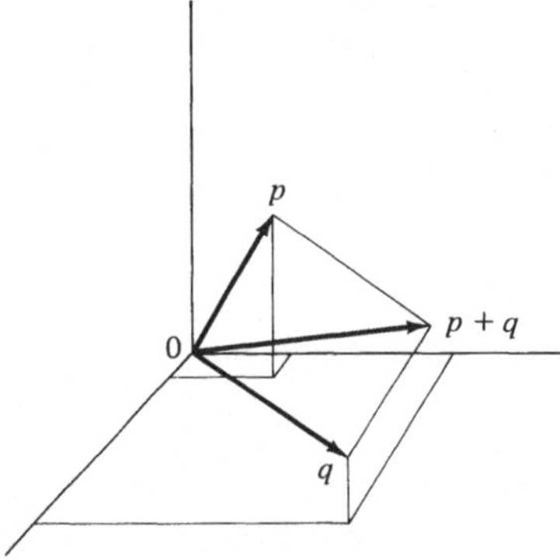

Figure 6.3

Concerning exterior multiplication we see that the points $(0, 0, 0)$, (p_1, p_2, p_3), and (ap_1, ap_2, ap_3) all lie on the same line. The length of the segment from $(0, 0, 0)$ to (ap_1, ap_2, qp_3) is $|a|$ times the length from $(0, 0, 0)$ to (p_1, p_2, p_3). Multiplying an arrow $\overrightarrow{0p}$ by a scalar a produces an arrow terminating on the same line but longer (or shorter) by a factor $|a|$ (see Figure 6.4).

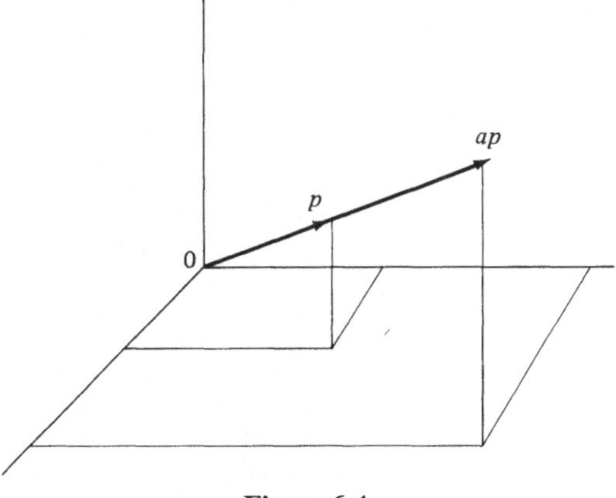

Figure 6.4

If $a > 0$ the arrow terminates on the same side of the line as does $\overrightarrow{0p}$. If $a < 0$ the direction is reversed.

A subspace of $\mathbb{R}^3$ such as $\{t(a_1, a_2, a_3) | t \in \mathbb{R}\}$ for some nonzero (a_1, a_2, a_3) consists of all possible scalar multiples of a single vector. The arrows of the subspace all terminate on a single line through the origin (see Figure 6.5).

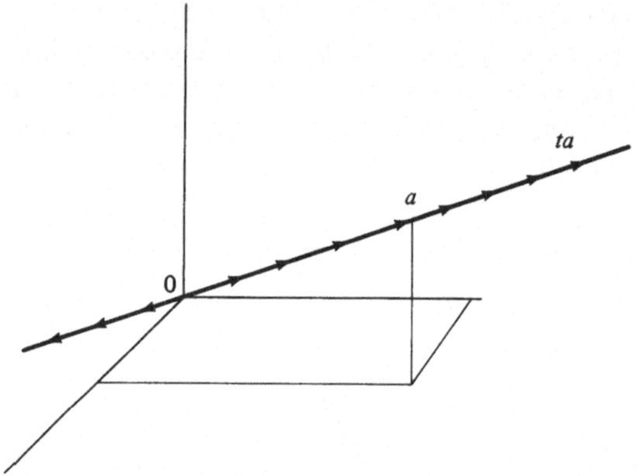

Figure 6.5

A subspace of $\mathbb{R}^3$ such as $\{t(a_1, a_2, a_3) + u(b_1, b_2, b_3) | t, u \in \mathbb{R}\}$, given that a and b are not on the same line, consists of all possible arrows terminating in a plane containing (0, 0, 0), (a_1, a_2, a_3) and (b_1, b_2, b_3) (see Figure 6.6). The discussion in this example is an intuitive one and is based upon a previous knowledge of geometry. The axioms or definitions used as starting points for lines, planes, parallelism are not made clear. For this reason nothing is really proved in this example. Nevertheless, it should be helpful to the reader, especially when he is familiar with the use of arrows as vectors.

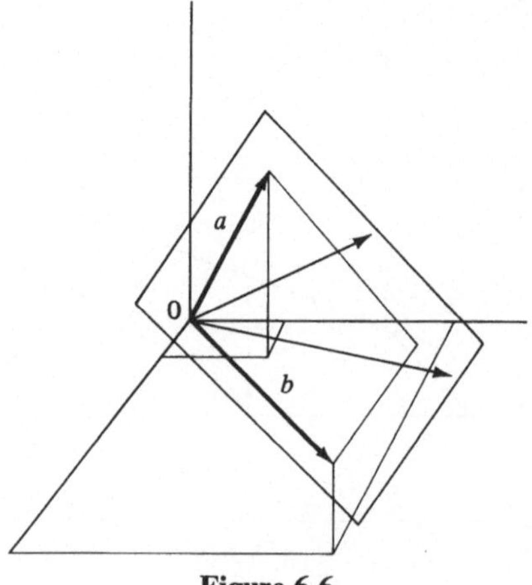

Figure 6.6

Questions

1. Which of the following subsets generate *proper* submodules of the entire module?
   (A) The subset $\{7, 13\}$ of the $\mathbb{Z}$-module $\mathbb{Z}$
   (B) The subset $\{2, 4, 8, 16, 32, \ldots\}$ of the $\mathbb{Z}$-module $\mathbb{Q}$
   (C) The subset $\{4, 8, 12, 16, 20, \ldots\}$ of the $\mathbb{Q}$-module $\mathbb{Q}$
   (D) The subset $(\mathbb{Q} \times \{0\}) \cup (\{0\} \times \mathbb{Q})$ of the $\mathbb{Z}$-module $\mathbb{Q} \times \mathbb{Q}$.
   (E) None generates a proper submodule.

2. Which of these statements are true?
   (A) If a subset of a module generates the entire module then the subset cannot be empty.
   (B) Every submodule S of a module M satisfies the inequality $\{\zeta\} \subseteq S \subseteq M$.
   (C) Two distinct subsets of M must generate two distinct submodules of M.
   (D) If S generates a submodule N of a module M then $M \cap N$ includes S.
   (E) None of the statements is true.

3. Which of the following are submodules of the $\mathbb{Q}$-module $\mathbb{R}$?
   (A) $\mathbb{Z}$
   (B) $\mathbb{Q}$
   (C) $\{a + b\pi | a, b \in \mathbb{Q}\}$
   (D) $\{p(\pi) | p(X) \in \mathbb{Q}[X]\}$.
   (E) None is a submodule.

4. Which of the following are submodules of the $\mathbb{R}$-module $\mathbb{R}^{\mathbb{N}}$?
   (A) $\mathbb{Q}^{\mathbb{N}}$
   (B) $\{(x_1, x_2, x_3, \ldots) | x_n \in \mathbb{Q}$ for all $n \in \mathbb{N}^+$ and $x_n = 0$ for all but a finite number of $n\}$
   (C) $\mathbb{R} \times \mathbb{R}$
   (D) $\mathbb{R}(X)$.
   (E) None is a submodule of $\mathbb{R}^{\mathbb{N}}$.

5. The set of all even degree polynomials with real coefficients together with the zero polynomial fails to be a vector space because
   (A) multiplication by a scalar (exterior multiplication) is not well defined (fails to be closed)
   (B) addition is not well defined (fails to be closed)
   (C) multiplication of polynomials is not well defined (fails to be closed)
   (D) negation is not well defined (fails to be closed).
   (E) The set of polynomials as described is a vector space.

6. Which of the following sets are submodules of the $\mathbb{R}$-module $\mathbb{R}^{[0,1]}$?
   (A) The constant functions: $\{f | f \in \mathbb{R}^{[0,1]}$ and $f(x) = c$ for some $c \in \mathbb{R}$ and all $x \in [0, 1]\}$
   (B) The polynomial functions: $\{f | f \in \mathbb{R}^{[0,1]}$ and $f(x) = a_0 + a_1x + \cdots + a_nx^n$ for some $a_0, a_1, \ldots, a_n \in \mathbb{R}\}$
   (C) All polynomial functions of even degree and the zero function
   (D) All functions in $\mathbb{R}^{[0,1]}$ such that $f(x) = f(1 - x)$ for all $x \in [0, 1]$.
   (E) None of the sets is a submodule of $\mathbb{R}^{[0,1]}$.

7. Which of the following sets are submodules of the $\mathbb{R}$-module $\mathbb{R}^{[0,1]}$?
   (A) All polynomial functions of degree one or less and the zero function.
   (B) All functions f in $\mathbb{R}^{[0,1]}$ such that $f(0) = f(1) = 0$

(C) All functions f in $\mathbb{R}^{[0,1]}$ such that $f(0) + f(1) = 0$
(D) All polynomial functions of degree two.
(E) None of the sets is a submodule of the given module.

8. The set of all real-valued sequences with real limits
(A) is a vector space over the field $\mathbb{R}$
(B) has a subspace of all rational-valued sequences with limits
(C) has a subspace of all real-valued sequences with limits
(D) has a subspace of all real-valued sequences with constant values (i.e., $s_n = c$ for all $n \in \mathbb{N}$ and some $c \in \mathbb{R}$).
(E) None of the alternatives completes a true sentence.

Exercises

1. Let $\langle G, +, \theta\rangle$ be a commutative group and H a subgroup of G. Prove that H is a submodule of the $\mathbb{Z}$-module G.

2. $\mathbb{R}^3$ is an $\mathbb{R}$-module. Describe $[\{(1, 0, 0)\}]$, $[\{(1, 0, 0), (0, 1, 0)\}]$, $[\{(1, 0, 0), (0, 1, 0), (0, 0, 1)\}]$.

3. For the $\mathbb{Z}$-module $\mathbb{R}$ and separately for the $\mathbb{R}$-module $\mathbb{R}$ describe each of the following submodules: $[\{1\}]$, $[\mathbb{Z}]$, $[\mathbb{Q}]$, $[\mathbb{R} - \mathbb{Q}]$.

4. Give an example of a subset of a module which is a subgroup but fails to be a submodule.

5. $\mathbb{Z}_4$ is both a $\mathbb{Z}_4$-module and a $\mathbb{Z}$-module. Describe how the two modules differ because of the different exterior multiplication. Is $\{4\mathbb{Z}, 2 + 4\mathbb{Z}\}$ a submodule in both cases?

6. $\mathbb{R}^{[a,b]}$ is an $\mathbb{R}$-module, the set of all real-valued functions defined on the closed unit interval $[a, b]$. Is $\{f \mid f \in \mathbb{R}^{[a,b]}$ and f is continuous on $[a, b]\}$ a submodule? You must use some basic calculus to answer this question. Is $\{f \mid f \in \mathbb{R}^{[a,b]}$ and f is differentiable on $[a, b]\}$ a submodule? How do the two submodules compare?

7. Let $p_0, p_1, \ldots, p_{n-1}$ belong to $\mathbb{R}^{[a,b]}$ and be continuous there. Define $p(x, y) = y^{(n)} + p_{n-1}(x)y^{(n-1)} + \cdots + p_1(x)y^{(1)} + p_0(x)y$, a polynomial of degree one in y and its derivatives $y^{(1)}, y^{(2)}, \ldots, y^{(n)}$. A solution $f(x)$ of the equation $p(x, y) = 0$ is a function f defined and n times differentiable on $[a, b]$ such that $p(x, f(x)) = 0$ for all $x \in [a, b]$. Let S be the set of all solutions to the equation $p(x, y) = 0$. Prove that S is a submodule of $\mathbb{R}^{[a,b]}$. S is called the solution space of the linear homogeneous differential equation $p(x, y) = 0$.

8. Let $\langle M, +, \cdot\rangle$ be an R-module with submodules L and N. Show that $L + N = \{x + y \mid x \in L, y \in N\}$ is the smallest submodule of M containing both L and N; i.e., show $L + N = [L \cup N]$.

9. Let M be and R-module with submodules L and N. Prove that if $L \cup N$ is a submodule then $L \subseteq N$ or $N \subseteq L$.

10. Let M be an R-module with subsets S and T. Prove $S \subseteq T$ implies $[S] \subseteq [T]$. Prove $[[S]] = [S]$.

11. Show that any R-module M has a proper nontrivial submodule or is generated by a single element.

Appendix 6A A method for solution of linear equations

In order to assist the solution of various problems that will arise in sections that follow we propose now a system of solving simultaneous linear equations. Our goal now is not a complete treatment of the solution of linear equations but only enough knowledge to proceed with the exercises. The exercises are designed, of course, to assist in understanding the text.

Rather than beginning with an abstract description of the process for solving we illustrate the method in use on several examples. The equations

$$\begin{aligned} X_1 + X_2 + X_3 &= 3 \\ X_1 \phantom{{}+X_2} - X_3 &= 1 \\ X_2 + 2X_3 &= 2 \end{aligned}$$

can be written in tabular form for brevity.

$$\begin{array}{ccc|c} 1 & 1 & 1 & 3 \\ 1 & 0 & -1 & 1 \\ 0 & 1 & 2 & 2. \end{array}$$

We propose to replace this system of linear equations by an equivalent system. An equivalent system of equations is one with precisely the same solutions as the former. Our first equivalent system for this example is obtained by multiplying the first equation through by -1

$$\begin{array}{ccc|c} -1 & -1 & -1 & -3 \end{array}$$

and adding the result to the second equation. The first and third equations are left unchanged. The new system is

$$\begin{array}{ccc|c} 1 & 1 & 1 & 3 \\ 0 & -1 & -2 & -2 \\ 0 & 1 & 2 & 2. \end{array}$$

Any solution of the old system will be a solution of this new system. The new system consists merely of the same equations, sums or multiples of equations in the old system. On the other hand, we can, in reverse, produce the old system from the new by adding the first equation to the second. Any solution of the new system must then be a solution of the old. The two systems are therefore equivalent.

We continue now to obtain a sequence of equivalent systems until we ultimately arrive at a system for which the solutions will be obvious. At each step we must take care to use only alterations that produce equivalent systems, operations that are reversible.

We now add 1 times equation two to equation three and leave equations one and two unchanged

$$\begin{array}{rrr|r} 1 & 1 & 1 & 3 \\ 0 & -1 & -2 & -2 \\ 0 & 0 & 0 & 0. \end{array}$$

And then add 1 times equation two to equation one.

$$\begin{array}{rrr|r} 1 & 0 & -1 & 1 \\ 0 & -1 & -2 & -2 \\ 0 & 0 & 0 & 0. \end{array}$$

Finally, we multiply equation two by -1.

$$\begin{array}{rrr|r} 1 & 0 & -1 & 1 \\ 0 & 1 & 2 & 2 \\ 0 & 0 & 0 & 0. \end{array}$$

This is our final form, called *row-reduced echelon form.* We observe the following things about the left side of the equations (to the left of the line):

All rows consisting entirely of zeros are at the bottom.
The first nonzero entry in any row is a 1, which we call an intitial 1.
Above and below in the same column as an initial 1 there are only zeros.
An initial 1 in any row must be further to the right than any initial 1 in a row above.

This constitutes a description of the row-reduced echelon form.

The system we have obtained in this example is

$$\begin{aligned} X_1 \qquad - \ X_3 &= 1 \\ X_2 + 2X_3 &= 2 \\ 0 &= 0 \end{aligned}$$

when written with the unknowns. Since the third equation is always satisfied an equivalent system is

$$\begin{aligned} X_1 &= 1 + \ X_3 \\ X_2 &= 2 - 2X_3. \end{aligned}$$

It is clear that we can obtain solutions for X_1 and X_2 for any value of X_3 whatsoever. Working with real numbers, for any t in $\mathbb{R}$

$$\begin{aligned} X_1 &= 1 + \ t \\ X_2 &= 2 - 2t \\ X_3 &= \qquad t \end{aligned}$$

will be a solution and all solutions will be of this form. The set of solutions is

$$\{(1 + t, 2 - 2t, t) | t \in \mathbb{R}\} \quad \text{or} \quad (1, 2, 0) + \{t(1, -2, 1) | t \in \mathbb{R}\}.$$

There is a strong connection between the form of this solution, the procedures we have used, and the module theory we are developing. We will in time develop these connections fully. Our goal at this moment, however, is merely to be able to solve simple equations that arise in the exercises. Besides listing when the equations are in row-reduced form we also list now the three elementary operations we have used to transform systems to equivalent systems:

Interchanging two equations
Adding a multiple of one equation to another equation
Multiplying one equation by a unit (an invertible element) of the ring containing the coefficients.

All of these elementary operations are reversible ones which lead to equivalent linear systems.

We now give a second example:

$$\begin{aligned} X_1 + X_2 - X_3 &= 1 \\ 2X_1 - X_2 + 3X_3 &= 3 \\ X_1 + 2X_2 - 2X_3 &= 2. \end{aligned}$$

$$\begin{array}{rrr|r} 1 & 1 & -1 & 1 \\ 2 & -1 & 3 & 3 \\ 1 & 2 & -2 & 2 \end{array}$$

(-2) times the first equation added to the second

$$\begin{array}{rrr|r} 1 & 1 & -1 & 1 \\ 0 & -3 & 5 & 1 \\ 1 & 2 & -2 & 2 \end{array}$$

(-1) times the first equation added to the third

$$\begin{array}{rrr|r} 1 & 1 & -1 & 1 \\ 0 & -3 & 5 & 1 \\ 0 & 1 & -1 & 1 \end{array}$$

interchange equation two and equation three

$$\begin{array}{rrr|r} 1 & 1 & -1 & 1 \\ 0 & 1 & -1 & 1 \\ 0 & -3 & 5 & 1 \end{array}$$

(3) times equation two added to equation three

$$\begin{array}{rrr|r} 1 & 1 & -1 & 1 \\ 0 & 1 & -1 & 1 \\ 0 & 0 & 2 & 4 \end{array}$$

(-1) times equation two added to equation one

$$\begin{array}{rrr|r} 1 & 0 & 0 & 0 \\ 0 & 1 & -1 & 1 \\ 0 & 0 & 2 & 4 \end{array}$$

multiply equation three by $\frac{1}{2}$

$$\begin{array}{rrr|r} 1 & 0 & 0 & 0 \\ 0 & 1 & -1 & 1 \\ 0 & 0 & 1 & 2 \end{array}$$

(1) times equation three added to equation two

$$\begin{array}{rrr|r} 1 & 0 & 0 & 0 \\ 0 & 1 & 0 & 3 \\ 0 & 0 & 1 & 2 \end{array}$$

The equations are now in row-reduced echelon form.

Writing the equations in full with the unknowns

$$\begin{aligned} X_1 \qquad\qquad &= 0 \\ X_2 \qquad &= 3 \\ X_3 &= 2. \end{aligned}$$

The set of solutions is the singleton set $\{(0, 3, 2)\}$.

An example in which the number of equations does not coincide with the number of unknowns is the following system:

$$\begin{aligned} X_1 + 2X_2 + X_3 + 5X_4 &= 5 \\ -2X_1 - 4X_2 - X_3 - 8X_4 &= -8 \\ X_1 + 2X_2 + 2X_3 + 7X_4 &= 7. \end{aligned}$$

$$\begin{array}{rrrr|r} 1 & 2 & 1 & 5 & 5 \\ -2 & -4 & -1 & -8 & -8 \\ 1 & 2 & 2 & 7 & 7 \\ \hline 1 & 2 & 1 & 5 & 5 \\ 0 & 0 & 1 & 2 & 2 \\ 1 & 2 & 2 & 7 & 7 \\ \hline 1 & 2 & 1 & 5 & 5 \\ 0 & 0 & 1 & 2 & 2 \\ 0 & 0 & 1 & 2 & 2 \\ \hline 1 & 2 & 1 & 5 & 5 \\ 0 & 0 & 1 & 2 & 2 \\ 0 & 0 & 0 & 0 & 0 \\ \hline 1 & 2 & 0 & 3 & 3 \\ 0 & 0 & 1 & 2 & 2 \\ 0 & 0 & 0 & 0 & 0. \end{array}$$

This is row-reduced echelon form. Written in full we have

$$\begin{aligned} X_1 + 2X_2 \qquad + 3X_4 &= 3 \\ X_3 + 2X_4 &= 2 \\ 0 \quad &= 0. \end{aligned}$$

This in turn can be written

$$\begin{aligned} X_1 &= 3 - 2X_2 - 3X_4 \\ X_3 &= 2 \qquad\quad - 2X_4. \end{aligned}$$

For any arbitrary values t, u assigned to X_2 and X_4 values of X_1 and values of X_3 are determined to provide a solution.

$$\begin{aligned} X_1 &= 3 - 2t - 3u \\ X_2 &= \phantom{3 - {}} t \\ X_3 &= 2 \phantom{{}- 2t} - 2u \\ X_4 &= u. \end{aligned}$$

The solution set is $\{(3 - 2t - 3u, t, 2 - 2u, u)|t, u \in \mathbb{R}\} = (3, 0, 2, 0) + \{t(-2, 1, 0, 0) + u(-3, 0, -2, 1)|t, u \in \mathbb{R}\}$.

Another example is the set of equations

$$\begin{aligned} X_1 + 3X_2 - X_3 &= 4 \\ X_1 + 2X_2 + X_3 &= 2 \\ 3X_1 + 7X_2 + X_3 &= 9. \end{aligned}$$

$$\begin{array}{rrr|r} 1 & 3 & -1 & 4 \\ 1 & 2 & 1 & 2 \\ 3 & 7 & 1 & 9 \end{array} \quad \begin{array}{rrr|r} 1 & 3 & -1 & 4 \\ 0 & -1 & 2 & -2 \\ 0 & -2 & 4 & -3 \end{array} \quad \begin{array}{rrr|r} 1 & 3 & -1 & 4 \\ 0 & 1 & -2 & 2 \\ 0 & 0 & 0 & 1 \end{array}$$

$$\begin{array}{rrr|r} 1 & 3 & -1 & 4 \\ 0 & -1 & 2 & -2 \\ 3 & 7 & 1 & 9 \end{array} \quad \begin{array}{rrr|r} 1 & 3 & -1 & 4 \\ 0 & 1 & -2 & 2 \\ 0 & -2 & 4 & -3 \end{array} \quad \begin{array}{rrr|r} 1 & 0 & 5 & -2 \\ 0 & 1 & -2 & 2 \\ 0 & 0 & 0 & 1. \end{array}$$

The final list is in row-reduced echelon form. The equations are

$$\begin{aligned} X_1 \phantom{{}+X_2} + 5X_3 &= -2 \\ X_2 - 2X_3 &= 2 \\ 0 &= 1. \end{aligned}$$

It is evident that regardless of what values are assigned to X_1, X_2, and X_3 the third equation and the system will never be satisfied. The set of solutions is the empty set.

We have, we hope, achieved here in this discussion our limited objective of assisting the reader to solve linear equations for the exercises. We have described the process and the end form to be found, the row-reduced echelon form. We have not formally demonstrated the achieveability of the row-reduced echelon form. In a later section we shall discuss again more fully the solution of linear equations on a more formal basis.

Exercises

1. Solve these equations using the method outlined in this section, finding an equivalent system in row-reduced echelon form.

(a) $X_1 + X_2 - X_3 = 4$
$X_2 - X_3 = 2$
$X_3 = 2.$

(b) $X_1 - X_2 + X_3 - X_4 = 1$
$X_1 + X_2 + X_3 + X_4 = 2.$

(c) $2X_1 - X_2 + X_3 = 1$
$3X_1 + X_2 + 2X_3 = 4$
$X_1 - 3X_2 = 2.$

(d) $3X_1 + X_2 - 4X_3 = 5.$

2. Each of the following two sets are submodules of the $\mathbb{R}$-module $\mathbb{R}^3$. Verify this.

$$\{t(1, 1, 2) + u(-1, 1, 2) | t, u \in \mathbb{R}\}$$
$$\{v(2, 1, -1) + w(2, -1, -2) | v, w \in \mathbb{R}\}$$

Find the intersection of the two submodules expressing the set in the same manner as the two given sets. [*Hint*: Set the two expressions equal, extract linear equations in t, u, v, w, and then solve the equations in the manner explained in the section.]

6.3 Quotient modules

In this section we see how, as with rings, it is possible to define a quotient structure of a module.

Before reading this section it is advisable to review the construction of the quotient ring in Section 2.6. If $\langle M, +, \zeta\rangle$ is a given R-module and A is a submodule of M then the cosets $\{x + A | x \in M\}$ form a quotient set M/A of M. The verification here closely follows that in Section 2.6 and is based upon the equivalence relation on $M: x \sim y$ if and only if $x - y \in A$. On this quotient set M/A the addition of cosets is defined according to the rule $(x + A) + (y + A) = x + y + A$. A, itself, is the neutral element for M/A and each coset $x + A$ has a negative $-x + A$. At this point the construction parts company with the ring construction of Section 2.6. Instead of a multiplication of cosets as we had in the ring construction we must define an exterior multiplication for the quotient module: $R \times M/A \to M/A$. We define $r(x + A) = rx + A$. It is trivial to confirm that this multiplication is independent of the representative x of the coset $x + A$. It also holds that

$$\begin{aligned}(r + s)(x + A) &= (r + s)x + A \\ r(x + A + y + A) &= r(x + A) + r(y + A) \\ (rs)(x + A) &= r(s(x + A)) \\ \nu(x + A) &= x + A.\end{aligned}$$

In summary we have

Theorem. *If $\langle M, +, \zeta\rangle$ is an R-module and A is a submodule then there exists a quotient module $\langle M/A, +, A\rangle$ with operations $(x + A) + (y + A) = x + y + A$ and $r(x + A) = rx + A$.*

The cosets of quotient modules are often called *linear varieties.*

EXAMPLES. We take the $\mathbb{Q}$-module $\mathbb{Q} \times \mathbb{Q}$ and consider the submodule $A = \{(s, s) | s \in \mathbb{Q}\}$. $(\mathbb{Q} \times \mathbb{Q})/A = \{(a, b) + A | (a, b) \in \mathbb{Q} \times \mathbb{Q}\}$. Two cosets $(a, b) + A$ and $(c, d) + A$ are equal if and only if $(a, b) - (c, d) \in A$ if and only if $(a - c, b - d) \in A$ if and only if $a - c = b - d$. For example, (2, 1) and (4, 3) determine the same coset. Representing $\mathbb{Q} \times \mathbb{Q}$ with a plane diagram each coset is a line parallel to A (see Figure 6.7).

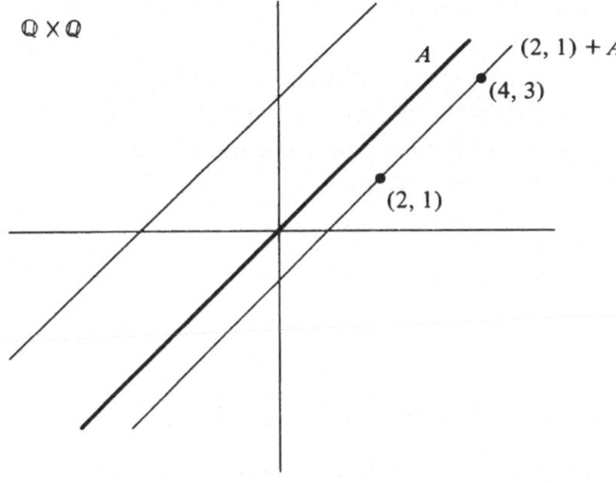

Figure 6.7

Another example is given by the submodule $3\mathbb{Z}$ of the $\mathbb{Z}$-module $\mathbb{Z}$. $\mathbb{Z}/3\mathbb{Z}$ has three members $3\mathbb{Z}, 1 + 3\mathbb{Z}, 2 + 3\mathbb{Z}$.

A third example is to take as defining submodule the submodule $A = \{t(1, 2, 3) | t \in \mathbb{R}\}$ of $\mathbb{R}^3$. The quotient module is the collection of cosets $\{(a_1, a_2, a_3) + A | (a_1, a_2, a_3) \in \mathbb{R}^3\}$ (see Figure 6.8). Other examples will be given in the exercises.

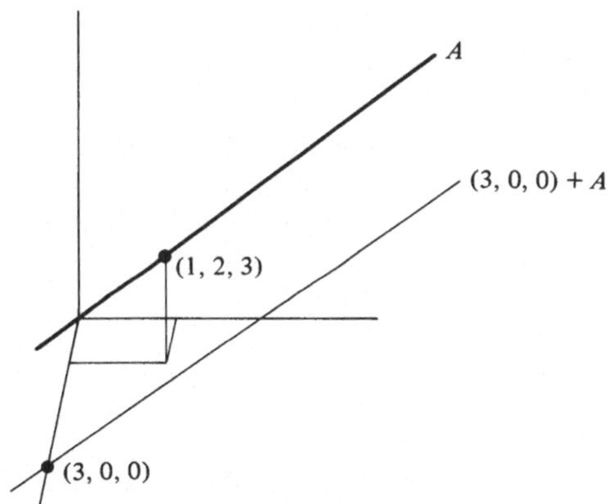

Figure 6.8

Questions

1. The quotient module M/A of the R-module M
   (A) is defined by any submodule A of M
   (B) contains an infinite number of vectors
   (C) has A as a submodule
   (D) is a submodule of M.
   (E) None of the alternatives completes a true sentence.

2. The $\mathbb{Q}$-module $\mathbb{R}$ differs from the $\mathbb{R}$-module $\mathbb{R}$ in that
(A) the former has a nontrivial proper submodule while the latter does not
(B) the latter is a vector space while the former is not
(C) $\{1\}$ generates the latter module while not the former
(D) $\mathbb{Z}$ is a submodule of the latter but not the former.
(E) None of the comparisons is true.

3. The quotient $\mathbb{R}$-module $\mathbb{R}^3/[(1, 1, 1)]$
(A) has only a finite number of cosets
(B) has finite cosets
(C) contains (2, 2, 2)
(D) contains $\{(s, s, s)|s \in \mathbb{R}\}$.
(E) None of the alternatives completes a true sentence.

4. Let A be a submodule of the R-module M. Which are true?
(A) $x + A = y + A$ implies $x = y$.
(B) $x + A = M$ for some x in M implies $M/A = \{M\}$.
(C) $A = \{\zeta\}$ implies M/A is equal to M.
(D) $R(M/A) = M/A$.
(E) None of the statements is true.

5. Let N and P be submodules of an R-module M. Then the linear varieties $x + N$ and $y + P$ are equal if and only if
(A) $x = y$ and $N = P$
(B) $x \in N$ and $y \in P$ and $N = P$
(C) $x - y \in N$ and $N = P$
(D) $(x + N) \cap (y + P) \neq \varnothing$ and $N = P$.
(E) None of the four alternatives completes a true sentence.

6. Which of the following are linear varieties of an R-module M?
(A) $\{x + N\}$ where N is a submodule of M
(B) $x + sa + tb$ where x, a, b belong to M and $s, t \in R$
(C) $\{x\}$ where $x \in M$
(D) $x + \{sa\}$ where $s \in R$, $a \in M$, $x \in M$.
(E) None is a linear variety of M.

Exercises

1. Show that $2\mathbb{Z}$, the set of all even integers, is a submodule of the $\mathbb{Z}$-module $\mathbb{Z}$. Describe the quotient module $\mathbb{Z}/2\mathbb{Z}$ including exterior multiplication.

2. $\mathbb{Z}$ is a submodule of the $\mathbb{Z}$-module $\mathbb{R}$. Describe $\mathbb{R}/\mathbb{Z}$. Prove that if $x \in \mathbb{R}$ then there exists x' in $[0, 1)$ such that $x + \mathbb{Z} = x' + \mathbb{Z}$.

3. Describe the $\mathbb{Z}$-module $\mathbb{R}/\mathbb{Q}$. How many cosets are there?

4. Show that if a linear variety (or coset) of a vector space contains two distinct multiples of the same vector then that variety is a subspace.

5. Let $\mathscr{L}$ stand for the $\mathbb{R}$-module of all real-valued sequences with real limits. Let $\mathscr{N}$ stand for all those members of $\mathscr{L}$ with limit equal to 0. We call $\mathscr{N}$ the set of null sequences. Prove that $\mathscr{N}$ is a submodule of $\mathscr{L}$. What equivalence relation on $\mathscr{L}$ is associated with the quotient module $\mathscr{L}/\mathscr{N}$?

6. Let $a = (a_1, a_2, a_3) \in \mathbb{R}^3$ and be nonzero. Show that $[\{a\}] = \{ta | t \in \mathbb{R}\}$. Show that $(X_1, X_2, X_3) \in [\{a\}]$ if and only if

$$X_1 = ta_1 \qquad X_2 = ta_2 \qquad X_3 = ta_3$$

for some $t \in \mathbb{R}$. Now let $c = (c_1, c_2, c_3) \in \mathbb{R}^3$. Show that $(X_1, X_2, X_3) \in c + [\{a\}] \in \mathbb{R}^3/[\{a\}]$ if and only if

$$X_1 = c_1 + ta_1 \qquad X_2 = c_2 + ta_2 \qquad X_3 = c_3 + ta_3$$

for some $t \in \mathbb{R}$. The coset $c + [\{a\}]$ is interpreted geometrically as the line through the point (c_1, c_2, c_3) with direction vector (a_1, a_2, a_3) (see Figure 6.9).

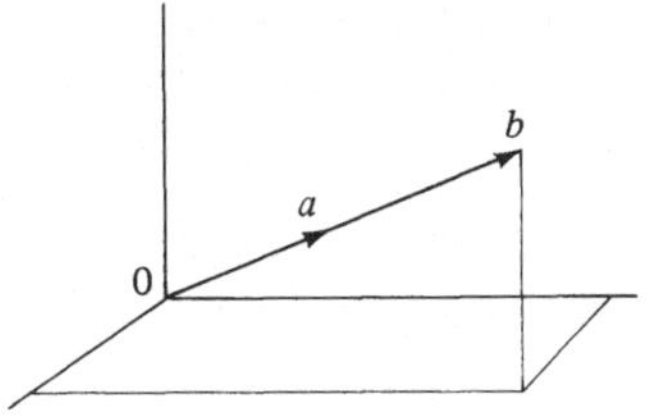

Figure 6.9

7. Let $a = (a_1, a_2, a_3)$ and $b = (b_1, b_2, b_3)$ belong to $\mathbb{R}^3$. Suppose also $ta + ub \neq (0, 0, 0)$ except for $t = u = 0$. This assumption means that vectors a and b do not have the same direction. Show that $[\{a, b\}] = \{ta + ub | t, u \in \mathbb{R}\}$. Show that $(X_1, X_2, X_3) \in [\{a, b\}]$ if and only if

$$X_1 = ta_1 + ub_1 \qquad X_2 = ta_2 + ub_2 \qquad X_3 = ta_3 + ub_3$$

for some $t, u \in \mathbb{R}$. Now let $(c_1, c_2, c_3) \in \mathbb{R}^3$. Show that $(X_1, X_2, X_3) \in c + [\{a, b\}] \in \mathbb{R}^3/[\{a, b\}]$ if and only if

$$X_1 = c_1 + ta_1 + ub_1 \qquad X_2 = c_2 + ta_2 + ub_2 \qquad X_3 = c_3 + ta_3 + ub_3$$

for some $t, u \in \mathbb{R}$. The coset $c + [\{a, b\}]$ is interpreted geometrically as the plane through the point (c_1, c_2, c_3) with vectors (a_1, a_2, a_3) and (b_1, b_2, b_3) parallel to the plane (see Figure 6.10).

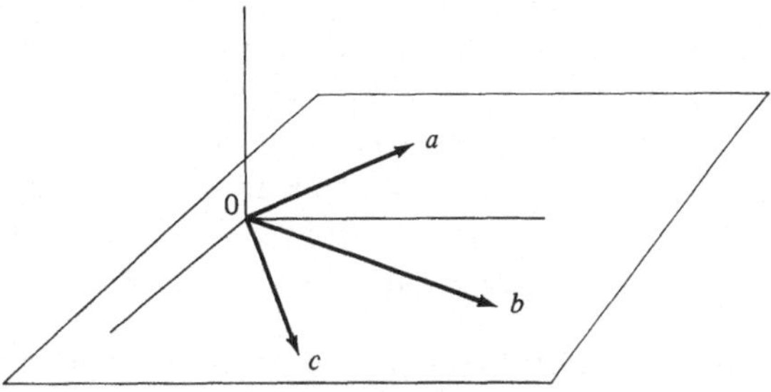

Figure 6.10

8. Using the results of Exercises 7 and 8 write the equations of a line containing the point (1, 2, 3) and having direction vector (7, 2, 1). Write the equations of a plane containing the point (1, 2, 3) and with direction given by the two vectors (6, 2, 1)

and (4, −1, 2). Write the equations of a plane containing the points (0, 1, −1), (4, 2, −3), (7, 2, 1). [*Hint*: One vector parallel to the plane is the vector (3, 0, 4) = (7, 2, 1) − (4, 2, −3) which joins the point (4, 2, −3) to (7, 2, 1). Use this as one of the direction vectors.]

6.4 Morphisms

In this section we define morphisms for modules, define kernel and range of a morphism, and explore structure-preserving properties of morphisms.

Definition. Let $\langle M, +, \zeta\rangle$ and $\langle M', +', \zeta'\rangle$ be two modules over a unitary ring R. We represent exterior multiplication on M and M' by $\boxdot$ and $\boxdot'$, respectively. $f: M \to M'$ is a *morphism* of modules if and only if f preserves the operations:

1. $f(x + y) = f(x) +' f(y)$ for all $x, y \in M$
2. $f(r \boxdot x) = r \boxdot' f(x)$ for all $r \in R$, $x \in M$
3. $f(\zeta) = \zeta'$.

The three properties in the definition are not independent. The third equation is implied by the second upon substituting $r = \theta$. It is therefore necessary and sufficient for $f: M \to M'$ to be a morphism that properties 1 and 2 hold. Property 1 is called *additivity* and property 2 is called *homogeneity*. A function which is both additive and homogeneous is called *linear*; the morphisms of modules are frequently called linear transformations (transformation, along with mapping and family, is merely another name for function).

Definition. As a morphism is injective, surjective, or bijective it is called respectively a *monomorphism*, an *epimorphism*, or an *isomorphism*. As with other structures a morphism of a module M into itself is called an *endomorphism* and an isomorphism of a module into itself is called an *automorphism*.

EXAMPLES. $f: \mathbb{Z} \times \mathbb{Z} \to \mathbb{Z}$ such that $f(r_1, r_2) = r_1 + r_2$ is a morphism of the $\mathbb{Z}$-modules $\mathbb{Z} \times \mathbb{Z}$ and $\mathbb{Z}$. The verification goes as follows. $f((r_1, r_2) + (s_1, s_2)) = f(r_1 + s_1, r_2 + s_2) = r_1 + s_1 + r_2 + s_2 = r_1 + r_2 + s_1 + s_2 = f(r_1, r_2) + f(s_1, s_2)$. $f(r(r_1, r_2)) = f(rr_1, rr_2) = rr_1 + rr_2 = r(r_1 + r_2) = rf(r_1, r_2)$. Moreover, this f is an epimorphism since for any $v \in \mathbb{Z}$, the codomain, there is a $(v, 0) \in \mathbb{Z} \times \mathbb{Z}$ such that $f(v, 0) = v + 0 = v$. This f is not, however, a monomorphism because $f(2, 1) = f(1, 2)$.

Associated with every morphism $f: M \to M'$ of two R-modules $\langle M, +, \zeta\rangle$ and $\langle M', +', \zeta'\rangle$ are two distinguished submodules.

Definition. Kernel $f = \{x \mid x \in M \text{ and } f(x) = \zeta'\}$. Range $f = \{f(x) \mid x \in M\}$.

Since kernel $f = f^{-1}(\{\zeta'\})$ and range $f = f(M)$ we can demonstrate that kernel f and range f are submodules by proving this more general result.

Theorem. *Let $f:M \to M'$ be a morphism of R-modules $\langle M, +, \zeta\rangle$ and $\langle M', +', \zeta'\rangle$. Then*

N is a submodule of M implies $f(N)$ is a submodule of M';
N' is a submodule of M' implies $f^{-1}(N')$ is a submodule of M.

PROOF. If N is a submodule of M then $\zeta \in N$ and so $f(\zeta) \in f(N)$. This proves $f(N) \neq \emptyset$. Let $y_1, y_2 \in f(N)$. $y_1 = f(x_1)$, $y_2 = f(x_2)$ for some $x_1, x_2 \in N$. $y_1 +' y_2 = f(x_1) +' f(x_2) = f(x_1 + x_2)$. But $x_1 + x_2 \in N$ since N is a submodule. Therefore $y_1 +' y_2 \in f(N)$. Let $r \in R$, $y \in f(N)$. $y = f(x)$ for $x \in N$. $ry = rf(x) = f(rx)$ with $rx \in N$. $f(N)$ is a submodule of M'.

Now assume N' is a submodule of M'. We consider $f^{-1}(N')$. Let $x_1, x_2 \in f^{-1}(N')$. $f(x_1), f(x_2) \in N'$. $f(x_1) +' f(x_2) \in N'$. $f(x_1 + x_2) = f(x_1) +' f(x_2)$. $f(x_1 + x_2) \in N'$. $x_1 + x_2 \in f^{-1}(N')$ proving $f^{-1}(N')$ closed under addition. Let $x \in f^{-1}(N')$, $r \in R$. $f(x) \in N'$. $rf(x) \in N'$. $f(rx) \in N'$. $rx \in f^{-1}(N')$ proving $f^{-1}(N')$ closed under scalar multiplication. ζ will belong to $f^{-1}(N')$ and $f^{-1}(N')$ will be nonempty if we can show $f(\zeta) \in N'$. But $f(\zeta) = f(\theta\zeta) = \theta f(\zeta) = \zeta' \in N'$. □

EXAMPLE. $f:\mathbb{Z} \times \mathbb{Z} \to \mathbb{Z}$ such that $f(x, y) = x + y$ is an epimorphism of the $\mathbb{Z}$-modules. Kernel $f = \{(x, y)|x + y = 0\} = \{(x, y)|y = -x\} = \{(x, -x)|x \in \mathbb{Z}\} = \{x(1, -1)|x \in \mathbb{Z}\}$. The range of f is all of the codomain $\mathbb{Z}$ since f is an epimorphism.

We now discuss the inverse function problem. We know that a function $f:M \to M'$ has an inverse function $f^{-1}:M' \to M$ if and only if f is a bijection. The question to be settled is whether or not f^{-1} is itself a morphism.

Theorem. *If $f:M \to M'$ is an isomorphism of the modules $\langle M, +, \zeta\rangle$ and $\langle M', +', \zeta'\rangle$ then $f^{-1}:M' \to M$ is an isomorphism.*

PROOF. If $f:M \to M'$ is a bijection then the inverse function $f^{-1}:M' \to M$ exists and also is a bijection. We must merely show that it is a morphism. Let $y', y'' \in M'$. There exist $x', x'' \in M$ such that $f(x') = y'$ and $f(x'') = y''$. $f(x' + x'') = f(x') +' f(x'') = y' +' y''$. This shows $x' + x'' \in f^{-1}(y' +' y'')$. Thus $f^{-1}(y' +' y'') = x' + x'' = f^{-1}(y') + f^{-1}(y'')$.

Now let $r \in R$ and $y' \in M'$. $f(x') = y'$ for some $x' \in M$. $ry' = rf(x') = f(rx')$. Therefore $f^{-1}(ry') = rx' = rf^{-1}(y')$. f^{-1} is therefore a morphism. □

Again we warn of the necessity for distinguishing between the inverse function $f^{-1}:M' \to M$ of $f:M \to M'$ and the inverse image $f^{-1}(B)$ of some subset B of M'. The inverse image is always defined whether $f:M \to M'$ is a bijection or not. The arguments for the inverse function $f^{-1}:M' \to M$ are members of M'. The arguments for the inverse image function $f^{-1}:\mathscr{P}(M') \to \mathscr{P}(M)$ are subsets of M'.

EXAMPLES. With $\mathbb{Z}$ as a $\mathbb{Z}$-module, $f:\mathbb{Z} \to \mathbb{Z}$ such that $f(x) = 4x$ defines a morphism with kernel $f = \{0\}$ and range $f = \{4x|x \in \mathbb{Z}\} = 4\mathbb{Z}$, a proper submodule of $\mathbb{Z}$. There is no inverse function f^{-1} but the inverse image of $8\mathbb{Z}$, $f^{-1}(8\mathbb{Z})$, for example, is $2\mathbb{Z}$.

We define $g:\mathbb{R}^{\mathbb{N}} \to \mathbb{R}^{\mathbb{N}}$ such that $g(x_0, x_1, x_2, \ldots) = (x_1, x_2, x_3, \ldots)$. The effect of the function on a sequence is to shift values one place to the left dropping the first term. One can verify that this function is a morphism of the $\mathbb{R}$-modules of infinite sequences. The kernel of g is $\{(x_0, 0, 0, \ldots)|x_0 \in \mathbb{R}\}$ whereas the range is $\mathbb{R}^{\mathbb{N}}$. This example is interesting because it shows that an endomorphism can be an epimorphism without being a monomorphism. The inverse function does not exist.

We now move to the fundamental morphism theorem for modules.

Theorem. *Let $\langle M, +, \zeta\rangle$ and $\langle M', +', \zeta'\rangle$ be R-modules and $f:M \to M'$ a morphism. Then there exist*

an epimorphism $\varphi:M \to M/\ker f$
a monomorphism $f':M/\ker f \to M'$ such that $f' \circ \varphi = f$.

Proof. The submodule kernel f of M defines a quotient module $M/\ker f$ and the surjection $\varphi:M \to M/\ker f$, taking each element x of M into the containing coset $x + \ker f$, is a morphism.

$$\begin{aligned}\varphi(x + y) &= x + y + \ker f = x + \ker f + y + \ker f\\ &= \varphi(x) + \varphi(y).\\ \varphi(rx) &= rx + \ker f = r(x + \ker f) = r\varphi(x).\end{aligned}$$

Likewise, the injection $f':M/\ker f \to M'$ such that $f'(x + \ker f) = f(x)$ is a morphism. These maps are all defined in Section 1.7; it is their morphism properties that need checking.

$$\begin{aligned}f'(x + \ker f + y + \ker f) &= f'(x + y + \ker f)\\ &= f(x + y) = f(x) + f(y)\\ &= f'(x + \ker f) + f'(y + \ker f).\end{aligned}$$

$f'(r(x + \ker f)) = f'(rx + \ker f) = f(rx) = rf(x) = rf'(x + \ker f)$. The reader should also consult the proof in Section 2.7 of the analogous theorem for rings. □

Corollary. *If $f:M \to M'$ is an R-module morphism then $M/\ker f$ is isomorphic with* range f.

Examples. For the morphism $f:\mathbb{Z} \times \mathbb{Z} \to \mathbb{Z}$ such that $f(x, y) = x + y$ we can conclude $(\mathbb{Z} \times \mathbb{Z})/\{n(1, -1)|n \in \mathbb{Z}\}$ is isomorphic with $\mathbb{Z}$.

From $f:\mathbb{Z} \to \mathbb{Z}$ such that $f(n) = 4n$ we can conclude that $\mathbb{Z}/\{0\}$ is isomorphic with $4\mathbb{Z}$.

From the morphism $f:\mathbb{R}^{\mathbb{N}} \to \mathbb{R}^{\mathbb{N}}$ such that $f(x_0, x_1, \ldots) = (x_1, x_2, x_3, \ldots)$ we conclude $\mathbb{R}^{\mathbb{N}}/\{(x, 0, 0, \ldots)|x \in \mathbb{R}\}$ is isomorphic with $\mathbb{R}^{\mathbb{N}}$.

We can characterize monomorphisms and epimorphisms in terms of kernel and range.

Theorem. *Let $f:M \to M'$ be a morphism of R-modules $\langle M, +, \zeta\rangle$ and $\langle M', +', \zeta'\rangle$. Then*

f is a monomorphism if and only if kernel $f = \{\zeta\}$;
f is an epimorphism if and only if range $f = M'$.

Proof. We must show first that injectivity of f is equivalent to kernel f being trivial. Suppose f to be injective. Let $x \in$ kernel f. $f(x) = \zeta'$. So also does $f(\zeta) = \zeta'$. $f(x) = f(\zeta)$. $x = \zeta$. Since $\zeta \in$ kernel f we have kernel $f = \{\zeta\}$. Conversely, let kernel $f = \{\zeta\}$. If $f(x) = f(y)$ then $f(x - y) = \zeta'$. $x - y \in$ kernel f. $x - y = \zeta$. $x = y$.

For the second part of the theorem f is a surjection if and only if $y \in M'$ implies there exists an $x \in M$ such that $f(x) = y$. However this must mean $f(M) = M'$. □

We continue our study of morphisms with a few special theorems.

Theorem. *Let $\langle M, +, \zeta\rangle$ be an R-module and L and N be two submodules so that $L \subseteq N$. Then M/N is isomorphic with $(M/L)/(N/L)$.*

Proof. The proof of this theorem relies upon the fundamental morphism theorem and shows some of the power of this theorem. From M/L to M/N we define a function f taking each coset of M/L into the coset which contains it. $f:M/L \to M/N$ such that $f(x + L) = x + N$. Since $L \subseteq N$ we have $x + L \subseteq x + N$. f is an epimorphism. The kernel of f is $\{x + L | f(x + L) = N\} = \{x + L | x + N = N\} = \{x + L | x \in N\} = N/L$. By the fundamental morphism theorem there exists an isomorphism $f':(M/L)/(N/L) \to M/N$. □

Theorem. *Let $\langle M, +, \zeta\rangle$ be an R-module with submodules L, N. Then $L/(L \cap N)$ is isomorphic with $(L + N)/N$.*

Proof. We define $f:L \to (L + N)/N$ such that $f(x) = x + N$. $x + N \in (L + N)/N$ for each $x \in L$ and f is surjective. N is a submodule of the module $L + N$ (consult Exercise 8 of Section 6.2). $f(x + y) = x + y + N = x + N + y + N = f(x) + f(y)$. $f(rx) = rx + N = r(x + N) = rf(x)$. Kernel $f = \{x | x \in L$ and $f(x) = N\} = \{x | x \in L$ and $x + N = N\} = \{x | x \in L$ and $x \in N\} = L \cap N$. The isomorphism follows from an application of the fundamental morphism theorem. □

Questions

1. Let $f:M \to M'$ be a mapping of $\mathbb{Z}$-modules. Which of these statements are true?
   (A) If $f(x + y) = f(x) + f(y)$ then f is a morphism.
   (B) If f is a morphism and $n \in \ker f$ then $n\mathbb{Z} \subseteq \ker f$.

(C) If f is an epimorphism then $f(M)$ is isomorphic with $\mathbb{Z}$.
(D) If f is a monomorphism then $f(x) = f(-x)$ for all $x \in M$.
(E) None of the statements is true.

2. Which of these mappings are monomorphisms?
(A) $f:\mathbb{Q} \times \mathbb{Q} \to \mathbb{Q} \times \mathbb{Q}$ such that $f(r_1, r_2) = (2r_1 - r_2, 4r_1 - 2r_2)$ where $\mathbb{Q} \times \mathbb{Q}$ is a $\mathbb{Z}$-module
(B) $f:\mathbb{R} \to \mathbb{R}$ such that $f(r) = 3r + 2$ where $\mathbb{R}$ is a $\mathbb{Q}$-module
(C) $f:\mathbb{Q} \to \mathbb{R}$ such that $f(r) = r$ where $\mathbb{Q}$, $\mathbb{R}$ are $\mathbb{Q}$-modules
(D) $f:\mathbb{R} \times \mathbb{R} \to \mathbb{R}^3$ such that $f(r_1, r_2) = (r_2, r_1, 0)$, $\mathbb{R}$-modules.
(E) None of the mappings is a monomorphism.

3. Which of the following are true for an endomorphism f?
(A) kernel $f \subseteq$ range f.
(B) range $f \subseteq$ domain f.
(C) kernel $f \subseteq$ kernel$(f \circ f)$.
(D) range $f \subseteq$ range$(f \circ f)$.
(E) None of the statements is true.

4. Which of the following statements are correct?
(A) Every epimorphism is additive, homogeneous, and surjective.
(B) Some injections are not monomorphisms.
(C) Every bijection is a monomorphism.
(D) No isomorphism is surjective.
(E) None of the statements is true.

5. Given that $f:M \to M'$ is a morphism of R-modules, which of the following sets fail to be submodules.
(A) $f^{-1}(\zeta')$
(B) $f(M)$
(C) $f^{-1}(M')$
(D) $f(\zeta)$.
(E) All four of the sets are submodules.

Exercises

1. Let $f:\mathbb{R}^2 \to \mathbb{R}^2$ such that $f(r_1, r_2) = (r_1 + 2r_2, 3r_1 + 6r_2)$. Show that f is an $\mathbb{R}$-module morphism. Show that the kernel of f is $\{t(-2, 1) | t \in \mathbb{R}\}$.

2. Let $f:\mathbb{R}^2 \to \mathbb{R}^2$ such that $f(r_1, r_2) = (r_1 + 2r_2, 3r_1 - 6r_2)$. Show that f is an $\mathbb{R}$-module morphism. Show that f is a monomorphism by showing kernel $f = \{(0, 0)\}$. Show that range $f = \mathbb{R}^2$. Conclude that f is an isomorphism.

3. Let $f:M \to M'$ be a morphism of R-modules. Show $f(\zeta) = \zeta'$. Show $f(-x) = -f(x)$ for all $x \in M$.

4. Let $f:M \to M'$ be a function from one R-module to another. Show if $f(rx + sy) = rf(x) + sf(y)$ for all $r, s \in R$, $x, y \in M$ then f is a morphism. Show also the converse.

5. Let S be a set with precisely one element. Let R be a unitary ring. Show that the R-modules R^S and R are isomorphic.

6. Let R be a commutative unitary ring. Argue in favor of R^R being an R-module. Let $\mathscr{E}(R)$ be the set of all *ring* endomorphisms of R. Show that $\mathscr{E}(R)$ is a *submodule* of the R-module R^R.

7. Let $\mathscr{L}$ be the $\mathbb{R}$-submodule of $\mathbb{R}^{\mathbb{N}}$ of all infinite sequences with real limits. Let $\mathscr{N}$ be the $\mathbb{R}$-submodule of $\mathbb{R}^{\mathbb{N}}$ of all infinite sequences with limit equal to zero. Show that $\mathscr{L}/\mathscr{N}$ is isomorphic to $\mathbb{R}$.

8. Let M, M' be isomorphic R-modules. Let N be a submodule of M which is isomorphic with a submodule N' of M'. Construct an isomorphism from M/N to M'/N'.

9. A member of the quotient module M/N of an R-module M by a submodule N we have called a coset or sometimes a linear variety. Let $f:M \to M'$ be a morphism. Show that if $a + N$ is a linear variety of M then $f(a + N)$ is a linear variety of M'. Show also that if $b + N'$ is a linear variety of M' then $f^{-1}(b + N')$ is a linear variety of M. [*Hint*: Do not reprove the first theorem of this section.]

10. Which of the following are morphisms, monomorphisms, epimorphisms, isomorphisms?
    (a) $f:\mathbb{R}^n \to \mathbb{R}$ such that $f(r_1, r_2, \ldots, r_n) = |r_1| + |r_2| + \cdots + |r_n|$.
    (b) $f:\mathbb{R}^n \to \mathbb{R}$ such that $f(r_1, r_2, \ldots, r_n) = r_1 + r_2 + \cdots + r_n$.
    (c) $f:\mathbb{R}^n \to \mathbb{R}$ such that $f(r_1, r_2, \ldots, r_n) = r_1^2 + r_2^2 + \cdots + r_n^2$.
    (d) $f:\mathbb{R}^n \to \mathbb{R}$ such that $f(r_1, r_2, \ldots, r_n) = r_1 + r_2 + \cdots + r_n + 2$.
    (e) $f:\mathbb{R}^{\mathbb{N}+} \to \mathbb{R}$ such that $f(r_1, r_2, \ldots) = r_1 + r_2 + \cdots$.
    (f) $\mathbf{C}[a, b] \to R$ such that $f(\varphi) = \int_a^b \varphi(t)e^{-t}\,dt$ where $\mathbf{C}[a, b]$ is the space of real-valued continuous functions defined on the closed interval $[a, b]$.

11. Let $G = \mathbb{Z} + \mathbb{Z}i = \{x + yi | x, y \in \mathbb{Z}\}$, $i = \sqrt{-1}$. G is called the set of Gaussian integers. Show that G is a $\mathbb{Z}$-module. Find all morphisms $f:G \to G$ such that $f(\mathbb{Z}) \subseteq \mathbb{Z}$ and $f(\mathbb{Z}i) \subseteq \mathbb{Z}i$.

12. Let M, M' be R-modules. We define a function $g:M \to M'$ to be an affine mapping if and only if there exist $b \in M'$ and a morphism $f:M \to M'$ such that $g(x) = f(x) + b$ for all $x \in M$. Now let g be such an affine mapping. Show that if V is a linear variety of M then $g(V)$ is a linear variety of M'. Show that if W is a linear variety of M then $g(V)$ is a linear variety of M'. Show that if W is a linear variety of M' then $g^{-1}(W)$ is a linear variety of M or the empty set.

6.5 Products and direct sums

We discuss the Cartesian product of modules, the direct sum of submodules, and the relationship between these two concepts.

From two given modules, as with rings, we can construct a new module by using the Cartesian product.

Theorem. *If $\langle M', +', \zeta' \rangle$ and $\langle M'', +'', \zeta'' \rangle$ are two given R-modules then $\langle M' \times M'', +, (\zeta', \zeta'') \rangle$ is an R-module with binary operations*

$$(x_1, x_2) + (y_1, y_2) = (x_1 +' y_1, x_2 +'' y_2)$$
$$r(x_1, x_2) = (rx_1, rx_2).$$

Proof. The exterior multiplication of the Cartesian product is achieved by using the exterior multiplication of M' in the first component and the exterior multiplication of M'' in the second component. The addition of the product is commutative and associative which follows from the commutativity and associativity of the addition in each component.

$$\begin{aligned}(x_1, x_2) + (y_1, y_2) &= (x_1 +' y_1, x_2 +'' y_2)\\ &= (y_1 +' x_1, y_2 +'' x_2)\\ &= (y_1, y_2) + (x_1, x_2).\\ (x_1, x_2) + [(y_1, y_2) + (z_1, z_2)] &= (x_1, x_2) + (y_1 +' z_1, y_2 +'' z_2)\\ &= (x_1 +' (y_1 +' z_1), x_2 +'' (y_2 +'' z_2))\\ &= ((x_1 +' y_1) +' z_1, (x_2 +'' y_2) +'' z_2)\\ &= (x_1 +' y_1, x_2 +'' y_2) + (z_1, z_2)\\ &= [(x_1, y_1) + (x_2, y_2)] + (z_1, z_2).\end{aligned}$$

It is very easy to verify similarly the other laws in the definition of the module; for example, $(rs)(x_1, x_2) = ((rs)x_1, (rs)x_2) = (r(sx_1), r(sx_2)) = r(sx_1, sx_2) = r(s(x_1, x_2))$. □

Example. $\langle \mathbb{Z} \times \mathbb{Z}_2, +, (0, 2\mathbb{Z})\rangle$ is the Cartesian product of the $\mathbb{Z}$-modules $\langle \mathbb{Z}, +, 0\rangle$ and $\langle \mathbb{Z}_2, +, 2\mathbb{Z}\rangle$. Some sample calculations are $(2, 1 + 2\mathbb{Z}) + (3, 1 + 2\mathbb{Z}) = (5, 2\mathbb{Z})$ and $4(3, 1 + 2\mathbb{Z}) = (12, 2\mathbb{Z})$.

The Cartesian product constructs a larger module from two given modules. Each of the two given modules is isomorphic to a submodule of the Cartesian product: $M' \times \{\zeta''\}$ and $\{\zeta') \times M''$. We now ask a reverse question. When is a given module the Cartesian product of two modules? Phrased in this manner the question is too restrictive. We introduce a new concept, the direct sum, to answer the question and show that the direct sum shares properties with the Cartesian product.

Definition. An R-module $\langle M, +, \zeta\rangle$ is the *direct sum* of two submodules M_1 and M_2 if and only if $M = M_1 + M_2$ and $M_1 \cap M_2 = \{\zeta\}$. If M is the direct sum of M_1 and M_2 then we write $M = M_1 \oplus M_2$.

Example. The $\mathbb{Z}$-module $\mathbb{Z}_6$ is the direct sum of two submodules: $\{6\mathbb{Z}, 2 + 6\mathbb{Z}, 4 + 6\mathbb{Z}\}$ and $\{6\mathbb{Z}, 3 + 6\mathbb{Z}\}$. Using the notation $\bar{0}, \bar{1}, \bar{2}, \bar{3}, \bar{4}, \bar{5}$ for the six cosets of $\mathbb{Z}_6$ we have $\mathbb{Z}_6 = \{\bar{0}, \bar{2}, \bar{4}\} + \{\bar{0}, \bar{3}\}$ because every element of $\mathbb{Z}_6$ can be written in the form $x_1 + x_2$ with $x_1 \in \{\bar{0}, \bar{2}, \bar{4}\}$ and $x_2 \in \{\bar{0}, \bar{3}\}$. $\bar{0} = \bar{0} + \bar{0}$; $\bar{1} = \bar{4} + \bar{3}$; $\bar{2} = \bar{2} + \bar{0}$; $\bar{3} = \bar{0} + \bar{3}$; $\bar{4} = \bar{4} + \bar{0}$; $\bar{5} = \bar{2} + \bar{3}$. Furthermore, $\{\bar{0}, \bar{2}, \bar{4}\} \cap \{\bar{0}, \bar{3}\} = \{\bar{0}\}$. Thus $\mathbb{Z}_6 = \{\bar{0}, \bar{2}, \bar{4}\} \oplus \{\bar{0}, \bar{3}\}$.

Example. $\mathbb{Z} \times \mathbb{Z} = \mathbb{Z} \times \{0\} \oplus \{0\} \times \mathbb{Z}$.

Another criterion for a module to be a direct sum is contained in this theorem.

Theorem. *An R-module M is the direct sum of two submodules M_1 and M_2 if and only if each element of M is uniquely expressible as a sum of an element in M_1 and an element in M_2.*

PROOF. We know, of course, that a module is a sum of M_1 and M_2 (not necessarily direct) if and only if every member of the module is expressible (not necessarily uniquely) as a sum of an element in M_1 and an element in M_2. We have therefore left to show that the directness of the sum is equivalent to the uniqueness of the expression. First, suppose there are two expressions for some element $x \in M$. $x = x_1 + x_2$ and $x = y_1 + y_2$ with x_1, y_1 in M_1 and x_2, y_2 in M_2. $x_1 + x_2 = y_1 + y_2$. $x_1 - y_1 = y_2 - x_2$. $x_1 - y_1 \in M_1$ and so does $y_2 - x_2$, its equal. But it is clear that $y_2 - x_2$ also belongs to M_2. Thus $y_2 - x_2$ belongs to $M_1 \cap M_2$. But assuming the sum to be direct we have $y_2 - x_2 = \zeta$. $y_2 = x_2$. So also does $x_1 = y_1$. This proves that if the sum is direct then the expression is unique. Conversely, we now suppose that the expression is unique. Let x be in $M_1 \cap M_2$. $x = x + \zeta$ and $x = \zeta + x$ are two sum expressions for x. By uniqueness, $x = \zeta$ and $\zeta = x$. $M_1 \cap M_2 = \{\zeta\}$. □

We now proceed to compare the Cartesian product with the direct sum. By $M \approx N$ we mean M is isomorphic with N.

Theorem. *Let the R-module $\langle M, +, \zeta\rangle$ be the direct sum of two submodules M_1 and M_2. Then M is isomorphic with $M_1 \times M_2$; i.e., $M_1 \oplus M_2 \approx M_1 \times M_2$.*

PROOF. To demonstrate that $M_1 \oplus M_2$ and $M_1 \times M_2$ are isomorphic we construct an isomorphism. Let $f: M_1 \times M_2 \to M_1 \oplus M_2$ such that $f(x_1, x_2) = x_1 + x_2$. First we verify that f is a morphism. $f((x_1, x_2) + (y_1, y_2)) = f(x_1 + y_1, x_2 + y_2) = x_1 + y_1 + x_2 + y_2 = x_1 + x_2 + y_1 + y_2 = f(x_1, x_2) + f(y_1, y_2)$. $f(r(x_1, x_2)) = f(rx_1, rx_2) = rx_1 + rx_2 = r(x_1 + x_2) = rf(x_1, x_2)$. We now verify that f is surjective. Let $x \in M$. $x = x_1 + x_2$ for some x_1 in M_1 and some x_2 in M_2. If we take $f(x_1, x_2) = x_1 + x_2 = x$ with (x_1, x_2) in $M_1 \times M_2$ we see that the mapping is a surjection. To see that the mapping is injective we look at the kernel. Kernel $f = \{(x_1, x_2) | x_1 + x_2 = \zeta\}$. Knowing $M_1 + M_2$ to be a direct sum and comparing $x_1 + x_2 = \zeta$ with $\zeta + \zeta = \zeta$ we conclude $x_1 = \zeta$ and $x_2 = \zeta$. Thus kernel $f = \{(\zeta, \zeta)\}$. □

In this last theorem M is not the Cartesian product of M_1 and M_2 but merely isomorphic with the Cartesian product. The elements of the Cartesian product must be ordered pairs of elements of M and therefore M is not the Cartesian product.

EXAMPLE. As before, $\mathbb{Z}_6 = \{\bar{0}, \bar{2}, \bar{4}\} \oplus \{\bar{0}, \bar{3}\}$. The Cartesian product of $\{\bar{0}, \bar{2}, \bar{4}\}$ and $\{\bar{0}, \bar{3}\}$ is $\{(\bar{0}, \bar{0}), (\bar{0}, \bar{3}), (\bar{2}, \bar{0}), (\bar{2}, \bar{3}), (\bar{4}, \bar{0}), (\bar{4}, \bar{3})\}$. The isomorphism described in the previous theorem, $f:\{\bar{0}, \bar{2}, \bar{4}\} \times \{\bar{0}, \bar{3}\} \to \mathbb{Z}_6$, has the following images:

$$\begin{array}{ll} (\bar{0}, \bar{0}) \mapsto \bar{0} + \bar{0} = \bar{0} & (\bar{2}, \bar{3}) \mapsto \bar{2} + \bar{3} = \bar{5} \\ (\bar{0}, \bar{3}) \mapsto \bar{0} + \bar{3} = \bar{3} & (\bar{4}, \bar{0}) \mapsto \bar{4} + \bar{0} = \bar{4} \\ (\bar{2}, \bar{0}) \mapsto \bar{2} + \bar{0} = \bar{2} & (\bar{4}, \bar{3}) \mapsto \bar{4} + \bar{3} = \bar{1}. \end{array}$$

Since $\{\bar{0}, \bar{2}, \bar{4}\} \approx \mathbb{Z}_3$ and $\{\bar{0}, \bar{3}\} \approx \mathbb{Z}_2$ we assert that $\mathbb{Z}_6 \approx \mathbb{Z}_3 \times \mathbb{Z}_2$ (consult Exercise 2).

We continue with a theorem showing some more of the nature of the direct sum.

Theorem. *Let M_1, M_2 be submodules of an R-module M such that $M = M_1 \oplus M_2$. Then $M_2 \approx M/M_1$.*

Proof. We show the existence of an isomorphism $g:M_2 \to M/M_1$. We define $g(x_2) = x_2 + M_1$. Each x_2 in M_2 has a uniquely defined coset in M/M_1 as its g image. We show g to be a morphism. $g(x_2 + y_2) = x_2 + y_2 + M_1 = x_2 + M_1 + y_2 + M_1 = g(x_2) + g(y_2)$. $g(rx_2) = rx_2 + M_1 = r(x_2 + M_1) = rg(x_2)$. Kernel $g = \{x_2 | x_2 \in M_2 \text{ and } g(x_2) = M_1\} = \{x_2 | x \in M_2 \text{ and } x_2 + M_1 = M_1\} = \{x_2 | x_2 \in M_2 \text{ and } x_2 \in M_1\} = M_1 \cap M_2 = \{\zeta\}$. Finally, let $x + M_1 \in M/M_1$. $x = x_1 + x_2$ for some $x_1 \in M_1$, $x_2 \in M_2$. $g(x_2) = x_2 + M_1 = x_1 + x_2 + M_1 = x + M_1$. We have showed g to be a surjection, an injection and a morphism. □

Example. As before $\mathbb{Z}_6 = \{\bar{0}, \bar{2}, \bar{4}\} \oplus \{\bar{0}, \bar{3}\}$. $\mathbb{Z}_6/\{\bar{0}, \bar{2}, \bar{4}\} \approx \{\bar{0}, \bar{3}\} \approx \mathbb{Z}_2$.

A word of caution is in order. If $M_2 = M/M_1$ it is not necessarily the case that $M \approx M_1 \oplus M_2$. For example, $\mathbb{Z}_4/\{\bar{0}, \bar{2}\} \approx \mathbb{Z}_2$, but $\mathbb{Z}_4$ is not the direct sum of proper submodules. The next theorem pins down this lack of a direct converse more precisely.

Theorem. *Let M be an R-module and P a submodule of M. Then there exists a submodule N of M such that $P \oplus N = M$ if and only if there exists a morphism $f:M \to P$ which is an extension of the identity $P \to P$.*

Proof. First, suppose there is a submodule N such that $P \oplus N = M$. Every element of x of M can be uniquely expressed as a sum $x_p + x_n$ with $x_p \in P$, $x_n \in N$. We define $f:P \oplus N \to P$ such that $f(x_p + x_n) = x_p$. f is an epimorphism which is the identity when restricted to P. By way of motivation we can notice that f is like the projection $p_1:P \times N \to P$ for f is really the composite of the natural isomorphism $P \oplus N \to P \times N$ and the projection $P \times N \to P$.

Conversely, suppose there exists a morphism $f:M \to P$ which extends the identity. It is then an epimorphism. Since $f':M/\ker f \to P$ is an isomorphism we propose to show that $M = P + \ker f$. Given x in M we have $x = f(x) + (x - f(x))$, $f(x) \in P$. Since f is the identity on P, $f(f(x)) = f(x)$ yielding $f(f(x) - x) = \zeta$ for all $x \in M$. Thus $x - f(x)$ belongs to kernel f. This shows x is the sum of $f(x)$ in P and a member of kernel f. $M = P + \ker f$. To show that the sum is direct let $x \in P \cap \ker f$. $x \in$ implies $f(x) = x$ and $x \in \text{kernel} f$ implies $f(x) = \zeta$. $x = \zeta$. $M = P \oplus \ker f$. □

Questions

1. If $\langle M', +', \zeta' \rangle$ and $\langle M'', +'', \zeta'' \rangle$ are R-modules then which of the following statements are true?
   (A) M' is a submodule of the Cartesian product $M' \times M''$.
   (B) M' is not a submodule of the Cartesian product $M' \times M''$ but is isomorphic to a submodule of $M' \times M''$.
   (C) M' is not, in general, isomorphic to a submodule of $M' \times M''$.
   (D) M' is never isomorphic to a submodule of $M' \times M''$.
(E) None is true.

2. The $\mathbb{Z}$-module $\mathbb{Z}_8$ is which of the following direct sums?
   (A) $\mathbb{Z}_2 \oplus \mathbb{Z}_4$
   (B) $(\mathbb{Z}_2 \oplus \mathbb{Z}_2) \oplus \mathbb{Z}_2$
   (C) $\mathbb{Z}_3 \oplus \mathbb{Z}_3$
   (D) $((\mathbb{Z}_2 \oplus \mathbb{Z}_2) \oplus \mathbb{Z}_2) \oplus \mathbb{Z}_2$
(E) $\mathbb{Z}_8$ is none of the direct sums listed.

3. If M_1 and M_2 are submodules of an R-module $\langle M, +, \zeta \rangle$ then which of the following are true?
   (A) $M_1 \oplus M_2 = M$ implies there exists an epimorphism $f: M_1 \times M_2 \to M$.
   (B) $M_1 \cap M_2 = \{\zeta\}$ implies there exists a monomorphism $f: M_1 \times M_2 \to M$.
   (C) There exists $x \in M$ with two distinct representations in $M_1 + M_2$ (i.e., $x = x_1 + x_2 = y_1 + y_2$ with $x_1, y_1 \in M_1$ and $x_2, y_2 \in M_2$ and ($x_1 \neq y_1$ or $x_2 \neq y_2$) implies every x in M has two distinct representations in $M_1 + M_2$.
   (D) $M_1 \subseteq M_2$ implies $M_1 + M_2 \neq M$.
(E) None of the statements is true.

4. Given $M = M_1 + M_2$, M_1 and M_2 submodules of an R-module M, the map $g: M_2 \to M/M_1$ such that $g(x_2) = x_2 + M_1$
   (A) maps each element of M_2 into its containing coset
   (B) is an epimorphism
   (C) is a monomorphism if and only if $M_1 \cap M_2 = \{\zeta\}$
   (D) is the identity if $M_2 \subseteq M_1$.
(E) None of the alternatives completes a true sentence.

Exercises

1. Show that the R-modules R^3 and $R^2 \times R$ are isomorphic.

2. Show that if M_1, M_2, N_1, N_2 are all R-modules and $M_1 \approx N_1$ and $M_2 \approx N_2$ then $M_1 \times M_2 \approx N_1 \times N_2$.

3. If M and N are R-modules prove that $(M \times N)/(\{\zeta\} \times N) \approx M$.

4. Let R be a unitary ring. Show that $R \times R$ is isomorphic (as an R-module) with R^S if S is a set with precisely two members.

5. Show that $R^2 = \{(r_1, \theta)|r_1 \in R\} \oplus \{(\theta, r_2)|r_2 \in R\}$ for any unitary ring R. The sum, of course, is an R-module direct sum.

6. Let M_1, M_2 be R-modules with submodules N_1, N_2 respectively. Show that $(M_1 \times M_2)/(N_1 \times N_2) \approx (M_1/N_1) \times (M_2/N_2)$.

7. Given R-modules M_1, M_2 show that the two projection functions $p_1: M_1 \times M_2 \to M_1$ such that $p_1(x_1, x_2) = x_1$ and $p_2: M_1 \times M_2 \to M_2$ such that $p_2(x_1, x_2) = x_2$ are epimorphisms.

8. If M_1, M_2 are submodules of an R-module M such that $M = M_1 \oplus M_2$ show that the two embedding mappings $q_1: M_1 \to M$ such that $q_1(x_1) = x_1$ and $q_2: M_2 \to M$ such that $q_2(x_2) = x_2$ are monomorphisms.

9. Let S, T be disjoint sets and R a unitary ring. Prove that there is an isomorphism between the two R-modules $R^S \times R^T$ and $R^{S \cup T}$.

10. Let N_1 and N_2 be submodules of some given module over a commutative unitary ring R. For any $r \in R$, show that rN_1 is also a submodule of M. Show that $r(N_1 \oplus N_2) = rN_1 \oplus rN_2$ for each $r \in R$.

11. Let M be a module over a commutative unitary ring R. Let $r \in R$. Show that the function $r\cdot\cdot: M \to M$ such that $r\cdot(x) = rx$ is an endomorphism of M. $r\cdot\cdot: M \to M$ is, of course, simply exterior multiplication by the ring element r.

12. Let M be a module over a commutative unitary ring R. Let $r \in R$. Denote the kernel of $r\cdot\cdot: M \to M$ by M_r. Show that if $M = N_1 \oplus N_2$ then $(N_1 \oplus N_2)_r = (N_1)_r \oplus (N_2)_r$.

13. Let M and M' be modules over an integral domain R. An element x in M has a nontrivial annihilator $r \in R$ if and only if $r \neq \theta$ and $rx = \zeta$. Show that if M and M' both have elements with nontrivial annihilators then the Cartesian product module $M \times M'$ also has an element with a nontrivial annihilator.

6.6 Families and matrices

In this section we develop family notation and discuss multiplication of matrices and the algebra of square matrices.

We introduce family notation to have a more unified view of matrices, sequences, and other functions; the family notation gives us an alternate way of writing functions. We also have in mind some notational problems with linear combinations of vectors which will occur in Section 6.7. The family notation will allow us to handle these accurately.

We have earlier seen in Section 6.1 that various function spaces form modules. We now talk about these spaces with the family notation.

Definition. A *family* $(x_i|i \in I)$ is a function with domain I and value x_i for each $i \in I$. I is called the index set of the family. $(x_i|i \in I)$ is also frequently written $(x_i)_{i \in I}$.

EXAMPLES.

$(x_i|i \in \mathbb{N}^+)$ is the sequence $(x_1, x_2, x_3, \ldots)$.
$(x_i|i \in \hat{3})$ is the triple (x_1, x_2, x_3).
$(x_i|i \in \mathbb{R})$ is the function $f: \mathbb{R} \to \mathbb{R}$ such that $f(i) = x_i$.
$(1/n|n \in \mathbb{N}^+)$ is the sequence $(1, \frac{1}{2}, \frac{1}{3}, \ldots)$.
$(1/n^2|n \in \hat{3})$ is the triple $(1, \frac{1}{4}, \frac{1}{9})$.
$(n^2|n \in \mathbb{R})$ is the function $f: \mathbb{R} \to \mathbb{R}$ such that $f(n) = n^2$.

One should distinguish carefully between the family $(x_i|i \in I)$ and the range of the family, $\{x_i|i \in I\}$. The first is the function itself and the second is the set of all values of the function. $((-1)^n|n \in \mathbb{N})$ is the sequence $(1, -1, 1, -1, \ldots)$ whereas $\{(-1)^n|n \in \mathbb{N}\}$ is the set $\{1, -1\}$.

We now list some of the collections of families we have considered. For a unitary ring R and a set I we have the R-module of all functions from I to R,

$$R^I = \{(x_i|i \in I)|x_i \in R\}.$$

This R-module can be made up from sequences, n-ples, or m-rowed, n-columned matrices.

$$R^{\mathbb{N}} = \{(x_i|i \in \mathbb{N})|x_i \in R\}.$$
$$R^{\hat{n}} = \{(x_i|i \in \hat{n})|x_i \in R\} = \{(x_1, x_2, \ldots, x_n)|x_i \in R\}.$$
$$R^{\hat{m} \times \hat{n}} = \{(x_{ij}|(i, j) \in \hat{m} \times \hat{n}\,|\,x_{ij} \in R\}.$$
$$= \left\{ \begin{pmatrix} x_{11} & x_{12} & \cdots & x_{1n} \\ x_{21} & x_{22} & \cdots & x_{2n} \\ \cdots & & & \\ x_{m1} & x_{m2} & \cdots & x_{mn} \end{pmatrix} \middle| x_{ij} \in R \right\}, \quad m, n \in \mathbb{N}.$$

We also find the family notation useful for families of sets:

$$(A_i|i \in I), \qquad (A_i|i \in \mathbb{N}), \qquad (A_i|i \in \hat{3}).$$

We can define for such families of sets the union and the intersection. $\bigcup(A_i|i \in I) = \bigcup_{i \in I} A_i = \bigcup\{x|x \in A_i \text{ for some } i \in I\} = \bigcup\{A_i|i \in I\}$. $\bigcap(A_i|i \in I) = \bigcap_{i \in I} A_i = \bigcap\{x|x \in A_i \text{ for all } i \in I\} = \bigcap\{A_i|i \in I\}$. For the intersection we must assume $I \neq \varnothing$.

The R-module of m-rowed, n-columned *matrices* with entries in R, $R^{\hat{m} \times \hat{n}}$, has its operations expressed as follows in family notation: $(x_{ij}|(i, j) \in \hat{m} \times \hat{n}) + (y_{ij}|(i, j) \in \hat{m} \times \hat{n}) = (x_{ij} + y_{ij}|(i, j)\ \hat{m} \times \hat{n})$. $r(x_{ij}|(i, j) \in \hat{m} \times \hat{n}) = (rx_{ij}|(i, j) \in \hat{m} \times \hat{n})$. With the number of rows and columns understood from context one frequently abbreviates to $(x_{ij}) + (y_{ij}) = (x_{ij} + y_{ij})$ and $r(x_{ij}) = (rx_{ij})$.

EXAMPLES. $(1, -3, 4) + (2, 7, 3) = (3, 4, 7)$. $2(1, 2) = (2, 4)$. $(1/n|n \in \mathbb{N}^+) + (-1/n^2|n \in \mathbb{N}^+) = ((n - 1)/n^2|n \in \mathbb{N}^+)$. $(2/(m + n)|(m, n) \in \hat{3} \times \hat{2}) = \begin{pmatrix} 1 & \frac{2}{3} \\ \frac{2}{3} & \frac{1}{2} \\ \frac{1}{2} & \frac{2}{5} \end{pmatrix}$.

We propose now to introduce another operation on matrices, called multiplication. We precede the formal definition with a few examples of multiplication.

$$\begin{pmatrix} 1 & 2 & 6 \\ 0 & 4 & 7 \end{pmatrix} \begin{pmatrix} 1 \\ -1 \\ 2 \end{pmatrix} = \begin{pmatrix} (1)(1) + (2)(-1) + (6)(2) \\ (0)(1) + (4)(-1) + (7)(2) \end{pmatrix} = \begin{pmatrix} 11 \\ 10 \end{pmatrix}.$$

$$\begin{pmatrix} 1 & 1 \\ 1 & 0 \end{pmatrix} \begin{pmatrix} 1 & 1 \\ 0 & 1 \end{pmatrix} = \begin{pmatrix} 1 & 2 \\ 1 & 1 \end{pmatrix}. \quad \begin{pmatrix} 1 & 1 \\ 0 & 1 \end{pmatrix} \begin{pmatrix} 1 & 1 \\ 1 & 0 \end{pmatrix} = \begin{pmatrix} 2 & 1 \\ 1 & 0 \end{pmatrix}.$$

Definition. We let R be a unitary ring. For $(x_{ij}) \in R^{m \times n}$ and $(y_{jk}) \in R^{n \times p}$, $m, n, p \in \mathbb{N}$, we define

$$(x_{ij})(y_{jk}) = (u_{ik}) \quad \text{where } u_{ik} = \sum_{j=1}^{n} x_{ij} y_{jk}.$$

In order for the product to be defined the number of columns of the first matrix must be the same as the number of rows of the second matrix. The entry in row p and column q of the product is computed using row p of the first matrix and column q of the second matrix. Two examples given before the definition show that matrix multiplication is noncommutative. Some properties of matrix multiplication are now listed.

Theorem. *Let R be a unitary ring. Matrix multiplication is associative. Matrix multiplication is left and right distributive with respect to matrix addition. If R is a commutative ring then*

$$r[(x_{ij})(y_{jk})] = [r(x_{ij})](y_{jk}) = (x_{ij})[r(y_{jk})]$$

for all $r \in R$ and matrices of the appropriate size with entries in R.

Proof. We show the associativity and leave the rest to the reader.

$$\begin{aligned}(x_{ij})[(y_{jk})(z_{kl})] &= (x_{ij})\left(\sum_{k=1}^{p} y_{jk} z_{kl}\right) = \left(\sum_{j=1}^{n} x_{ij} \sum_{k=1}^{p} y_{jk} z_{kl}\right) \\ &= \left(\sum_{j=1}^{n} \sum_{k=1}^{p} x_{ij}(y_{jk} z_{kl})\right) = \left(\sum_{k=1}^{p} \sum_{j=1}^{n} (x_{ij} y_{jk}) z_{kl}\right) \\ &= \left(\sum_{j=1}^{n} x_{ij} y_{jk}\right)(z_{kl}) = [(x_{ij})(y_{jk})](z_{kl}).\end{aligned}$$ □

In order to have a structure closed under matrix multiplication it is necessary that the matrices be square: the number of rows and the number of columns of the matrices are equal. We now introduce a new name for a structure describing the properties of square matrices.

Definition. Let $\langle M, +, \zeta\rangle$ be a module over a commutative unitary ring $\langle R, +, \cdot, \theta, v\rangle$. Let there also be given an associative product $M \times M \to M$. Furthermore, let

$$\begin{aligned} x(y + z) &= xy + xz && \text{for all } x, y, z \in M \\ (y + z)x &= yx + zx && \text{for all } x, y, z \in M \\ r(xy) &= (rx)y = x(ry) && \text{for all } r \in R, \quad x, y \in M. \end{aligned}$$

Then we call $\langle M, +, \cdot, \zeta\rangle$ an *R-algebra* over the commutative unitary ring $\langle R, +, \cdot, \theta, v\rangle$.

It immediately follows that

Theorem. *For any unitary commutative ring $\langle R, +, \cdot, \theta, v\rangle$, the R-module of square matrices $R^{n\times n}$ is an R-algebra.*

Definition. If every element of an R-algebra $\langle M, +, \cdot, \zeta\rangle$ except ζ has a multiplicative inverse then M is called a *division algebra.*

There are fascinating results that the only division algebras over the ring of real numbers consist of the real numbers themselves, the complex numbers, and an algebra called the quaternions. If one drops the requirement that the multiplication be associative then there is one more called the Cayley algebra. These results are beyond the scope of this book.

QUESTIONS

1. Which of the following statements are true?
   (A) $\text{Range}(x_i|i \in I) = \{x_i|i \in I\}$.
   (B) $\text{Domain}(x_i|i \in I) = I$.
   (C) $x_i = x_j$ and $i \neq j$; $i, j \in I$ imply the family $(x_i|i \in I)$ is not an injection.
   (D) $\text{Codomain}(x_i|i \in I) = I$.
   (E) None of the statements is true.

2. Which of the following statements are true?
   (A) $(n^2|n \in \mathbb{N}) \subseteq (n^2|n \in \mathbb{Q})$.
   (B) $\text{Index}(n^2|n \in \mathbb{N}) \subseteq \mathbb{Q}$.
   (C) $\bigcap(A_i|i \in \mathbb{N}) \subseteq \bigcup(A_i|i \in \mathbb{N})$.
   (D) $\bigcup(A_i|i \in \mathbb{N}) = \bigcup \text{range}(A_i|i \in \mathbb{N})$.
   (E) None of the statements is true.

3. Which of these statements are true if $m \neq n$?
   (A) $\mathbb{Q}^{m\times n}$ is a $\mathbb{Q}$-module.
   (B) $\mathbb{Q}^{m\times n}$ is a $\mathbb{Q}$-vector space.
   (C) $\mathbb{Q}^{m\times n}$ is a $\mathbb{Q}$-algebra.
   (D) $\mathbb{Q}^{m\times n}$ is a $\mathbb{Q}$-division algebra.
   (E) None of the statements is true.

4. Which of these statements are true of the complex numbers, $\mathbb{C}$? (Cf. Section 2.5, Exercise 15.)
   (A) $\mathbb{C}$ is an $\mathbb{R}$-module.
   (B) $\mathbb{C}$ is an $\mathbb{R}$-vector space.
   (C) $\mathbb{C}$ is an $\mathbb{R}$-algebra.
   (D) $\mathbb{C}$ is an $\mathbb{R}$-division algebra.
 (E) None of the statements is true.

5. Which of these statements are true? R is a unitary commutative ring.
   (A) $R^{n \times n}$ is a unitary algebra over R.
   (B) $R^{n \times n}$ is not a division algebra over R if $n > 1$.
   (C) $R^{n \times n}$ is a division algebra over R if $n = 1$ and R is a field.
   (D) $R^{n \times n}$ is a noncommutative algebra over R if $n > 1$.
 (E) None of the statements is true.

Exercises

1. Compute these products:

$$(1 \quad 0 \quad 2)\begin{pmatrix} 6 & -1 & 2 \\ 5 & 0 & 4 \\ 1 & 3 & 2 \end{pmatrix} \qquad \begin{pmatrix} 1 & 2 \\ -1 & 4 \end{pmatrix}\begin{pmatrix} 6 & 0 \\ 2 & 7 \end{pmatrix}$$

$$(1 \quad 0 \quad 2 \quad 3)\begin{pmatrix} 6 \\ 0 \\ 2 \\ 4 \end{pmatrix} \qquad (x \quad y \quad z)\begin{pmatrix} 2 & 6 & 3 \\ 5 & 0 & 4 \\ 4 & 2 & 1 \end{pmatrix}.$$

2. Show that matrix multiplication is left distributive with respect to matrix addition.

3. Show that both of the following matrix equations are equivalent to the same simultaneous linear equations.

$$(x \quad y \quad z)\begin{pmatrix} 2 & 6 & 3 \\ -1 & 0 & -2 \\ 4 & 2 & 1 \end{pmatrix} = (7 \quad 1 \quad 2). \qquad \begin{pmatrix} 2 & -1 & 4 \\ 6 & 0 & 2 \\ 3 & -2 & 1 \end{pmatrix}\begin{pmatrix} x \\ y \\ z \end{pmatrix} = \begin{pmatrix} 7 \\ 1 \\ 2 \end{pmatrix}.$$

4. Let δ_{ij} be the Kronecker delta symbol: $\delta_{ij} = 1$ if $i = j$ and $\delta_{ij} = 0$ if $i \neq j$. Let R be a commutative unitary ring. For a matrix $(x_{ij} | (i, j) \in m \times n)$ with entries in R show that $(\delta_{hi}\nu | (h, i) \in m \times m)$ is a left neutral element of multiplication and $(\delta_{jk}\nu | (j, k) \in n \times n)$ is a right neutral element of multiplication. For example,

$$\begin{pmatrix} 1 & 2 & 3 \\ 4 & 5 & 6 \end{pmatrix}\begin{pmatrix} 1 & 0 & 0 \\ 0 & 1 & 0 \\ 0 & 0 & 1 \end{pmatrix} = \begin{pmatrix} 1 & 2 & 3 \\ 4 & 5 & 6 \end{pmatrix}.$$

5. Find in $\mathbb{Q}^{2 \times 2}$ the multiplicative inverse of the matrix $\begin{pmatrix} 1 & 2 \\ 3 & 4 \end{pmatrix}$.

6. Let R be a unitary ring and I be an infinite set. Define $M = \{(x_i | i \in I) | x_i \in R$ and $x_i = \theta$ for all except for a finite number of $i\}$. Show that M is a submodule of the R-module R^I.

7. Show that $R[X]$ is an R-algebra for any commutative unitary ring R.

8. Find $A^2 - 10A + 24$ when $A = \begin{pmatrix} 6 & 2 \\ 1 & 4 \end{pmatrix}$ and $\delta = \begin{pmatrix} 1 & 0 \\ 0 & 1 \end{pmatrix}$.

9. Find $(A - 6\delta)(A - 4\delta)$ when $A = \begin{pmatrix} 6 & 2 \\ 1 & 4 \end{pmatrix}$.

10. A generalization of the Cartesian product of two modules is the Cartesian product of any family of R-modules, $(M_i|i \in I)$. $\times(M_i|i \in I) = \times_{i \in I} M_i = \{(x_i|i \in I)|x_i \in M_i$ for all $i \in I\}$. With an addition, $(x_i|i \in I) + (y_i|i \in I) = (x_i + y_i|i \in I)$, and an exterior multiplication, $r(x_i|i \in I) = (rx_i|i \in I)$, the Cartesian product of R-modules is itself an R-module. Verify this result. An important submodule of $\times(M_i|i \in I)$ is $\times^w(M_i|i \in I)$, the weak Cartesian product, consisting of all families which have the value ζ except for a finite number of values. $\times^w(M_i|i \in I) = \{(x_i|i \in I)|x_i \in M_i$ and $x_i = \zeta$, except for a finite number of values$\}$. Show that $\times^w(M_i|i \in I)$ is a submodule of $\times(M_i|i \in I)$.

11. The direct sum of a family $(M_i|i \in I)$ of submodules of an R-module M is defined as follows: $M = \bigoplus(M_i|i \in I)$ if and only if $M = \sum^w (M_i|i \in I) = \sum^w_{i \in I} M_i$ and $M_j \cap [\bigcup_{i \neq j} M_i] = \{\zeta\}$. Show that $\bigoplus(M_i|i \in I)$ and $\times^w(M_i|i \in I)$ are isomorphic.

12. Given a matrix $A = (A_{ij}|(i, j) \in m \times n)$ with m rows and n columns, the matrix obtained from this given matrix by interchanging rows for columns and columns for rows is called the *transpose* of A and is written A^*. $A^* = (A_{ij}|(i, j) \in m \times n)^* = (B_{ji}|(j, i) \in n \times m)$ with $B_{ji} = A_{ij}$. For example, $\begin{pmatrix} 1 & 2 & 3 \\ 4 & 5 & 6 \end{pmatrix}^* = \begin{pmatrix} 1 & 4 \\ 2 & 5 \\ 3 & 6 \end{pmatrix}$. Show that $(A + B)^* = A^* + B^*$; $(AB)^* = B^*A^*$.

6.7 Bases

In this section we discuss families of vectors which are linearly independent, which generate submodules and which are bases. We also give a function space model for a module with a basis.

A linear combination of vectors is a finite sum of multiples of the vectors; for example, $7(1, 3, -5) + 2(1, 4, 2) - 3(2, 0, 4)$ is a linear combination of the vectors $(1, 3, -5)$, $(1, 4, 2)$, $(2, 0, 4)$. For a given family of vectors $((1, 3, -5), (1, 4, 2), (2, 0, 4))$ there are many linear combinations

$$r_1(1, 3, -5) + r_2(1, 4, 2) + r_3(2, 0, 4)$$

depending upon which ring scalars r_1, r_2, r_3 are chosen. We can form linear combinations also from infinite families provided we use only a finite number of vectors in any one sum.

Definition. A *linear combination* of a family $(x_i|i \in I)$ of vectors or elements of an R-module $\langle M, +, \zeta\rangle$ is a sum $\sum^w_{i \in I} r_i x_i$ in which $r_i \in R$ and all but a finite number of the coefficients r_i are zero.

It will be understood that if $r_i = \theta$ then the term $r_i x_i$ will not appear in the sum. If all r_i are zero then the sum is taken by definition to be ζ. In any sum there are only a finite number of nonzero terms and therefore the sum is finite and well defined. This understanding for sums of infinite families will save us complication of notation. We have included the superscript w on the sum to remind us that the sum is finite and all but a finite number of the terms are zero. The symbol stands for the word *weak* and the sum is called a *weak sum.*

EXAMPLE. Let $e_1 = (1, 0, 0, \ldots)$, $e_2 = (0, 1, 0, 0, \ldots)$, $e_3 = (0, 0, 1, 0, \ldots)$, ... $(e_j | j \in \mathbb{N}^+)$ is an infinite family of infinite sequences. Some linear combinations are, for example, $6(0,1,0,0,\ldots)+3(0,0,1,0,0,\ldots)-2(0,0,0,0,1,0,\ldots)$ and $2(1,0,0,\ldots) + 4(0,0,0,0,0,0,0,1,0,\ldots)$. A linear combination of the family $(e_i | i \in \mathbb{N}^+)$, $\sum_{j \in \mathbb{N}^+}^w r_j e_j$, always has all save a finite number of terms equal to zero; $r_j = \theta$ for all but a finite number of $j \in \mathbb{N}^+$.

We can and do consider the set of all linear combinations of a given family of vectors $(x_i | i \in I)$ of a module M

$$\left\{ \sum_{i \in I}^{w} r_i x_i \middle| r \in R \right\}.$$

This set is a submodule of M.

Theorem. *Let $\langle M, +, \zeta \rangle$ be an R-module and $(x_i | i \in I)$ be a family of vectors in M. Then the set of all linear combinations of the given family is a submodule of M and is the submodule generated by the set $\{x_i | i \in I\}$.*

PROOF. The choice $r_i = \theta$ for all $i \in I$ produces the linear combination $\sum_{i \in I}^w \theta x_i = \zeta$. The sum of two linear combinations $\sum_{i \in I}^w r_i x_i$ and $\sum_{i \in I}^w s_i x_i$ is the vector $\sum_{i \in I}^w (r_i + s_i) x_i$, another linear combination. If $s \in R$ and $\sum_{i \in I}^w r_i x_i$ is a linear combination then so also is $s(\sum_{i \in I}^w r_i x_i) = \sum_{i \in I}^w (sr_i) x_i$. Thus the set of linear combinations is a submodule of M. Since one possible choice of scalars is $r_k = v$ and $r_j = \theta$ for all $j \neq k$ we must have x_k in the submodule for every $k \in I$. The submodule then contains all vectors in the given family. On the other hand, if any submodule of M contains the vectors $\{x_i | i \in I\}$ it must contain every linear combination of those vectors. The set of linear combinations is $\{x_i | i \in I\}$. □

We are then led to this natural definition.

Definition. We define the submodule *generated* by a family $(x_i | i \in I)$ to be the submodule generated by the range of the family: $\{x_i | i \in I\}$.

We note that the use of families permits repetitions of a vector while in a set there can be no repetitions of vectors. This is to say, in a given set a vector cannot belong twice. It is either a member of a set or it is not a member of the set. On the other hand, a family may have the same value for several different arguments. The question of repeated vectors becomes important when we discuss linear independence.

Definition. A family $(x_i|i \in I)$ of vectors in an R-module M is *linearly dependent* if and only if there exists a family of scalars $(r_i|r \in I)$ in R, not all zero, such that $\sum_{i\in I}^{w} r_i x_i = \zeta$. It is equivalent that a family is *linearly independent* (not linearly dependent) if and only if for every family of scalars $(r_i|i \in I)$ in R, $\sum_{i\in I}^{w} r_i x_i = \zeta$ implies $r_i = \theta$ for all $i \in I$.

EXAMPLES. In $\mathbb{R}^3$, $((1, 1, 0), (0, 1, 0), (0, 1, 1))$ is a linearly independent family because $r_1(1, 1, 0) + r_2(0, 1, 0) + r_3(0, 1, 1) = (0, 0, 0)$ implies

$$(r_1, r_1 + r_2, r_3) = (0, 0, 0).$$

This implies $r_1 = 0$, $r_1 + r_2 = 0$, $r_3 = 0$ which implies $r_1 = 0$, $r_2 = 0$, $r_3 = 0$.

In $R^{\mathbb{N}^+}$, $(e_j|j \in \mathbb{N}^+)$ is a linearly independent family. Suppose $\sum_{i\in I}^{w} r_i e_i = \zeta$. All but a finite number of $r_i = \theta$. Let $r_{i_1}, r_{i_2}, \ldots, r_{i_k}$ be the finite number of coefficients not known to be zero. $r_{i_1}e_{i_1} + r_{i_2}e_{i_2} + \cdots + r_{i_k}e_{i_k} = (\theta, \theta, \ldots)$. Since e_{i_1} has a v in place number i and elsewhere zero we get $r_{i_1} = \theta$. Similarly, $r_{i_2} = \theta, \ldots, r_{i_k} = \theta$. Hence all $r_i = \theta$. This establishes the linear independence.

An example of a dependent family is $((1, 2), (3, 4), (-1, 5))$ in $\mathbb{R}^2$. Assuming $r_1(1, 2) + r_2(3, 4) + r_3(-1, 5) = (0, 0)$ in $\mathbb{R}^2$ yields $(r_1 + 3r_2 - r_3, 2r_1 + 4r_2 + 5r_3) = (0, 0)$. This yields $r_1 + 3r_2 - r_3 = 0$ and $2r_1 + 4r_2 + 5r_3 = 0$. If we can prove r_1, r_2, r_3 must all be zero then the family is linearly independent. On the other hand, if we can find nonzero solutions for r_1, r_2, r_3 then the family is linearly dependent. Leaving all details to the reader we simply note one nonzero solution is $r_1 = -\frac{19}{2}$, $r_2 = \frac{7}{2}$, $r_3 = 1$.

$$-\tfrac{19}{2}(1, 2) + \tfrac{7}{2}(3, 4) + (1)(-1, 5) = (0, 0).$$

EXAMPLE. We now return to and examine the example $\mathscr{D}_0$ of directed line segments emanating from the origin for the intuitive meaning of linear independence. $\mathscr{D}_0$ is the example discussed in Section 6.2. We offer in this discussion only intuitive geometrical arguments; our goal here is an intuitive one and the discussion is not meant to be part of the logical structure of the section'

A family of a single vector is linearly independent if and only if the vector is not the zero vector $\overrightarrow{00}$. This is true because multiplying any nonzero arrow by a nonzero scalar produces an arrow of nonzero length. Two distinct nonzero arrows are linearly independent if and only if they do not terminate on the same line. This is true because they terminate on the same line if and only if their coordinates (projections on the axes) are proportional. $\overrightarrow{0a} = r\overrightarrow{0b}$ for some nonzero $r \in \mathbb{R}$. This is equivalent to $(1)\overrightarrow{0a} - (r)\overrightarrow{0b} = \overrightarrow{00}$ and linear dependence (see Figure 6.9).

A family of three distinct nonzero vectors (arrows) is linearly independent if and only if the three directed line segments are not contained in the same plane. First assume the three arrows are dependent. $r\overrightarrow{0a} + s\overrightarrow{0b} + t\overrightarrow{0c} = \overrightarrow{00}$ with at least one of the numbers r, s, t being nonzero. For convenience, we assume it is the $t \neq 0$. Then $\overrightarrow{0c} = (-r/t)\overrightarrow{0a} + (-s/t)\overrightarrow{0b}$. Multiples of $\overrightarrow{0a}$ lie on the line containing $\overrightarrow{0a}$; multiples of $\overrightarrow{0b}$ lie on the line containing $\overrightarrow{0b}$. Sums of $\overrightarrow{0a}$ and $\overrightarrow{0b}$ or multiples of $\overrightarrow{0a}$ and multiples of $\overrightarrow{0b}$ lie in the plane determined by 0, a, and b. $\overrightarrow{0c}$ lies in the plane determined by 0, a, and b (see Figure 6.10).

Conversely, assume $\overrightarrow{0c}$ lies in the plane of $\overrightarrow{0a}$ and $\overrightarrow{0b}$. Construct in this plane parallels through c to $\overrightarrow{0b}$ and to $\overrightarrow{0a}$. (We assume $\overrightarrow{0b}$ and $\overrightarrow{0a}$ are not themselves collinear for in that case obviously the three arrows are linearly dependent.) The parallels intersect the lines through $\overrightarrow{0a}$ and through $\overrightarrow{0b}$ respectively at a' and b'. $\overrightarrow{0b'}$ is a multiple of $\overrightarrow{0b}$, say, $t\overrightarrow{0b}$, and $\overrightarrow{0a'}$ is a multiple of $\overrightarrow{0a}$, say $s\overrightarrow{0a}$. We then have $\overrightarrow{0c} = \overrightarrow{0a'} + \overrightarrow{0b'} = s\overrightarrow{0a} + t\overrightarrow{0b}$. This immediately yields linear dependence (see Figure 6.11).

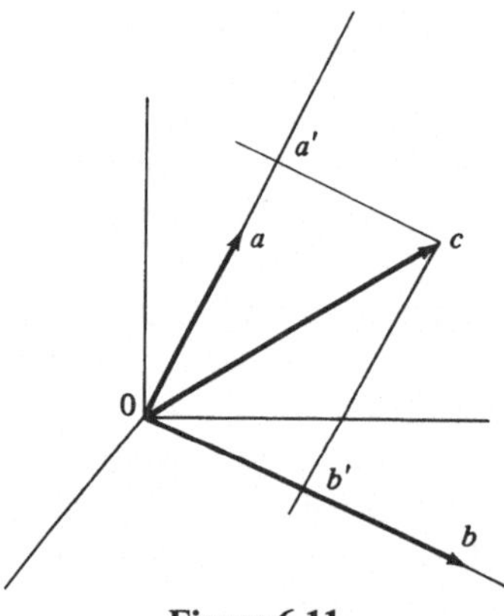

Figure 6.11

As a final part of this intuitive example we show how any four vectors are linearly dependent. Let $\overrightarrow{0a}$, $\overrightarrow{0b}$, $\overrightarrow{0c}$, $\overrightarrow{0d}$ be any four distinct arrows, no two lying in the same line and no three lying in the same plane. By what we have argued earlier we now need consider only this most general situation. Through d construct a line parallel to the line of $\overrightarrow{0c}$. Let this line intersect the plane $0ab$ in d'. Parallel to the plane of 0, a, and b construct a plane through d. Let c' be the intersection of this plane with the line containing $\overrightarrow{0c}$. $\overrightarrow{0d} = \overrightarrow{0c'} + \overrightarrow{0d'}$. $\overrightarrow{0c'} = t\overrightarrow{0c}$ for some $t \in \mathbb{R}$. $\overrightarrow{0d'} = r\overrightarrow{0a} + s\overrightarrow{0b}$ for some $r, s \in \mathbb{R}$. Thus $\overrightarrow{0d} = r\overrightarrow{0a} + s\overrightarrow{0b} + t\overrightarrow{0c}$ (see Figure 6.12).

Now we prove a theorem about linear independence.

Theorem. *Let $\langle M, +, \zeta \rangle$ be an R-module. Let $(x_i | i \in I)$ be a family of M and let $J \subseteq I$. If the family $(x_i | i \in I)$ is linearly independent then the sub-*

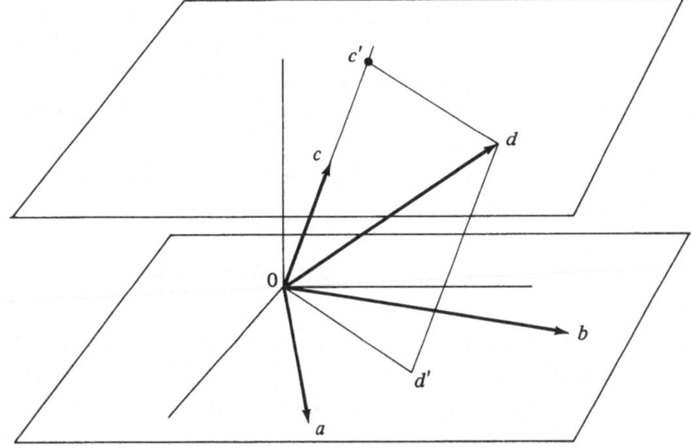

Figure 6.12

family $(x_i|i \in J)$ *is also linearly independent. If the subfamily* $(x_i|i \in J)$ *is linearly dependent then the entire family* $(x_i|i \in I)$ *is linearly dependent.*

PROOF. The two statements to be proved are equivalent. We prove the second. Suppose there exist scalars $(r_i|i \in J)$, not all zero, so that $\sum_{i \in J}^{w} r_i x_i = \zeta$. $\sum_{i \in I-J}^{w} \theta x_i = \zeta$. $\sum_{i \in J}^{w} r_i x_i + \sum_{i \in I-J}^{w} \theta x_i = \zeta$. Thus $(x_i|i \in I)$ is linearly dependent. □

For some similar theorems we refer the reader to Exercises 20 and 21.

Definition. A family $(x_i|i \in I)$ of an R-module M which is both linearly independent and generates M is a *basis* for M. One also calls the set $\{x_i|i \in I\}$ a basis as well as the family $(x_i|i \in I)$.

EXAMPLES. A basis for $\mathbb{R}^3$ is $((1, 0, 0), (0, 1, 0), (0, 0, 1))$. Another basis for $\mathbb{R}^3$ is $((1,1,0), (0,1,0), (0,1,1))$. A basis for $\mathbb{R}[X]$ is the family $(1, X, X^2, X^3, \ldots)$.

A basis for $\mathbb{Q}$ as a $\mathbb{Q}$-module consists of any nonzero element. However, a basis for $\mathbb{Q}$ as a $\mathbb{Z}$-module does not exist. In order for a set to generate $\mathbb{Q}$ it must generate (at least) all fractions of the form $1/2^n$, $n = 1, 2, 3, \ldots$. No multiple or linear combination of $1/2^m$ will produce $1/2^{m+1}$, yet $1/2^m$ and $1/2^{m+1}$ are linearly dependent. Hence it is not possible to find a linearly independent family from $\mathbb{Q}$ which generates $\mathbb{Q}$.

No finite nontrivial commutative group $\langle G, +, \zeta \rangle$ conceived of as a $\mathbb{Z}$-module can have a basis. For, if $x \in G$ and $x \neq \zeta$ then $nx = \zeta$ if $n = \operatorname{crd} G$ (the number of elements in the set G). This result is proved in the chapter on groups. We actually can give a brief argument here to establish that some multiple of x must be zero. If all multiples of x were distinct then the group would be infinite, a contradiction. There must be, therefore, two distinct multiples of x which are equal, say, $mx = px$. Then $(m - p)x = \zeta$ and $m - p \neq 0$. Thus every singleton subset of G is linearly dependent. Any larger family must be linearly dependent also. The only linearly independent family is $\varnothing$. But this cannot generate a group of two or more elements.

We now move towards proving that every module with a basis is isomorphic to a function space.

Definition. A module with a basis is called a *free module.*

We again turn to that submodule of R^I consisting of all functions which have all but a finite number of values zero. $(R^I)^w = \{f|f:I \to R \text{ and } f(i) = \theta$ for all but a finite number of $i\} = \{(x_i|i \in I)|x_i \in R$ and $x_i = \theta$ for all except a finite number of i in $I\}$.

Theorem. *Let R be a unitary ring and I an arbitrary set. Then $(R^I)^w$ is a free module.*

Proof. The basis for $(R^I)^w$ is simply the family $(e_i|i \in I)$ we have considered before. $e_i:I \to R$ such that $e_i(n) = \theta$ if $n \neq i$ and $= v$ if $n = i$. We begin by proving the family to be linearly independent. Suppose $\sum_{i\in I}^{w} r_i e_i = z$.

$$\sum_{i\in I}^{w} r_i e_i(n) = z(n) = \theta \quad \text{for any } n \in I.$$

$$\sum_{i\neq n}^{w} r_i e_i(n) + r_n e_n(n) = \theta.$$

$$\sum_{i\neq n}^{w} r_i \theta + r_n = \theta \qquad r_n = \theta \quad \text{for any } n \in I.$$

To show that $(e_i|i \in I)$ generates $(R^I)^w$ let $f \in (R^I)^w$. $f(n) = \theta$ for all save a finite number of $n \in I$. The sum $\sum_{i\in I}^{w} f(i)e_i$ is a finite linear combination in $(R^I)^w$. But

$$\begin{aligned} \sum_{i\in I}^{w} f(i)e_i(n) &= \sum_{i\neq n}^{w} f(i)e_i(n) + f(n)e_n(n) \\ &= \sum_{i\neq n}^{w} f(i)\theta + f(n)v \\ &= f(n) \quad \text{for each } n \in I. \end{aligned}$$

Hence, $f = \sum_{i\in I}^{w} f(i)e_i$. □

Examples. In these examples we choose several different index sets I, look at the module $(R^I)^w$, and see what is the basis $(e_i|i \in I)$. If $I = \hat{3}$ then $(\mathbb{R}^3)^w = \mathbb{R}^3 = \{(r_1, r_2, r_3)|r_1, r_2, r_3 \in \mathbb{R}\}$. $e_1 = (1, 0, 0)$, $e_2 = (0, 1, 0)$, $e_3 = (0, 0, 1)$. If $I = \mathbb{N}^+$ then $(\mathbb{R}^{\mathbb{N}^+})^w = \{(r_1, r_2, r_3, \ldots)|r_i \in \mathbb{R}$ for all $i \in I$ and all but a finite number of $r_i = 0\}$. $e_1 = (1, 0, 0, \ldots)$, $e_2 = (0, 1, 0, 0, \ldots)$, $e_3 = (0, 0, 1, 0, 0, \ldots)$, etc.

Definition. The basis $(e_i|i \in I)$ for the function space $(R^I)^w$ such that $e_i : I \to R$ such that

$$e_i(n) \begin{cases} = \theta & \text{if } n \neq i \\ = v & \text{if } n = i \end{cases}$$

we call the *standard basis* for the space.

We shall defer until the next section showing that every free module is isomorphic to a function space module of the form $(R^I)^w$. The theorem falls naturally into the material of the next section and serves to illustrate that material.

Questions

1. Which of the following completes a true sentence? A family $(x_i|i \in I)$ of a module M is linearly independent if and only if
   (A) there exist scalars $(r_i|i \in I)$ such that $\sum_{i \in I}^{w} r_i x_i = \zeta$
   (B) for all ring elements $(r_i|i \in I)$ we have $\sum_{i \in I}^{w} r_i x_i = \zeta$
   (C) all scalars are zero ($r_i = \theta$ for all $i \in I$) implies $\sum_{i \in I}^{w} r_i x_i = \zeta$
   (D) there are no scalars $(r_i|i \in I)$ such that $\sum_{i \in I}^{w} r_i x_i = \zeta$.
   (E) None of the alternatives is true.

2. Which of these are true?
   (A) If a subfamily $(x_i|i \in J)$ of a family $(x_i|i \in I)$, $J \subseteq I$, is linearly independent then the entire family is linearly independent.
   (B) A linearly dependent family contains at least one value.
   (C) Every family contains at least one linearly independent subfamily.
   (D) Every linearly dependent family contains a finite linearly dependent subfamily.
   (E) None of the sentences is true.

3. Which of these statements are true?
   (A) If $(x_i|i \in J)$ is a subfamily of the family $(x_i|i \in I)$, $J \subseteq I$, then the submodule $[(x_i|i \in J)]$ generated by $(x_i|i \in J)$ is included in the submodule $[(x_i|i \in I)]$ generated by $(x_i|i \in I)$.
   (B) If $(x_i|i \in I)$ is a nonempty family ($I \neq \varnothing$) then $[(x_i|i \in I)] \neq \{\zeta\}$.
   (C) If J is a proper subset of I and $[(x_i|i \in I)] = [(x_i|i \in J)]$ then $(x_i|i \in I)$ is a linearly dependent family.
   (D) Every linearly independent family $(x_i|i \in I)$ has a finite subfamily.
   (E) None of the statements is true.

4. Which of these statements are true?
   (A) For any unitary ring R, R is a free R-module.
   (B) If S is any subring of R, a unitary ring, then S is an R-module.

(C) $\mathbb{Q}[X]$ is a free $\mathbb{Z}$-module.
(D) $\mathbb{Z}[X]$ is a free $\mathbb{Z}$-module.
(E) None of the statements is true.

5. Let S be a subset of a K-vector space M.
(A) $[[S]] = [S]$.
(B) $[S] = \{\sum_{i=1}^{n} r_i x_i | r_1, \ldots, r_n \in K; x_1, \ldots, x_n \in M\}$ whenever S is nonempty.
(C) $[S] = M$ implies S is a basis of M.
(D) $[S] = \{N | N \text{ is a subspace of } M \text{ and } S \subseteq N\}$.
(E) None of the statements is true.

Exercises

1. Prove that the family $((1, 1, 0), (1, 0, 1), (0, 1, 1))$ is a basis for the $\mathbb{R}$-module $\mathbb{R}^3$.

2. Prove that the family $((1, 1, 0, 0), (0, 1, 1, 0), (0, 0, 1, 1))$ is a linearly independent family of the $\mathbb{Z}$-module $\mathbb{Z}^4$. Prove that the family is not a basis for $\mathbb{Z}^4$.

3. Prove that $((1, 1, 0, 0), (0, 1, 1, 0), (0, 0, 1, 1), (1, 0, 0, 1))$ is not a basis for the $\mathbb{Z}$-module $\mathbb{Z}^4$.

4. Prove that $((1, 1, 0, 0), (0, 1, 1, 0), (0, 0, 1, 1), (0, 0, 0, 1))$ is a basis for the $\mathbb{Z}$-module $\mathbb{Z}^4$.

5. Prove that $((2, 2, 0, 0), (0, 2, 2, 0), (0, 0, 2, 2), (0, 0, 0, 2))$ is a basis for the $\mathbb{Q}$-module $\mathbb{Q}^4$ but not for the $\mathbb{Z}$-module $\mathbb{Z}^4$.

6. Prove that the $\mathbb{Z}$-module $\mathbb{Z}$ has a basis.

7. Prove that in the $\mathbb{Z}$-module $\mathbb{Z}_6$ there are no linearly independent families except the empty one and therefore $\mathbb{Z}_6$ has no basis.

8. Let

$$d_i(x) \begin{cases} = 1 & \text{if } x = i \text{ or } x = i + 1 \\ = 0 & \text{if } x \neq i \text{ and } x \neq i + 1 \text{ for all } x \in \mathbb{N}. \end{cases}$$

Does $(d_i | i \in \mathbb{N})$ generate the $\mathbb{Z}$-module $(\mathbb{Z}^{\mathbb{N}})^w$? Is the family linearly independent?

9. Does the $\mathbb{Z}$-module $\mathbb{Z}_4$ have a basis?

10. Let S be the set of all polynomials with real coefficients with degree less than or equal to three and also the zero polynomial. Show that $\{X - 1, X^2 + 1, 2\}$ is a linearly independent set. Adjoin one more polynomial to the set to produce a basis.

11. Is the family $((2, 1, 0), (4, 2, 1), (3, 3, 0))$ linearly independent in the $\mathbb{Z}$-module $\mathbb{Z}^3$? Does it generate $\mathbb{Z}^3$? Is it a basis?

12. If any family $(x_i | i \in I)$ in an R-module M is linearly dependent then there exists a finite subfamily of $(x_i | i \in I)$ which is linearly dependent. Prove.

13. If $(u_1, u_2, \ldots, u_n)$, $n \geqslant 3$, is a basis for an R-module M which of the following are also bases?
(a) $(u_1, u_1 + u_2, u_1 + u_2 + u_3, \ldots, u_1 + u_2 + \cdots + u_n)$
(b) $(u_1 + u_2, u_2 + u_3, u_3 + u_4, \ldots, u_{n-1} + u_n, u_n)$

(c) $(u_1 + u_2, u_2 + u_3, u_3 + u_4, \ldots, u_{n-1} + u_n, u_n + u_1)$
In case (c) is the family a basis if R is a field (M is a vector space)?

14. Prove that a family $(x_i|i \in I)$ of and R-module M is linearly independent if and only if every permutation of the family is also linearly independent.

15. Let $(x_i|i \in I)$ be a given family in an R-module M. Suppose there exists in M a vector a such that $a = \sum_{i \in I}^{w} r_i x_i$ and $a = \sum_{i \in I}^{w} s_i x_i$ with $(r_i|i \in I) \neq (s_i|i \in I)$. Prove that the family is linearly dependent.

16. We have seen that $(e_i|i \in \mathbb{N}^+)$ is a basis for $(R^{\mathbb{N}^+})^w$ and have called it the standard basis. Prove that $(e_i|i \in \mathbb{N}^+)$ fails to be a basis for $\mathbb{R}^{\mathbb{N}^+}$.

17. Let the family $(x_i|i \in I)$ generate the module M. Suppose each element x_i, $i \in I$, can be expressed as a linear combination of the family $(y_j|j \in J)$. Show that M is generated by $(y_j|j \in J)$.

18. Let $(x_1, x_2, \ldots, x_n)$ be a finite basis for the R-module M. Prove $M = [x_1] \oplus [x_2] \oplus \cdots \oplus [x_n]$ (cf. Section 6.6, Exercise 11).

19. Let $(x_i|i \in I)$ be a basis for the R-module M. Prove $M = \bigoplus_{i \in I} [x_i]$.

20. Show that if any family $(x_i|i \in I)$ of an R-module M has two values identical (say, $x_j = x_k$ for $j \neq k$) then the family is linearly dependent.

21. Show that if any value of the family $(x_i|i \in I)$ is zero (say, $x_k = \zeta$ for some $k \in I$) then the family is linearly dependent in the R-module M.

6.8 The coordinate morphism

We define coordinates, the coordinate morphism and use this function to characterize R-modules in terms of function spaces.

We begin with the definition of coordinate morphism.

Definition. Given a family $(x_i|i \in I)$ of vectors in an R-module M we define the *coordinate morphism* to be the mapping

$$L:(R^I)^w \to M \quad \text{such that } L(r_i|i \in I) = \sum_{i \in I}^{w} r_i x_i.$$

Each value of the coordinate morphism is a linear combination of the given family $(x_i|i \in I)$. Each argument is the family of coefficients, scalars, which appear in the linear combination yielding the value of the function. Since we have showed in the last section that $(R^I)^w$ is an R-module and M is given to be an R-module then we have a map from one R-module to another R-module. The map L depends, of course, upon which given family $(x_i|i \in I)$ is used to compute the linear combinations. In cases where several families are used to define coordinate morphisms we shall use subscripts to distinguish between the two morphisms: a family $(x_i|i \in I)$ defines $L_x(r_i|i \in I) = \sum_{i \in I}^{w} r_i x_i$, the family $(y_i|i \in I)$ defines $L_y(r_i|i \in I) = \sum_{i \in I}^{w} r_i y_i$. The families might also differ in their index sets.

But we have not yet justified the use of the word *morphism*.

Theorem. *Given a family $(x_i i \in I)$ of vectors or elements in an R-module M the coordinate mapping $L:(R^I)^w \to M$ such that $L(r_i|i \in I) = \sum_{i\in I}^w r_i x_i$ is a morphism.*

PROOF. $L((r_i|i \in I) + (s_i|i \in I)) = L(r_i + s_i|i \in I) = \sum_{i\in I}^w (r_i + s_i)x_i = \sum_{i\in I}^w r_i x_i + \sum_{i\in I}^w s_i x_i = L(r_i|i \in I) + L(s_i|i \in I)$. $L(s(r_i|i \in I)) = L(sr_i|i \in I) = \sum_{i\in I}^w (sr_i)x_i = s(\sum_{i\in I}^w r_i x_i) = sL(r_i|i \in I)$. □

EXAMPLES. The family $((1, 0, 0), (0, 1, 0), (0, 0, 1))$ defines a morphism $L:(\mathbb{R}^3)^w \to \mathbb{R}^3$ such that $L(r_1, r_2, r_3) = r_1(1, 0, 0) + r_2(0, 1, 0) + r_3(0, 0, 1)$.

The family $((1, 1, 0), (1, 0, 1), (0, 1, 1))$ defines another coordinate morphism $L(\mathbb{R}^3)^w \to \mathbb{R}^3$ such that $L(r_1, r_2, r_3) = r_1(1, 1, 0) + r_2(1, 0, 1) + r_3(0, 1, 1)$. It is a different example from the first.

The family $(e_i|i \in \mathbb{N}^+)$ defines a morphism $L:(\mathbb{R}^{\mathbb{N}^+})^w \to \mathbb{R}^{\mathbb{N}^+}$ such that $L(r_i|i \in \mathbb{N}^+) = \sum_{i\in \mathbb{N}^+}^w r_i e_i$. In the family of coefficients $(r_i|i \in \mathbb{N}^+)$ all save a finite number of values must be zero.

We now wish to move on to understand better the properties of the coordinate morphism. That the map $L:(R^I)^w \to M$ is a surjection means that for every $x \in M$ there exists a family of coefficients $(r_i|i \in I)$ such that $L(r_i|i \in I) = x$. The mapping $L:(R^I)^w \to M$ is an injection if and only if $L(r_i|i \in I) = L(s_i|i \in I)$ implies $(r_i|i \in I) = (s_i|i \in I)$. We move from these facts to a theorem.

Theorem. *Let M be a module over a unitary ring R. The family $(x_i|i \in I)$ is linearly independent if and only if the associated coordinate morphism L is a monomorphism. The family $(x_i|i \in I)$ generates M if and only if the associated coordinate morphism L is an epimorphism.*

PROOF. $(x_i|i \in I)$ is linearly independent if and only if $\sum_{i\in I}^w r_i x_i = \zeta$ implies $(r_i|i \in I) = (\theta|i \in I)$ if and only if $L(r_i|i \in I) = \zeta$ implies $(r_i|i \in I) = (\theta|i \in I)$ if and only if kernel $L = \{(\theta|i \in I)\}$ if and only if L is a monomorphism. $(x_i|i \in I)$ generates M if and only if $x \in M$ implies $\sum_{i\in I}^w r_i x_i = x$ for some $(r_i|i \in I)$ in $(R^I)^w$ if and only if $x \in M$ implies $L(r_i|i \in I) = x$ for some $(r_i|i \in I)$ in $(R^I)^w$ if and only if L is an epimorphism. □

Corollary. *The family $(x_i|i \in I)$ is a basis for M if and only if the coordinate morphism $L:(R^I)^w \to M$ is an isomorphism.*

Corollary. *Every free R-module M is isomorphic to $(R^I)^w$ for some set I. A module isomorphic to some function space $(R^I)^w$ is free.*

PROOF. If M is a free R-module then there exists a basis $(x_i|i \in I)$. The coordinate morphism is an isomorphism between $(R^I)^w$ and M. On the other hand, if $(R^I)^w$ is isomorphic to M then the family of images of the standard basis $(e_i|i \in I)$ in $(R^I)^w$ is a basis in M.

We now investigate the relationships among morphism, generation, and linear independence, and how the properties of generation and linear independence are preserved by morphisms.

Theorem. *Let $f:M \to M'$ be an epimorphism between R-modules $\langle M, +, \zeta\rangle$ and $\langle M', +', \zeta'\rangle$. Then:*

(a) *$(x_i|i \in I)$ generates M implies $(f(x_i)|i \in I)$ generates M'.*
(b) *If $(y_i|i \in I)$ is a linearly independent family of M' then there exists a linearly independent family $(x_i|i \in I)$ of M such that $y_i = f(x_i)$ for all $i \in I$.*

Let $f:M \to M'$ be a monomorphism between R-modules $\langle M, +, \zeta\rangle$ and $\langle M', +', \zeta'\rangle$. Then:

(a) *If $(x_i|i \in I)$ is a linearly independent family of M then $(f(x_i)|i \in I)$ is a linearly independent family of M'.*
(b) *If $(x_i|i \in I)$ is a family of M such that $(f(x_i)|i \in I)$ generates M' then $(x_i|i \in I)$ generates M.*

Proof. We offer a proof of part (a) of the second statement. We leave the the other parts to the reader. Let $f:M \to M'$ be a monomorphism and $(x_i|i \in I)$ a linearly independent family of M. We must show that $(f(x_i)|i \in I)$ is a linearly independent family. Suppose $\sum_{i\in I}^{w} r_i f(x_i) = \zeta'$. $f(\sum_{i\in I}^{w} r_i x_i) = \zeta'$. $\sum_{i\in I}^{w} r_i x_i \in \text{kernel } f$. $\sum_{i\in I}^{w} r_i x_i = \zeta$ since kernel $f = \{\zeta\}$. But $(x_i|i \in I)$ linearly independent implies all $r_i = \theta$. □

Corollary. *The image or preimage of a basis under an isomorphism is also a basis.*

Proof. This result follows from combining the parts of the theorem. □

We have previously demonstrated that every free R-module, that is, every module with a basis, is isomorphic to $(R^I)^w$ for some set I, the index set of the basis. Every R-module with a basis is essentially a space of functions taking values in the ring R. The character of this function space depends only upon the index set or size of the basis and not upon the basis itself. A consequence of this is that any two modules over the same ring R are isomorphic when they have bases of the same size. There is essentially only one free R-module of a given size.

Theorem. *If $\langle M, +, \zeta\rangle$ and $\langle M', +', \zeta'\rangle$ are free R-modules with bases of the same size then M and M' are isomorphic.*

PROOF. Let $(x_i | i \in I)$ be a basis for M and let $(y_j | j \in J)$ be a basis for M'. We express the fact that the bases have the same size by postulating that there exists a bijection $\varphi: I \to J$, a one-to-one correspondence between the index sets. This bijection φ defines in a natural way an isomorphism $\Phi: (R^J)^w \to (R^I)^w$ such that $\Phi(f) = f \circ \varphi$.

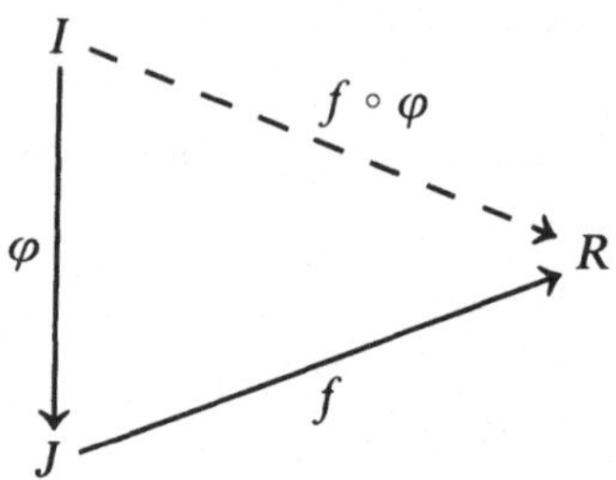

We verify that Φ is an isomorphism. $\Phi(f + g) = (f + g) \circ \varphi = f \circ \varphi + g \circ \varphi = \Phi(f) + \Phi(g)$. $\Phi(rf) = (rf) \circ \varphi = r(f \circ \varphi) = r\Phi(f)$. Let $\Phi(f) = \Phi(g)$. $f \circ \varphi = g \circ \varphi$. $f \circ \varphi \circ \varphi^{-1} = g \circ \varphi \circ \varphi^{-1}$. $f = g$. Let $g \in (R^I)^w$. Then $\Phi(g \circ \varphi^{-1}) = g \circ \varphi^{-1} \circ \varphi = g$. Φ is a morphism which is injective and surjective and is therefore an isomorphism.

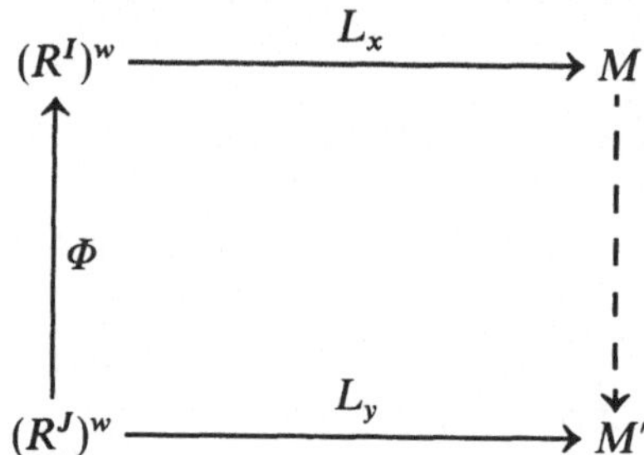

To show that M and M' are isomorphic we compose three isomorphisms. $L_y \circ \Phi^{-1} \circ L_x^{-1}: M \to M'$. Consult the accompanying diagram. □

We will later have something to say about the possibility of a module's having bases of different size. We turn now to the possibility of a module's having no basis at all. What can be said about such modules and their representations as function spaces? Certainly every module has a generating family, the module itself at worst. We now show that every module with a generating family is isomorphic to a quotient module of a function space.

Theorem. *If M is an R-module with a generating family $(x_i | i \in I)$ then there exists a submodule N of $(R^I)^w$ such that M is isomorphic with $(R^I)^w/N$.*

PROOF. The coordinate morphism associated with the generating family $(x_i | i \in I)$, namely, $L: (R^I)^w \to M$ is an epimorphism. Using the fundamental morphism theorem there exists an isomorphism $L': (R^I)^w \to M$ where $N = \text{kernel } L$. □

In general when L is not a monomorphism the kernel N will be nontrivial. Cosets of $(R^I)^w$ will then be involved.

EXAMPLE. $(X^n|n \in \mathbb{N})$ is a family of polynomials which generates the polynomial ring $\mathbb{Z}_4[X]$, a $\mathbb{Z}$-module. $L(r_i|i \in \mathbb{N}) = \sum_{i\in\mathbb{N}}^w r_iX^i$ is the coordinate morphism. It is an epimorphism. Kernel $L = \{(r_i|i \in \mathbb{N})|\sum_{i\in\mathbb{N}}^w r_iX^i = \overline{0}\} = \{(r_i|i \in \mathbb{N})|r_i \in 4\mathbb{Z}$ for all $i \in \mathbb{N}\} = ((4\mathbb{Z})^{\mathbb{N}})^w$. Then $\mathbb{Z}_4[X]$, a $\mathbb{Z}$-module without a basis, is isomorphic to a quotient module of the function space $(\mathbb{Z}^{\mathbb{N}})^w$, namely the quotient module $(\mathbb{Z}^{\mathbb{N}})^w/((4\mathbb{Z})^{\mathbb{N}})^w$.

To each basis for a given module M will be associated a different coordinate morphism. We can actually relate two bases, $(x_i|i \in I)$ and $(x'_j|j \in J)$, for the same R-module M by a set of equations. We can do this because each basis element of one basis is necessarily expressible as a linear combination of the members of the other basis. For one,

$$x_j = \sum_{i\in I}^w E_{ij}x'_i \quad \text{for each } j \in J.$$

Since the linear combinations are unique, to know each family of scalars $(E_{ij}|i \in I)$ is to know each x_j and vice versa. Each family $(E_{ij}|i \in I)$ of scalars is zero for all but a finite number of i in I.

We now categorize three particularly simple relationships between pairs of bases. We call these three simple relationships elementary change of bases. We assume each basis has the same number of elements and we therefore index by the same set.

I. The interchange of two basis elements. Let p, q be two indices in I, $p \neq q$. Let $(x_i|i \in I)$ be the first basis and $(u_j|j \in I)$ be the second.

$$\begin{aligned} u_i &= x_i \quad \text{for } i \in I, i \neq p, i \neq q \\ u_p &= x_q \\ u_q &= x_p. \end{aligned}$$

II. The adding of a multiple of one basis element to another. Let p, q be two indices in I, $p \neq q$. Let $r \in R$. We define a basis $(u_j|j \in I)$ in terms of a basis $(x_i|i \in I)$.

$$\begin{aligned} u_i &= x_i \quad \text{for all } i \in I, i \neq q \\ u_q &= x_q - rx_p. \end{aligned}$$

III. The multiplying of one basis element by a nonzero invertible constant. Let p be an index in I. Assume $s \in R$ and s^{-1} is also in R. We define a basis $(u_j|j \in I)$ in terms of a basis $(x_i|i \in I)$.

$$\begin{aligned} u_i &= x_i \quad \text{for } i \in I, i \neq p \\ u_p &= s^{-1}x_p. \end{aligned}$$

We must verify that the transitions described do define new bases.

Theorem. *Let M be an R-module. Let $(x_i|i \in I)$ be a given basis. Then $(u_i|i \in I)$ as described in* (I), (II), *and* (III) *above is a basis.*

Proof. For part I suppose $\sum_{i\in I}^{w} r_i u_i = \zeta$. $\sum_{i\neq p,q}^{w} r_i u_i + r_p u_p + r_q u_q = \zeta$. $\sum_{i\neq p,q}^{w} r_i x_i + r_p x_q + r_q x_p = \zeta$. $r_i = \theta$ for all $i \in I$. $(u_i|i \in I)$ is linearly independent. Now suppose $x \in M$. $x = \sum_{i\in I}^{w} r_i x_i$ for some $r_i \in R$. Therefore, $x = \sum_{i\neq p,q}^{w} r_i u_i + r_p u_q + r_q u_p$. $(u_i|i \in I)$ generates M. The other two parts are proved similarly. □

Corollary. *Let M be an R-module. The inverses of the three changes of bases described above are*

I. $x_i = u_i, \quad i \in I, i \neq p, i \neq q$
$x_p = u_q$
$x_q = u_p$.

II. $x_i = u_i, \quad i \in I, i \neq q$
$x_q = u_q + r u_p$

III. $x_i = u_i, \quad i \in I, i \neq p$
$x_p = s u_p$.

Questions

1. If the coordinate morphism L defined by a family $(x_i|i \in I)$ of an R-module M is an epimorphism then
   (A) there may be two distinct linear combinations of $(x_i|i \in I)$ yielding a given vector x in M
   (B) there may be no linear combination of $(x_i|i \in I)$ yielding a given vector x in M
   (C) kernel L consists of the set of all coordinates of ζ
   (D) L is an injection.
   (E) None of the alternatives completes a true sentence.

2. Which of the following are true?
   (A) If the coordinate morphism L_x defined by the family $(x_i|i \in I)$ is an isomorphism then L_x^{-1} is also an isomorphism.
   (B) If L_x and L_y are two coordinate morphisms for M defined by bases $(x_i|i \in I)$ and $(y_i|i \in I)$ then $L_y \circ L_x^{-1}$ is an automorphism of M.
   (C) If f is an automorphism of M and L is a coordinate isomorphism for M then $f \circ L$ is another coordinate isomorphism.
   (D) If L_x and L_y are two coordinate epimorphisms for M then $L_y \circ L_x$ is another coordinate epimorphism for M.
   (E) None of the sentences is true.

3. Which of these statements are true?
   (A) The image $(f(x_i)|i \in I)$ of a linearly independent family $(x_i|i \in I)$ under an injection $f: M \to M'$ of R-modules is linearly independent.
   (B) The image $(f(x_i)|i \in I)$ of a basis $(x_i|i \in I)$ under a monomorphism $f: M \to M'$ of R-modules is a generating family for M'.
   (C) Any preimage $(x_i|i \in I)$ of a basis $(y_i|i \in I)$, $y_i = f(x_i)$, under an epimorphism $f: M \to M'$ of R-modules is a linearly independent family of M.

(D) If $(x_i|i \in I)$ is a basis of M and $f:M \to M'$ is an epimorphism of R-modules then $(f(x_i)|i \in I)$ is a basis for M'.
(E) None of the sentences is true.

4. Which of these sentences are true?
   (A) The function space $(R^I)^w$ is a *proper* submodule of R^I whenever I is infinite and R is a unitary ring.
   (B) $(e_i|i \in I)$, the standard basis, is a basis for $(R^I)^w$, but not for R^I, for every index set I.
   (C) The vector $(v|i \in I)$ is a member of R^I, but is not a member of $(R^I)^w$.
   (D) No linear combination of $(e_i|i \in I)$, the standard basis, will yield $(v|i \in I)$.
   (E) None of the sentences is true.

EXERCISES

1. Let L be the coordinate morphism defined by the basis $((1, 1, 0, 0), (0, 1, 1, 0), (0, 0, 1, 1), (0, 0, 0, 1))$ of the $\mathbb{Z}$-module $\mathbb{Z}^4$. What is $L(r_1, r_2, r_3, r_4)$? $L(2, 4, -1, 3)$? $L^{-1}(1, 2, 3, 4)$? What vector has coordinates (r_1, r_2, r_3, r_4)? What vector has coordinates $(2, 4, -1, 3)$? What are the coordinates of the vector $(1, 2, 3, 4)$?

2. Show that (sine, cosine) is a linearly independent family of two functions in the $\mathbb{R}$-vector space $\mathbb{R}^{[0, \pi]}$. If f is a member of the space $[\{\text{sine, cosine}\}]$ and $f(\pi/4) = f(\pi/6) = 1$ then find $L^{-1}(f)$.

3. Let p be a prime natural number. Show that the $\mathbb{Z}$-module $\mathbb{Z}_p$ is not free. Find a generating family $(x_i|i \in I)$ for $\mathbb{Z}_p$. Find the submodule N of $(\mathbb{Z}^I)^w$ so that $\mathbb{Z}_p$ is isomorphic to $(\mathbb{Z}^I)^w/N$.

4. Give an example of a morphism $f:M \to M'$ and a family $(x_i|i \in I)$ of M such that $(x_i|i \in I)$ generates M and $(f(x_i)|i \in I)$ fails to generate M'.

5. Give an example of a morphism $f:M \to M'$ and a linearly independent family $(x_i|i \in I)$ such that $(f(x_i)|i \in I)$ is linearly dependent.

6. Let M be the $\mathbb{R}$-module $\mathbb{R}^2$ and $(x_i|i \in \hat{3})$ be the family $((1, 0), (1, 1), (0, 1))$. What is L with respect to this family? Find kernel L. Find $L^{-1}(2, 3)$.

7. Show that the $\mathbb{Z}$-module $\mathbb{Q}$ has no basis. Find a nontrivial (not $\mathbb{Q}$ itself) family which generates $\mathbb{Q}$.

8. Consider the family $((2, 1, 0), (4, 2, 1), (3, 3, 0))$ in the $\mathbb{Z}$-module $\mathbb{Z}^3$. What is L for this family? Is L a monomorphism? an epimorphism?

9. Prove that if any family $(x_i|i \in I)$ of an R-module M is linearly dependent then there exists a finite subfamily of $(x_i|i \in I)$ which is also linearly dependent.

10. Let $((1, 0), (1, 1), (0, 1))$ be a family in $\mathbb{R}^2$. Find L for this family. Find kernel L and also find $L^{-1}(2, 3)$.

6.9 Morphisms and bases, kernel, and range

We show that a morphism is completely determined by its behavior on the basis elements of the domain and prove theorems relating to the size of the kernel, range, and domain of a morphism.

Theorem. *Let M, M' be R-modules. Let $(x_i|i \in I)$ be a basis for M and $(y_i|i \in I)$ be an arbitrary family of M' of the same size as the basis of M. Then there is one and only one morphism $f:M \to M'$ with values $f(x_i) = y_i$, $i \in I$.*

PROOF. Every vector in M is uniquely representable as a linear combination of the basis elements $(x_i|i \in I)$. We define a mapping $f:M \to M'$ such that $f(\sum_{i\in I}^{w} r_i x_i) = \sum_{i\in I}^{w} r_i y_i$. Since for any $x \in M$ the family $(r_i|i \in I)$ is uniquely determined then the mapping f is uniquely determined also. It is clear that $f(x_i) = y_i$ for every $i \in I$. That f is a morphism is easily verified; we leave this verification to the reader. We show f is unique. Let $g:M \to M'$ be any (other) morphism such that $g(x_i) = y_i$ for all $i \in I$. $f(x_i) = g(x_i)$ for all $i \in I$. $r_i f(x_i) = r_i g(x_i)$ for any $i \in I$, $r_i \in R$. $\sum_{i\in I}^{w} r_i f(x_i) = \sum_{i\in I}^{w} r_i g(x_i)$. $\sum_{i\in I}^{w} f(r_i x_i) = \sum_{i\in I}^{w} g(r_i x_i)$. $f(\sum_{i\in I}^{w} r_i x_i) = g(\sum_{i\in I}^{w} r_i x_i)$. $f(x) = g(x)$ for all $x \in M$. $f = g$. □

EXAMPLE. We let $I = \hat{3}$ and $M = M' = \mathbb{R}^{\hat{3}}$. $(x_i|i \in \hat{3}) = ((1, 0, 0), (0, 1, 0), (0, 0, 1))$. $(y_i|i \in \hat{3}) = ((2, 1, 2), (1, 1, 0), (1, 0, 2))$. $f:\mathbb{R}^3 \to \mathbb{R}^3$ such that $f(1, 0, 0) = (2, 1, 2)$, $f(0, 1, 0) = (1, 1, 0)$, $f(0, 0, 1) = (1, 0, 2)$. $f(r_1(1, 0, 0) + r_2(0, 1, 0) + r_3(0, 0, 1)) = r_1(2, 1, 2) + r_2(1, 1, 0) + r_3(1, 0, 2)$. $f(r_1, r_2, r_3) = (2r_1 + r_2 + r_3, r_1 + r_2, 2r_1 + 2r_3)$.

To determine the behavior of a morphism of modules it is only necessary to know the behavior of the morphism on the basis elements of the domain.

We now begin theorems which relate the size of the kernel with the size of the range of a morphism.

Lemma. *Let $f:M \to M'$ be a morphism of R-modules. Let $(x_i|i \in J)$ generate* kernel f. *Then there exists a superfamily $(x_i|i \in I)$, $J \subseteq I$, such that $(x_i|i \in I)$ generates M.*

PROOF. Let $I' = M -$ kernel f. The family $(x_i|i \in J \cup I')$ generates M where $x_i = i$ if $i \in I'$. □

In this proof the elements of the complement of $[\{x_i|i \in J\}]$ in M are adjoined to $(x_i|i \in J)$ to produce a family generating all of M. Of course, this generation of M is done inefficiently; many fewer elements would, in general, suffice. The point is that it can be done. Having showed that it is possible to extend any family generating the kernel to a family generating all of M we now discuss generating the range.

Lemma. *Let $f:M \to M'$ be a morphism of R-modules. Let $(x_i|i \in J)$ generate* kernel f *and $(x_i|i \in I)$ generate M and suppose $J \subseteq I$. Then $(f(x_i)|i \in I - J)$ generates* range f.

PROOF. Let y be in range f. $y = f(x)$ for some x in M. $x = \sum_{i\in I}^{w} r_i x_i$ for some r_i in R. $x = \sum_{i\in I-J}^{w} r_i x_i + \sum_{i\in J}^{w} r_i x_i$. $f(x) = \sum_{i\in I-J}^{w} r_i f(x_i) + \sum_{i\in J}^{w} r_i f(x_i) = \sum_{i\in I-J}^{w} r_i f(x_i) + \zeta'$. $y = \sum_{i\in I-J}^{w} r_i f(x_i)$. □

We now include linear independence in our hypothesis for our main theorem.

Theorem. *Let $f:M \to M'$ be a morphism of R-modules. Let $(x_i|i \in J)$ be a basis for* kernel f *and $(x_i|i \in I)$ be a basis for M with $J \subseteq I$. Then $(f(x_i)|i \in I - J)$ is a basis for* range f.

Proof. Because of the lemmas we need only prove $(f(x_i)|i \in I - J)$ is linearly independent. Suppose $\sum_{i \in I-J}^{w} r_i f(x_i) = \zeta'$. $f(\sum_{i \in I-J}^{w} r_i x_i) = \zeta'$. $\sum_{i \in I-J}^{w} r_i x_i \in \text{kernel} f$. $\sum_{i \in I-J}^{w} r_i x_i = \sum_{i \in J}^{w} s_i x_i$. $\sum_{i \in I-J}^{w} r_i x_i + \sum_{i \in J}^{w} (-s_i) x_i = \zeta$. By linear independence of $(x_i|i \in I)$ we have $r_i = \theta$ for all $i \in I - J$ and $-s_i = \theta$ for all $i \in J$. Hence $r_i = \theta$ for all $i \in I - J$. □

This theorem shows that the number of elements in the basis for the domain (index set I) equals the number of elements in the basis for the kernel (index set J) plus the number of elements in the basis for the range (index set $I - J$). It is not always possible, having found a basis for the kernel of a morphism, to extend this basis to a basis for the entire space, at least for modules. For vector spaces it is always possible to make such an extension. An example of the former is the following. Let $g:\mathbb{Z} \to \mathbb{Z}/2\mathbb{Z}$ where $g(n) = \bar{0}$ or $\bar{1}$ according to whether the remainder upon dividing n by 2 is 0 or 1. g is a morphism with kernel $2\mathbb{Z}$. $2\mathbb{Z}$ has a basis consisting of the singleton 2. The basis (2) for $2\mathbb{Z}$ cannot be extended to a basis for the entire module $\mathbb{Z}$ since the only bases for $\mathbb{Z}$ are the singletons (1) and (-1).

Example. We consider the morphism $f:\mathbb{R}^3 \to \mathbb{R}^2$ such that $f(r_1, r_2, r_3) = (r_1 + r_2, 2r_1 - r_3)$. The kernel of f is $\{(r_1, r_2, r_3)|(r_1 + r_2, 2r_1 - r_3) = (0, 0)\} = \{(r_1, r_2, r_3)|r_2 = -r_1 \text{ and } r_3 = 2r_1\} = \{(t, -t, 2t)|t \in R\} = \{(t(1, -1, 2)|t \in R\}$. A basis for the kernel is the singleton family $((1, -1, 2))$. A superfamily of it is the family $((1, 0, 0), (0, 1, 0), (1, -1, 2))$ which is a basis for $\mathbb{R}^3$, the domain of the morphism. The family $(f(1, 0, 0), f(0, 1, 0)$ must therefore be a basis for the range of f. The family is $((1, 1), (2, 0))$. The number of members of the basis for the domain is 3, the number of members of the basis for the kernel is 1 and the number of members of the basis for the range is 2.

Questions

1. Which of the following statements are true?
   (A) If a vector belongs to the kernel of a module morphism $f:M \to M'$ then the vector is ζ.
   (B) The morphism $z:M \to M'$ such that $z(x) = \zeta'$ for all $x \in M$ can never have kernel equaling $\{\zeta\}$.
   (C) Kernel $f \subseteq$ range f if $f:M \to M$ is an endomorphism.
   (D) Kernel $f \cap$ range $f = \varnothing$ if and only if $f:M \to M'$ is a monomorphism.
   (E) None of the statements is true.

2. Let $f:M \to M'$ be a morphism of R-modules. Which of the following are true?
   (A) If $(x_1, x_2, \ldots, x_n)$ is a basis of M and $(y_1, y_2, \ldots, y_n)$ is a basis for M' and $f(x_i) = y_i$, $i = 1, 2, \ldots, n$, then f is an isomorphism.
   (B) If $(x_1, x_2, \ldots, x_n)$ is a basis of M and $(f(x_1), f(x_2), \ldots, f(x_n))$ is a basis of M' then f is an isomorphism.
   (C) If $(x_1, x_2, \ldots, x_n)$ is a basis of M and $f(x_1) = f(x_2)$ then kernel $f \neq \{\zeta\}$ and $x_2 - x_1 \in$ kernel f.
   (D) If $(x_1, x_2, \ldots, x_n)$ is a basis for M and $x_1, x_2, \ldots, x_n \in$ kernel f then $f(x) = \zeta'$ for all $x \in M$.
   (E) None of the statements is true.

3. Let $f:M \to M'$ be a morphism of R-modules. Which of the following statements are true?
   (A) If $(x_1, x_2, \ldots, x_n)$ is a basis for M and $x_1 \in$ kernel f then a basis for range f must have strictly fewer than n members.
   (B) If $(x_1, x_2, \ldots, x_n)$ is a basis for M then $(f(x_1), f(x_2), \ldots, f(x_n))$ generates range f.
   (C) If $(x_1, x_2, \ldots, x_n)$ is a basis for M and kernel $f \neq \{\zeta\}$ then $(f(x_1), f(x_2), \ldots, f(x_n))$ cannot be a basis for range f.
   (D) If $(x_1, x_2, \ldots, x_n)$ is a basis for M and $(f(x_1), f(x_2), \ldots, f(x_n))$ is linearly dependent then kernel $f \neq \{\zeta\}$.
   (E) None of the statements is true.

4. Let $f:M \to M'$ be a morphism of R-modules. Which of these statements are true?
   (A) If $(x_1, x_2, \ldots, x_n)$ is a basis for M and $g:M' \to M$ such that $g(f(x_i)) = x_i$, $i = 1, 2, \ldots, n$, then f is an isomorphism.
   (B) If $(x_1, x_2, \ldots, x_n)$ is a basis for M and $(y_1, y_2, \ldots, y_n)$ is a basis for M' and $g:M' \to M$ such that $g(y_i) = x_i$, $i = 1, 2, \ldots, n$, then kernel $f = \{\zeta\}$.
   (C) If $(x_1, x_2, \ldots, x_k)$ is a basis for kernel f and $(x_1, x_2, \ldots, x_k, x_{k+1}, \ldots, x_n)$ is a basis for M and $k < n$ then range $f \neq \{\zeta'\}$.
   (D) Range $f \neq \varnothing$.
   (E) None of the statements is true.

Exercises

1. Let $f:\mathbb{R}^3 \to \mathbb{R}^3$ such that $f(r_1, r_2, r_3) = (r_1 + 2r_2, 3r_1 + 6r_2, 7r_2 + 14r_3)$. Find the kernel of f and a basis for the kernel of f, given that one exists. Find a basis for the range of f. Is f a monomorphism? If f an epimorphism?

2. For the function $g:\mathbb{R}^3 \to \mathbb{R}^2$ such that $g(s_1, s_2, s_3) = (s_1 + s_2 + s_3, s_1 - s_2 + s_3)$ follow the instructions of Exercise 1.

3. For the function $h:\mathbb{R}^3 \to \mathbb{R}^4$ such that $h(r_1, r_2, r_3) = (3r_1 + r_2 - r_3, r_1 + 5r_2 + 2r_3, 2r_1 + 12r_2 + 5r_3, -r_1 + r_2 + r_3)$ follow the instructions of Exercise 1.

4. Find a morphism $f:\mathbb{R}^3 \to \mathbb{R}^2$ such that $(1, 1, 1) \in$ kernel f and $(1, 1) \in$ range f. Is this morphism unique?

5. Show that $\{t(1, 2) | t \in \mathbb{Z}\}$ is a submodule of $\mathbb{Z}^2$, a $\mathbb{Z}$-module. What is a basis for this submodule? Can you find a morphism from $\mathbb{Z}^2$ to $\mathbb{Z}^2$ having precisely this submodule as its kernel?

6. ((1, 1, 0), (1, 0, 1), (0, 1, 1)) is a basis for $\mathbb{Z}^3$ and ((1, 0), (1, 1), (0, 1)) is a family in $\mathbb{Z}^2$ with the same index set $\hat{3}$. According to the first theorem of this section there is one and only one morphism $f:\mathbb{Z}^3 \to \mathbb{Z}^2$ such that $f(1, 1, 0) = (1, 0)$, $f(1, 0, 1) = (1, 1)$, $f(0, 1, 1) = (0, 1)$. Find $f(s_1, s_2, s_3)$ for an arbitrary (s_1, s_2, s_3) in $\mathbb{Z}^3$. Find kernel f. Find range f.

7. Let $f:M \to M'$ be a morphism of R-modules. Let $(x_i|i \in J)$ be a basis for kernel f and $(x_i|i \in I)$ also be given as a basis for M so that $J \subseteq I$. Show that the following three R-modules are isomorphic: M, $\ker f \times \text{range } f$, $\ker f \times (M/\ker f)$.

8. Let R be a nontrivial commutative unitary ring. Show that the R-modules R and $R \times R$ cannot be isomorphic.

9. If $(x_1, x_2, \ldots, x_n)$ and $(y_1, y_2, \ldots, y_n)$ are bases of R-modules M and M', respectively, prove that any morphism $f:M \to M'$ such that $f(x_i) = y_i$, $i = 1, 2, \ldots, n$, is an isomorphism. Is f unique?

10. What is the inverse of the morphism in problem 9? Is the inverse an isomorphism?

11. Let $f:M \to M'$ be a morphism of R-modules. Let $(x_i|i \in J)$ be a basis for kernel f and $(x_i|i \in K)$, $J \subseteq K$, be a linearly independent family of M. Show $(f(x_i)|i \in K - J)$ is a linearly independent family of M'.

6.10 Vector spaces

In this section we discuss results on bases and dimension that are especially available to vector spaces.

By restricting our attention to vector spaces, modules over division rings, and fields, we can demonstrate some rather strong results. We will give our proofs for commutative division rings, i.e., fields. The first result is that every vector space has a basis, a fact not true of modules in general.

Theorem. *Let M be a K-vector space. Let $(x_i|i \in J)$ be a linearly family of M. Then there exists a superfamily $(x_i|i \in I)$, $J \subseteq I$, of the given family which is a basis for M.*

Proof. The proof of this theorem requires some techniques of set theory more profound than those used in chapter one. For this reason we place a proof in Appendix 6B at the end of this section. We suggest that the reader accept the theorem as an axiom and defer the proof; it is common in undergraduate texts to assume the equivalent of this result. We have included the proof for the sake of the reader who wishes to pursue the matter further. □

Corollary. *If M is a K-vector space then M has a basis.*

Proof. The empty set is a linearly independent family of M and by the theorem $\emptyset$ can be extended to a basis for M. This extension is the required basis of M. □

Corollary. *If M is a K-vector space and N is a subspace of M then there exists a basis $(x_i|i \in J)$ for N and a superfamily $(x_i|i \in I)$, $J \subseteq I$, which is a basis for all of M.*

PROOF. As a vector space N has a basis. This basis of N can, as a linearly independent family of M, be extended to a basis for M. □

It is important to note that a basis for M, the entire space, does not necessarily have a subfamily which is a basis for the subspace N. For example, $((1, 0), (0, 1))$ is a basis for $\mathbb{R}^2$ yet the subspace $\{t(1, 1)|t \in \mathbb{R}\}$ has neither $((1, 0))$ nor $((0, 1))$ as a basis. One can, however, begin with the basis $((1, 1))$ for the subspace and extend this basis to $((1, 1), (1, 0))$, a basis for the entire space.

We now prove a sequence of results about vector spaces and finite bases which do not depend upon the general theorem just discussed.

Theorem. *Let M be a K-vector space where K is a field. A finite family $(x_1, x_2, \ldots, x_n)$, $n \geqslant 1$, is linearly dependent if and only if some x_k, $k \geqslant 2$, is a linear combination of $(x_1, x_2, \ldots, x_{k-1})$ or $x_1 = \zeta$.*

PROOF. Suppose first that $(x_1, x_2, \ldots, x_n)$ is linearly dependent. Then there exist $r_1, r_2, \ldots, r_n$ in K, not all zero, such that $r_1x_1 + r_2x_2 + \cdots + r_nx_n = \zeta$. Let r_k be the nonzero coefficient with largest subscript. $r_1x_1 + r_2x_2 + \cdots + r_kx_k = \zeta$ with $r_k \neq \theta$. If $k > 1$ then $x_k = -r_k^{-1}r_1x_1 - r_k^{-1}r_2x_2 - \cdots - r_k^{-1}r_{k-1}x_{k-1}$ is a solution of the equation for x_k. Since K is a field the inverse r_k^{-1} belongs to K and x_k is a linear combination of the preceding $x_1, x_2, \ldots, x_{k-1}$. If $k = 1$ we have $r_1x_1 = \zeta$ with $r_1 \neq \theta$. Again the existence of the inverse r_1^{-1} in K proves $x_1 = \zeta$.

For the converse, if $x_1 = \zeta$ then certainly the family is dependent. If $x_k = s_1x_1 + s_2x_2 + \cdots + s_{k-1}x_{k-1}$ with $1 < k \leqslant n$ then $s_1x_1 + s_2x_2 + \cdots + s_{k-1}x_{k-1} + (-\nu)x_k = \zeta$ proves $(x_1, x_2, \ldots, x_k)$ is linearly dependent. $(x_1, x_2, \ldots, x_n)$ is also linearly dependent. □

Theorem. *Let M be a K-vector space. If M has a finite, nonempty basis $(x_1, x_2, \ldots, x_n)$ and $(y_1, y_2, \ldots, y_m)$ is a linearly independent family then there exist $y_{m+1}, y_{m+2}, \ldots, y_{m+p}$, $0 \leqslant p \leqslant n$, such that $(y_1, y_2, \ldots, y_m, y_{m+1}, \ldots, y_{m+p})$ is a basis for M.*

PROOF. Consider the family $(y_1, y_2, \ldots, y_m, x_1, x_2, \ldots, x_n)$. It certainly generates M. If this family is linearly dependent either $y_1 = \zeta$ or some vector in the family is a linear combination of the preceding vectors. As $y_1 \neq \zeta$ and no "y" can be a linear combination of preceding y's we have that some x_k is a linear combination of the preceding vectors $y_1, y_2, \ldots, y_m, x_1, \ldots, x_{k-1}$. The family $(y_1, y_2, \ldots, y_m, x_1, \ldots, x_{k-1}, x_{k+1}, \ldots, x_n)$ still generates M. If this family is linearly independent then the theorem is proved. If not then the deletion process is repeated. After at most n steps the process must terminate because the y's alone are linearly independent. Denote the final linearly independent family which generates M by $(y_1, y_2, \ldots, y_m, y_{m+1}, \ldots, y_{m+p})$ in which $y_{m+1}, \ldots, y_{m+p}$ stand for the nondeleted x's. □

This previous theorem tells us that if a vector space is known to have a finite basis then any linearly independent family can be extended to a basis which is also finite. This result is similar to the previously expounded theorem that every linearly independent family in a vector space can be extended to a basis. The proof here, however, assumes the existence of a finite basis for the space. We now move towards proving that every basis has the same number of elements.

Theorem. *Let M be a K-vector space. If the finite family $(x_1, x_2, \ldots, x_n)$ generates M and the finite family $(y_1, y_2, \ldots, y_n)$ is linearly independent in M then $m \geqslant n$.*

Proof. Consider $(y_n, x_1, x_2, \ldots, x_m)$. This family generates M because $(x_1, x_2, \ldots, x_m)$ generates M. The augmented family is linearly dependent because y_n is a linear combination of $x_1, x_2, \ldots, x_m$. There is a first x_i which is a linear combination of the previous $y_n, x_1, x_2, \ldots, x_{i-1}$. Note $y_n \neq \zeta$. We obtain $(y_n, x_1, x_2, \ldots, x_{i-1}, x_{i+1}, \ldots, x_m)$ which again generates M. In essence, we have replaced an "x" by a "y". For this reason this is often called the replacement or exchange theorem.

Now consider $(y_{n-1}, y_n, x_1, x_2, \ldots, x_{i-1}, x_{i+1}, \ldots, x_m)$ which must be linearly dependent. There is a first "x" which is a linear combination of the preceding vectors. Delete this vector and continue the argument as before. By this process the x's cannot be used up before the y's because otherwise some smaller subfamily of the y's than $(y_1, y_2, \ldots, y_n)$ would generate M making $(y_1, y_2, \ldots, y_n)$ a linearly dependent family. The number of x's is the same as or exceeds the number of y's. $m \geqslant n$. □

Corollary. *Let M be a K-vector space. If $(x_1, x_2, \ldots, x_m)$ and $(y_1, y_2, \ldots, y_n)$ are finite bases for M then $m = n$.*

Proof. $(x_1, x_2, \ldots, x_m)$ generates M and $(y_1, y_2, \ldots, y_n)$ is linearly independent show $m \geqslant n$. $(y_1, y_2, \ldots, y_n)$ generates M and $(x_1, x_2, \ldots, x_m)$ linearly independent show $n \geqslant m$. $m = n$. □

Corollary. *Let M be a K-vector space with a finite basis $(x_1, x_2, \ldots, x_m)$ and let $(y_j \mid j \in J)$ be another basis. Then J cannot be infinite.*

Proof. If $(y_j \mid j \in J)$ were an infinite family it would contain a finite subfamily with $m + 1$ members which is linearly independent. The m members of the given finite basis $(x_1, x_2, \ldots, x_m)$ generate M. $m \geqslant m + 1$ is a contradiction. □

Since a vector space cannot have at the same time a finite basis and an infinite basis the only remaining comparison is between two infinite bases.

Theorem. *Let M be a K-vector space. Then any two bases have the same number of members.*

PROOF. We leave this remaining comparison between two infinite bases to Appendix 6C at the end of this section. It is entirely appropriate to accept the result as an axiom. □

Because every basis for a given vector space has the same size we can assign to each vector space a number, the number of elements in any basis. We call this number the dimension of the vector space.

Definition. The *dimension* of a vector space is the number of elements in any basis of the vector space.

We will also use the word dimension for the number of elements in a basis of an R-module over a commutative unitary ring when that R-module has a basis. Since not every R-module has a basis not every R-module will have a dimension. Every vector space does have a dimension. We give a proof in Appendix 6D at the end of this section that the number of elements in a basis of a free module over a commutative unitary ring is also an invariant; each basis has the same number of elements as any other basis. This justifies the use of the word dimension for free modules over commutative unitary rings.

EXAMPLE. $\mathbb{R}^3$ has the standard basis $(e_1, e_2, e_3) = ((1, 0, 0), (0, 1, 0), (0, 0, 1))$. Other bases such as $((1, 1, 0), (1, 0, 1), (0, 1, 1))$ also have three members. $\mathbb{R}^3$ is an $\mathbb{R}$-vector space of dimension 3. $\mathbb{Q}[X]$ is a $\mathbb{Q}$-vector space with basis $(1, X, X^2, \ldots) = (X^j | j \in \mathbb{N})$. This basis has the same number of elements as $\mathbb{N}$, an infinite set. We denote the number of members of $\mathbb{N}$ with the symbol ω. The vector space $(\mathbb{R}^{\mathbb{N}})^w$ has the basis $(e_i | i \in \mathbb{N})$ which has ω members also. R, itself, as a vector space over the field R has a basis (1) and therefore has dimension 1. The trivial vector space $\{\zeta\}$ over any field has $\varnothing$ as its basis. The dimension of the trivial vector space is therefore 0.

For finite dimensional vector spaces we have some more particular results.

Theorem. *Any family with $n + 1$ or more vectors in a finite n-dimensional vector space must be linearly dependent.*

PROOF. The number of vectors in a generating family must equal or exceed the number of vectors in a linearly independent family. If a family of $n + 1$ vectors were linearly independent we would have $n + 1 \leqslant n$, manifestly false. □

Theorem. *In an n-dimensional vector space, n finite, any linearly independent family* $(x_1, x_2, \ldots, x_n)$ *is a basis.*

PROOF. An n-dimensional vector space M has a basis of n members. We must show that the family $(x_1, x_2, \ldots, x_n)$ generates the vector space. Let x be in M and consider the family $(x_1, x_2, \ldots, x_n, x)$. Since this family has $n + 1$ members it must be linearly dependent. There must be a first value which is a linear combination of the preceding vectors. This first value must be x because of the linear independence of $(x_1, x_2, \ldots, x_n)$. □

Theorem. *Any finite generating family of n members must be a basis in an n-dimensional vector space.*

PROOF. Let $(x_1, x_2, \ldots, x_n)$ be any generating family. If this set is not linearly independent there is a subfamily of $n - 1$ members which still generates M. A generating family of $n - 1$ members and a linearly independent family (the given basis) in the same space of n members is a contradiction. The generating family is therefore also linearly independent. □

EXAMPLE. Any family of three members which generates $\mathbb{R}^3$ must be a basis for the $\mathbb{R}$-vector space $\mathbb{R}^3$. Any family of three members which is linearly independent is also a basis for $\mathbb{R}^3$. It is only necessary to establish one of the two conditions (linear independence and generation) once the finite dimension of the space is known.

We now return to some earlier material on kernel and range and strengthen the theory with our vector space results.

Theorem. *Let* $f: M \to M'$ *be a morphism of K-vector spaces. Then* $\dim M = \dim \text{kernel} f + \dim \text{range} f$.

PROOF. This is the theorem of the previous section restated in terms of dimension. We know there exists a basis $(x_i | i \in J)$ for the subspace, kernel f, of M. This basis can be extended to a superfamily $(x_i | i \in I)$ of M, $J \subseteq I$. By the theorem of Section 6.9 $(f(x_i) | i \in I - J)$ is a basis for range f. Since $I = J \cup (I - J)$ we have $\text{crd}\, I = \text{crd}\, J + \text{crd}(I - J)$. But $\text{crd}\, I = \dim M$. $\text{Crd}\, J = \dim \ker f$. $\text{Crd}(I - J) = \dim \text{range} f$. (Crd in this theorem is an abbreviation for cardinal number. The cardinal number of a set, we recall from Section 4.1, is the number of members in the set.) □

Definition. If $f: M \to M'$ is a morphism of K-vector spaces then *rank* $f = \dim \text{range} f$ and *nullity* $f = \dim \text{kernel} f$.

The previous theorem can be stated as

$$\dim \text{domain} f = \text{nullity} f + \text{rank} f.$$

Example. $f: \mathbb{R}^3 \to \mathbb{R}^3$ such that $f(r_1, r_2, r_3) = (r_1 + r_2, r_2 + r_3, 0)$ has kernel $f = \{(r_1, r_2, r_3) | r_1 + r_2 = 0 \text{ and } r_2 + r_3 = 0\} = \{(r_1, r_2, r_3) | r_1 = -t, r_2 = t, r_3 = -t\} = \{(-t, t, -t) | t \in \mathbb{R}\} = \{t(-1, 1, -1) | t \in \mathbb{R}\}$. A basis for kernel f is $((-1, 1, -1))$. Nullity $f = 1$. Dimension domain $f = 3$. Therefore rank f must be 2.

Example. In $\mathbb{R}^3$ the subspaces of dimension 0 are only $\{(0, 0, 0)\}$. The subspaces of dimension 1 are $\{t(a_1, a_2, a_3) | t \in \mathbb{R}\}$ where (a_1, a_2, a_3) is a nonzero vector. These subspaces of dimension 1 are lines through the origin. (See Figure 6.13). The subspaces of dimension 2 are $\{t(a_1, a_2, a_3) + u(b_1, b_2, b_3) | t, u \in \mathbb{R}\}$ where a and b are linearly independent vectors. These subspaces of dimension 2 are planes through the origin. (See Figure 6.14). The only subspace of dimension 3 is the entire space $\mathbb{R}^3$. Linear varieties of dimension 0 are $(c_1, c_2, c_3) + \{(0, 0, 0)\} = \{(c_1, c_2, c_3)\}$, singleton points. Linear varieties of dimension 1 are $(c_1, c_2, c_3) + \{t(a_1, a_2, a_3) | t \in \mathbb{R}\}$,

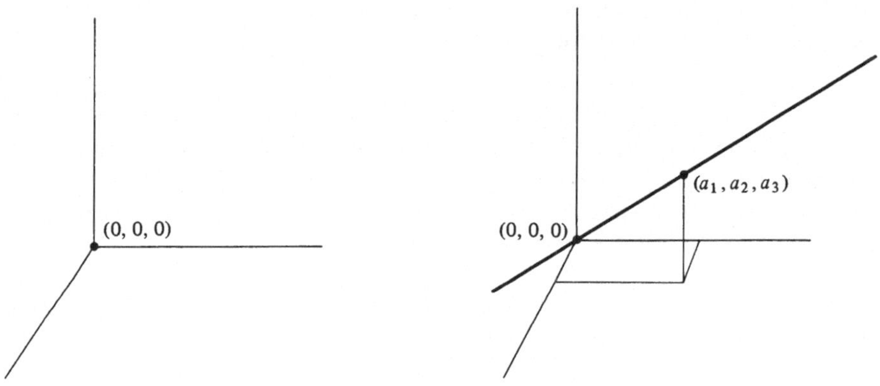

Figure 6.13

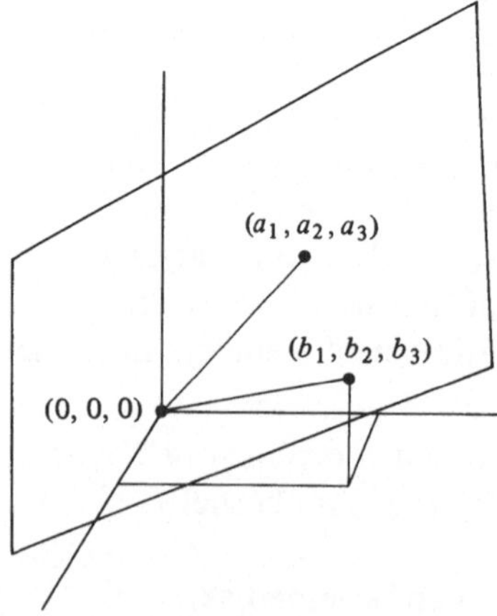

Figure 6.14

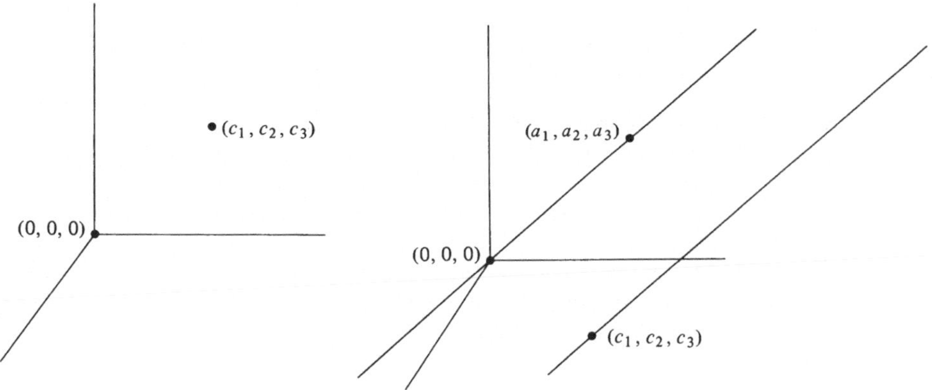

Figure 6.15

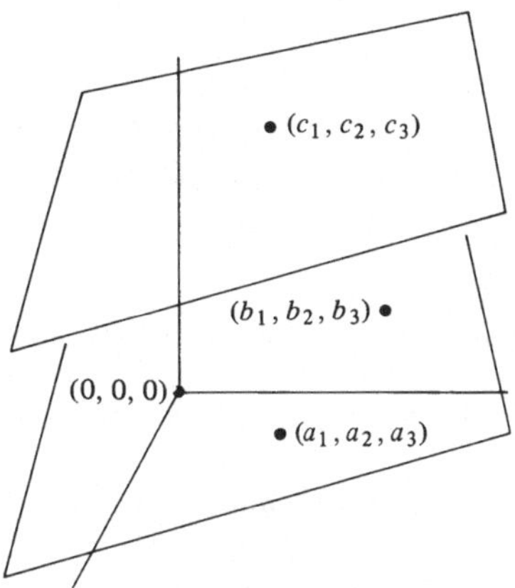

Figure 6.16

lines translated from the origin by (c_1, c_2, c_3) (see Figure 6.15). Linear varieties of dimension 2 are of the form $(c_1, c_2, c_3) + \{t(a_1, a_2, a_3) + u(b_1, b_2, b_3) | t, u \in \mathbb{R}\}$, planes translated from the origin by (c_1, c_2, c_3) (see Figure 6.16).

QUESTIONS

1. Which of these modules are vector spaces?
   (A) the $\mathbb{Z}$-module $\mathbb{Q}$
   (B) the $\mathbb{Q}$-module $\mathbb{R}$
   (C) the $\mathbb{Q}$-module $\mathbb{Q}[X]$
   (D) the $\mathbb{Z}$-module $(\mathbb{Q}^{\mathbb{N}})^w$.
(E) None of the listed modules is a vector space.

2. Which of these statements are true?
   (A) If a module has a basis then it is a vector space.
   (B) If a module is a vector space then it has a basis.

(C) Every module has a generating family.
(D) For every module there is at least one linearly independent set or family.
(E) None of the statements is true.

3. Which of the statements about *modules* are true?
(A) If a family is linearly dependent then at least one vector in the family can be expressed as a linear combination of the others.
(B) If a subfamily of a given family of vectors is linearly dependent then the entire family is linearly dependent.
(C) A singleton family, a family with one member, can be linearly dependent without necessarily being the zero vector.
(D) Every vector space is a module.
(E) None of the statements is true.

4. The set $\Delta = \{(t, t) | t \in \mathbb{R}\}$ is a subspace of the $\mathbb{R}$-vector space $\mathbb{R}^2$. The family (e_1, e_2) is the standard basis for $\mathbb{R}^2$, yet no subfamily of (e_1, e_2) is a basis for the subspace Δ. Which of these statements are true?
(A) Not every subspace of $\mathbb{R}^2$ need have a basis.
(B) Δ is not really a subspace of $\mathbb{R}^2$.
(C) The entire family $((1, 0), (0, 1))$ is a basis for Δ as well as $\mathbb{R}^2$.
(D) A basis for $\mathbb{R}^2$ need not have a subfamily which is a basis for Δ.
(E) None of the statements is true.

5. Which of these statements are true?
(A) The number of linearly independent vectors in a vector space must exceed or equal the number of vectors in a generating family.
(B) A generating family in a module must be either linearly independent or linearly dependent.
(C) If a basis for a vector space is infinite then every finite family in the space is linearly independent.
(D) No family in a module can be linearly dependent unless some finite subfamily of the given family is linearly dependent.
(E) None of the statements is true.

6. Let us say that a vector space allows home runs if one can find at least four distinct bases for the space. How many different (nonisomorphic) vector spaces over $\mathbb{R}$, the real numbers, do not allow home runs?
(A) 0
(B) 1
(C) 4
(D) 2^4
(E) an infinite number.

7. If M is a vector space generated by n vectors, $n \in \mathbb{N}$, which are linearly independent then
(A) $\dim M \leqslant n$
(B) $\dim M < n$
(C) $n > 0$
(D) $n = 0$.
(E) None of the alternatives completes a true sentence.

8. If N and P are *distinct* subspaces of a vector space M over a field K then
(A) $\dim N + \dim P = \dim(N + P)$
(B) $\dim N + \dim P < \dim(N + P)$
(C) $\dim(N \times P) = (\dim N)(\dim P)$
(D) $\dim N + \dim P \geqslant 1$.
(E) None of the alternatives completes a true sentence.

Exercises

1. Let $f:\mathbb{Q}^3 \to \mathbb{Q}^3$ such that $f(r_1, r_2, r_3) = (r_1 + 2r_2, r_1 + r_2, 2r_1 + 3r_2 + r_3)$. Find kernel f and range f and bases for each subspace.

2. Find a morphism from $\mathbb{R}^2$ to $\mathbb{R}^2$ with the smallest possible kernel containing the set $\{(\frac{1}{2}, 3), (2, 12)\}$.

3. Find a morphism f from $\mathbb{R}^3$ to $\mathbb{R}^3$ which has range precisely the space $[(2, 1, 1), (3, 2, 0)]$. $f(s_1, s_2, s_3) = ?$.

4. Show that a morphism $f:\mathbb{R}^3 \to \mathbb{R}^3$ with trivial kernel must be an epimorphism.

5. Let $\mathbf{C}[a, b]$ be the set of all real-valued functions defined and continuous on the real closed interval $[a, b]$. Verify that $\mathbf{C}[a, b]$ is a submodule of the $\mathbb{R}$-module $\mathbb{R}^{[a,b]}$. Consider the integral as a function $\mathbf{C}[a, b] \to \mathbb{R}$ with value $\int_a^b f(t)\,dt$. Prove that $\int_a^b:\mathbf{C}[a, b] \to \mathbb{R}$ is a morphism. Prove $\int_a^b$ is not a monomorphism by showing that the kernel is nontrivial. [*Hint*: Construct a nontrivial member of the kernel. Prove that $\int_a^b$ is an epimorphism.]

6. Denote the derivative of a function f by Df. The derivative of a polynomial $p(X) = a_0 + a_1X + \cdots + a_nX^n$ is $Dp(X) = a_1 + 2a_2X + \cdots + na_nX^{n-1}$. Prove $D:\mathbb{R}[X] \to \mathbb{R}[X]$ is a morphism. Find kernel D. Is D a monomorphism? Is D an epimorphism?

7. Show that if N is a subspace of a vector space M and the subspace has the same dimension as M then $N = M$. Show also that if N is a submodule of a module M that the same result may not hold.

8. Let M' and M'' be subspaces of a K-vector space. Show that $\dim(M' + M'') + \dim(M' \cap M'') = \dim M' + \dim M''$. [*Hint*: Choose first a basis for $M' \cap M''$ and extend separately to obtain bases for M' and for M''.]

9. Let $f:M \to M'$ be a morphism of K-vector spaces. Prove that these three conditions are equivalent: f is a monomorphism, nullity of f is zero, rank $f = \dim M$.

10. Let M' and M'' be K-vector spaces. Prove $\dim(M' \times M'') = \dim M' + \dim M''$.

11. Let N be a subspace of a K-vector space M. Prove $\dim M = \dim M/N + \dim N$.

12. Let $(x_1, x_2, \ldots, x_n)$, a finite family of vectors, generate a K-vector space M. Prove that there exists a subfamily of $(x_1, x_2, \ldots, x_n)$ which is a basis of M.

13. Let $f:\mathbb{R}^3 \to \mathbb{R}^4$ such that $f(s_1, s_2, s_3) = (s_2 + 2s_3, s_1 - s_2, s_1 + 3s_2 - s_3, s_1 + s_3)$. Find a basis for kernel f, extend to a basis for domain f and find a basis for range f. Verify dim domain f = nullity f + rank f.

14. The family $((39, 51, 3), (13, 17, 1), (26, 34, 2), (52, 68, 4))$ is linearly dependent. In any linearly dependent family either the first vector is zero or there is a first vector in the family which is a linear combination of the preceding ones. Apply this theorem to this example.

15. Show that if one value of a family $(x_i|i \in I)$ is a multiple of another value (say $x_j = rx_k$ for some $r \in K$, j, $k \in I$) in a K-vector space M then the family is linearly dependent.

16. Find all the morphisms $f:\mathbb{R}^3 \to \mathbb{R}^3$ such that $(1, 1, 1)$ and $(1, 2, 3)$ both belong to the kernel of f. What are the ranks of these morphisms?

17. Let $f:M \to M'$ be a vector space morphism. Show that dim kernel $f \leqslant$ dim domain f and dim range $f \leqslant \min\{\text{dim domain } f, \text{dim codomain } f\}$.

18. Let M, M' be K-vector spaces of the same finite dimension. Let $f:M \to M'$ be a monomorphism. Prove f is an isomorphism.

19. Let M, M' be K-vector spaces of the same finite dimension. Let $f:M \to M'$ be an epimorphism. Prove f is an isomorphism.

20. Let M, M' be K-vector spaces and let $f:M \to M'$ be an isomorphism. Prove M and M' must have the same dimension.

21. Give a K-vector space M and a subspace N find a vector space M' and a morphism $f:M \to M'$ such that kernel $f = N$.

22. Given a K-vector space M' and a subspace N' find a vector space M and a morphism $f:M \to M'$ such that range $f = N'$.

23. Let $M = \{t(1, 1, 2, 2) + u(0, 2, 1, 0)|t, u \in \mathbb{R}\}$ and $N = \{v(1, 0, 2, 2) + w(2, 1, 3, 4)|v, w \in \mathbb{R}\}$ be two subspaces of $\mathbb{R}^4$. Find a basis for $M \cap N$. Extend this basis to a basis for M. Also extend the basis of $M \cap N$ to a basis for N. Verify the equation $\dim(M + N) + \dim(M \cap N) = \dim M + \dim N$.

24. If $(a_1, a_2, \ldots, a_n)$ is any nonzero vector in $\mathbb{R}^n$, $n \geqslant 2$, prove there is a nontrivial morphism $f:\mathbb{R}^n \to \mathbb{R}$ such that $f(a_1, a_2, \ldots, a_n) = 0$.

25. The dimension of a coset or linear variety $a + N$ of a K-vector space M is defined to be the dimension of the subspace N. An affine mapping, we recall, is a mapping taking linear varieties into linear varieties. If $g:M \to M'$ is an affine mapping of the K-vector spaces M, M' show that $\dim(a + N) = \dim g(a + N)$ whenever g is injective. Show that if g is also a bijection then $\dim(b + N') = \dim g^{-1}(b + N')$ for any variety $b + N'$ in M'.

26. This is a long exercise in which some of the more obvious geometrical results of $\mathbb{R}^3$ are given vector space definitions.

Definitions

A line is a one dimensional linear variety of $\mathbb{R}^3$.

A plane is a two dimensional linear variety of $\mathbb{R}^3$.

Lines are parallel if and only if they are linear varieties with the same subspace.

Planes are parallel if and only if they are linear varieties with the same subspace.

Prove these theorems:

Two planes are parallel or intersect in a line.

Two lines are disjoint or intersect in a point or are identical.

Two parallel lines cannot intersect in a single point.
Two lines parallel to a third line are parallel to each other.
Through a given point there is one and only one line parallel to a given line.
A line and a plane are disjoint or intersect in a single point or the line lies in the plane.

27. We now define the term *independent* which is similar to, but not the same as *linear independent*. Let M be an R-module. A subset A of M is *independent* if and only if $[A] \neq [A - \{x\}]$ for all x in A. In other words, a subset is independent if and only if no proper subset generates the same subspace: it is a minimal generating set. Prove that a subset A of a K-vector space is independent if and only if A is linearly independent.

28. Show that in the $\mathbb{Z}$-module $\mathbb{Z}^2$ the set $\{(3, 0), (2, 0), (0, 1)\}$ is independent yet not linearly independent.

29. Show that a set S is a basis for a K-vector space M if and only if S is a maximal linearly independent set (S is linearly independent and $S \cup \{x\}$ is linearly dependent for any $x \in M - S$).

Appendix 6B The existence of a basis for a vector space

Demonstrating that every vector space has at least one basis requires the use of somewhat advanced techniques from set theory. In an undergraduate course one can reasonably take the existence of a basis for a vector space as an axiom leaving for another day the question of whether there exist vector spaces without bases. The techniques used here can also be used to pursue details of the theorem at the end of Section 1.6. We intend here to give a proof of the existence of a basis using Zorn's lemma, a theorem of set theory equivalent to the axiom of choice and to the well-ordering theorem. The theorem we intend to prove is simple in conception; consider the set of all linearly independent families in the vector space, find a maximal one and prove it generates the entire space. The use of the set theoretic technique is to assert the existence of the maximal linearly independent family.

Definition. A collection of families, $\mathscr{C}$, is a chain provided the collection is totally ordered; that is, given any two families in the collection one is a subfamily of the other.

Definition. A family B is maximal in a collection $\mathscr{X}$ if and only if there is no family D in $\mathscr{X}$ which properly extends B.

Zorn's lemma. *If $\mathscr{X}$ is a nonempty collection of families such that every nonempty chain $\mathscr{C}$ of families from $\mathscr{X}$ has the property $\bigcup\mathscr{C} \in \mathscr{X}$ then $\mathscr{X}$ contains a maximal family.*

Theorem. *If M is a K-vector space over a field (or division ring) K and $(x_i|i \in I)$ is any linearly independent family of M then there exists a basis of M which extends $(x_i|i \in I)$.*

PROOF. Denote by $\mathscr{X}$ the set of all linearly independent families of M which have $(x_i|i \in I)$ as a subfamily. The collection is nonempty for it includes $(x_i|i \in I)$ itself. Let $\mathscr{C}$ be any nonempty chain of sets in $\mathscr{X}$. We now show $\bigcup\mathscr{C}$ is a linearly independent family of M, i.e., belongs to $\mathscr{X}$. Suppose $r_1x_1' + r_2x_2' + \cdots + r_nx_n'$ is a linear combination of elements $x_1', x_2', \ldots, x_n'$ which are values of the family $\bigcup\mathscr{C}$ and is equal to ζ. $x_1' \in \bigcup\mathscr{C}$ implies x_1' is a value in C_1 for some family C_1 in $\mathscr{C}$. x_2' is a value in C_2 for some C_2 in $\mathscr{C}$. Since the collection $\mathscr{C}$ is a chain either $C_1 \subseteq C_2$ or $C_2 \subseteq C_1$. Thus both x_1' and x_2' are values in the bigger of C_1 and C_2. Inductively, there is a set $C = \max\{C_1, C_2, \ldots, C_n\}$ in $\mathscr{C}$ in which all the $x_1', x_2', \ldots, x_n'$ are values. As C is a linearly independent family $r_1x_1' + r_2x_2' + \cdots + r_nx_n' = \zeta$ implies all $r_1, r_2, \ldots, r_n$ are zero. $\bigcup\mathscr{C}$ is a linearly independent family. Since $\mathscr{C}$ is not empty there is at least one family D in $\mathscr{C}$. $(x_i|i \in I) \subseteq D \subseteq \bigcup\mathscr{C}$. We conclude $\bigcup\mathscr{C} \in \mathscr{X}$.

Having fulfilled the hypothesis of Zorn's lemma we may conclude $\mathscr{X}$, the collection of all linearly independent families of M which extend $(x_i|i \in I)$, contains a maximal family B. By Exercise 29 of Section 6.10 this maximal linearly independent family generates M. □

Appendix 6C Equicardinality of infinite bases of a vector space

We wish to argue that if $(x_i|i \in I)$ and $(y_j|j \in J)$ are both infinite bases of a vector space $\langle M, +, \zeta\rangle$ over a field K then $\operatorname{crd} I = \operatorname{crd} J$. This is to say that the bases have the same number of elements.

Each basis element y_j can be expressed as a linear combination of the basis family $(x_i|i \in I)$. Since the family $(y_j|j \in J)$ is linearly independent we wish to count the number of linearly independent combinations of the family $(x_i|i \in I)$. The actual number of y_j, the cardinality of J, must be smaller than or equal to the number of linearly independent combinations of $(x_i|i \in I)$ we can form which result in a linearly independent set. We count the number of linear combinations we can form so that the collection of linear combinations remains linearly independent. Of linear combinations of length 1 we can choose any one of the vectors x_i, $i \in I$, making n choices where $n = \operatorname{crd} I$. We cannot choose two multiples of the same x_i for then our collection would be linearly dependent. Of length 2 there are at most two linear combinations of a given two vectors which are linearly independent. We can choose a pair of vectors from $(x_i|i \in I)$ in fewer than n^2 ways. Thus of length 2 one has at most $2n^2$ linearly independent linear combinations. Considering all possible finite lengths we can have at most $n + 2n^2 + 3n^3 + 4n^4 + \cdots$ linearly independent combinations of $(x_i|i \in I)$. Now n is an infinite cardinal number so that any finite power of n or finite multiple of n is again simply n. Thus the number of linearly independent combinations of $(x_i|i \in I)$ we can form is less than or equal to n. Since the number is clearly at least as large as n it must be exactly n. We have showed that if crd $I = n$ then crd $J = n$ also.

Appendix 6D Dimension of a module over a commutative unitary ring

The assignment of a dimension to a vector space was possible because the number of elements in any basis was the same as the number of elements in any other basis. We now prove a theorem that shows that the size of bases for modules over commutative unitary rings is also invariant. This makes it possible to define a dimension for free modules over commutative unitary rings. Of course, not every module has a basis and these modules will have no dimension. The $\mathbb{Z}$-module $\mathbb{Q}$, for example, is without dimension.

Theorem. *Let M be a module over a commutative unitary ring R with bases $(x_i|i \in I)$ and $(y_j|j \in J)$. Then* crd $I =$ crd J.

PROOF. With an ideal A of R we shall also consider the submodule $[AM]$ generated by all ideal multiples of M and the corresponding quotient R-module $M/[AM]$. We see that $M/[AM]$ is not only an R-module but is also an (R/A)-module with the following definition of exterior multiplication: $(r/A)\, x/[AM] = rx/[AM]$, $r/A \in R/A$, $x/[AM] \in M/[AM]$. We verify that the exterior multiplication is well-defined. Let $r/A = s/A$. $r - s \in A$. $(r - s)x \in [AM]$. $rx - sx \in [AM]$. $rx/[AM] = sx/[AM]$. $(r/A)\, x/[AM] = (s/A)x/[AM]$.

The next stage in this proof is to show that $(x_i/[AM]|i \in I)$ is a basis for the (R/A)-module $M/[AM]$. Let $\sum_{i \in I}^{w} (r_i/A)x_i/[AM]$ be a linear combination of the family and set it equal to the zero, $\zeta/[AM]$. Then $\sum_{i \in I}^{w} r_i x_i/[AM] = \zeta/[AM]$. $(\sum_{i \in I}^{w} r_i x_i)/[AM] = \zeta/[AM]$. $\sum_{i \in I}^{w} r_i x_i \in [AM]$. There exist s_i in A such that $\sum_{i \in I}^{w} r_i x_i = \sum_{i \in I}^{w} s_i x_i$. $\sum_{i \in I}^{w} (r_i - s_i)x_i = \zeta$. $r_i - s_i = \theta$ for all $i \in I$. $r_i = s_i$ for all $i \in I$. Since $s_i \in A$ we must also have $r_i \in A$. $r_i/A = \theta/A$ for all $i \in I$. Thus the family is linearly independent.

To show that the family generates the module let $x/[AM] \in M/[AM]$. Then $x \in M$ and $x = \sum_{i \in I}^{w} r_i x_i$ for some $r_i \in R$. $x/[AM] = (\sum_{i \in I}^{w} r_i x_i)/[AM] = \sum_{i \in I}^{w} r_i x_i/[AM] = \sum_{i \in I}^{w} (r_i/A)x_i/[AM]$. Therefore $(x_i/[AM]|i \in I)$ generates the (R/A)-module $M/[AM]$.

Having established the previous results for an arbitrary ideal A of R we now choose a maximal ideal A of the commutative unitary ring R (cf. Section 2.8). R/A is then a field. The R/A-module $M/[AM]$ is a vector space with a basis, the family $(x_i/[AM]|i \in I)$. So also is the family $(y_i/[AM]|i \in J)$, as $(y_i|i \in J)$ is also a given basis family for M. We then have two bases for the vector space $M/[AM]$ over the field R/A. The cardinality or size of I and J must be the same. □

One easily establishes that a proper subspace of a vector space has smaller dimension than the entire space. However, $2\mathbb{Z}$ is a proper submodule of the $\mathbb{Z}$-module $\mathbb{Z}$. The submodule has the basis (2) and the entire module has the basis (1). The dimension of the proper submodule and the dimension of the entire module are both 1. The basis for the submodule cannot be extended to a basis for the entire space. We offer the following theorem.

Theorem. *Let M be a module over a principal ideal domain R and let M have a finite basis. If N is a submodule of M then N also has a basis and* $\dim N \leqslant \dim M$.

PROOF. Let M have the basis $(x_i | i \in m)$. Let $N_j = N \cap [(x_1, x_2, \ldots, x_j)]$, $j = 0, 1, \ldots, m$. $N_0 = \{\zeta\}$ and $N_m = N$. $N_1 = N \cap [x_1]$ is a submodule of x_1. Every member of N_1 is a multiple of x_1. The multiples themselves (members of R) form an ideal of R of which a_1 is the generator. Thus $N_1 = [a_1 x_1]$ for some a_1 in R. N_1 is either $\{\zeta\}$ or has a basis of one element. To run an induction assume N_k has a basis and the dimension of N_k is $\leqslant k$. Let $A = \{a | a \in R$ and there exist $x \in N, b_1, b_2, \ldots, b_k \in R$ such that $x = b_1 x_1 + \cdots + b_k x_k + a x_{k+1}\}$. A is an ideal of R. Let a_{k+1} be a generator of A. If $a_{k+1} = \theta$ then $N_{k+1} = N_k$ and the induction is complete. If $a_{k+1} \neq \theta$ then let $w \in N_{k+1}$ such that $w = a_1 x_1 + \cdots + a_k x_k + a_{k+1} x_{k+1}$. If x is any member of N_{k+1} there exists $c \in R$ such that $x - cw \in N_k$. $N = N_k + [w]$. But $N_k \cap [w] = \{\zeta\}$. $N = N_k \oplus [w]$. w is an additional basis element and the induction is complete. □

The previous theorem can also be proved for nonfinite bases.

For modules without bases it becomes important to assign them a finite quality whenever possible, that of being finitely generated. A module is *finitely generated* if it is generated by a finite family.

Corollary. *Let M be a finitely generated module over a principal ideal domain R. If N is a submodule of M then N is also finitely generated.*

PROOF. Let M be generated by the finite family $(y_1, y_2, \ldots, y_m)$. Let $\varphi: R^m \to M$ such that $\varphi(e_j) = y_j$, $j = 1, 2, \ldots, m$, and $(e_j | j \in \hat{m})$ is the standard basis. φ is an epimorphism. The preimage $\varphi^{-1}(N)$ is a submodule of R^m, the free module, and therefore $\varphi^{-1}(N)$ must also have a finite basis $(d_1, d_2, \ldots, d_k)$, $k \leqslant m$. $[\varphi(d_1), \varphi(d_2), \ldots, \varphi(d_k)] = N$. □

Linear algebra: The module of morphisms 7

In this chapter we continue our development of linear algebra. We study the module of morphisms, the module made from collecting together all the morphisms from one module to another. The coordinate morphism provides the tie between the module of morphisms and the module of m by n matrices. The collective structure of the module of morphisms is further extended through composition to produce the algebra of endomorphisms and the corresponding algebra of square matrices. Through the use of the matrix of a morphism and the coordinates of a vector we calculate the images of morphisms in the traditional way.

Through change of basis we explore the relation of equivalence of matrices and search for bases which produce exceptionally simple matrix representation of morphisms: canonical forms. We further link the transpose of a matrix to the dual of a morphism and use this tool in the study of the rank of a matrix.

In the section on linear equations we study the complete solution of linear equations justifying the earlier methods introduced in Appendix 6A. We show that every matrix with entries in a field is row equivalent to a *unique* matrix in row-reduced echelon form.

The section on determinants is highly detailed and is based upon permutations. It is the traditional constructive approach to determinants. A reader preferring perhaps a more elegant approach can consult [3, p. 98]. One of our motives here is to give the student a working knowledge of permutations and the associated kinds of manipulative mathematics.

7.1 $\mathscr{L}(M, M')$, the module of morphisms

In this section we investigate the module structure of the set of all morphisms from one module M to another module M', both over a commutative unitary ring R. We define the matrix of a morphism and set up the fundamental correspondence between morphisms and matrices.

Definition. Let M, M' be modules over R, a commutative unitary ring. By $\mathscr{L}(M, M')$ we mean the set of all morphisms from M to M'.

$\mathscr{L}(M, M')$ is a set of functions, a subset of the set M'^M of all functions from M to M'. The additive group structure, $\langle M', +, \zeta' \rangle$, immediately gives us an additive group of functions, $\langle M'^M, +, z \rangle$, with operations

$$(f + g)(x) = f(x) + g(x) \quad \text{for all } x \in M,$$
$$z(x) = \zeta' \quad \text{for all } x \in M.$$

We do not repeat details here of this construction; consult Section 6.1 for a similar discussion in detail. The exterior multiplication on M' from the ring R induces an exterior multiplication on M'^M,

$$(rf)(x) = rf(x) \quad \text{for all } r \in R, x \in M.$$

In summary, $\langle M'^M, +, z \rangle$ is an R-module. It is not, however, the module of our principal interest. This is the submodule $\mathscr{L}(M, M')$.

Theorem. *Let M, M' be modules over a commutative unitary ring R. $\langle \mathscr{L}(M, M'), +, z \rangle$ is an R-module.*

PROOF. We must establish closure of addition, exterior multiplication. We must show $\mathscr{L}(M, M')$ is nonempty. This will show $\mathscr{L}(M, M')$ is a submodule.

Let f, g belong to $\mathscr{L}(M, M')$. Both functions are then linear. We now show $f + g$ is also linear. $(f + g)(u + v) = f(u + v) + g(u + v) = f(u) + f(v) + g(u) + g(v) = f(u) + g(u) + f(v) + g(v) = (f + g)(u) + (f + g)(v)$. $(f + g)(ru) = f(ru) + g(ru) = rf(u) + rg(u) = r(f(u) + g(u)) = r(f + g)(u)$. $f + g \in \mathscr{L}(M, M')$.

Let $s \in R$, $f \in \mathscr{L}(M, M')$. $(sf)(u + v) = sf(u + v) = s(f(u) + f(v)) = sf(u) + sf(v) = (sf)(u) + (sf)(v)$. To prove the homogeneity of sf we must use the commutativity of R. $(sf)(ru) = sf(ru) = s(rf(u)) = (sr)f(u) = (rs)f(u) = r(sf(u)) = r(sf)(u)$. $sf \in \mathscr{L}(M, M')$.

Finally, z belongs to $\mathscr{L}(M, M')$. $z(u + v) = \zeta' = \zeta' + \zeta' = z(u) + z(v)$. $z(ru) = \zeta' = r\zeta' = rz(u)$. $\langle \mathscr{L}(M, M'), +, z \rangle$ is an R-module.

Having established the module status of $\mathscr{L}(M, M')$ we begin a discussion which will lead ultimately to finding the dimension of $\mathscr{L}(M, M')$ when M and M' are free. We accomplish this aim by the construction of a basis for $\mathscr{L}(M, M')$. We restrict our attention to modules which have finite bases.

Theorem. *Let M, M' be modules over a commutative unitary ring R. Let $(x_j | j \in n)$ be a basis for M and $(y_i | i \in m)$ be a basis for M'. The family of* *morphisms* $(e_{ij} | i \in m, j \in n)$ *such that*

$$\begin{aligned} e_{ij}(x_q) &= \zeta' \quad \text{if } q \neq j \\ &= y_i \quad \text{if } q = j, q = 1, 2, \ldots, n, \end{aligned}$$

is a basis for $\mathscr{L}(M, M')$.

PROOF. Since a morphism is uniquely defined by its behavior on the basis elements of its domain, each e_{ij} is completely and uniquely defined:

$$\begin{aligned} e_{ij}(x) &= e_{ij}\left(\sum_{k=1}^{n} X_k x_k\right) = \sum_{k=1}^{n} X_k e_{ij}(x_k) \\ &= \sum_{k \neq j} X_k \zeta' + X_j y_i = X_j y_i. \end{aligned}$$

To complete the theorem we must show $(e_{ij} | i \in m, j \in n)$ is a linearly independent family which generates $\mathscr{L}(M, M')$. We begin by letting $\sum_{(i,j) \in m \times n} r_{ij} e_{ij} = z$.

$$\begin{aligned} \left(\sum_{(i,j) \in m \times n} r_{ij} e_{ij}\right)(x_q) &= z(x_q) \quad \text{for any } q = 1, 2, \ldots, n. \\ \sum_{(i,j) \in m \times n} r_{ij} e_{ij}(x_q) &= \zeta'. \end{aligned}$$

Since $e_{ij}(x_q) = \zeta'$ for all $j \neq q$ we drop from the sum all terms with $j \neq q$.

$$\begin{aligned} \sum_{i \in m} r_{iq} e_{iq}(x_q) &= \zeta' \quad \text{for any } q = 1, 2, \ldots, n. \\ \sum_{i \in m} r_{iq} y_i &= \zeta'. \end{aligned}$$

We know $(y_i | i \in m)$ to be a linearly independent family of M'.

$$r_{iq} = \theta \quad \text{for all } i \in m \text{ and all } q \in n.$$

$(e_{ij} | i \in m, j \in n)$ is therefore a linearly independent family.

To show that $(e_{ij} | i \in m, j \in n)$ generates $\mathscr{L}(M, M')$ we must show that any morphism $f: M \to M'$ can be written as a linear combination of the e_{ij}. The coefficients will, naturally, depend upon f. For each $j \in n$ we have $f(x_j)$ an element of M'. As a member of M', $f(x_j)$ can be expressed as a linear combination of the family $(y_i | i \in m)$. There exist coefficients in R, $\mu(f)_{1j}, \mu(f)_{2j}, \ldots, \mu(f)_{mj}$ such that

$$f(x_j) = \sum_{i=1}^{m} \mu(f)_{ij} y_i, \quad j = 1, 2, \ldots, n.$$

These equations are the defining equations for the matrix of f with respect to the bases $(x_j | j \in n)$, $(y_i | i \in m)$. The use of the notation $\mu(f)_{ij}$ indicates the dependence of the coefficients upon f and stands for the matrix of f. The matrix of f is the family of coefficients $(\mu(f)_{ij}|i \in m, j \in n) = \mu(f)$. We now propose to show that $f = \sum_{(i,j)\in m\times n} \mu(f)_{ij}e_{ij}$; the matrix of f provides the coefficients for the expression of f in terms of the basis family $(e_{ij}|i \in m, j \in n)$. Both f and $\sum_{i\in m, j\in n} \mu(f)_{ij}e_{ij}$ are morphisms in $\mathscr{L}(M, M')$. We wish to show that they are equal. Two morphisms in $\mathscr{L}(M, M')$ are equal, of course, if they agree upon the basis elements $(x_j|j \in n)$ of the domain. For any $q \in n$,

$$\begin{aligned}\Bigl(\sum_{i\in m, j\in n} \mu(f)_{ij}e_{ij}\Bigr)(x_q) &= \sum_{i\in m, j\in n} \mu(f)_{ij}e_{ij}(x_q)\\ &= \sum_{i\in m} \mu(f)_{iq}e_{iq}(x_q)\\ &= \sum_{i\in m} \mu(f)_{iq}y_i = f(x_q).\end{aligned}$$

The last equality follows from the defining equations of $\mu(f)$. □

The proof of the theorem is complete but we repeat now for emphasis the definition of the matrix of f which occurred naturally in the theorem.

Definition. Let M, M' be modules over a commutative unitary ring R. Let $(x_j|j \in n)$, $(y_i|i \in m)$ be bases for M, M' respectively. $\mu(f)$, *the matrix of* f, is the family $(\mu(f)_{ij}|i \in m, j \in n)$ defined by the equations

$$f(x_j) = \sum_{i\in m} \mu(f)_{ij}y_i, \quad j \in n.$$

We refer to these equations as *the defining equations of the matrix of* f.

We have showed that $\mathscr{L}(M, M')$ will have a basis $(e_{ij}|i \in m, j \in n)$ of mn members when M has a basis of n members and M' has a basis of m members. If M is of dimension n and M' is of dimension m then $\mathscr{L}(M, M')$ is of dimension mn. In proving the theorem we have linked a matrix from R to every morphism of $\mathscr{L}(M, M')$ and every choice of bases. The matrix, $\mu(f)$, associated with f depends very much upon the bases chosen in M and M'. A change of bases, as we shall see, affects the matrix.

Before continuing with the development of these concepts we show an example.

EXAMPLE. $f: \mathbb{R}^3 \to \mathbb{R}^2$ such that $f(r_1, r_2, r_3) = (r_1 + 2r_2 + 3r_3, r_1 - r_2)$ is a morphism in $\mathscr{L}(\mathbb{R}^3, \mathbb{R}^2)$. A basis for $\mathbb{R}^3$ is the standard one $((1, 0, 0), (0, 1, 0), (0, 0, 1))$ and for $\mathbb{R}^2$ also the standard one $((1, 0), (0, 1))$. The defining equations for the matrix of f are

$$\begin{aligned} f(1,0,0) &= (1,1) &&= 1(1,0) + 1(0,1)\\ f(0,1,0) &= (2,-1) &&= 2(1,0) + (-1)(0,1)\\ f(0,0,1) &= (3,0) &&= 3(1,0) + 0(0,1).\end{aligned}$$

The matrix of f with respect to the given bases is

$$\begin{pmatrix} 1 & 2 & 3 \\ 1 & -1 & 0 \end{pmatrix}.$$

Since the matrix values yield the coefficients in the linear combination of f in terms of the family $(e_{ij} | i \in 2, j \in 3)$ we can write $f = e_{11} + 2e_{12} + 3e_{13} + e_{21} - e_{22}$. The dimension of $\mathscr{L}(\mathbb{R}^3, \mathbb{R}^2)$ is $2 \cdot 3 = 6$.

It is to be noticed that when the defining equations for the matrix are written the rows and columns are transposed in the array.

$$\begin{aligned}
f(x_1) &= \mu(f)_{11} y_1 + \mu(f)_{21} y_2 + \cdots + \mu(f)_{m1} y_m \\
f(x_2) &= \mu(f)_{12} y_1 + \mu(f)_{22} y_2 + \cdots + \mu(f)_{m2} y_m \\
&\cdots \\
f(x_n) &= \mu(f)_{1n} y_1 + \mu(f)_{2n} y_2 + \cdots + \mu(f)_{mn} y_m.
\end{aligned}$$

It will be necessary then when extracting the coefficients from the defining equations to interchange the positions of rows and columns to obtain the matrix $\mu(f) =$

$$\begin{pmatrix} \mu(f)_{11} & \mu(f)_{12} & \cdots & \mu(f)_{1n} \\ \mu(f)_{21} & \mu(f)_{22} & \cdots & \mu(f)_{2n} \\ \cdots & & & \\ \mu(f)_{m1} & \mu(f)_{m2} & \cdots & \mu(f)_{mn} \end{pmatrix}.$$

A morphism from a module of dimension n to a module of dimension m leads to a matrix which has m rows and n columns.

A further example and an observation of interest is to find the matrices that correspond to the e_{ij} themselves.

$$\begin{aligned}
e_{ij}(x_1) &= \theta y_1 + \theta y_2 + \cdots + \theta y_i + \cdots + \theta y_m \\
e_{ij}(x_2) &= \theta y_1 + \theta y_2 + \cdots + \theta y_i + \cdots + \theta y_m \\
e_{ij}(x_3) &= \theta y_1 + \theta y_2 + \cdots + \theta y_i + \cdots + \theta y_m \\
&\cdots \\
e_{ij}(x_j) &= \theta y_1 + \theta y_2 + \cdots + v y_i + \cdots + \theta y_m \\
&\cdots \\
e_{ij}(x_n) &= \theta y_1 + \theta y_2 + \cdots + \theta y_i + \cdots + \theta y_m.
\end{aligned}$$

$$\mu(e_{ij}) = \begin{pmatrix} \theta & \theta & \cdots & \theta & \cdots & \theta \\ \theta & \theta & \cdots & \theta & \cdots & \theta \\ \cdots & & & & & \\ \theta & \theta & \cdots & v & \cdots & \theta \\ \cdots & & & & & \\ \theta & \theta & \cdots & \theta & \cdots & \theta \end{pmatrix} \leftarrow \text{row } i$$

$\uparrow$ column j

To show how the matrix of f depends upon the choice of bases for M and M' we calculate the matrix of f, the same morphism as before in the example, $f(r_1, r_2, r_3) = (r_1 + 2r_2 + 3r_3, r_1 - r_2)$, with different bases. We let the bases be $((1, 1, 0), (1, 0, 1), (0, 1, 1))$ for $\mathbb{R}^3$ and $((1, 1), (0, 1))$ for $\mathbb{R}^2$. Any two bases could serve as an example. We now calculate the defining equations for the matrix for these bases. We know from substitution

$$\begin{aligned} f(1, 1, 0) &= (3, 0) \\ f(1, 0, 1) &= (4, 1) \\ f(0, 1, 1) &= (5, -1). \end{aligned}$$

To complete the calculation we must express each of these elements as a linear combination of the basis elements $((1, 1), (0, 1))$. We look for coefficients such that

$$\begin{aligned} (3, 0) &= r_{11}(1, 1) + r_{21}(0, 1) \\ (4, 1) &= r_{12}(1, 1) + r_{22}(0, 1) \\ (5, -1) &= r_{13}(1, 1) + r_{23}(0, 1). \end{aligned}$$

Solutions for these equations are

$$\begin{aligned} r_{11} &= 3 & r_{21} &= -3 \\ r_{12} &= 4 & r_{22} &= -3 \\ r_{13} &= 5 & r_{23} &= -6. \end{aligned}$$

The matrix of f with respect to this pair of bases is

$$\begin{pmatrix} 3 & 4 & 5 \\ -3 & -3 & -6 \end{pmatrix}.$$

The appearance of a basis $(e_{ij}|i \in m, j \in n)$ for the R-module $\mathscr{L}(M, M')$ allows us to speak of the coordinate isomorphism $L: R^{m \times n} \to \mathscr{L}(M, M')$ associated with this basis. We wish to compare this mapping with the just defined matrix mapping $\mu: \mathscr{L}(M, M') \to R^{m \times n}$ which assigns to each morphism f its matrix $\mu(f)$.

Theorem. *Let M, M' be modules over a commutative unitary ring R. Let $(x_j|j \in n)$ be a basis for M and $(y_i|i \in m)$ be a basis for M'. Then the coordinate isomorphism $L: R^{m \times n} \to \mathscr{L}(M, M')$ and the matrix mapping $\mu: \mathscr{L}(M, M') \to R^{m \times n}$ are inverse functions. The matrix mapping is also an isomorphism.*

PROOF. $\mu(L(r_{ij}|i \in m, j \in n)) = \mu(\sum_{i \in m, j \in n} r_{ij}e_{ij}) = (r_{ij}|i \in m, j \in n)$. $\mu \circ L = I$, the identity. $L(\mu(f)) = L(\mu_{ij}(f)|i \in m, j \in n)$ where $f = \sum_{i \in m, j \in n} \mu(f)_{ij}e_{ij}$. $L(\mu(f)) = f$. $L \circ \mu = I$. L and μ are inverse functions. Since L is an isomorphism so also is μ. □

Establishing that μ is an isomorphism yields for us standard rules about matrices such as the matrix of the sum of two morphisms is the sum of the matrices and the matrix of the scalar multiple of a morphism is the scalar times the matrix of the morphism.

QUESTIONS

1. Let R be a commutative unitary ring and M, M' be modules with finite bases. Which of the following statements are true?
   (A) $\mu(f + g) = \mu(f) + \mu(g)$ for all $f, g \in \mathscr{L}(M, M')$.
   (B) $\mu(rf) = r\mu(f)$ for all $r \in R, f \in \mathscr{L}(M, M')$.
   (C) $\mu(f) = \mu(g)$ implies $f = g$ for all $f, g \in \mathscr{L}(M, M')$.
   (D) For each $(r_{ij}|i \in m, j \in n)$ in $R^{m \times n}$ there exists an f in $\mathscr{L}(M, M')$ such that $\mu(f) = (r_{ij}|i \in m, j \in n)$.
   (E) None of the statements is true.

2. $\mathscr{L}(M, M')$ is not an R-module for noncommutative rings R because
   (A) $\mathscr{L}(M, M')$ is not closed under addition
   (B) the function rf for $r \in R, f \in \mathscr{L}(M, M')$ is not well defined
   (C) the function rf fails to be additive in some cases
   (D) $f + g$ is not necessarily homogeneous.
   (E) None of the alternatives completes a true sentence.

3. If the finite dimension of a module M is n and that of M' is m, both over a commutative unitary ring R, then
   (A) $\mathscr{L}(M, M')$ may fail to have a basis, but if $\mathscr{L}(M, M')$ does have a basis it is finite
   (B) $\mathscr{L}(M, M')$ may have an infinite basis because there is no guarantee that all sums are finite
   (C) $\mathscr{L}(M, M')$ has dimension mn, but $\mathscr{L}(M, M')$ does not necessarily have a basis
   (D) $\mathscr{L}(M, M')$ has finite dimension.
   (E) None of the alternatives completes a true sentence.

4. Which of the following are the defining equations for the matrix of a morphism?
   (A) $(r_{ij}|i \in m, j \in n) \in R^{m \times n}$ if and only if $r_{ij} \in R$ for all $i \in m, j \in n$.
   (B) $(r_{ij}|i \in m, j \in n) + (s_{ij}|i \in m, j \in n) = (r_{ij} + s_{ij}|i \in m, j \in n)$.
   (C) $\mu(rf + sg) = r\mu(f) + s\mu(g)$ for all $r, s \in R, f, g \in \mathscr{L}(M, M')$.
   (D) $f(x_j) = \sum_{i=1}^{m} \mu(f)_{ij} y_i, j = 1, 2, \ldots, n$.
   (E) None of the above equations is the defining equations for the matrix of a morphism.

EXERCISES

1. Let $f: \mathbb{R}^3 \to \mathbb{R}^3$ such that $f(s_1, s_2, s_3) = (2s_1 + s_2 + s_3, s_1 + s_2, 2s_1 + 2s_3)$. Find the matrix, $\mu(f)$, associated with the standard bases in both domain and codomain. Express f as a linear combination of the family of morphisms $(e_{ij}|i \in 3, j \in 3)$.

2. Let $g: \mathbb{R}^3 \to \mathbb{R}^3$ such that $g = -2e_{11} + 4e_{12} - e_{13} + 19e_{22} + 7e_{31} - 4e_{33}$. What is $g(s_1, s_2, s_3)$? Find $\mu(g)$.

3. Let $f:\mathbb{Q}^2 \to \mathbb{Q}^3$ such that $f(s_1, s_2) = (2s_1 + s_2, s_1, s_1 - 3s_2)$, $g:\mathbb{Q}^2 \to \mathbb{Q}^3$ such that $g(s_1, s_2) = (3s_1, 4s_1 + 3s_2, s_1)$. Using the standard bases for both $\mathbb{Q}^2$ and $\mathbb{Q}^3$ find $\mu(f)$ and $\mu(g)$. Find $f + g$ in $\mathscr{L}(\mathbb{Q}^2, \mathbb{Q}^3)$. Find $\mu(f + g)$. Verify $\mu(f + g) = \mu(f) + \mu(g)$.

4. Using the same functions defined in Exercise 3 find the matrices of f and g with respect to bases $((1, 2), (-1, 3))$ and $((1, 1, 0), (0, 1, 1), (2, 1, 4))$.

5. Let $f:\mathbb{R}^2 \to \mathbb{R}^3$ be a morphism and have matrix

$$\begin{pmatrix} 6 & 1 \\ 2 & 0 \\ 0 & 4 \end{pmatrix}$$

with respect to the standard bases. Find bases for $\mathbb{R}^2$ and $\mathbb{R}^3$ so that the matrix of f with respect to these bases is

$$\begin{pmatrix} 1 & 0 \\ 0 & 1 \\ 0 & 0 \end{pmatrix}.$$

6. Let the standard bases be given for the vector spaces $\mathbb{R}^3$ and $\mathbb{R}^2$. Find the matrix of e_{12}, a morphism of $\mathscr{L}(\mathbb{R}^3, \mathbb{R}^2)$.

7. Let M, M' be vector spaces over a field K. Let M have a finite basis of n members and M' have a finite basis of m members. Prove $\kappa(f) + \kappa(g) \leqslant n + \kappa(f + g)$ and $\rho(f + g) \leqslant \rho(f) + \rho(g)$ where κ stands for nullity and ρ stands for rank.

8. In the text we have discussed the matrix of a morphism when each module has a finite basis. The defining equations for the matrix of a morphism are $f(x_j) = \sum_{i \in m} \mu(f)_{ij} y_i, j \in n$. Correspondingly for infinite bases the defining equations for a (infinite) matrix would be $f(x_j) = \sum_{i \in I}^{w} \mu(f)_{ij} y_i, j \in J$.
   (a) Let $f:(\mathbb{R}^{\mathbb{N}^+})^w \to (\mathbb{R}^{\mathbb{N}^+})^w$ such that $f(s_k|k \in \mathbb{N}^+) = (s_k + s_{k+1}|k \in \mathbb{N}^+)$. With respect to the standard basis $(e_i|i \in \mathbb{N}^+)$ in both domain and codomain find $\mu(f)$, the matrix of f.
   (b) Let $g:(\mathbb{R}^{\mathbb{N}^+})^w \to (\mathbb{R}^{\mathbb{N}^+})^w$ such that $g(s_k|k \in \mathbb{N}^+) = (s_k + s_{k+1} + s_{k+2} + \cdots |k \in \mathbb{N}^+)$. Find $\mu(g)$ with respect to the standard bases in both domain and codomain. Notice that in every column of $\mu(g)$ there are only a finite number of nonzero entries. Contrast this with the rows of $\mu(g)$. What can you say of the sum $\sum_{(i, j) \in I \times J} \mu(f)_{ij} e_{ij}$ and of the family $(e_{ij}|i, j) \in I \times J)$ as a basis for $\mathscr{L}(M, M')$?

9. Let the $\mathbb{Z}$-modules $\mathbb{Z}^n$ and $\mathbb{Z}^m$ be given and $f:\mathbb{Z}^n \to \mathbb{Z}^m$ such that $f(s_1, s_2, \ldots, s_n) = (\sum_{j=1}^{n} a_{1j}s_j, \sum_{j=1}^{n} a_{2j}a_j, \ldots, \sum_{j=1}^{n} a_{mj}s_j)$ where a_{ij} are constants, members of $\mathbb{Z}$. Find the matrix of f with respect to the standard bases.

10. Give an example of a vector space endomorphism which is a monomorphism but not an epimorphism. Give an example of a vector space endomorphism which is an epimorphism but not a monomorphism.

11. Let g be a given endomorphism of a vector space M. Define a function $\varphi:\mathscr{L}(M, M) \to \mathscr{L}(M, M)$ such that $\varphi(f) = g \circ f$. Prove that φ is itself a morphism. Can every member of $\mathscr{L}(M, M)$ be obtained as an image of φ?

7.2 Composition of morphisms, the endomorphism algebra $\mathscr{E}$(M)

In this section we consider composition of morphisms and its relation to multiplication of matrices. We construct the algebra of endomorphisms of a module.

We begin with a theorem on composition.

Theorem. *Let M, M', M'' be modules over a commutative unitary ring R. Let $(x_k | k \in p)$, $(y_j | j \in n)$, $(z_i | i \in m)$ be finite bases for M, M', M'', respectively. Let $f \in \mathscr{L}(M, M')$, $g \in \mathscr{L}(M', M'')$. Then $g \circ f \in \mathscr{L}(M, M'')$ and $\mu(g \circ f) = \mu(g)\mu(f)$.*

PROOF. It is quite routine to verify that the composition of two morphisms is a morphism. We turn therefore to the more difficult part of the conclusion, computing the matrix. The defining equations of the matrices $\mu(f)$ and $\mu(g)$ are

$$f(x_k) = \sum_{j \in n} \mu(f)_{jk} y_j, \quad k \in p,$$
$$g(y_j) = \sum_{i \in m} \mu(g)_{ij} z_i, \quad j \in n.$$

To find the matrix of $g \circ f$ we compute its defining equations

$$\begin{aligned}(g \circ f)(x_k) &= g(f(x_k)) = g\left(\sum_{j \in n} \mu(f)_{jk} y_j\right) \\ &= \sum_{j \in n} \mu(f)_{jk} g(y_j) \\ &= \sum_{j \in n} \mu(f)_{jk} \sum_{i \in m} \mu(g)_{ij} z_i \\ &= \sum_{i \in m} \left[\sum_{j \in n} \mu(g)_{ij} \mu(f)_{jk}\right] z_i.\end{aligned}$$

Comparing this result with the defining equations of $\mu(g \circ f)$,

$$(g \circ f)(x_k) = \sum_{i \in m} \mu(g \circ f)_{ik} z_i,$$

we have, because coefficients are unique,

$$\mu(g \circ f)_{ik} = \sum_{j \in n} \mu(g)_{ij} \mu(f)_{jk}.$$

This, however, is a matrix product (cf. Section 6.6),

$$(\mu(g \circ f)_{ik} | i \in m, k \in p) = (\mu(g)_{ij} | i \in m, j \in n)(\mu(f)_{jk} | j \in n, k \in p).$$
$$\mu(g \circ f) = \mu(g)\mu(f).$$

The matrix of the composition of two morphisms is the product of the matrices. □

We now give an example to illustrate the previous theorem.

EXAMPLE. Let $\mathbb{R}^3$, $\mathbb{R}^2$, $\mathbb{R}^4$ be given with the standard bases. Let $f:\mathbb{R}^3 \to \mathbb{R}^2$ such that $f(r_1, r_2, r_3) = (r_1 + r_2, r_1 + r_3)$ and $g:\mathbb{R}^2 \to \mathbb{R}^4$ such that $g(s_1, s_2) = (s_1 + s_2, 2s_1 + 4s_2, s_1 - s_2, 2s_1 + s_2)$. The matrices of f and g are found from the defining equations of the matrices.

$$\begin{aligned} f(1, 0, 0) &= (1, 1) = 1(1, 0) + 1(0, 1), \\ f(0, 1, 0) &= (1, 0) = 1(1, 0) + 0(0, 1), \\ f(0, 0, 1) &= (0, 1) = 0(1, 0) + 1(0, 1). \end{aligned}$$

$$g(1, 0) = (1, 2, 1, 2) = 1(1, 0, 0, 0) + 2(0, 1, 0, 0) + 1(0, 0, 1, 0) + 2(0, 0, 0, 1),$$

$$g(0, 1) = (1, 4, -1, 1) = 1(1, 0, 0, 0) + 4(0, 1, 0, 0) - 1(0, 0, 1, 0) + 1(0, 0, 0, 1).$$

$$\mu(f) = \begin{pmatrix} 1 & 1 & 0 \\ 1 & 0 & 1 \end{pmatrix}. \qquad \mu(g) = \begin{pmatrix} 1 & 1 \\ 2 & 4 \\ 1 & -1 \\ 2 & 1 \end{pmatrix}.$$

The composition of f and g, $g \circ f:\mathbb{R}^3 \to \mathbb{R}^4$ is $g(s_1, s_2) = g(f(r_1, r_2, r_3)) = (r_1 + r_2 + r_1 + r_3, 2(r_1 + r_2) + 4(r_1 + r_3), r_1 + r_2 - (r_1 + r_3), 2(r_1 + r_2) + r_1 + r_3) = (2r_1 + r_2 + r_3, 6r_1 + 2r_2 + 4r_3, r_2 - r_3, 3r_1 + 2r_2 + r_3)$.

The defining equations for the matrix of $g \circ f$ are

$$\begin{aligned} (g \circ f)(1, 0, 0) &= 2(1, 0, 0, 0) + 6(0, 1, 0, 0) + 0(0, 0, 1, 0) + 3(0, 0, 0, 1), \\ (g \circ f)(0, 1, 0) &= 1(1, 0, 0, 0) + 2(0, 1, 0, 0) + 1(0, 0, 1, 0) + 2(0, 0, 0, 1), \\ (g \circ f)(0, 0, 1) &= 1(1, 0, 0, 0) + 4(0, 1, 0, 0) - 1(0, 0, 1, 0) + 1(0, 0, 0, 1). \end{aligned}$$

$$\mu(g \circ f) = \begin{pmatrix} 2 & 1 & 1 \\ 6 & 2 & 4 \\ 0 & 1 & -1 \\ 3 & 2 & 1 \end{pmatrix}.$$

In accordance with the theorem

$$\mu(g)\mu(f) = \begin{pmatrix} 1 & 1 \\ 2 & 4 \\ 1 & -1 \\ 2 & 1 \end{pmatrix} \begin{pmatrix} 1 & 1 & 0 \\ 1 & 0 & 1 \end{pmatrix} = \begin{pmatrix} 2 & 1 & 1 \\ 6 & 2 & 4 \\ 0 & 1 & -1 \\ 3 & 2 & 1 \end{pmatrix}.$$

Given $f, g \in \mathscr{L}(M, M')$ it is not possible to compose f and g to produce $g \circ f$ unless domain g equals codomain f. If we desire $\mathscr{L}(M, M')$ to be closed under composition then we must have $M = M'$. We are led to consider $\mathscr{L}(M, M)$, the set of endomorphisms of the module M, in order to have a structure closed under addition, R-exterior multiplication, and composition. We denote the module of endomorphisms of M by $\mathscr{E}(M)$; $\mathscr{E}(M) = \mathscr{L}(M, M)$.

Theorem. *Let R be a commutative unitary ring and M an R-module. Then $\langle \mathscr{E}(M), +, \circ, z, I\rangle$ is a unitary R-algebra.*

Proof. We know that $\mathscr{E}(M) = \mathscr{L}(M, M)$ is an R-module. The second binary operation for $\mathscr{E}(M)$ is composition. The composition of two morphisms is another morphism. With respect to composition on $\mathscr{E}(M)$, the neutral element is I, the identity function on M. The following properties will establish $\mathscr{E}(M)$ to be an R-algebra:

$$h \circ (g \circ f) = (h \circ g) \circ f \quad \text{for all } f, g, h \in \mathscr{E}(M),$$
$$(g + h) \circ f = g \circ f + h \circ f \quad \text{for all } f, g, h \in \mathscr{E}(M),$$
$$h \circ (g + f) = h \circ g + h \circ f \quad \text{for all } f, g, h \in \mathscr{E}(M).$$
$$r(f \circ g) = (rf) \circ g = f \circ (rg) \quad \text{for all } r \in R, f, g \in \mathscr{E}(M),$$
$$f \circ I = I \circ f = f \quad \text{for all } f \in \mathscr{E}(M).$$

We discuss now the verification of these properties. That composition of functions of any kind is associative we established in chapter one. The next equation follows simply from the definition of addition of functions. $((g + h) \circ f)(x) = (g + h)(f(x)) = g(f(x)) + h(f(x)) = (g \circ f)(x) + (h \circ f)(x) = ((g \circ f) + (h \circ f))(x)$. The left distributive law, unlike the right, depends upon morphism properties as well as function properties. $(h \circ (f + g))(x) = h((f + g)(x)) = h(f(x) + g(x)) = h(f(x)) + h(g(x)) = (h \circ f)(x) + (h \circ g)(x) = ((h \circ f) + (h \circ g))(x)$. The next property relates R-exterior multiplication to the composition. $(r(f \circ g))(x) = r(f(g(x))) = rf(g(x)) = (rf)(g(x)) = ((rf) \circ g)(x)$ and $(r(f \circ g))(x) = r(f(g(x))) = f(rg(x)) = f((rg)(x)) = (f \circ (rg))(x)$. The property of the identity function we have used often before. □

Because $\langle \mathscr{E}(M), +, \circ, z, I\rangle$ is a ring of vectors (not the ring R) it is often called the ring of endomorphisms of M. Ring, however, does not completely describe the situation because it makes no mention of the R-exterior multiplication. An algebra is a ring, but with another ring (of scalars) "behind" the ring of vectors.

Example. $\mathscr{E}(\mathbb{R}^2)$, the ring of endomorphisms of the $\mathbb{R}$-vector space $\mathbb{R}^2$ includes f and g such that $f(r_1, r_2) = (r_1 + r_2, r_1 - r_2)$ and $g(s_1, s_2) = (2s_1 - s_2, s_1 + 2s_2)$. Then $(f + g)(r_1, r_2) = (3r_1, 2r_1 + r_2)$, $(g \circ f)(r_1, r_2) = (r_1 + 3r_2, 3r_1)$, $(f \circ g)(r_1, r_2) = (3r_1 + r_2, r_1 + r_2)$.

The isomorphisms of M into M are called, as usual, automorphisms. We represent *the set of automorphisms of M* with the symbol $\mathscr{A}(M)$.

Theorem. *Let M be a module over a commutative unitary ring R. $\langle \mathscr{A}(M), \circ, I\rangle$ is a group.*

PROOF. The composition of two automorphisms is also an automorphism. I is an automorphism and every automorphism has an inverse which is also an automorphism. □

We now show how the extra properties of the structure of $\mathscr{E}(M)$ carry over to the matrices.

Theorem. *Let M be a module over a commutative unitary ring R and let $(x_i|i \in n)$ be a finite basis for M. Then the endomorphism algebra $\mathscr{E}(M)$ and the matrix algebra $R^{n \times n}$ are isomorphic.*

PROOF. With respect to the basis $(x_i|i \in n)$, used for both domain and codomain, we have already showed that the matrix mapping $\mu:\mathscr{L}(M, M) \to R^{n \times n}$ is a module isomorphism. We have also proved that $\mu(g \circ f) = \mu(g)\mu(f)$. This additional formula shows that the composition of the algebra $\mathscr{E}(M)$ is preserved in the matrix multiplication of $R^{n \times n}$. The matrix mapping $\mu:\mathscr{E}(M) \to R^{n \times n}$ is an algebra isomorphism. This isomorphism does depend upon the basis choice. □

Because the algebra of square matrices and the algebra of endomorphisms are isomorphic the invertible matrices are simply the matrices of invertible morphisms under the matrix mapping. But we know that an endomorphism is invertible if and only if it is an automorphism. A matrix is invertible if and only if it is the matrix of an automorphism.

Detailed examples of calculations with matrices and their inverses related to automorphisms will be given in the exercises.

QUESTIONS

1. The equation $\mu(g \circ f) = \mu(g)\mu(f)$ of this section
   (A) shows that matrices can be multiplicatively cancelled
   (B) shows that composition of morphisms is preserved in multiplication of matrices
   (C) will show that $\mu(f^2) = \mu(f)^2$ by setting $g = f$
   (D) shows that the matrix of z, the zero morphism, cannot have a multiplicative inverse.
   (E) None of the alternatives completes a true sentence.

2. Which of the following are true?
   (A) An epimorphism of a finite dimensional module into itself is an automorphism.
   (B) A monomorphism of a finite dimensional module into itself is an automorphism.
   (C) Every isomorphism of a finite dimensional module into itself is an automorphism.
   (D) Any endomorphism f of a module M satisfies the relation $f(\text{kernel } f) \subseteq \text{kernel } f$.
   (E) None of the sentences is true.

3. Which of the following are true?

(A) $(a_{ij}|i \in m, j \in n)(b_{ik}|i \in n, k \in m) = (\sum_{i \in m} a_{ij}b_{ki}|j \in n, k \in n)$.

(B) $(a_{ij}|i \in m, j \in n)(b_j|j \in n) = (\sum_{j \in n} a_{ij}b_j|i \in m)$.

(C) $(a_{ij}|i \in m, j \in n)(\delta_{jk}|j \in n, k \in n) = (a_{ik}|i \in m, k \in n)$.

(D) $(a_{ij}|i \in m, j \in n)(b_{jk}|j \in n, k \in p) = (\sum_{j \in n} a_{ij}b_{jk}|i \in m, k \in p)$.

(E) None of the products is correct.

4. Which of these statements are true?

(A) It is possible to compose two nonzero endomorphisms of a module M to produce the zero endomorphism, z.

(B) It is possible to multiply two nonzero matrices to produce the zero matrix.

(C) Every nonzero endomorphism of a vector space M has an inverse.

(D) If $f \circ g = f \circ h$ for endomorphisms f, g, h of a vector space M and f has no inverse then $g \neq h$.

(E) None of the statements is true.

Exercises

1. Prove that the composition of two morphisms is a morphism.

2. Prove that the composition of two automorphisms is an automorphism.

3. Let M be a free module over a commutative unitary ring R and $(x_j|j \in n)$ a finite basis for M. Prove the following:

(a) $\mu(f)$ has an inverse in $R^{n \times n}$ if and only if f has an inverse in $\mathscr{E}(M)$ if and only if $f \in \mathscr{A}(M)$.

(b) If f^{-1} exists then $\mu(f)^{-1} = \mu(f^{-1})$.

(c) $\mu(I) = \delta = (\delta_{ij}|i \in n, j \in n)$. Recall $\delta_{ij} = \theta$ if $i \neq j$ and equals v if $i = j$.

4. Let f, $g \in \mathscr{E}(\mathbb{R}^3)$ such that $f(r_1, r_2, r_3) = (2r_1 - r_2, 2r_1 + r_2, r_1 + r_2)$ and $g(s_1, s_2, s_3) = (s_1 + s_2 + s_3, s_1 - s_2 + 2s_3, 3s_1 - s_2 + 5s_3)$. Using the standard basis for $\mathbb{R}^3$ find $\mu(f)$, $\mu(g)$, $\mu(g \circ f)$ and verify that $\mu(g \circ f) = \mu(g)\mu(f)$. Prove f is an automorphism. Find f^{-1}. Find $\mu(f^{-1})$. Verify that $\mu(f)\mu(f^{-1}) = \mu(f^{-1})\mu(f) = \delta$. Show that g is not an automorphism. Is $g \circ f$ an automorphism?

5. Given the matrix

$$\begin{pmatrix} 1 & 2 & 1 \\ 3 & 0 & 2 \\ 1 & -1 & 2 \end{pmatrix}$$

use the following procedure to find its inverse. Regard the matrix as the matrix of an endomorphism f of $\mathbb{R}^3$ with respect to the standard basis. Determine the formula for the endomorphism f. See whether f is an automorphism. If f is an automorphism find its inverse automorphism f^{-1}. Find the matrix of f^{-1}. Verify that the matrix found is indeed the inverse matrix of the given matrix.

6. Let M, M', M'' be finite dimensional vector spaces over a field K. Let $\dim M = p$, $\dim M' = n$, $\dim M'' = m$. Let $f \in \mathscr{L}(M, M')$, $g \in \mathscr{L}(M', M'')$. Let ρ stand for rank and κ stand for nullity. Show

(a) $\rho(g \circ f) \leqslant \min\{\rho(f), \rho(g)\}$

(b) $\kappa(g \circ f) \geqslant \max\{\kappa(f), \kappa(g)\}$ if $p = n$

(c) $\ker(g \circ f) = \ker f \oplus [\ker(g \circ f) - \ker f]$

(d) $\kappa(g \circ f) \leqslant \kappa(f) + \kappa(g)$
(e) $\rho(f) + \rho(g) \leqslant \rho(g \circ f) + n$.

7. Let f be a nonzero endomorphism of a vector space M. Let there exist a number n such that f^n is the zero morphism ($f^2 = f \circ f$, $f^{j+1} = f^j \circ f$). Show
(a) there exists a smallest natural number m so that $f^m = z$
(b) there exists an $a \in M$ such that $f^{m-1}(a) \neq \zeta$
(c) $(a, f(a), \ldots, f^{m-1}(a))$ is a linearly independent family of M
(d) $f([a, f(a), \ldots, f^{m-1}(a)]) \subseteq [a, f(a), \ldots, f^{m-1}(a)]$.

7.3 Matrix calculation of morphisms

In this section we calculate morphisms by matrix multiplication.

We will be able to calculate the values of a morphisms by multiplying the coordinates of a vector by the matrix of the morphism.

$$\begin{pmatrix} \mu(f)_{11} & \mu(f)_{12} & \cdots & \mu(f)_{1n} \\ \mu(f)_{21} & \mu(f)_{22} & \cdots & \mu(f)_{2n} \\ \cdots & & & \\ \mu(f)_{m1} & \mu(f)_{m2} & \cdots & \mu(f)_{mn} \end{pmatrix} \begin{pmatrix} X_1 \\ X_2 \\ \cdots \\ X_n \end{pmatrix}.$$

Before stating precisely our theorem we repeat a definition from Section 7.1. For a morphism $f: M \to M'$ and finite bases $(x_j | j \in n)$ for M, $(y_i | i \in m)$ for M' the defining equations of the matrix $\mu(f)$ are

$$f(x_j) = \sum_{i \in m} \mu(f)_{ij} y_i, \quad j \in n.$$

In finding the values of a morphism in matrix form we shall use these equations.

Theorem. *Let M, M' be modules over a commutative unitary ring R. Let $f: M \to M'$ be a morphism. If $f(\sum_{j=1}^{n} X_j x_j) = \sum_{i=1}^{m} Y_i y_i$ then*

$$\begin{pmatrix} \mu(f)_{11} & \mu(f)_{12} & \cdots & \mu(f)_{1n} \\ \mu(f)_{21} & \mu(f)_{22} & \cdots & \mu(f)_{2n} \\ \cdots & & & \\ \mu(f)_{m1} & \mu(f)_{m2} & \cdots & \mu(f)_{mn} \end{pmatrix} \begin{pmatrix} X_1 \\ X_2 \\ \cdots \\ X_n \end{pmatrix} = \begin{pmatrix} Y_1 \\ Y_2 \\ \cdots \\ Y_m \end{pmatrix}.$$

In more abbreviated form the last equation is $\mu(f)X = Y$.

PROOF. $f(\sum_{j=1}^{n} X_j x_j) = \sum_{j=1}^{n} X_j f(x_j) = \sum_{j=1}^{n} X_j \sum_{i=1}^{m} \mu(f)_{ij} y_i = \sum_{i=1}^{m} (\sum_{j=1}^{n} \mu(f)_{ij} X_j) y_i$. A comparison with $\sum_{i=1}^{m} Y_i y_i$ yields $Y_i = \sum_{j=1}^{n} \mu(f)_{ij} X_j$ for each $i \in m$ because representations in terms of the basis $(y_i | i \in m)$ must be unique. $Y = (Y_i | i \in m) = (\sum_{j \in n} \mu(f)_{ij} X_j | i \in m) = (\mu(f)_{ij} | i \in m, j \in n)(X_j | j \in n) = \mu(f)X$ by the rule for multiplication of matrices. □

We wish now to observe that this theorem makes correspond two functions, one, the morphism f from M to M' and, two, the multiplication by the matrix $\mu(f)$ taking R^n to R^m, the domains for the coordinates of M and M'.

We locate these two functions on a diagram.

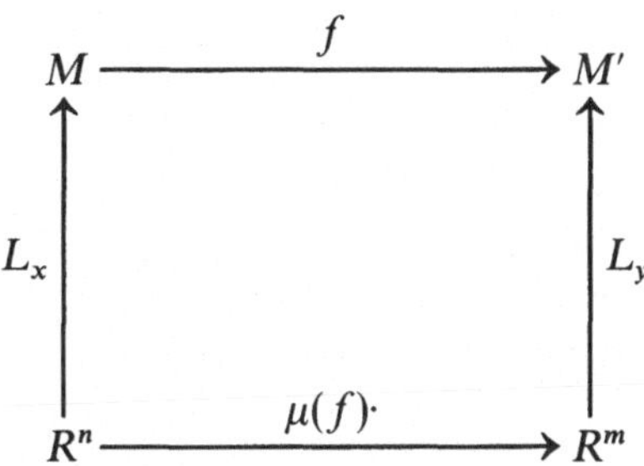

L_x and L_y are the coordinate morphisms, now hopefully quite familiar. We have denoted the function *left multiplication by the matrix* $\mu(f)$ with the symbol $\mu(f)\cdot$. We here are making a distinction between the matrix $\mu(f)$ and the function of left multiplication by $\mu(f)$. The diagram leads one to investigate the equation $\mu(f)\cdot = L_y^{-1} \circ f \circ L_x$. This assertion is correct and is essentially the content of the theorem.

$$\begin{aligned}(L_y^{-1} \circ f \circ L_x)(X) &= L_y^{-1}(f(L_x(X_j|j \in n))) = L_y^{-1}\left(f\left(\sum_{j=1}^{n} X_j x_j\right)\right) \\ &= L_y^{-1}\left(\sum_{j=1}^{n} X_j f(x_j)\right) = L_y^{-1}\left(\sum_{j=1}^{n} X_j \sum_{i=1}^{m} \mu(f)_{ij} y_i\right) \\ &= L_y^{-1}\left(\sum_{i=1}^{m}\left(\sum_{j=1}^{n} \mu(f)_{ij} X_j\right) y_i\right) = \left(\sum_{j=1}^{n} \mu(f)_{ij} X_j | i \in m\right) \\ &= \mu(f)(X_j|j \in n) = \mu(f)X.\end{aligned}$$

Therefore, $L_y^{-1} \circ f \circ L_x = \mu(f)\cdot$.

Example. Let $f: \mathbb{R}^3 \to \mathbb{R}^3$ such that $f(r_1, r_2, r_3) = (2r_1 + r_2 + r_3, r_1 + r_2, 2r_1 + 2r_3)$. Using the standard bases in both domain and codomain $\mu(f)$ is

$$\begin{pmatrix} 2 & 1 & 1 \\ 1 & 1 & 0 \\ 2 & 0 & 2 \end{pmatrix}.$$

The equation relating the coordinates is

$$\begin{pmatrix} Y_1 \\ Y_2 \\ Y_3 \end{pmatrix} = \begin{pmatrix} 2 & 1 & 1 \\ 1 & 1 & 0 \\ 2 & 0 & 2 \end{pmatrix}\begin{pmatrix} X_1 \\ X_2 \\ X_3 \end{pmatrix}.$$

If we wish, for example, to find the coordinates of $f(x)$ when x has coordinates $(-1, 6, 3)$ then we calculate

$$\begin{pmatrix} 2 & 1 & 1 \\ 1 & 1 & 0 \\ 2 & 0 & 2 \end{pmatrix}\begin{pmatrix} -1 \\ 6 \\ 3 \end{pmatrix}$$

which is

$$\begin{pmatrix} 7 \\ 5 \\ 4 \end{pmatrix}.$$

Left multiplication by the matrix of f, $\mu(f)$, to compute the coordinate image of f could have been as well right multiplication. For example, in the calculation of the previous example we can write

$$(7 \quad 5 \quad 4) = (-1 \quad 6 \quad 3)\begin{pmatrix} 2 & 1 & 2 \\ 1 & 1 & 0 \\ 1 & 0 & 2 \end{pmatrix}.$$

More generally,

$$(Y_1 \quad Y_2 \quad \cdots \quad Y_m) = (X_1 \quad X_2 \quad \cdots \quad X_n)\begin{pmatrix} \mu(f)_{11} & \mu(f)_{21} & \cdots & \mu(f)_{m1} \\ \mu(f)_{12} & \mu(f)_{22} & \cdots & \mu(f)_{m2} \\ \cdots & & & \\ \mu(f)_{1n} & \mu(f)_{2n} & \cdots & \mu(f)_{mn} \end{pmatrix}$$

if and only if

$$(Y_1 \quad Y_2 \quad \cdots \quad Y_m) = \left(\sum_{j=1}^{n} X_j\mu(f)_{1j}, \ldots, \sum_{j=1}^{n} X_j\mu(f)_{mj}\right)$$

if and only if

$$\begin{pmatrix} Y_1 \\ Y_2 \\ \cdots \\ Y_m \end{pmatrix} = \begin{pmatrix} \sum_{j=1}^{n} \mu(f)_{1j}X_j \\ \sum_{j=1}^{n} \mu(f)_{2j}X_j \\ \cdots \\ \sum_{j=1}^{n} \mu(f)_{mj}X_j \end{pmatrix}$$

if and only if

$$\begin{pmatrix} Y_1 \\ Y_2 \\ \cdots \\ Y_m \end{pmatrix} = \begin{pmatrix} \mu(f)_{11} & \mu(f)_{12} & \cdots & \mu(f)_{1n} \\ \mu(f)_{21} & \mu(f)_{22} & \cdots & \mu(f)_{2n} \\ \cdots & & & \\ \mu(f)_{m1} & \mu(f)_{m2} & \cdots & \mu(f)_{mn} \end{pmatrix}\begin{pmatrix} X_1 \\ X_2 \\ \cdots \\ X_n \end{pmatrix}.$$

And more briefly from the two ends of the equation sequence, $Y^* = X^*\mu(f)^*$ if and only if $Y^* = (\mu(f)X)^*$ if and only if $Y = \mu(f)X$.

The decision to use left multiplication of matrices or right multiplication of matrices is a matter of taste. It is also a matter for choice whether to write function values $f(x)$ as we have done or to write xf as do many linear algebra texts. A reader of mathematics must be prepared for and accept many notational variations.

Since we shall use the *transpose matrix* $\mu(f)^*$ again we now note some of its properties in a theorem. The *transpose* A^* of a matrix A is a matrix C where $C_{ji} = A_{ij}$ for all $j \in n$, $i \in m$.

Theorem. *Let R be a commutative unitary ring. The transpose function $-^*: R^{m \times n} \to R^{n \times m}$ with value A^* for each $A \in R^{m \times n}$ enjoys these properties: $(A + B)^* = A^* + B^*$, $\theta^* = \theta$, $(rA)^* = rA^*$ for $r \in R$, $A, B \in R^{m \times n}$. Moreover, when a product AB is defined then also is defined B^*A^* and $(AB)^* = B^*A^*$. The function $-^*: R^{n \times n} \to R^{n \times n}$ is an algebra isomorphism (actually anti-isomorphism).*

PROOF. The easy verifications here are left to the reader. The order of matrix multiplication is reversed under transposition; hence, we have called transposition an anti-isomorphism. □

In Section 7.2 we showed that $\mu(g \circ f) = \mu(g)\mu(f)$. We now offer a diagram of the involved functions including the matrix multiplication of coordinates. We hope the diagrams facilitate a better understanding of the functions involved.

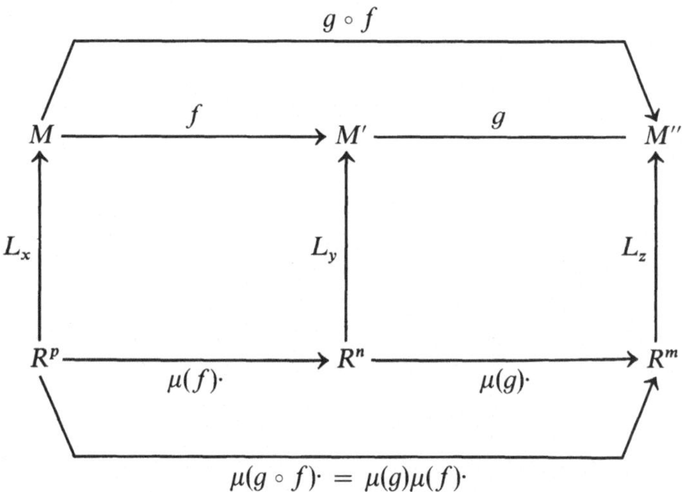

For the diagram it is understood that we are assuming the existence of bases for M, M' and M'' which are respectively $(x_k | k \in p)$, $(y_j | j \in n)$ and $(z_i | i \in m)$. The diagram illustrates the equation $\mu(g \circ f)\cdot = L_z^{-1} \circ (g \circ f) \circ L_x = L_z^{-1} \circ g \circ L_y \circ L_y^{-1} \circ g \circ L_x = \mu(g)\cdot \circ \mu(f)\cdot$.

The matrix of a morphism depends, of course, upon the bases used in M and M' to compute the matrix. A different basis choice will result in a different matrix. So also does the choice of basis affect the isomorphism L between the space of coordinates and the module. We put again the question about how to choose bases in M and M' to produce a simple matrix. We offer one answer in the form of a theorem based upon the fundamental theorem of Section 6.9.

Theorem. *Let M, M' be finite dimensional K-vector spaces and $f: M \to M'$ a morphism. Then there exist bases $(x_j | j \in n)$ and $(y_i | \in m)$ for M and M' with $n = \dim M$, $m = \dim M'$ such that $\mu(f)_{ij} = \delta_{ij}$ for $i \leqslant \operatorname{rank} f$, $j \leqslant \operatorname{rank} f$ and $\mu(f)_{ij} = \theta$ otherwise.*

PROOF. In less compact notation the matrix $\mu(f)$ described in the conclusion of the theorem is

$$\mu(f) = \begin{pmatrix} v & \theta & \theta & \cdots & \theta & \theta & \cdots & \theta \\ \theta & v & \theta & \cdots & \theta & \theta & \cdots & \theta \\ \theta & \theta & v & \cdots & \theta & \theta & \cdots & \theta \\ \cdots & & & & & & & \\ \theta & \theta & \theta & \cdots & v & \theta & \cdots & \theta \\ \theta & \theta & \theta & \cdots & \theta & \theta & \cdots & \theta \\ \cdots & & & & & & & \\ \theta & \theta & \theta & \cdots & \theta & \theta & \cdots & \theta \end{pmatrix}$$

with precisely rank f entries equal to v.

The theorem is a simple consequence of the theorem proving $n = \operatorname{nullity} f + \operatorname{rank} f$. A basis $(x_{\rho+1}, x_{\rho+2}, \ldots, x_n)$ is chosen for kernel f, a subspace of M, where $\rho = \operatorname{rank} f$. These basis elements for the kernel are numbered at the end so that columns of all zeros will be located on the extreme right of the matrix. The basis $(x_{\rho+1}, x_{\rho+2}, \ldots, x_n)$ for the subspace is extended to a basis $(x_1, x_2, \ldots, x_\rho, x_{\rho+1}, x_{\rho+2}, \ldots, x_n)$ for all of the space M. The extending basis elements are listed first, remembering that any permutation of a basis is also a basis. For M' we first choose $(f(x_1), f(x_2), \ldots, f(x_\rho))$ which is a basis for range f, a subspace of M'. This basis is then extended, if necessary, to a basis $(f(x_1), f(x_2), \ldots, f(x_\rho), y_{\rho+1}, \ldots, y_m)$ of M' in a purely arbitrary manner. The defining equations of the matrix of f are then

$$\begin{aligned} f(x_1) &= vf(x_1) + \theta f(x_2) + \cdots + \theta f(x_\rho) + \theta y_{\rho+1} + \cdots + \theta y_m \\ f(x_2) &= \theta f(x_1) + vf(x_2) + \cdots + \theta f(x_\rho) + \theta y_{\rho+1} + \cdots + \theta y_m \\ &\cdots \\ f(x_\rho) &= \theta f(x_1) + \theta f(x_2) + \cdots + vf(x_\rho) + \theta y_{\rho+1} + \cdots + \theta y_m \\ f(x_{\rho+1}) &= \theta f(x_1) + \theta f(x_2) + \cdots + \theta f(x_\rho) + \theta y_{\rho+1} + \cdots + \theta y_m \\ &\cdots \\ f(x_n) &= \theta f(x_1) + \theta f(x_2) + \cdots + \theta f(x_\rho) + \theta y_{\rho+1} + \cdots + \theta y_m. \end{aligned}$$

The matrix is now read off the defining equations. □

The simplicity of this matrix comes from the fact that f takes the first basis element of M into the first basis element of M', the second basis element of M into the second basis element of M', and so forth until the rank is exhausted, then the rest of the basis elements of M into the zero vector of M'. Thus by choosing precisely the right bases the morphism has become particularly simple viewed in matrix form. One could say that if a morphism of vector spaces looks complicated then one must have basically the wrong point of view. The problem of choosing bases to produce a simple matrix will be revisited in Chapter 10. This last theorem leads us naturally into the next section for a complete discussion of change of basis.

Questions

1. Let M, M' be modules over a commutative unitary ring R. Let $(x_j | j \in n)$ and $(y_i | i \in m)$ be finite bases for M and M'. Let $f:M \to M'$ be a morphism. Which of the following statements are true?
   (A) $f:M \to M'$ is an isomorphism if and only if $\mu(f)\cdot : R^n \to R^m$ is an isomorphism.
   (B) $f:M \to M'$ is an isomorphism if and only if dim M = dim M'.
   (C) Dim M = dim M' only if L_x and L_y are both isomorphisms.
   (D) If f^{-1} exists (as a function) then $f^{-1} = L_x \circ \mu(f)^{-1}\cdot \circ L_y^{-1}$.
(E) None of the statements is true.

2. Let M, M' be modules over a commutative unitary ring R. Let $(x_j | j \in n)$ and $(y_i | i \in m)$ be finite bases for M and M'. Let $f:M \to M'$ be a morphism. Which of the following are true?
   (A) If $f(x_j) = \sum_{i \in m} A_{ij} y_i, j \in n$, then $A = \mu(f)$.
   (B) If $\mu(f)_{ij} = \delta_{ij}$ for all i, j and $n \leqslant m$ then f is a monomorphism.
   (C) If $\mu(f)_{ij} = \delta_{ij}$ and $m < n$ then f is an epimorphism.
   (D) If $\mu(f)_{ij} = \delta_{ij}$ and $m = n$ then f is an isomorphism.
(E) None of the statements is true.

3. Which of these sentences are true?
   (A) $(\mu(f)^*)^{-1} = (\mu(f)^{-1})^*$ for all isomorphisms f.
   (B) $(\mu(g)\mu(f))^* = \mu(g)^*\mu(f)^*$ when both products are defined.
   (C) $\mu(g \circ f)^* = \mu(g)^*\mu(f)^*$ when the composition and product are defined.
   (D) $\delta^* = \delta$ for the n by n identity matrix δ.
(E) None of the statements is true.

4. Let $f:M \to M'$ be a morphism of finite dimensional K-vector spaces. Which of these statements are true?
   (A) If a basis element x_j belongs to kernel f then the *column* $(\mu(f)_{1j}, \mu(f)_{2j}, \ldots, \mu(f)_{mj})$ of $\mu(f)$ consists entirely of zeros.
   (B) The morphism f cannot have a matrix with more than $n - \rho$ columns entirely zero (n = dim domain, ρ = rank f).
   (C) $\mu(f - g) = \mu(f) - \mu(g)$ for any morphism $g:M \to M'$.
   (D) $\sigma \leqslant m$ and $\rho \leqslant n$ where ρ = rank f, n = dim domain f, m = dim codomain f.
(E) None of the statements is satisfactory.

Exercises

1. Let $f:\mathbb{R}^3 \to \mathbb{R}^2$ be given such that $f(r_1, r_2, r_3) = (r_1 - r_2, r_3)$. Find $\mu(f)$ with respect to the standard bases for $\mathbb{R}^3$ and $\mathbb{R}^2$. Calculate directly $f(1, 2, 3)$ from the given formula for f. Calculate $f(1, 2, 3)$ by matrix multiplication. Compare the two results.

2. Using the same formula for f as given in Exercise 1 compute the matrix of f with respect to the bases $((1, 1, 0), (1, 0, 1), (0, 1, 1))$ and $((1, 1), (0, 1))$. Calculate the coordinates of $(1, 2, 3)$ with respect to the new bases (requires solving some linear equations). Multiply by the matrix of f to obtain the coordinates of $f(1, 2, 3)$. Verify that these are the coordinates of $f(1, 2, 3)$.

3. Let $f:\mathbb{Q}^2 \to \mathbb{Q}^3$ such that $f(r_1, r_2) = (2r_1 + r_2, r_1, r_1 - 3r_2)$ and $g:\mathbb{Q}^2 \to \mathbb{Q}^3$ such that $g(r_1, r_2) = (3r_1, 4r_1 + 3r_2, r_1)$. Find with respect to the standard bases, $\mu(f)$ and $\mu(g)$. Calculate using $\mu(f)$ and $\mu(g)$ the values of $f(1, 2)$ and $g(1, 2)$. Find $\mu(f + g)$ and calculate the value of $(f + g)(1, 2)$.

4. Let $f:\mathbb{Q}^6 \to \mathbb{Q}^3$ such that $f(r_1, r_2, r_3, r_4, r_5, r_6) = (r_1 + r_2, r_3 + r_4, r_5 + r_6)$. Find $\mu(f)$ with respect to the standard bases. Find $\mu(f)^*$, the transpose of $\mu(f)$. Find $f(-5, 5, 1, 2, -1, 2)$ first using left multiplication by $\mu(f)$ and then by using right multiplication by $\mu(f)^*$.

5. Let $f:\mathbb{R}^2 \to \mathbb{R}^3$ such that $f(r_1, r_2) = (r_1 + r_2, r_1 - r_2, 2r_1 + r_2)$ and $g:\mathbb{R}^3 \to \mathbb{R}^2$ such that $g(s_1, s_2, s_3) = (2s_1 - s_2, 3s_3 - s_2)$. Find $g \circ f$. Find with respect to the standard bases the matrices $\mu(f)$, $\mu(g)$, $\mu(g \circ f)$. Verify $\mu(g \circ f) = \mu(g)\mu(f)$.

6. Let $f:\mathbb{R}^2 \to \mathbb{R}^3$ such that $f(r_1, r_2) = (r_1 + r_2, r_1 - r_2, 2r_1 + r_2)$. Find bases for the domain and the codomain of f so that the matrix of f is a matrix of only 0's and 1's and the number of 1's denotes the rank of f.

7. Give examples from some $\mathscr{E}(M)$ to show that there exist endomorphisms f, g such that $f \circ g = g \circ f$ and there also exist endomorphisms f, g such that $f \circ g \neq g \circ f$. [*Hint*: Use $\mathscr{E}(\mathbb{R}^2)$ and look at matrices.]

8. Let M be a K-vector space. For endomorphisms $f, g \in \mathscr{E}(M)$ show that this implication is not always true
$$f \circ g = z \quad \text{implies } g \circ f = z.$$

9. Suppose it be given that $f:\mathbb{R}^3 \to \mathbb{R}^3$ is an endomorphism and that $f(1, 1, 1) = (0, 1, 0)$, $f(2, 1, 0) = (2, 1, 1)$, $f(3, 1, 0) = (3, 0, 1)$. Find the matrix of f with respect to the standard basis of $\mathbb{R}^3$.

10. Let $f:\mathbb{R}^3 \to \mathbb{R}^3$ be given such that $f(0, 0, 2) = (0, 0, -2)$, $f(\frac{1}{2}, \frac{1}{2}, 1) = (0, 1, -1)$, $f(0, -1, 3) = (\frac{1}{2}, -\frac{1}{2}, 3)$. What is the matrix of f with respect to the standard bases? What is the rank of f?

11. Prove directly from the definition of matrix transpose that $(BA)^* = A^*B^*$. Prove that A is nonsingular (has an inverse) if and only if A^* is nonsingular.

12. Show that the set of matrices $\left\{\begin{pmatrix} a & b \\ -b & a \end{pmatrix} \middle| \, a, b \in \mathbb{R}\right\}$ is a field. Shorten your argument as much as possible by using the results from the text.

13. Let A, B be matrices with entries in a field K. Show that $\operatorname{rank}(A + B) \leqslant \operatorname{rank} A + \operatorname{rank} B$ when the sum is defined.

14. Let A, B be matrices with entries in a field K and let the product BA be defined. Show that $\operatorname{rank} A + \operatorname{rank} B - n \leqslant \operatorname{rank} BA \leqslant \min\{\operatorname{rank} A, \operatorname{rank} B\}$ where n = number of rows of A.

7.4 Change of basis

In this section we discuss change of basis in a module, computation with the change of basis matrix, elementary changes of basis, and a connection of change of basis with solving linear equations.

The point of view used to find the matrix of a morphism can be turned to find the matrix of change of basis. *The change of basis matrix* is that matrix by which one multiplies the coordinates with respect to one basis of a vector to produce the coordinates with respect to a different basis. We find this matrix by finding the matrix of the identity morphism with respect to the two bases in question.

Theorem. *Let M be a module over a commutative unitary ring R. Let $(x_j | j \in n)$ and $(u_j | j \in n)$ be two finite bases for M. Then the coordinates U of a vector x with respect to the basis $(u_j | j \in n)$ are related to the coordinates X with respect to the basis $(x_j | j \in n)$ by the equation*

$$U = \mu_{xu}(I)X.$$

Moreover, the change of basis matrix $\mu_{xu}(I)$ is an invertible matrix.

PROOF. $\mu_{xu}(I)$ is the change of basis matrix because multiplying the $(x_j | j \in n)$ coordinates of a vector by $\mu_{xu}(I)$ produces the $(u_j | j \in n)$ coordinates.

We consider the identity isomorphism $I: M \to M$ using the basis $(x_j | j \in n)$ for the domain and the basis $(u_j | j \in n)$ for the codomain. The situation is summarized in this diagram.

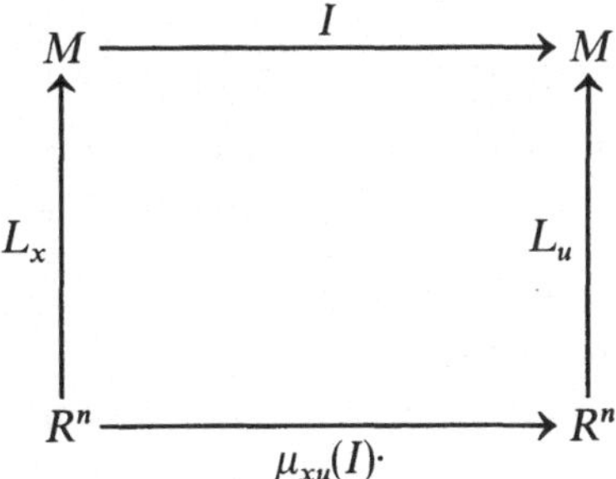

The matrix $\mu_{xu}(I)$ has defining equations

$$I(x_j) = \sum_{i \in n} \mu_{xu}(I)_{ij} u_i, \quad i \in n.$$

But since $I(x_j) = x_j$ these equations are simply

$$x_j = \sum_{i \in n} \mu_{xu}(I)_{ij} u_i, \quad i \in n.$$

By the principal theorem of Section 7.3 the coordinates X of the vector x are related to the coordinates U of $x = I(x)$ by the equation

$$U = \mu_{xu}(I)X.$$

As $\mu_{xu}(I)$ is the matrix of an isomorphism it is invertible. □

Every change of basis matrix is invertible and conversely every invertible n by n matrix can define a change of basis on an n-dimensional free module. For simplicity we let $\mu_{xu}(I) = P$ and make our diagram of change of basis as follows.

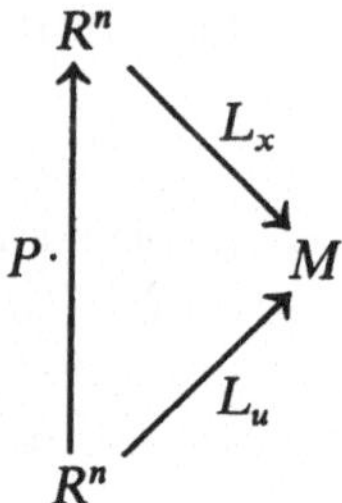

As shown in the diagram, to change the "old" coordinates X to the "new" coordinates U we multiply X by an invertible n by n matrix P on the left.

$$U = PX.$$

$P\cdot : R^n \to R^n$ is, of course, an isomorphism.

Example. Two basis for $\mathbb{Q}^3$ are $x = ((1, 0, 0), (0, 1, 0), (0, 0, 1))$ and $u = ((1, 1, 0), (1, 0, 1), (0, 1, 1))$.

$$\begin{aligned}
(1, 0, 0) &= P_{11}(1, 1, 0) + P_{21}(1, 0, 1) + P_{31}(0, 1, 1)\\
(0, 1, 0) &= P_{12}(1, 1, 0) + P_{22}(1, 0, 1) + P_{32}(0, 1, 1)\\
(0, 0, 1) &= P_{13}(1, 1, 0) + P_{23}(1, 0, 1) + P_{33}(0, 1, 1)
\end{aligned}$$

are the defining equations for the change of basis matrix, $\mu_{xu}(I) = P$. We must solve for the various entries, $P_{11}, P_{12}, \ldots, P_{33}$, to know the matrix P. This requires solution of linear equations, for example,

$$(1, 0, 0) = (P_{11} + P_{21}, P_{11} + P_{31}, P_{21} + P_{31}).$$

$$\begin{aligned}
P_{11} + P_{21} \qquad\qquad &= 1\\
P_{11} \qquad\qquad + P_{31} &= 0\\
P_{21} + P_{31} &= 0.
\end{aligned}$$

These equations are solved by reduction to echelon form.

$$\left(\begin{array}{ccc|c} 1 & 1 & 0 & 1 \\ 1 & 0 & 1 & 0 \\ 0 & 1 & 1 & 0 \end{array}\right) \quad \left(\begin{array}{ccc|c} 1 & 0 & 1 & 0 \\ 0 & -1 & 1 & -1 \\ 0 & 0 & 2 & -1 \end{array}\right) \quad \left(\begin{array}{ccc|c} 1 & 0 & 1 & 0 \\ 0 & 1 & 0 & \frac{1}{2} \\ 0 & 0 & 1 & -\frac{1}{2} \end{array}\right)$$

$$\left(\begin{array}{ccc|c} 1 & 1 & 0 & 1 \\ 0 & -1 & 1 & -1 \\ 0 & 1 & 1 & 0 \end{array}\right) \quad \left(\begin{array}{ccc|c} 1 & 0 & 1 & 0 \\ 0 & 1 & -1 & 1 \\ 0 & 0 & 2 & -1 \end{array}\right) \quad \left(\begin{array}{ccc|c} 1 & 0 & 0 & \frac{1}{2} \\ 0 & 1 & 0 & \frac{1}{2} \\ 0 & 0 & 1 & -\frac{1}{2} \end{array}\right)$$

$$\left(\begin{array}{ccc|c} 1 & 1 & 0 & 1 \\ 0 & -1 & 1 & -1 \\ 0 & 0 & 2 & -1 \end{array}\right) \quad \left(\begin{array}{ccc|c} 1 & 0 & 1 & 0 \\ 0 & 1 & -1 & 1 \\ 0 & 0 & 1 & -\frac{1}{2} \end{array}\right) \quad \begin{array}{l} P_{11} = \frac{1}{2} \\ P_{21} = \frac{1}{2} \\ P_{31} = -\frac{1}{2}. \end{array}$$

Thus,

$$(1, 0, 0) = \tfrac{1}{2}(1, 1, 0) + \tfrac{1}{2}(1, 0, 1) - \tfrac{1}{2}(0, 1, 1).$$

In a similar manner one can find

$$\begin{aligned} (0, 1, 0) &= \tfrac{1}{2}(1, 1, 0) - \tfrac{1}{2}(1, 0, 1) + \tfrac{1}{2}(0, 1, 1) \\ (0, 0, 1) &= -\tfrac{1}{2}(1, 1, 0) + \tfrac{1}{2}(1, 0, 1) + \tfrac{1}{2}(0, 1, 1). \end{aligned}$$

The matrix $\mu_{xu}(I) = P$ is then

$$\begin{pmatrix} \frac{1}{2} & \frac{1}{2} & -\frac{1}{2} \\ \frac{1}{2} & -\frac{1}{2} & \frac{1}{2} \\ -\frac{1}{2} & \frac{1}{2} & \frac{1}{2} \end{pmatrix}.$$

On the other hand the matrix taking u coordinates to x coordinates is gotten from the defining equations,

$$\begin{aligned} (1, 1, 0) &= 1(1, 0, 0) + 1(0, 1, 0) + 0(0, 0, 1), \\ (1, 0, 1) &= 1(1, 0, 0) + 0(0, 1, 0) + 1(0, 0, 1), \\ (0, 1, 1) &= 0(1, 0, 0) + 1(0, 1, 0) + 1(0, 0, 1). \end{aligned}$$

$$\mu_{ux}(I) = \begin{pmatrix} 1 & 1 & 0 \\ 1 & 0 & 1 \\ 0 & 1 & 1 \end{pmatrix}.$$

The two matrices,

$$\begin{pmatrix} \frac{1}{2} & \frac{1}{2} & -\frac{1}{2} \\ \frac{1}{2} & -\frac{1}{2} & \frac{1}{2} \\ -\frac{1}{2} & \frac{1}{2} & \frac{1}{2} \end{pmatrix} \text{ and } \begin{pmatrix} 1 & 1 & 0 \\ 1 & 0 & 1 \\ 0 & 1 & 1 \end{pmatrix},$$

are inverses of each other. The two change of basis equations are $U = PX$ and $X = P^{-1}U$,

$$\begin{pmatrix} U_1 \\ U_2 \\ U_3 \end{pmatrix} = \begin{pmatrix} \frac{1}{2} & \frac{1}{2} & -\frac{1}{2} \\ \frac{1}{2} & -\frac{1}{2} & \frac{1}{2} \\ -\frac{1}{2} & \frac{1}{2} & \frac{1}{2} \end{pmatrix} \begin{pmatrix} X_1 \\ X_2 \\ X_3 \end{pmatrix} \quad \text{and} \quad \begin{pmatrix} X_1 \\ X_2 \\ X_3 \end{pmatrix} = \begin{pmatrix} 1 & 1 & 0 \\ 1 & 0 & 1 \\ 0 & 1 & 1 \end{pmatrix} \begin{pmatrix} U_1 \\ U_2 \\ U_3 \end{pmatrix}.$$

Having seen the change of basis matrix we now observe how the matrix of a morphism is altered by changes of basis in the modules.

Theorem. *Let M and M' be modules over a commutative unitary ring R. Let $(x_j|j \in n)$ and $(u_j|j \in n)$ be finite bases for M and let $(y_i|i \in m)$ and $(v_i|i \in m)$ be finite bases for M'. Then the matrices of f and the change of basis matrices are related by the equation*

$$\mu_{uv}(f) = \mu_{yv}(I_{M'})\mu_{xy}(f)\mu_{xu}(I_M)^{-1}.$$

PROOF. The following diagram illustrates the mappings involved.

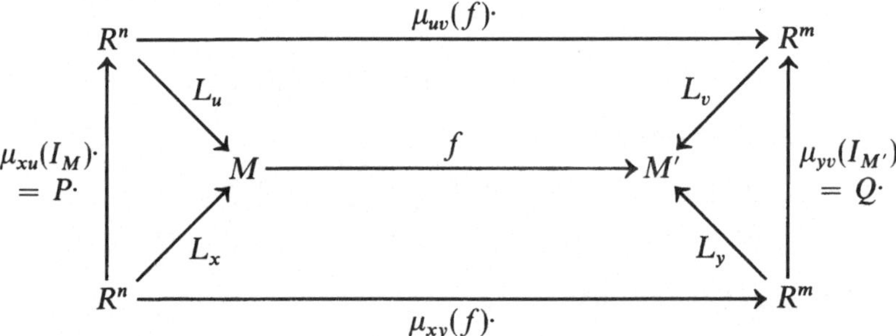

We know from our earlier theorems:

$$\mu_{uv}(f)\cdot = L_v^{-1} \circ f \circ L_u; \qquad P\cdot = \mu_{xu}(I_M)\cdot = L_u^{-1} \circ L_x;$$
$$\mu_{xy}(f)\cdot = L_y^{-1} \circ f \circ L_x; \qquad Q\cdot = \mu_{yv}(I_{M'})\cdot = L_v^{-1} \circ L_y.$$

We combine these equations.

$$\begin{aligned}\mu_{uv}(f)\cdot &= L_v^{-1} \circ f \circ L_u = L_v^{-1} \circ (L_y \circ \mu_{xy}(f)\cdot \circ L_x^{-1}) \circ L_u \\ &= (L_v^{-1} \circ L_y) \circ \mu_{xy}(f)\cdot \circ (L_x^{-1} \circ L_u) \\ &= \mu_{yv}(I_{M'})\cdot \circ \mu_{xy}(f)\cdot \circ \mu_{xu}(I_M)^{-1}\cdot \\ &= Q\cdot \circ \mu_{xy}(f)\cdot \circ P^{-1}\cdot = Q\mu_{xy}(f)P^{-1}\cdot.\end{aligned}$$

It is to be carefully noticed that the change of basis matrices P and Q are always invertible. The matrices of f are invertible if and only if f happens to be an isomorphism. □

EXAMPLE. Let $f:\mathbb{Q}^3 \to \mathbb{Q}^2$ such that $f(r_1, r_2, r_3) = (r_1 + 2r_2 + r_3, r_1 - r_2 + r_3)$. For $\mathbb{Q}^3$ we use the two bases $((1, 0, 0), (0, 1, 0), (0, 0, 1)) = x$ and $((1, 1, 0), (1, 0, 1), (0, 1, 1)) = u$. For $\mathbb{Q}^2$, the codomain, we use the bases $y = ((1, 0), (0, 1))$ and $v = ((1, 1), (0, 1))$. From the defining equations

$$\begin{aligned}(1, 0) &= 1(1, 1) + (-1)(0, 1), \\ (0, 1) &= 0(1, 1) + \quad 1(0, 1),\end{aligned}$$

we have the coordinate equation with matrix Q, $V = QY$,

$$\begin{pmatrix} V_1 \\ V_2 \end{pmatrix} = \begin{pmatrix} 1 & 0 \\ -1 & 1 \end{pmatrix} \begin{pmatrix} Y_1 \\ Y_2 \end{pmatrix}.$$

We have the matrix P from the previous example all worked out.

$$\mu_{xy}(f) = \begin{pmatrix} 1 & 2 & 1 \\ 1 & -1 & 1 \end{pmatrix}.$$

$$\mu_{uv}(f) = \mu_{yv}(I_{M'})\mu_{xy}(f)\mu_{xu}(I_M)^{-1} = Q\mu_{xy}(f)P^{-1}.$$

$$\mu_{uv}(f) = \begin{pmatrix} 1 & 0 \\ -1 & 1 \end{pmatrix} \begin{pmatrix} 1 & 2 & 1 \\ 1 & -1 & 1 \end{pmatrix} \begin{pmatrix} \frac{1}{2} & \frac{1}{2} & -\frac{1}{2} \\ \frac{1}{2} & -\frac{1}{2} & \frac{1}{2} \\ -\frac{1}{2} & \frac{1}{2} & \frac{1}{2} \end{pmatrix}$$

$$= \begin{pmatrix} 1 & 0 \\ -1 & 1 \end{pmatrix} \begin{pmatrix} 1 & 2 & 1 \\ 1 & -1 & 1 \end{pmatrix} \begin{pmatrix} 1 & 1 & 0 \\ 1 & 0 & 1 \\ 0 & 1 & 1 \end{pmatrix} = \begin{pmatrix} 3 & 2 & 3 \\ -3 & 0 & -3 \end{pmatrix}.$$

We now discuss elementary change of basis matrices. In Section 6.8 we discussed three elementary changes of basis:

I. The interchange of two basis elements
II. The adding of a multiple of one basis element to another basis element
III. The multiplying of one basis element by an invertible scalar.

Each of these changes has a corresponding change of basis matrix which we find from the defining equations of the change of basis matrix. Noting carefully the form of these defining equations, already written out for us in Section 6.8, we recopy them here in detail for clarity in all three cases.

I.

$$\begin{aligned} x_i &= u_i, \quad i \neq p, q \\ x_p &= u_q \\ x_q &= u_p \end{aligned}$$

becomes in detail to show all the coefficients

$$\begin{aligned} x_1 &= vu_1 + \theta u_2 + \cdots + \theta u_p + \cdots + \theta u_q + \cdots + \theta u_n \\ x_2 &= \theta u_1 + vu_2 + \cdots + \theta u_p + \cdots + \theta u_q + \cdots + \theta u_n \\ &\cdots \\ x_p &= \theta u_1 + \theta u_2 + \cdots + \theta u_p + \cdots + vu_q + \cdots + \theta u_n \\ &\cdots \\ x_q &= \theta u_1 + \theta u_2 + \cdots + vu_p + \cdots + \theta u_q + \cdots + \theta u_n \\ &\cdots \\ x_n &= \theta u_1 + \theta u_2 + \cdots + \theta u_p + \cdots + \theta u_q + \cdots + vu_n. \end{aligned}$$

From these equations we obtain the matrix $\mu_{xu}(I)$ which we call in this case $E(p, q)$.

$$E(p, q) = \begin{pmatrix} v & \theta & \cdots & \theta & \cdots & \theta & \cdots & \theta \\ \theta & v & \cdots & \theta & \cdots & \theta & \cdots & \theta \\ \cdots \\ \theta & \theta & \cdots & \theta & \cdots & v & \cdots & \theta \\ \cdots \\ \theta & \theta & \cdots & v & \cdots & \theta & \cdots & \theta \\ \cdots \\ \theta & \theta & \cdots & \theta & \cdots & \theta & \cdots & v \end{pmatrix} \begin{matrix} \\ \\ \\ \text{row } p \\ \\ \leftarrow \text{row } q \\ \\ \\ \end{matrix}$$

$$\uparrow \text{ column } p \qquad \uparrow \text{ column } q$$

This matrix $E(p, q)$ looks just like the n by n identity matrix δ except for having rows p and q interchanged. The inverse of $E(p, q)$ found from the inverse equations is also $E(p, q)$.

II.

$$x_i = u_i, i \neq q$$
$$x_q = u_q + ru_p.$$

$$\begin{aligned} x_1 &= vu_1 + \theta u_2 + \cdots + \theta u_p + \cdots + \theta u_q + \cdots + \theta u_n \\ x_2 &= \theta u_1 + vu_2 + \cdots + \theta u_p + \cdots + \theta u_q + \cdots + \theta u_n \\ &\cdots \\ x_p &= \theta u_1 + \theta u_2 + \cdots + vu_p + \cdots + \theta u_q + \cdots + \theta u_n \\ &\cdots \\ x_q &= \theta u_1 + \theta u_2 + \cdots + ru_p + \cdots + vu_q + \cdots + \theta u_n \\ &\cdots \\ x_n &= \theta u_1 + \theta u_2 + \cdots + \theta u_p + \cdots + \theta u_q + \cdots + vu_n. \end{aligned}$$

These defining equations yield the matrix $\mu_{xu}(I)$, which we name in this case $E(r, q; p)$.

$$E(r, q; p) = \begin{pmatrix} v & \theta & \cdots & \theta & \cdots & \theta & \cdots & \theta \\ \theta & v & \cdots & \theta & \cdots & \theta & \cdots & \theta \\ \cdots \\ \theta & \theta & \cdots & v & \cdots & r & \cdots & \theta \\ \cdots \\ \theta & \theta & \cdots & \theta & \cdots & v & \cdots & \theta \\ \cdots \\ \theta & \theta & \cdots & \theta & \cdots & \theta & \cdots & v \end{pmatrix} \begin{matrix} \\ \\ \\ \leftarrow \text{row } p \\ \\ \leftarrow \text{row } q \\ \\ \\ \end{matrix}$$

$$\uparrow \text{ column } p \qquad \uparrow \text{ column } q$$

This elementary change of basis matrix differs from the identity matrix in having r times row q added to row p. From the equations of u in terms of x the inverse can be determined to be $E(-r, q; p)$.

III.

$$x_i = u_i, \quad i \neq p$$
$$x_p = su_p, \quad s \text{ invertible.}$$

$$\begin{aligned}
x_1 &= vu_1 + \theta u_2 + \cdots + \theta u_p + \cdots + \theta u_n \\
x_2 &= \theta u_1 + vu_2 + \cdots + \theta u_p + \cdots + \theta u_n \\
&\cdots \\
x_p &= \theta u_1 + \theta u_2 + \cdots + su_p + \cdots + \theta u_n \\
&\cdots \\
x_n &= \theta u_1 + \theta u_2 + \cdots + \theta u_p + \cdots + vu_n.
\end{aligned}$$

These defining equations yield a change of basis matrix we call $E(s; p)$.

$$E(s; p) = \begin{pmatrix} v & \theta & \cdots & \theta & \cdots & \theta \\ \theta & v & \cdots & \theta & \cdots & \theta \\ \cdots \\ \theta & \theta & \cdots & s & \cdots & \theta \\ \cdots \\ \theta & \theta & \cdots & \theta & \cdots & v \end{pmatrix}. \leftarrow \text{row } p$$

$\uparrow$
column p

This matrix differs from the identity matrix in having row p multiplied by the scalar s, a unit of R. The inverse of $E(s; p)$ can be found from the inverse equations to be $E(s^{-1}; p)$.

In all three cases $E(p, q)$, $E(r, q; p)$, $E(s; p)$, of the elementary change of basis matrices, the "new" coordinates U with respect to the new basis u are related to the "old" coordinates X with respect to the old basis x by the matrix equations

I. $U = E(p, q)X$
II. $U = E(r, q; p)X$
III. $U = E(s; p)X$.

An application of such elementary change of basis matrices can be made to the problem of solving linear equations. We illustrate this for the case when the coefficients of the equations lie in a field. We have earlier illustrated the method of solving simultaneous linear equations by reduction to row-reduced echelon form. Three elementary operations on linear equations were presented at that time.

I. Interchange of two equations
II. Adding a multiple of one equation to another
III. Multiplying an equation by an invertible constant.

Beginning with a system of m linear equations in n unknowns with coefficients in a field

$$\begin{aligned} A_{11}X_1 + A_{12}X_2 + \cdots + A_{1n}X_n &= B_1 \\ A_{21}X_1 + A_{22}X_2 + \cdots + A_{2n}X_n &= B_2 \\ \cdots \\ A_{m1}X_1 + A_{m2}X_2 + \cdots + A_{mn}X_n &= B_m \end{aligned}$$

we put the equations in matrix form.

$$\begin{pmatrix} \sum_{j\in n} A_{1j}X_j \\ \sum_{j\in n} A_{2j}X_j \\ \cdots \\ \sum_{j\in n} A_{mj}X_j \end{pmatrix} = \begin{pmatrix} B_1 \\ B_2 \\ \cdots \\ B_m \end{pmatrix}.$$

The left side factors yielding

$$\begin{pmatrix} A_{11} & A_{21} & \cdots & A_{1n} \\ A_{21} & A_{22} & \cdots & A_{2n} \\ \cdots \\ A_{m1} & A_{m2} & \cdots & A_{mn} \end{pmatrix} \begin{pmatrix} X_1 \\ X_2 \\ \cdots \\ X_n \end{pmatrix} = \begin{pmatrix} B_1 \\ B_2 \\ \cdots \\ B_m \end{pmatrix}.$$

The system of m linear equations in m unknowns is equivalent to the matrix equation $AX = B$ as written above. Each of the three elementary transformations or operations on the system of linear equations can be accomplished by left multiplication on both sides of the matrix equation by elementary change of basis matrices of types I, II, III. A matrix is in row-reduced echelon form if and only if

1. rows entirely zero are below any row with some nonzero entries
2. the first nonzero entry in any row is a v, called an initial v
3. above and below every initial v occur only zeros
4. any initial v in any row must be to the right of any initial v in any row above.

Thus by procedures like those outlined in our earlier presentation on linear equations matrices can be reduced to row-reduced echelon form by left multiplication by elementary change of basis matrices of types I, II, and III. In both equivalent situations we have, however, only given informal directions on reducing to the row-reduced echelon form. That we can always obtain such a form and that it is unique will be settled in Section 7.6 later. We wish also in this later section to explain more fully the relationship between the linear equations and the vector spaces.

EXAMPLE. The equations

$$\begin{aligned} X_1 + X_2 + X_3 &= 3 \\ X_1 - X_3 &= 1 \\ X_2 + 2X_3 &= 2 \end{aligned}$$

are put into matrix form

$$\begin{pmatrix} X_1 + X_2 + X_3 \\ X_1 \qquad - X_3 \\ \qquad X_2 + 2X_3 \end{pmatrix} = \begin{pmatrix} 3 \\ 1 \\ 2 \end{pmatrix}.$$

$$\begin{pmatrix} 1 & 1 & 1 \\ 1 & 0 & -1 \\ 0 & 1 & 2 \end{pmatrix} \begin{pmatrix} X_1 \\ X_2 \\ X_3 \end{pmatrix} = \begin{pmatrix} 3 \\ 1 \\ 2 \end{pmatrix}.$$

We multiply on the left by $E(-1, 1; 2)$.

$$\begin{pmatrix} 1 & 0 & 0 \\ -1 & 1 & 0 \\ 0 & 0 & 1 \end{pmatrix} \begin{pmatrix} 1 & 1 & 1 \\ 1 & 0 & -1 \\ 0 & 1 & 2 \end{pmatrix} \begin{pmatrix} X_1 \\ X_2 \\ X_3 \end{pmatrix} = \begin{pmatrix} 1 & 0 & 0 \\ -1 & 1 & 0 \\ 0 & 0 & 1 \end{pmatrix} \begin{pmatrix} 3 \\ 1 \\ 2 \end{pmatrix}.$$

$$\begin{pmatrix} 1 & 1 & 1 \\ 0 & -1 & -2 \\ 0 & 1 & 2 \end{pmatrix} \begin{pmatrix} X_1 \\ X_2 \\ X_3 \end{pmatrix} = \begin{pmatrix} 3 \\ -2 \\ 2 \end{pmatrix}.$$

We then multiply by $E(-1, 2; 3)$ to get

$$\begin{pmatrix} 1 & 1 & 1 \\ 0 & -1 & -2 \\ 0 & 0 & 0 \end{pmatrix} \begin{pmatrix} X_1 \\ X_2 \\ X_3 \end{pmatrix} = \begin{pmatrix} 3 \\ -2 \\ 0 \end{pmatrix}.$$

Then by $E(-1; 2)$ yielding

$$\begin{pmatrix} 1 & 1 & 1 \\ 0 & 1 & 2 \\ 0 & 0 & 0 \end{pmatrix} \begin{pmatrix} X_1 \\ X_2 \\ X_3 \end{pmatrix} = \begin{pmatrix} 3 \\ 2 \\ 0 \end{pmatrix}.$$

Next by $E(-1, 2; 1)$ yielding

$$\begin{pmatrix} 1 & 0 & -1 \\ 0 & 1 & 2 \\ 0 & 0 & 0 \end{pmatrix} \begin{pmatrix} X_1 \\ X_2 \\ X_3 \end{pmatrix} = \begin{pmatrix} 1 \\ 2 \\ 0 \end{pmatrix}.$$

This is equivalent to the system

$$\begin{aligned} X_1 \qquad - X_3 &= 1 \\ X_2 + 2X_3 &= 2 \\ 0 &= 0. \end{aligned}$$

This example was also worked out in the earlier section on linear equation solutions. The reader should assure himself that the same work is performed in both cases.

Questions

1. Which of the following statements are true?
   (A) Every change of basis matrix is invertible.
   (B) Every change of basis matrix is square.
   (C) Every square invertible matrix is a change of basis matrix for some pair of bases.
   (D) Every square invertible matrix is the matrix of some automorphism.
   (E) None of the sentences is true.

2. Which of the following equations are true? We presume the notation of this section.
   (A) $U = PX$ and $V = QY$ and $Y = AX$ imply $V = QAPU$.
   (B) $\mu_{uv}(f) = \mu_{ux}(I)\mu_{xy}(f)\mu_{yv}(I)$.
   (C) $L_u^{-1} \circ L_x = \mu_{ux}(I)$.
   (D) $L_x^{-1} \circ L_u = \mu_{ux}(I)$.
   (E) None of the statements is true.

3. Which of the following matrices are in row-reduced echelon form?
   (A) $\begin{pmatrix} 0 & 1 & 0 \\ 0 & 0 & 1 \\ 0 & 0 & 0 \end{pmatrix}$ (B) $\begin{pmatrix} 0 & 0 & 0 \\ 1 & 0 & 0 \\ 0 & 0 & 1 \end{pmatrix}$ (C) $\begin{pmatrix} 1 & 2 & 3 \\ 0 & 0 & 0 \\ 0 & 0 & 0 \end{pmatrix}$ (D) $\begin{pmatrix} 1 & 0 & 1 \\ 0 & 0 & 0 \\ 0 & 0 & 0 \end{pmatrix}$.
   (E) None of the matrices is in row-reduced echelon form.

4. Which of the following matrices can be change of basis matrices?
   (A) $\begin{pmatrix} 1 & 1 \\ 1 & 1 \end{pmatrix}$ (B) $\begin{pmatrix} 1 & 2 \\ 2 & 1 \end{pmatrix}$ (C) $\begin{pmatrix} 1 & 0 \\ 1 & 0 \end{pmatrix}$ (D) $\begin{pmatrix} 1 & 2 \\ 3 & 4 \end{pmatrix}$.
   (E) None can be change of basis matrices.

Exercises

1. Write down the defining equations for each of the three elementary change of basis matrices $E(p, q)$, $E(r, q; p)$, $E(s; p)$. Find the defining equations and the matrices of the inverses of each of the three elementary changes of basis.

2. Given the vector spaces $\mathbb{Q}^2$ and $\mathbb{Q}^3$ first with the standard bases and then with the new bases $((1, 2), (-1, 3))$ and $((1, 1, 0), (0, 1, 1), (2, 1, 4))$, find the change of basis matrices $\mu_{xu}(I)$ and $\mu_{yv}(I)$ for both spaces. For the morphism $f:\mathbb{Q}^2 \to \mathbb{Q}^3$ such that $f(r_1, r_2) = (2r_1 + r_2, r_1, r_1 - 3r_2)$ find $\mu_{xy}(f)$ and $\mu_{uv}(f)$ verifying the theorem of this section.

3. Let $f:\mathbb{R}^3 \to \mathbb{R}^2$ such that $f(r_1, r_2, r_3) = (r_1 + r_2 + r_3, r_1 - r_2 + r_3)$. Find the matrix of f, $\mu_{xy}(f)$, using the standard bases in both spaces. Find $\mu_{uv}(f)$ using for u the basis $((1, 1, 0), (1, 0, 1), (0, 1, 1))$ and for v the basis $((1, 1), (0, 1))$. Find both change of basis matrices and check the equation $\mu_{uv}(f) = \mu_{yv}(I)\mu_{xy}(f)\mu_{xu}(I)^{-1}$.

4. Let R be a field. We define two matrices A and B in $R^{m\times n}$ to be equivalent matrices if and only if A and B are both matrices of some one morphism $f:R^n \to R^m$. Note that the choice of bases to produce A and B will usually be different. Prove that this relation is an equivalence relation on $R^{m\times n}$. Show that any matrix A is equivalent to some matrix C where $C_{ij} = \delta_{ij}$, for $1 \leqslant i, j \leqslant \rho$ and $C_{ij} = \theta$ otherwise. ρ is the rank of the morphism of which A is the matrix. [*Hint*: See Section 7.3.]

5. Let $g:\mathbb{R}^4 \to \mathbb{R}^3$ such that $g(r_1, r_2, r_3, r_4) = (r_1 - r_2, r_2 - r_3, r_3 - r_4)$. Find bases for $\mathbb{R}^4$ and $\mathbb{R}^3$ such that the matrix of g has the form of C described in Exercise 4.

6. Let (x_1, x_2, x_3) be a given basis for $\mathbb{R}^3$. Show that $(x_1 + x_2, x_2 + x_3, x_3 + x_1)$ is also a basis for $\mathbb{R}^3$. What is the change of basis matrix?

7. Let R be a field. We define A, B in $R^{n\times n}$ to be similar matrices if and only if they are both matrices for some one endomorphism f in $\mathscr{E}(R^n)$. This is to say A will be the matrix of f for some basis choice and B will be the matrix of f for some other basis choice for R^n. The same basis must be used for both domain and codomain. Prove that this relation of similarity is an equivalence relation on $R^{n\times n}$.

8. Let $f:M \to M'$ be a morphism of the free modules M, M' over a commutative unitary ring R. Let $(x_j|j \in n)$ and $(y_i|i \in m)$ be bases for M and M' respectively. We do not assume the bases to be finite. Show that the equation $Y = \mu_{xy}(f)X$ still holds where X and Y are the coordinates of $x \in M$ and $f(x) \in M'$.

9. If the matrix of $f:\mathbb{Q}^3 \to \mathbb{Q}^2$ with respect to the bases $((1, 1, 0), (1, 0, 1), (0, 1, 1))$ and $((2, 1), (1, 2))$ is

$$\begin{pmatrix} 1 & 2 & 4 \\ 3 & 1 & 1 \end{pmatrix}$$

then what is the matrix of the morphism f with respect to the standard bases?

7.5 The dual space

The dual of a module is defined and the dual of a morphism is defined and correlated with the transpose of a matrix. The rank of a matrix is defined, characterized, and correlated with the rank of a morphism.

Definition. Let M be a module over a commutative unitary ring R. The R-module $\mathscr{L}(M, R)$ is called the *dual module* or *dual space* of M and is denoted by M^*.

We now find a basis for M^* in terms of a given basis for M.

Theorem. *Let M be a module over a commutative unitary ring R. If $(x_j|j \in n)$ is a finite basis for M then there exists a basis $(x^i|i \in n)$ for M^* such that $x^i(x_j) = \delta_{ij}\nu$ for all $i, j \in n$.*

PROOF. (ν), the singleton family of the unity of R, is a basis for the ring R considered as a module over itself. For the space $\mathscr{L}(M, R)$ we have the basis $(e_{1j}|j \in n)$ such that $e_{1j}(x_q) = \nu$ if $q = j$ and $= \theta$ if $q \neq j$. We simply rename this basis so that we write e_{1j} as x^j. We note that the superscript is not a power of x, obviously. Each x^i belongs to $\mathscr{L}(M, R) = M^*$ and the family $(x^i|i \in n)$ is a basis for M^*. The basis has $1 \cdot n = n$ members. □

For the finite dimensional module, the dual module M^* has the dimension of M. M and M^* are therefore isomorphic modules. Once again we remind the reader that not all modules have bases and the theorem only concerns modules with given finite bases.

As well as the dual of a module we have also the dual of a morphism.

Definition. Given that M, M' are modules over a commutative unitary ring R and that $f:M \to M'$ is a morphism we define the *dual morphism* of f to be the mapping $f^*:(M')^* \to M^*$ such that $f^*(y) = y \circ f$.

Whereas the original morphism f is directed from M to M', the dual morphism f^* is directed from the dual, M'^*, to the dual, M^*. One might write the dual morphism f^* as the function $- \circ f$ understanding the notation to mean the function with value $y \circ f$ at y. This diagram may help in remembering the definition of the dual morphism.

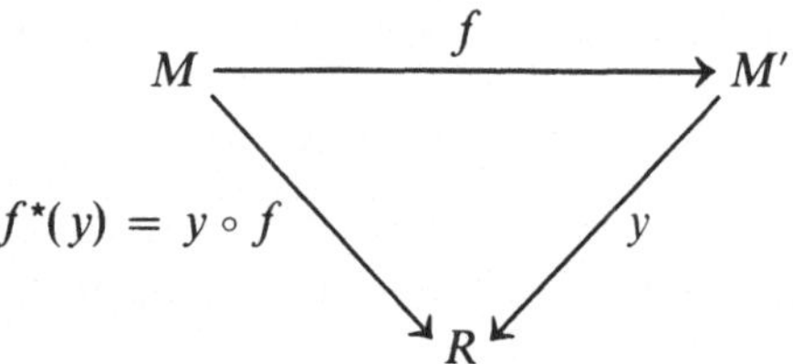

We must now verify that the dual "morphism" is indeed a morphism.

Theorem. *Let M and M' be modules over a commutative unitary ring R and $f:M \to M'$ a morphism. Then $f^*:M'^* \to M^*$ is also a morphism. Moreover, if f is an isomorphism so also is f^* an isomorphism.*

PROOF. $f^*(x + y) = (x + y) \circ f = x \circ f + y \circ f = f^*(x) + f^*(y)$. $f^*(ry) = (ry) \circ f = r(y \circ f) = rf^*(y)$. f^* is a morphism. If f is an isomorphism so also is $f^{-1}:M' \to M$ an isomorphism. Given $y \in M^*$ there exists in M'^* the linear mapping $y \circ f^{-1}$. $f^*(y \circ f^{-1}) = y \circ f^{-1} \circ f = y$ proves f^* is surjective. If $f^*(x) = f^*(y)$ then we have $x \circ f = y \circ f$. Composing on the right with f^{-1} yields $x = y$ proving f^* injective. □

Notationally, we can represent the operation of taking duals of modules and morphisms as $(f:M \to M')^* = (f^*:M'^* \to M^*)$. In the exercises of this section we shall develop some further properties of the dual. Now, however, we move immediately to the very important fundamental relationship between the matrix of a morphism and the matrix of the dual morphism.

Theorem. *Let M and M' be modules over a commutative unitary ring R and $f:M \to M'$ be a morphism. Let $(x_j|j \in n)$ and $(y_i|i \in m)$ be finite bases for M and M', respectively. Then $\mu(f^*)$, the matrix of the dual of f with respect to the dual bases $(y^i|i \in m)$ of M'^* and $(x^j|j \in n)$ of M^*, is the* *transpose of the matrix $\mu(f)$ with respect to the bases $(x_j|j \in n)$ of M and $(y_i|i \in m)$ of M'.*

PROOF. $\mathscr{L}(M, M')$ has the basis $(e_{ij}|i \in m, j \in n)$ where $e_{ij}(x_q) = \delta_{iq} y_i$. $\mathscr{L}(M'^*, M^*)$ has a basis we denote by $(e^{lk}|l \in n, k \in m)$ with $e^{lk}(y_p) = \delta_{kp} x_l$. Both f and f^* are expressible as unique linear combinations of basis elements.

$$f = \sum_{i \in m, j \in n} \mu_{xy}(f)_{ij} e_{ij}; \qquad f^* = \sum_{l \in n, k \in m} \mu_{y^*x^*}(f^*)_{lk} e^{lk}.$$

The definition of f^* asserts $f^*(y) = y \circ f$ for all $y \in M'^*$. In particular, $f^*(y^p) = y^p \circ f$ for all $p \in m$. Into this last equation we substitute both of the sums for f^* and f.

$$\sum_{(l,k) \in n \times m} \mu_{y^*x^*}(f^*)_{lk} e^{lk}(y^p) = y^p \circ \sum_{(i,j) \in m \times n} \mu_{xy}(f)_{ij} e_{ij} \quad \text{for all } p \in m.$$

$$\sum_{(l,k) \in n \times m} \mu_{y^*x^*}(f^*)_{lk}\, \delta_{kp} x^1 = \sum_{(i,j) \in m \times n} \mu_{xy}(f)_{ij} y^p \circ e_{ij}, \quad p \in m.$$

Applying both functions to the basis element x_q we have

$$\sum_{(l,k) \in n \times m} \mu_{y^*x^*}(f^*)_{lk}\, \delta_{kp} x^l(x_q) = \sum_{(i,j) \in m \times n} \mu_{xy}(f)_{ij} y^p(e_{ij}(x_q)) \quad \text{for all } q \in n, p \in m.$$

$$\sum_{(l,k) \in n \times m} \mu_{y^*x^*}(f)_{lk}\, \delta_{kp}\, \delta_{lq} v = \sum_{(i,j) \in m \times n} \mu_{xy}(f)_{ij} y^p(\delta_{jq} y_i) \quad \text{for all } q \in n, p \in m.$$

Remembering the Kronecker delta to be 1 when the subscripts coincide and zero otherwise we begin dropping terms from the sums in zero cases.

$$\mu_{y^*x^*}(f^*)_{qp} = \sum_{(i,j) \in m \times n} \mu_{xy}(f)_{ij}\, \delta_{jq}\, \delta_{pi} v \quad \text{for all } q \in n, p \in m,$$

$$= \mu_{xy}(f)_{pq} \quad \text{for all } p \in m, q \in n.$$

This proves $\mu_{y^*x^*}(f^*) = \mu_{xy}(f)^*$. The matrix of f^* is the transpose of the matrix of f. □

EXAMPLE. The morphism $f:\mathbb{R}^3 \to \mathbb{R}^2$ such that $f(r_1, r_2, r_3) = (2r_1 + r_2, r_1 + r_3)$ has $\mu(f) = \begin{pmatrix} 2 & 1 & 0 \\ 1 & 0 & -1 \end{pmatrix}$ with respect to the standard bases in $\mathbb{R}^3$ and $\mathbb{R}^2$. The matrix of $f^*:(\mathbb{R}^2)^* \to (\mathbb{R}^3)^*$ with respect to the standard dual bases is

$$\begin{pmatrix} 2 & 1 \\ 1 & 0 \\ 0 & -1 \end{pmatrix}.$$

A sample member of $(\mathbb{R}^2)^*$ is the mapping $y:\mathbb{R}^2 \to \mathbb{R}$ such that $y(r_1, r_2) = 2r_1 + 3r_2$. $y = 2e^1 + 3e^2$ where e^1 and e^2 are the members of the standard dual basis. The coordinates of y are then

$$\begin{pmatrix} 2 \\ 3 \end{pmatrix}.$$

The coordinates of $f^*(y)$ are

$$\begin{pmatrix} 2 & 1 \\ 1 & 0 \\ 0 & -1 \end{pmatrix} \begin{pmatrix} 2 \\ 3 \end{pmatrix} = \begin{pmatrix} 7 \\ 2 \\ -3 \end{pmatrix}.$$

$f^*(y)$ is $7e^1 + 2e^2 - 3e^3$ in $(\mathbb{R}^3)^*$.

EXAMPLE. The element $y:\mathbb{R}^3 \to \mathbb{R}$ of $(\mathbb{R}^3)^*$ with values $y(r_1, r_2, r_3) = 2r_1 - 3r_2 + r_3$ is, of course, a morphism. With respect to the standard bases (e_1, e_2, e_3) and (1) the morphism has a matrix $(2 \quad -3 \quad 1)$. The coordinate equation is

$$(Y_1) = (2 \quad -3 \quad 1) \begin{pmatrix} X_1 \\ X_2 \\ X_3 \end{pmatrix}.$$

In the same manner that the dual M^* of a given module M exists, so also does M^* itself have a dual M^{**}. And M^{**} has a dual, M^{***}. The process continues inductively. We are able, however, to show that for finite dimensional modules M and M^{**} can be identified.

Theorem. *Let M be a module over a commutative unitary ring R and $(x_j \mid j \in n)$ a finite basis for M. Then $\varphi:M \to M^{**}$ such that $\varphi(x)$ is the following member of M^{**},*

$$\varphi(x):M^* \to R \text{ with } \varphi(x)(f) = f(x)$$

is an isomorphism.

PROOF. We first show φ to be additive. $(\varphi(x + u) - \varphi(x) - \varphi(u))(f) = \varphi(x + u)(f) - \varphi(x)(f) - \varphi(u)(f) = f(x + u) - f(x) - f(u) = \theta$ for all $f \in M^*$ and x, $u \in M$. $\varphi(x + u) - \varphi(x) - \varphi(u)$ is the zero map of M^{**}. $\varphi(x + u) = \varphi(x) + \varphi(u)$. $(\varphi(rx) - r\varphi(x))(f) = \varphi(rx)(f) - r\varphi(x)(f) = f(rx) - rf(x) = \theta$ for all $f \in M^*$, $r \in R$, $x \in M$. $\varphi(rx) = r\varphi(x)$.

Suppose now $\varphi(x) = \varphi(u)$. $\varphi(x)(f) = \varphi(u)(f)$ for all $f \in M^*$. $f(x) = f(u)$ for all $f \in M^*$. In particular, $x^j(x) = x^j(u)$ for all $j \in n$. $x^j(\sum_{i=1}^n X_i x_i) = x^j(\sum_{i=1}^n U_i x_i)$. $\sum_{i=1}^n X_i x^j(x_i) = \sum_{i=1}^n U_i x^j(x_i)$. $\sum_{i=1}^n X_i\, \delta_{ji} v = \sum_{i=1}^n U_i\, \delta_{ji} v$. $X_j = U_j$ for all $j \in n$. $\sum_{j=1}^n X_j x_j = \sum_{j=1}^n U_j x_j$. $x = u$. φ is a monomorphism.

The dimensions of M, M^*, M^{**} are all equal to n. Range φ is also of dimension n since nullity $\varphi = 0$ and range φ is a submodule of M^{**}. In case M and M^{**} are vector spaces this is enough to prove range $\varphi = M^{**}$. For modules, however, we argue as follows. Let the basis of M^{**} which is dual to the basis $(x^i \mid i \in n)$ of M^* be denoted by $(\tilde{x}_i \mid i \in n)$. $\tilde{x}_k(x^i) = \delta_{ki} v$ by the dual basis definition. To show that φ is surjective let $u \in M^{**}$ and be expressed in terms of the basis for M^{**}: $u = \sum_{k=1}^n U_k \tilde{x}_k$ for some $U_1, U_2, \ldots, U_n \in R$. We are prepared now to show that $\varphi(\sum_{k=1}^n U_k x_k) = u$.

$$\begin{aligned}(\varphi(\textstyle\sum_{k=1}^n U_k x_k) - u)(x^j) &= (\textstyle\sum_{k=1}^n U_k \varphi(x_k) - \sum_{k=1}^n U_k \tilde{x}_k)(x^j)\\ &= \textstyle\sum_{k=1}^n U_k[\varphi(x_k)(x^j) - \tilde{x}_k(x^j)]\\ &= \textstyle\sum_{k=1}^n U_k[x^j(x_k) - \tilde{x}_k(x^j)]\\ &= \textstyle\sum_{k=1}^n U_k(\delta_{jk} - \delta_{kj})v = \theta.\end{aligned}$$

Since $\varphi(\sum_{k=1}^n U_k x_k) - u$ is a morphism from M^* to R which is zero on all the basis elements of M^* it must be the zero function from M^* to R. □

The next definitions and theorems are for the purpose of comparing the sizes of the ranges and kernels of f and f^*. This will give valuable information about the matrix of f and the matrix of f^*.

Definition. The *annihilator* of a submodule A of a module M over a commutative unitary ring R is the following subset of M^*. Anh $A = \{y \mid y \in M^*$ and $y(x) = \theta$ for all $x \in A\}$.

Theorem. *Let M be a module over a commutative unitary ring R and A a submodule of M. Then* anh A *is a submodule of M^*.*

Proof. Let $y, v \in \text{anh } A$. $y(x) = \theta$, $v(x) = \theta$ for all $x \in A$. $(y + v)(x) = \theta$ for all $x \in A$. $y + v \in \text{anh } A$. Let $r \in R$ and $y \in \text{anh } A$. $(ry)(x) = ry(x) = r\theta = \theta$ for all $x \in A$. $ry \in \text{anh } A$. Also clearly z, the zero mapping, belongs to anh A. Anh A is a submodule of M^*. □

Theorem. *If M is a finite dimensional vector space over a field K and A is any subspace of M then*

$$\dim A + \dim \text{anh } A = \dim M.$$

Proof. Let $(x_1, x_2, \ldots, x_k)$ be a basis for A and extend this to a basis $(x_1, \ldots, x_k, x_{k+1}, \ldots, x_n)$ for M. This is to include the possibilities of $k = 0$ with a basis $\varnothing$ for A and also of $k = n$ with basis $(x_1, \ldots, x_n)$ for A. We wish now to establish for the cases $k \neq n$ that $(x^{k+1}, x^{k+2}, \ldots, x^n)$ is a basis for anh A. The family, as a subfamily of the dual basis, is linearly independent. We now show that it generates anh A. Let $y \in \text{anh } A$. Because $y \in M^*$ we have $y = \sum_{j \in n} c_j x^j$ for some $c_j \in K$. But $\theta = y(x_p) = \sum_{j \in n} c_j x^j(x_p) = \sum_{j \in n} c_j \delta_{jp} v = c_p$ for all $p = 1, 2, \ldots, k$ because $x_p \in A$ when $p = 1, 2, \ldots, k$. Since all coefficients $c_1, c_2, \ldots, c_k$ are zero $y = c_{k+1}x^{k+1} + \cdots + c_n x^n$. □

This theorem was proved only for vector spaces and the results will apply only to matrices with entries in a field.

In a symmetrical manner to the annihilator of a subspace of M if we begin with a subspace B of the dual space M^* we have a subspace anh B

of M^{**}, $\{u | u \in M^{**} \text{ and } u(y) = \theta \text{ for all } y \in M^*\}$. Furthermore, dim B + dim anh B = dim M^{**}. Using the natural isomorphism $\varphi: M \to M^{**}$ discussed in an earlier theorem for finite dimensional spaces we may take anh B to be identified with a subspace of M.

$$\begin{aligned} \text{Anh } B &= \{\varphi(x) | x \in M \text{ and } \varphi(x)(y) = \theta \text{ for all } y \in M^*\} \\ &= \{x | x \in M \text{ and } y(x) = \theta \text{ for all } y \in M^*\}. \end{aligned}$$

We now prove the theorem which shows that rank f = rank f^*.

Theorem. *Let M and M' be finite dimensional vector spaces over a field K and let $f: M \to M'$ be a morphism. Then*

$$\begin{aligned} \text{kernel } f^* &= \text{anh range } f \\ \text{range } f^* &= \text{anh kernel } f \\ \text{rank } f^* &= \text{rank } f. \end{aligned}$$

PROOF. Kernel $f^* = \{y | y \in M'^* \text{ and } f^*(y) = \zeta^*\} = \{y | y \in M'^* \text{ and } y \circ f = \zeta^*\}$. But $y \circ f$ is the zero vector of M^* if and only if $(y \circ f)(x) = y(f(x)) = \theta$ for all $x \in M$ if and only if y annihilates the range of f. Kernel f^* = anh range f. Now let dim $M = n$ and dim $M' = m$, rank f = dim range $f = \rho$. Then dim anh range f = dim M' − dim range $f = n - \rho$. By part one of this theorem dim kernel $f^* = n - \rho$ also. Dim range f^* = dim M'^* − dim kernel $f^* = n - (n - \rho) = \rho$ which proves part three. To prove part two begin with $u \in$ range f^*. $u = f^*(y)$ for some $y \in M'^*$. $u(x) = (f^*(y))(x) = (y \circ f)(x) = y(f(x)) = y(\zeta') = \theta$ for all $x \in$ kernel f. We have therefore, range $f^* \subseteq$ anh kernel f. Since dim range f^* = dim range $f = n$ − dim kernel f = dim anh kernel f we have range f^* = anh kernel f. □

The next theorem applies the result to matrices.

Theorem. *Let $(a_{ij} | i \in m, j \in n)$ be a matrix with entries in a field K. Then there exists a natural number ρ such that $\rho \leqslant m$ and $\rho \leqslant n$ and ρ is the maximum number of linearly independent rows of the matrix in K^n and ρ is the maximum number of linearly independent columns of the matrix in K^m.*

PROOF. Choose K-vector spaces M and M' and bases $(x_j | j \in n)$ and $(y_i | i \in m)$ for M and M' respectively. There always exist such spaces, K^n and K^m, for example. Let f be the unique morphism $M \to M'$ such that $\mu(f) = a$, the given matrix. Let ρ be the rank of f. ρ is then $\leqslant m$ because rank f = dim range $f \leqslant$ dim $M' = m$. ρ is also the maximum number of linearly independent vectors in the family $(f(x_j) | j \in n)$, a generating family for range f. Because of the equation dim ker f + dim range f = dim $M = n$ we have ρ = dim range $f \leqslant n$.

$f(x_j) = \sum_{i \in m} \mu(f)_{ij} y_i = \sum_{i \in m} a_{ij} y_i$ implies each $f(x_j)$ has coordinates $(a_{1j}, a_{2j}, \ldots, a_{mj})$ in M' with respect to the basis $(y_i | i \in m)$. This family of coordinates is column number j in the matrix $(a_{ij} | i \in m, j \in n)$ and is a vector in K^m. The maximum number of linearly independent columns of matrix a in K^m is the same as ρ, the maximum number of linearly independent vectors in the family $(f(x_j)|j \in n)$ because the coordinate isomorphism L_y preserves linear independence.

Alternatively, ρ is also the dimension of the range of f^* which has matrix a^*, the transpose of a. The maximum number of linearly independent columns of a^* must be ρ which is dim range f^*. However, the columns of a^* are the rows of a. Hence the maximum number of linearly independent rows in K^n of a is also ρ. □

We are now in a position to define the rank of a matrix.

Definition. The *rank* of a matrix with entries in a field K is the maximum number of linearly independent rows or the maximum number of linearly independent columns.

We will deal with the rank of a matrix with entries in a ring in Chapter 10.

QUESTIONS

1. Which of the following statements are true?
   (A) Every member of the dual R-module M^* is a morphism.
   (B) Dim $M^* = n$, finite, implies the K-vector space M has dimension n.
   (C) No member of the dual module M^* can be a constant function.
   (D) For every module M, $M \neq M^*$.
   (E) None of the statements is true.

2. Which of the following statements about the dual morphism $f^*: M'^* \to M^*$ of R-modules are true?
   (A) $f^*(y) = f \circ y$.
   (B) If f^{-1} is the inverse of f then $(f^{-1})^*$ is the inverse of f^*.
   (C) f is an isomorphism implies f^* is also an isomorphism.
   (D) $f^* \in \mathscr{L}(M, R)$.
   (E) None of the statements is true.

3. Which of the following statements are true?
   (A) The annihilator of a subspace has the same dimension as the subspace.
   (B) The intersection of a subspace and its annihilator is sometimes a zero dimensional subspace.
   (C) Kernel f = anh range f for every morphism f.
   (D) $y \in$ anh A implies $y(x) = \theta$ for all $x \in M$.
   (E) None of the alternatives is true.

4. Which of the following statements are true?
   (A) The maximum number of linearly independent columns of a matrix (with entries in a field) is the same as the rank of its morphism.
   (B) The maximum number of linearly independent rows of a matrix (with entries in a field) is the same as the maximum number of linearly independent columns.

(C) If $f: M \to M'$ is a vector space morphism then rank f = the minimum number of generating vectors for range f in the family $(f(x_1), \ldots, f(x_n))$ where $(x_1, \ldots, x_n)$ is a basis of M.
(D) If $f: M \to M'$ is a vector space isomorphism then the columns of the matrix of f form a linearly independent family in K^m.
(E) None of the statements is true.

Exercises

1. Let $y \in (\mathbb{R}^3)^*$, the dual of the $\mathbb{R}$-module $\mathbb{R}^3$. Suppose the values of y are given by $y(r_1, r_2, r_3) = 5r_1 - 2r_2 + r_3$. Using the standard bases for $\mathbb{R}^3$ and $\mathbb{R}$ find the matrix of the morphism y. Write the matrix equation for computing values of y. Write y in terms of the dual basis.

2. For module M with any infinite basis $(x_j | j \in J)$ show that the dual family $(x^i | i \in J)$ is a linearly independent family. Show that the family fails to generate M^*.

3. For any morphisms f, g in $\mathscr{L}(M, M')$ show that $(f + g)^* = f^* + g^*$. For any morphism f in $\mathscr{L}(M, M')$ and $r \in R$ show that $(rf)^* = rf^*$.

4. If $f \in \mathscr{L}(M, M')$ and $g \in \mathscr{L}(M', M'')$ then show that $(g \circ f)^* = f^* \circ g^*$.

5. If M and M' have finite bases then $\mathscr{L}(M, M')$ and $\mathscr{L}(M'^*, M^*)$ are isomorphic. What is the dimension of the spaces?

6. Let M, M' be R-modules with finite bases, R a commutative unitary ring. For M'^* and M^* we use the dual bases of the given finite bases. Let $f, g \in \mathscr{L}(M, M')$ and $r \in R$. Prove these formulas:

$$(\mu(f) + \mu(g))^* = \mu(f + g)^* = \mu(f^* + g^*) = \mu(f)^* + \mu(g)^*.$$
$$\mu(rf)^* = \mu((rf)^*) = \mu(rf^*) = r\mu(f^*).$$

7. Let M, M', and M'' be R-modules with finite bases and utilize the appropriate dual bases in the dual spaces. Let $f \in \mathscr{L}(M, M')$ and $g \in \mathscr{L}(M', M'')$. Prove $(\mu(g)\mu(f))^* = \mu(f)^*\mu(g)^*$.

8. The taking of the transpose of a matrix is a mapping from $R^{m \times n}$ to $R^{n \times m}$. What can you say about this mapping with respect to the module structure of $R^{m \times n}$? What of the algebra structure of $R^{n \times n}$ in case $m = n$?

9. Discuss this diagram:

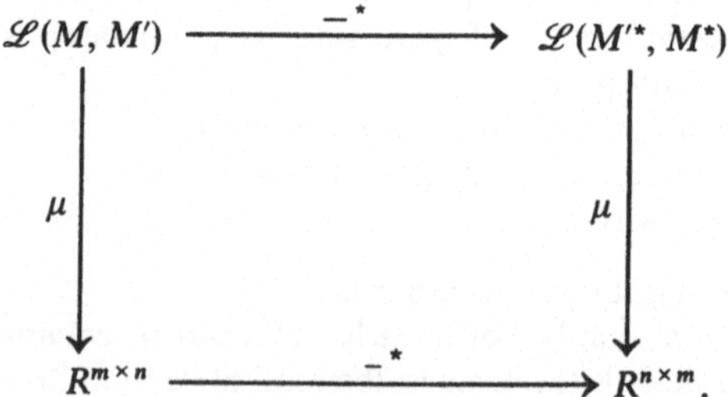

10. Let M be a module over a commutative unitary ring R with finite basis $(x_j | j \in n)$. Let $(x^j | j \in n)$ be the dual basis for M^*. Prove $(\varphi(x_j) | j \in n)$ is the basis for M^{**} which

is dual to $(x^j | j \in n)$. [*Note*: $\varphi: M \to M^{**}$ is the mapping defined in a theorem of this section.]

11. Show that in case the module M has an infinite basis then the mapping φ (see Exercise 10) is still a monomorphism.

12. Find the dual basis in $(\mathbb{R}^3)^*$ of the basis $((1, 1, 0), (1, 0, 1), (0, 1, 1))$ of $\mathbb{R}^3$.

13. Find all the linear functionals (morphisms) y in $(\mathbb{R}^3)^*$ which agree in value on the vectors $(2, 1, 4)$, $(0, 6, 2)$, and $(1, 0, 1)$.

14. Given the linear functional y in $(\mathbb{R}^3)^*$ such that $y(r_1, r_2, r_3) = 3r_1 - 2r_2 + r_3$, find a basis for the annihilator of $[y]$.

15. Let A and C be subspaces of a finite dimensional vector space M. Prove $A \subseteq C$ if and only if $\operatorname{anh} C \subseteq \operatorname{anh} A$.

16. Let x and x' be members of the K-vector space M such that $y(x) = \theta$ implies $y(x') = \theta$ for all $y \in M^*$. Show that x' is a multiple of x.

17. The set of all functions, real-valued, continuous on the real unit interval $[0, 1]$ is an $\mathbb{R}$-vector space. Show that the function $y: \mathbf{C}[0, 1] \to \mathbb{R}$ such that $y(f) = \int_0^1 f(t)\, dt$ is a member of $\mathbf{C}[0, 1]^*$.

18. If M is a nontrivial vector space then M^* is also nontrivial.

19. The subspace $\{t(1, 2, 3) | t \in R\}$ of $\mathbb{R}^3$ has a single vector $(1, 2, 3)$ as a basis. Find the subspace of $(\mathbb{R}^3)^*$ which is the annihilator of the given subspace. Find a basis for the annihilator.

20. If $x \in \mathbb{R}^n$ and $x \neq (0, 0, \ldots, 0)$ find a member of $(\mathbb{R}^n)^*$ such that $f(x) \neq 0$.

21. Let M be a finite dimensional vector space and y be a member of M^*. What is the dimension of the subspace $\{x | x \in M \text{ and } y(x) = \theta\}$ of M? If y is a nonzero linear functional (a member of M^*) and k an arbitrary member of the field K is there a vector x such that $y(x) = k$?

22. Let $((1, 1, 1), (0, 1, 1), (1, 1, 0))$ be a basis for $\mathbb{R}^3$. Find the dual basis in $(\mathbb{R}^3)^*$. Find a basis for the annihilator of the subspace $[(1, 1, 1), (0, 1, 1)]$.

7.6 Linear equations

In this section we prove theorems on the existence and uniqueness of solutions of systems of linear equations. We show that every matrix is row equivalent to a unique matrix in row-reduced echelon form.

Earlier in Appendix 6A we have discussed the solution of linear equations. We wish now to return to the subject and to use our knowledge of vector spaces as an aid to a deeper understanding. It is in this vector space setting that we are able to discuss efficiently the existence and nature of solutions.

The system of m linear equations in n unknowns

$$\begin{aligned} A_{11}X_1 + A_{12}X_2 + \cdots + A_{1n}X_n &= Y_1 \\ A_{21}X_1 + A_{22}X_2 + \cdots + A_{2n}X_n &= Y_2 \\ &\cdots \\ A_{m1}X_1 + A_{m2}X_2 + \cdots + A_{mn}X_n &= Y_m \end{aligned}$$

with coefficients in a field K is entirely equivalent to the matrix equation

$$\begin{pmatrix} A_{11} & A_{12} & \cdots & A_{1n} \\ A_{21} & A_{22} & \cdots & A_{2n} \\ \cdots & & & \\ A_{m1} & A_{m2} & \cdots & A_{mn} \end{pmatrix} \begin{pmatrix} X_1 \\ X_2 \\ \cdots \\ X_n \end{pmatrix} = \begin{pmatrix} Y_1 \\ Y_2 \\ \cdots \\ Y_m \end{pmatrix}.$$

A solution

$$\begin{pmatrix} X_1 \\ X_2 \\ \cdots \\ X_n \end{pmatrix}$$

to the matrix equation is a solution $X_1, X_2, \ldots, X_n$ to the m linear equations and vice versa. We abbreviate, as usual, the matrix with A and the matrices of X's and Y's with X and Y. We also indicate the matrix A augmented with the column matrix Y by $A:Y$. This is to say

$$A:Y = \begin{pmatrix} A_{11} & A_{12} & \cdots & A_{1n} & Y_1 \\ A_{21} & A_{22} & \cdots & A_{2n} & Y_2 \\ \cdots & & & & \\ A_{m1} & A_{m2} & \cdots & A_{mn} & Y_m \end{pmatrix}.$$

First we treat existence.

Theorem. *Let A be an m by n matrix with entries in a field K. Let $Y_1, Y_2, \ldots, Y_m \in K$. The equation $AX = Y$ has at least one solution if and only if* rank $A:Y$ = rank A.

PROOF. Choose K-vector spaces M and M' with finite bases of n and m members, respectively, say $(x_j \mid j \in n)$ and $(y_i \mid i \in m)$. There exists a unique morphism $f: M \to M'$ with matrix $\mu(f) = A$ (with respect to the chosen bases). The equation $AX = Y$ has a solution if and only if there exists a vector $x \in M$, $x = \sum_{j \in n} X_j x_j$, such that $f(x) = y$, $y = \sum_{i \in m} Y_i y_i$, if and only if $y \in \operatorname{range} f$. We continue with a list of equivalent conditions. $y \in \operatorname{range} f$. y belongs to the subspace generated by $(f(x_j) \mid j \in n)$. $\operatorname{Dim}[f(x_1), f(x_2), \ldots, f(x_n), y] = \dim[f(x_1), \ldots, f(x_n)]$. The maximum number of linearly independent vectors in the family $(f(x_1), \ldots, f(x_n), y)$ equals the maximum number of linearly independent vectors in the family $(f(x_1), \ldots, f(x_n))$. The maximum number of linearly independent columns (in K^m) of $A:Y$ equals the maximum number of linearly independent columns (in K^m) of A. Rank $A:Y$ = rank A (Rank A is, of course, equal to ρ, the rank of the morphism f). □

A system of linear equations with at least one solution is called *consistent* while a system with no solution is called *inconsistent*. The just proven theorem says that a system is consistent if and only if rank A = rank $A:Y$. We are able to say with this theorem which systems have solutions and which do not. For example, a system of homogeneous equations, i.e., equations for which $Y_1 = Y_2 = \cdots = Y_m = \theta$, must have at least one solution because rank A = rank $A:\theta$. Of course, this is hardly startling news, since it is obvious that $X_1 = X_2 = \cdots = X_n = \theta$ is a solution for a homogeneous system.

We now state and prove a theorem about uniqueness of solutions.

Theorem. *Let A be an m by n matrix with entries in a field K. Let $Y_1, Y_2, \ldots, Y_m$ belong to K. There at most one solution for $AX = Y$ if and only if* rank $A = n$.

PROOF. As with the previous theorem we place the problem in a vector space setting by choosing K-vector spaces M and M' with finite bases $(x_j \mid j \in n)$ and $(y_i \mid i \in m)$. We wish to show there cannot be two distinct solutions X, X' such that $AX = Y$ and $AX' = Y$. It is enough to show $AX = AX'$ implies $X = X'$. But $AX = AX'$ implies $X = X'$ if and only if $f(x) = f(x')$ implies $x = x'$ (x and x' are the vectors for which X and X' are the coordinates; $x = \sum_{j \in n} X_j x_j$, $x' = \sum_{j \in n} X'_j x_j$). $f(x) = f(x')$ implies $x = x'$ if and only if f is a monomorphism if and only if kernel $f = \{\zeta\}$ if and only if nullity $f = 0$ if and only if rank $f = n$ if and only if rank $A = n$. □

EXAMPLES. The system

$$\begin{aligned} X_1 + X_2 + 2X_3 + 2X_4 &= 0 \\ 2X_1 + X_2 + 3X_4 &= 0 \\ 3X_1 + 2X_2 + 2X_3 + 5X_4 &= 1 \end{aligned}$$

has a matrix of rank 2 because the rows (1, 1, 2, 2), (2, 1, 0, 3), (3, 2, 2, 5) are linearly dependent vectors in $\mathbb{R}^4$ whereas (1, 1, 2, 2), (2, 1, 0, 3) are linearly independent. When augmented by the column

$$\begin{pmatrix} 0 \\ 0 \\ 1 \end{pmatrix}$$

the resulting 3 by 5 matrix has rank 3. Using the notation of the theorem rank $A = 2$, rank $A:Y = 3$, $n = 4$. Because rank $A:Y \neq$ rank A there are no solutions. Because rank $A \neq 4$ even if there were solutions they would not be unique (!?).

The system

$$\begin{aligned} X_1 + X_2 + 2X_3 &= 0 \\ 2X_1 + X_2 &= 0 \\ 3X_1 + 2X_2 + 2X_3 &= 0 \end{aligned}$$

must have solutions because augmenting the matrix A by a column of zeros cannot change the rank. The solutions are not unique because rank $A = 2$ and $n = 3$.

In general we can note that if $m < n$, the number of equations is smaller than the number of unknowns, then the solutions cannot be unique because $\rho \leqslant m < n$.

We have now answered the question of existence and uniqueness of solutions for $AX = Y$ in terms of the rank of the matrices $A\!:\!Y$ and A. We turn to explore the nature of these solutions.

Theorem. *Let A be an m by n matrix with entries in a field K. Then the solutions of the homogeneous equation $AX = \theta$ form a subspace of K^n of dimension $n - \rho$, where $\rho =$ rank A.*

PROOF. Let M and M' be K-vector spaces of dimension n and m, finite, with bases $(x_j | j \in n)$ and $(y_i | i \in m)$. Let $f\!: M \to M'$ be the morphism with matrix A. X is a solution of the equation $AX = \theta$ if and only if $f(x) = \zeta'$ $(x = \sum_{j \in n} X_j x_j)$ if and only if x belongs to kernel f. Hence the solutions X to $AX = \theta$ form the kernel of the mapping $A\cdot$. We denote this subspace by N. Dim $N =$ nullity $A\cdot =$ nullity $f = n -$ rank $f = n -$ rank A. □

For consistent systems, that is, linear systems of equations with a least one solution, we have this result.

Theorem. *Suppose A is an m by n matrix with coefficients in a field K and $Y \in K^m$. Suppose in addition the equation $AX = Y$ has at least one solution $\tilde{X}$. Then the set of all solutions to $AX = Y$ is a coset $\tilde{X} + N$ of K^n/N where N is the $n -$ rank A dimensional subspace of K^n which is the set of solutions to $AX = \theta$.*

PROOF. We again choose vector spaces M, M' of finite dimension n, m with bases $(x_j | j \in n)$, $(y_i | i \in m)$. Corresponding under the coordinate morphism to the set of solutions $\{X | AX = Y \text{ and } X \in K^n\}$ is the set $\{x | f(x) = y \text{ and } x \in M\} = f^{-1}(y)$. $f^{-1}(y) \neq \varnothing$ since $\tilde{x} \in f^{-1}(y)$; $\tilde{x} = \sum_{j \in n} \tilde{X}_j x_j$. $f^{-1}(y)$ is the inverse image under a morphism of the coset $y + \{\zeta'\}$ of M' and must itself be a coset of M with respect to some subspace. In fact, the following equation holds:

$$f^{-1}(y) = \tilde{x} + f^{-1}(\zeta') = \tilde{x} + \text{kernel } f.$$

We leave the proof of this equation to the reader. We denote by N the coordinates of the vectors in kernel f, that is, $\{X | AX = \theta\}$. Corresponding to the set $\tilde{x}$ + kernel f is the set of coordinate $\tilde{X} + N$. Dimension $N =$ dim ker $f = n -$ rank $f = n -$ rank A. □

EXAMPLE. The matrix equation

$$\begin{pmatrix} 2 & -1 & 2 \\ 1 & 1 & 0 \\ 3 & 0 & 2 \end{pmatrix} \begin{pmatrix} X_1 \\ X_2 \\ X_3 \end{pmatrix} = \begin{pmatrix} 6 \\ 3 \\ 9 \end{pmatrix}$$

has rank $A = 2$, rank $A:Y = 2$ and $n = 3$. There exist solutions and they are not unique. Actually solving the equations by finding row-reduced echelon form gives the solutions $X_1 = (-\frac{2}{3})t + 3$, $X_2 = (-\frac{2}{3})t$, $X_3 = t$. In set form we find $\{((-\frac{2}{3})t + 3, (-\frac{2}{3})t, t) | t \in \mathbb{R}\} = (3, 0, 0) + \{t(-\frac{2}{3}, -\frac{2}{3}, 1) | t \in \mathbb{R}\}$, a coset of $\mathbb{R}^3$ determined by a dimension 1 subspace (see Figure 7.1).

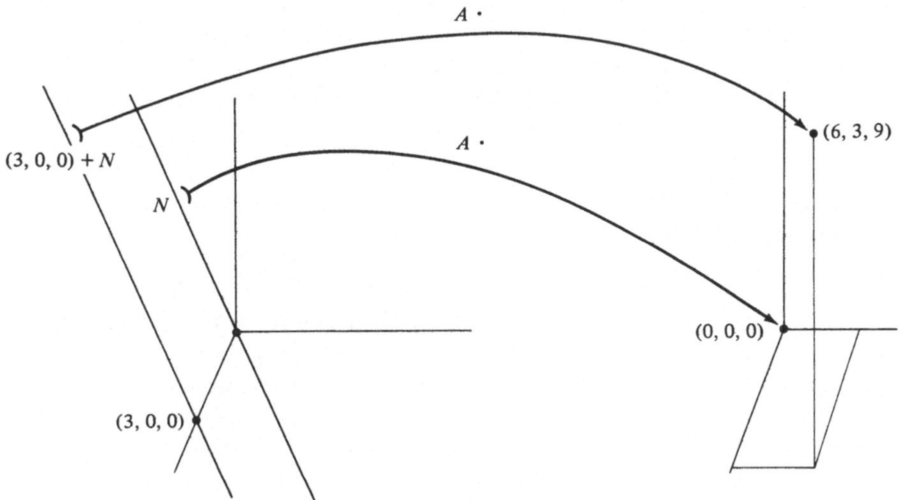

Figure 7.1

In our earlier intuitive introduction to the solution of linear equation in Appendix 6A we introduced the method of reducing to row-reduced echelon form. We put this method on an intuitive basis in order to enable us to be able to solve simple equations as we moved through a study of modules and vector spaces. We now return to this problem to show that every matrix can be reduced to row-reduced echelon form. We first give a slightly more formal definition of the row-reduced echelon form.

Definition. A matrix A with m rows and n columns with entries in a field K is in *row-reduced echelon form* if and only if

1. If row i is not all zeros then the first nonzero entry in row i is in column k_i and is the value v ($i = 1, 2, \ldots, m$)

2. There is an integer ρ, $0 \leqslant \rho \leqslant m$, such that rows $\rho + 1, \rho + 2, \ldots, m$ are all zeros, rows $1, 2, \ldots, \rho$ are not all zeros and $k_1 < k_2 < \cdots < k_\rho$
3. The only nonzero entry in column k_i is the v in place (i, k_i), $i = 1, 2, \ldots, \rho$.

We follow this definition with a definition of when two matrices are row equivalent.

Definition. m by n matrices, A and B, with entries in a field K are *row equivalent* if and only if there exists an invertible m by m matrix Q with entries in K such that $B = QA$.

The elementary row transformations or operations used earlier in the solution of linear equations are all equivalent to left multiplication by elementary change of basis matrices of type I, II, or III. All three are invertible square matrices. This definition of row equivalence thus includes multiplication by one or any finite number of such change of basis matrices.

We now show that every matrix can be reduced to a matrix in row-reduced echelon form. More precisely,

Theorem. *Every m by n matrix with entries in a field K is row equivalent to a matrix in row-reduced echelon form.*

Proof. If a matrix has all entries zero then it is already in row-reduced echelon form. If not all entries are zero then denote with k_1 the column number of the first column which is not zero in every row $1, 2, \ldots, m$. By interchanging rows (multiplying on the left by an elementary change of basis matrix of type I) we can produce a row equivalent matrix with a nonzero entry in place $(1, k_1)$. By multiplication by a change of basis matrix of type III we produce a row equivalent matrix with v in place $(1, k_1)$. We then use elementary change of basis matrices of type II to produce a matrix with v in place $(1, k_1)$ and zeros in the rest of column k_1. By use of elementary change of basis matrices of type I any all zero rows can be placed below any row not entirely zero. This begins our induction.

We now assume that our result holds up to and including row l. We then consider row l, $1 \leqslant l \leqslant m$. If there are v's in places $(1, k_1), (2, k_2), \ldots, (\rho, k_\rho)$, $\rho \leqslant l$ and $k_1 < k_2 < \cdots < k_\rho$, zeros in column k_i except at place (i, k_i), $i = 1, 2, \ldots, \rho$ and rows $\rho + 1$, $\rho + 2, \ldots, l$, $l + 1, \ldots, m$ are entirely zero then the matrix is in row-reduced echelon form. The other alternative is that there are v's in places $(1, k_1)$, $(2, k_2), \ldots, (l, k_l)$ and $k_1 < k_2 < \cdots < k_l$, zeros in column k_i except at place (i, k_i), $i = 1, \ldots, l$ and all rows entirely zero are below any rows not all zeros. Let k_{l+1} be the number of the first column with a nonzero entry in any row numbered $l + 1, l + 2, \ldots, m$. Shift, using an elementary change of basis matrix of type I, the nonzero entry to place $(l + 1, k_{l+1})$. Obtain an entry v in place $(l + 1, k_{l+1})$ using an elementary change of basis matrix of type III. Produce a matrix with zeros in column k_{l+1} except at place $(l + 1, k_{l+1})$ by repeated use of elementary change of basis matrices of type II. Shift all completely zero rows below any rows not completely zero by using elementary change of basis matrices of type I. This completes the induction. □

Having showed that every matrix is row equivalent to a matrix in row-reduced echelon form we show that the rank of a matrix in row-reduced echelon form is precisely the number of initial v's that occur in the form. Since matrices that are row equivalent necessarily have the same rank, reducing a matrix to row-reduced echelon form will provide an efficient way of finding its rank.

Theorem. *An m by n matrix in row-reduced echelon form with v's in places* $(1, k_1), (2, k_2), \ldots, (\rho, k_\rho)$ *has* rank ρ.

Proof. Regarding the matrix as the matrix of a morphism between two vector spaces we note all of the matrices for the same morphism have the same rank regardless of basis changes. Multiplication by change of basis matrices on the left are due to change of bases in the codomain (second) vector space. Hence all row equivalent matrices have the same rank. So also do all equivalent matrices produced by changes of basis in the domain vector space. Beginning with a matrix in row-reduced echelon form and using changes of basis in the domain space, multiplying on the right by change of basis matrices, we can produce a matrix with v's in places $(1, k_1)$, $(2, k_2), \ldots, (\rho, k_\rho)$ and zeros elsewhere. This is easily accomplished by use of elementary change of basis matrices of type II on the columns (right multiplication). The remaining matrix is obviously of rank ρ. The original row-reduced echelon matrix must also have been of rank ρ. □

We have showed that every matrix is row equivalent to a matrix in row-reduced echelon form and consequently every system of linear equations

$$\begin{aligned} A_{11}X_1 + A_{12}X_2 + \cdots + A_{1n}X_n &= Y_1 \\ A_{21}X_1 + A_{22}X_2 + \cdots + A_{2n}X_n &= Y_2 \\ \cdots \\ A_{m1}X_1 + A_{m2}X_2 + \cdots + A_{mn}X_n &= Y_m \end{aligned}$$

with coefficients in a field K can be replaced by an equivalent system

$$\begin{aligned} X_{k_1} + A'_{1k_1+1}X_{k_1+1} + \cdots + \theta X_{k_2} + A'_{1k_2+1}X_{k_2+1} + \cdots + \theta X_{k_\rho} + \cdots &= Y'_1 \\ X_{k_2} + A'_{2k_2+1}X_{k_2+1} + \cdots + \theta X_{k_\rho} + \cdots &= Y'_2 \\ \cdots \\ X_{k_\rho} + \cdots &= Y'_\rho \\ \theta &= Y'_{\rho+1} \\ \cdots \\ \theta &= Y'_m \end{aligned}$$

in which we can solve for ρ of the unknowns, namely $X_{k_1}, X_{k_2}, \ldots, X_{k_\rho}$ in terms of the remaining unknowns and $Y'_1, Y'_2, \ldots, Y'_m$. That $Y'_{\rho+1}, \ldots, Y'_m$ are zero is a necessary and sufficient condition for the consistency of the system. The remaining $n - \rho$ unknowns may be assigned arbitrary values yielding a set of solutions which is a coset of K^n/N formed by a subspace N of K^n of dimensional $n - \rho$.

Once we know every matrix is row equivalent to a matrix in row-reduced echelon form, that every system of linear equations with coefficients in a field can be solved by using the row-reduced echelon form, there remains the problem of uniqueness of the form. We wish then to show there is only one matrix in row-reduced echelon form to which a matrix is row equivalent.

Theorem. *Two row-reduced echelon matrices which are row equivalent are identical.*

PROOF. This theorem shows that a matrix can be reduced to a unique row-reduced echelon matrix, that variations in procedure will not result in producing a different row-reduced echelon matrix. Assume then that there are two matrices V' and V'' both in row-reduced echelon form and both row equivalent to a matrix A. $Q'A = V'$ and $Q''A = V''$ for some invertible, change of basis matrices Q', Q''. A is an m by n matrix representing some morphism f with respect to some choice of bases $(x_j | j \in n)$ and $(y_i | i \in m)$ for an n dimensional vector space domain and an m dimensional vector space codomain. Now let $(v'_i | i \in m)$ be a basis for the codomain so that V' is the matrix of f with respect to the pair of bases $(x_j | j \in n)$ and $(v'_i | i \in m)$. Let $(v''_i | i \in m)$ be a basis for the codomain so that V'' is the matrix of f with respect to the pair of bases $(x_j | j \in n)$ and $(v''_i | i \in m)$.

The finite sequence of places for the first nonzero entry, v, in each row of V' is $(1, k'_1), (2, k'_2), \ldots, (\rho, k'_\rho)$ and for V'' is $(1, k''_1), (2, k''_2), \ldots, (\rho, k''_\rho)$. The number ρ must be the same in both cases because ρ is the rank of the matrix and the ranks of row equivalent matrices are the same. It further follows that the sequence of places for the two matrices V' and V'' must be identical. Otherwise, let them differ first in row i:$(i, k'_i) \neq (i, k''_i)$. For notational convenience, assume $k'_i < k''_i$. Then according to matrix V', $\dim[f(x_1), f(x_2), \ldots, f(x_{k'_i})] = i$ yet according to matrix V'', $\dim[f(x_1), f(x_2), \ldots, f(x_{k'_i})] = i - 1$. This contradiction show that the two sequences of places are identical.

By use of the matrices V' and V'' we see that $f(x_{k'_i}) = v'_i$ for $i = 1, 2, \ldots, \rho$ and $f(x_{k''_i}) = v''_i$ for $i = 1, 2, \ldots, \rho$. Hence $v'_i = v''_i$ for $i = 1, 2, \ldots, \rho$. We know furthermore that rows $\rho + 1, \rho + 2, \ldots, m$ of both V' and V'' are all entirely zero from the definition of row-reduced echelon form. The defining equations for the matrices V' and V'' are then

$$f(x_j) = V'_{1j}v'_1 + V'_{2j}v'_2 + \cdots + V'_{\rho j}v'_\rho + \theta v'_{\rho+1} + \cdots + \theta v'_m$$
$$f(x_j) = V''_{1j}v''_1 + V''_{2j}v''_2 + \cdots + V''_{\rho j}v''_\rho + \theta v''_{\rho+1} + \cdots + \theta v''_m$$

with $j = 1, 2, \ldots, n$. We know $V'_{ij} = V''_{ij}$, $i = 1, 2, \ldots, \rho$, $j = 1, 2, \ldots, n$ because $v'_i = v''_i$, $i = 1, 2, \ldots, \rho$, and because linear combinations in terms of basis elements are unique. Thus we know $V' = V''$. □

Questions

1. Let A be an m by n matrix with entries in a field K. Let Y be an m by 1 matrix with entries in K. The matrix equation $AX = Y$ has a solution
   (A) if and only if A is an invertible matrix
   (B) if Y has all zero entries
   (C) if Y belongs to the range of the matrix mapping $A\cdot$
   (D) if the matrix mapping $A\cdot : K^n \to K^m$ is an epimorphism.
   (E) None of the alternatives completes a true sentence.

2. Let A be an m by n matrix with entries in a field K. Let Y be an m by 1 matrix with entries in K. The matrix equation $AX = Y$
   (A) has at most one solution if $A\cdot : K^n \to K^m$ is a monomorphism
   (B) has at least one solution if $m < n$
   (C) has no solutions if the number $m > n$
   (D) has solutions implies rank $A : Y$ = rank A.
   (E) None of the alternatives completes a true sentence.

3. Let A be an m by n matrix with entries in a field K. Let Y be an m by 1 matrix with entries in K. Which of these statements are true?
   (A) $AX = \theta$ has at least one solution implies $AX = Y$ has a solution.
   (B) $AX = \theta$ has more than one solution implies $AX = Y$ has more than one solution.
   (C) $AX = Y$ has more than one solution implies $AX = \theta$ has more than one solution.
   (D) If rank $A = n$ then $AX = \theta$ has only trivial solutions ($X = \theta$).
   (E) None of the statements is true.

4. Let A be an m by n matrix with entries in a field K. Let Y be an m by 1 matrix with entries in K. Which of these statements are true?
   (A) The set of solutions for $AX = Y$ is a subspace of K^m.
   (B) The set of solutions for $AX = \theta$ is a subspace of K^n.
   (C) The set of solutions for $AX = Y$ is the set $(A\cdot)^{-1}(Y)$.
   (D) The set of solutions for $AX = Y$ is a coset $\tilde{X} + N$ where $\tilde{X}$ is a solution and N is a subspace of K^n.
   (E) None of the statements is true.

5. Let A be an m by n matrix with entries in a field K. Which of the following statements are correct?
   (A) A is row equivalent to some matrix containing only θ's and ν's.
   (B) Any matrix row equivalent to A has the same rank as A.
   (C) Any m by n matrix with entries in K with the same rank as A must be row equivalent to A.
   (D) $AX = Y$ and $BX = Y$ have the same solution if and only if A and B are row equivalent.
   (E) None of the alternatives is true.

6. The subset $\{(X_1, X_2, X_3) | A_1X_1 + A_2X_2 + A_3X_3 = B\}$ of $\mathbb{R}^3$ is
  (A) a linear variety of $\mathbb{R}^3$ for any choice of A_1, A_2, A_3, B
  (B) a linear variety of dimension two, if some A_1, A_2, A_3, is not 0
  (C) a linear variety of dimension 1 for some possible choices of A_1, A_2, A_3, B
  (D) a linear variety of dimension 3 for some possible choices of A_1, A_2, A_3, B.
(E) None of the alternatives completes a true sentence.

7. The subset $\{(X_1, X_2, X_3) | A_{11}X_1 + A_{12}X_2 + A_{13}X_3 = B_1$ and $A_{21}X_1 + A_{22}X_2 + A_{23}X_3 = B_2\}$ of $\mathbb{R}^3$ is
  (A) a linear variety of dimension 1 for all possible choices of $A_{11}, A_{12}, \ldots, A_{23}, B_1, B_2$
  (B) a linear variety of dimension 2 for some possible choices of $A_{11}, A_{12}, \ldots, A_{23}, B_1, B_2$
  (C) a subspace of dimension 1 if $B_1 = B_2 = 0$ and rank $\begin{pmatrix} A_{11} & A_{12} & A_{13} \\ A_{21} & A_{22} & A_{23} \end{pmatrix} = 1$
  (D) possibly empty.
(E) None of the alternatives completes a true sentence.

8. In the $\mathbb{R}$-vector space $\mathbb{R}^4$
  (A) two planes (2 dimensional linear varieties) may be disjoint yet not parallel
  (B) two hyperplanes (3 dimensional linear varieties) are either parallel or intersect in a line
  (C) two planes can be parallel to the same hyperplane yet not parallel to each other
  (D) Sixteen lines can be parallel to the same hyperplane yet no two parallel.
(E) None of the alternatives completes a true sentence.

Exercises

1. Solve each of the following systems of linear equations using the row-reduced echelon form. The coefficients are presumed in $\mathbb{Q}$.

(a)

$$\begin{aligned} X_3 - 2X_4 &= -5 \\ X_1 + 6X_2 + 4X_3 + 6X_4 &= -15 \\ X_1 + 6X_2 + 2X_3 - 3X_4 &= -7. \end{aligned}$$

(b)

$$\begin{aligned} X_1 + X_2 &= 0 \\ X_2 + X_3 &= 0 \\ X_3 + X_4 &= 0 \\ X_4 + X_1 &= 0. \end{aligned}$$

(c)

$$2X_1 + 3X_2 + 4X_3 = 5.$$

(d)

$$\begin{aligned} 2X_1 + 5X_3 &= 4X_2 + 2X_4 \\ X_1 + 3X_3 &= X_2 - X_4. \end{aligned}$$

(e)

$$\begin{aligned} X_1 &= 5 + t \\ X_2 &= -1 + 3t \\ X_3 &= 2t \\ X_1 &= -3 - u \\ X_2 &= -5 + u \\ X_3 &= 4 + 2u. \end{aligned}$$

[*Hint for* (e): The five unknowns are X_1, X_2, X_3, t, u.]

(f)

$$\begin{aligned} X_1 - X_2 + X_3 &= 4 \\ -X_1 + X_2 - X_3 &= 1 \\ 2X_1 - X_2 + X_3 &= 3. \end{aligned}$$

2. Prove that an n by n matrix with entries in a field K is invertible if and only if it is row equivalent to the n by n identity matrix.

3. Prove that an n by n matrix with entries in a field K is invertible if and only if it is the product of elementary change of basis matrices.

4. Prove that two m by n matrices with entries in a field K are equivalent if and only if they have the same rank.

5. Solve the following equations in the field of real numbers.

(a)

$$\begin{aligned} X_1 + 3X_2 - X_3 &= 4 \\ X_1 + 2X_2 + X_3 &= 2 \\ 3X_1 + 7X_2 + X_3 &= 9. \end{aligned}$$

(b)

$$\begin{aligned} 2X_1 + X_2 - X_3 &= 3 \\ X_1 + X_2 &= 2 \\ X_1 - X_3 &= 1. \end{aligned}$$

6. Find all solutions of

$$\begin{aligned} 2X_1 - 3X_2 + X_3 &= 5 \\ X_1 + 2X_2 + 3X_3 &= 1. \end{aligned}$$

Assume the coefficients in $\mathbb{Q}$.

7. Solve for X_1, X_2, X_3 using the row reduction method over the field $\mathbb{R}$. The answer will depend on a in $\mathbb{R}$.

$$\begin{aligned} aX_1 + aX_3 &= 4a \\ X_2 + 2X_3 &= -2 \\ aX_1 + 6X_2 + 3X_3 &= 1. \end{aligned}$$

8. Is the set $\{(X_1, X_2, X_3) | X_1 + 2X_2 + X_3 = 6 \text{ and } 7X_1 + 3X_2 - X_3 = 4 \text{ and } 4X_1 - 3X_2 - 4X_3 = -14\}$ a subspace or a coset of a subspace of $\mathbb{R}^3$? If so what is its dimension?

9. Is there a matrix A such that

$$A \begin{pmatrix} 3 & -1 \\ 1 & 2 \\ 0 & 1 \end{pmatrix} = \begin{pmatrix} 3 & -1 \\ 1 & 0 \end{pmatrix}?$$

Is such an A unique if it exists?

10. Write $\{(X_1, X_2, X_3)|X_1 - 2X_2 + X_3 = 5\}$ as a coset of a subspace of $\mathbb{R}^3$.

11. Solve the equations

$$\begin{aligned} 2X_1 - 3X_2 + X_3 &= 9 \\ X_1 + X_2 - X_3 &= 4 \end{aligned}$$

in the field $\mathbb{Q}$. Also solve these equations

$$\begin{aligned} 5X_1 - 2X_2 + 3X_3 &= 3 \\ X_1 + 4X_2 - X_3 &= 9 \\ 4X_1 + 5X_2 \qquad &= 15. \end{aligned}$$

12. For what real values of Y_1, Y_2, Y_3 do the following equations have a solution and what is the solution?

$$\begin{aligned} X_1 - 2X_2 + X_3 &= Y_1 \\ 2X_1 + X_2 + X_3 &= Y_2 \\ 5X_2 - X_3 &= Y_3. \end{aligned}$$

13. For what real values of c do the following equations have a real solution?

$$\begin{aligned} X_1 + \qquad 2X_2 &= 1 \\ 2X_1 + (1 + c)X_2 &= -1 \\ (1 - c)X_1 + \qquad 3X_2 &= 2. \end{aligned}$$

14. If a square matrix A with entries in a field K has a left inverse or a right inverse then that left inverse or that right inverse is an inverse of A. Prove.

15. Write an alternative proof to that given in Section 7.5 for the equality of row rank and column rank of a matrix A with the following argument. Let A' be the row-reduced echelon form for A. Column rank of A = dim of subspace generated by the columns of A = dim range $A\cdot = n -$ dim ker $A\cdot = n - \dim\{X|AX = \theta\} = n - \dim\{X|A'X = \theta\} = n - (n -$ number of initial ones in $A')$ = number of initial ones in A' = dim of subspace generated by the rows of A' = dim subspace generated by the rows of A = row rank of A.

7.7 Determinants

In this section we explore some further properties of permutations then define the determinant. We discuss the relationship between the determinant and the rank of a matrix.

We begin by classifying all permutations either even or odd according to the number of inversions produced by the permutation in an original arrangement of a set. This we do by defining the sign of a permutation in $\mathfrak{S}_n$, $n \geqslant 2$, as

$$\varepsilon(\sigma) = \prod_{1 \leqslant i < j \leqslant n} \frac{X_{\sigma(j)} - X_{\sigma(i)}}{X_j - X_i}.$$

This somewhat complicated looking symbol is called an *alternator*; its meaning will become clearer in the examples which follow. The upper case pi is the product symbol. The purpose of the alternator is to count the number of inversions of order between pairs of arguments produced by a permutation. Such a count can be accomplished without the use of this symbol if one prefers. For $n = 2$, the alternator is

$$\varepsilon(\sigma) = \frac{X_{\sigma(2)} - X_{\sigma(1)}}{X_2 - X_1}.$$

For $n = 3$, the alternator is

$$\varepsilon(\sigma) = \frac{(X_{\sigma(3)} - X_{\sigma(2)})(X_{\sigma(3)} - X_{\sigma(1)})(X_{\sigma(2)} - X_{\sigma(1)})}{(X_3 - X_2)(X_3 - X_1)(X_2 - X_1)}.$$

And for $n = 4$, the alternator is

$$\varepsilon(\sigma) = \frac{(X_{\sigma(4)} - X_{\sigma(3)})(X_{\sigma(4)} - X_{\sigma(2)})(X_{\sigma(4)} - X_{\sigma(1)})}{(X_4 - X_3)(X_4 - X_2)(X_4 - X_1)} \cdot \frac{(X_{\sigma(3)} - X_{\sigma(2)})(X_{\sigma(3)} - X_{\sigma(1)})(X_{\sigma(2)} - X_{\sigma(1)})}{(X_3 - X_2)(X_3 - X_1)(X_2 - X_1)}.$$

EXAMPLES. In Section 1.8 we worked out the six permutations of $\mathfrak{S}_3$. For several of these six permutations we now compute their sign. We use the notation of Section 1.8.

$$\begin{aligned}
\varepsilon(\sigma_4) &= \frac{(X_{\sigma_4(3)} - X_{\sigma_4(2)})(X_{\sigma_4(3)} - X_{\sigma_4(1)})(X_{\sigma_4(2)} - X_{\sigma_4(1)})}{(X_3 - X_2)(X_3 - X_1)(X_2 - X_1)} \\
&= \frac{(X_1 - X_2)(X_1 - X_3)(X_2 - X_3)}{(X_3 - X_2)(X_3 - X_1)(X_2 - X_1)} \\
&= (-1)(-1)(-1) = -1. \\
\varepsilon(\sigma_5) &= \frac{(X_{\sigma_5(3)} - X_{\sigma_5(2)})(X_{\sigma_5(3)} - X_{\sigma_5(1)})(X_{\sigma_5(2)} - X_{\sigma_5(1)})}{(X_3 - X_2)(X_3 - X_1)(X_2 - X_1)} \\
&= \frac{(X_2 - X_1)(X_2 - X_3)(X_1 - X_3)}{(X_3 - X_2)(X_3 - X_1)(X_2 - X_1)} \\
&= (-1)(-1) = 1.
\end{aligned}$$

One can continue and verify that of the six permutations in $\mathfrak{S}_3$, $\sigma_1, \sigma_3, \sigma_5$ have sign 1, and $\sigma_2, \sigma_4, \sigma_6$ have sign -1. Permutations with sign 1 are called *even permutations* (ignore the nonmatching subscripts) and permutations with sign -1 are called *odd permutations.* If we were more committed to this notation we would arrange for the even permutations to have even subscripts. We intend, however, to introduce soon a different notation for permutations.

We now show that ε preserves the composition of $\mathfrak{S}_n$ in the multiplication of $\{1, -1\}$.

Theorem. $\varepsilon(\tau \circ \sigma) = \varepsilon(\tau)\varepsilon(\sigma)$ *for any* σ, τ *in* $\mathfrak{S}_n$, $n \geqslant 2$.

PROOF. Given any pair of natural numbers i, j such that $1 \leqslant i < j \leqslant n$ there exist i', j' in $\{1, 2, \ldots, n\}$ such that $\sigma(i') = i$ and $\sigma(j') = j$ because σ is a surjection. $i' \neq j'$ because σ is an injection. Therefore, in the alternator there is a term $X_j - X_i = X_{\sigma(j')} - X_{\sigma(i')}$ or a term $X_i - X_j = X_{\sigma(i')} - X_{\sigma(j')}$ in the numerator. The term $X_j - X_i$ in the denominator cancels into the numerator leaving 1 or -1 in place of the two terms. On the other hand, given any $X_{\sigma(j)} - X_{\sigma(i)}$ in the numerator there exist $i'', j'' \in \{1, 2, \ldots, n\}$ such that $\sigma(i) = i''$ and $\sigma(j) = j''$. If $i'' < j''$ then $X_{\sigma(j)} - X_{\sigma(i)}$ is cancelled by $X_{j''} - X_{i''}$ yielding the factor 1. If $i'' > j''$ then $X_{\sigma(j)} - X_{\sigma(i)}$ is cancelled by $X_{i''} - X_{j''}$ leaving the factor -1. $\varepsilon(\sigma)$ is then the product of members of $\{1, -1\}$ and is therefore 1 or -1.

$$\begin{aligned}
\varepsilon(\tau \circ \sigma) &= \prod_{1 \leqslant i < j \leqslant n} \frac{X_{\tau \circ \sigma(j)} - X_{\tau \circ \sigma(i)}}{X_j - X_i} \\
&= \prod_{1 \leqslant i < j \leqslant n} \frac{X_{\tau \circ \sigma(j)} - X_{\tau \circ \sigma(i)}}{X_{\sigma(j)} - X_{\sigma(i)}} \frac{X_{\sigma(j)} - X_{\sigma(i)}}{X_j - X_i} \\
&= \prod_{1 \leqslant i < j \leqslant n} \frac{X_{\tau \circ \sigma(j)} - X_{\tau \circ \sigma(i)}}{X_{\sigma(j)} - X_{\sigma(i)}} \prod_{1 \leqslant i < j \leqslant n} \frac{X_{\sigma(j)} - X_{\sigma(i)}}{X_j - X_i} \\
&= \prod_{i' \neq j',\, 1 \leqslant i' \leqslant n,\, 1 \leqslant j' \leqslant n} \frac{X_{\tau(j')} - X_{\tau(i')}}{X_{j'} - X_{i'}} \varepsilon(\sigma) \\
&= \prod_{1 \leqslant k < l \leqslant n} \frac{X_{\tau(l)} - X_{\tau(k)}}{X_l - X_k} \varepsilon(\sigma) = \varepsilon(\tau)\varepsilon(\sigma).
\end{aligned}$$

□

Definition. A permutation of a set which leaves all elements of the set fixt except for two which it interchanges we cell a *transposition.*

Theorem. *The sign $\varepsilon(\tau)$ of a transposition τ in $\mathfrak{S}_n$ is -1.*

PROOF. To compute the sign of the transposition we count the number of -1's in the alternator. Let a transposition τ interchange the indices k and l and leave all others fixt. For convenience we take $k < l$. For indices i such that $k < l < i$ we have terms in the alternator

$$\frac{(X_{\tau(i)} - X_{\tau(l)})(X_{\tau(i)} - X_{\tau(k)})}{(X_i - X_l)(X_i - X_k)} = \frac{(X_i - X_k)(X_i - X_l)}{(X_i - X_l)(X_i - X_k)} = 1.$$

For indices i such that $k < i < l$ we have factors in the alternator

$$\frac{(X_{\tau(l)} - X_{\tau(i)})(X_{\tau(i)} - X_{\tau(k)})}{(X_l - X_i)(X_i - X_k)} = \frac{(X_k - X_i)(X_i - X_l)}{(X_l - X_i)(X_i - X_k)} = 1.$$

For indices i such that $i < k < l$ we have terms

$$\frac{(X_{\tau(l)} - X_{\tau(i)})(X_{\tau(k)} - X_{\tau(i)})}{(X_l - X_i)(X_k - X_i)} = \frac{(X_k - X_i)(X_l - X_i)}{(X_l - X_i)(X_k - X_i)} = 1.$$

For pairs of indices not involving k or l there are also terms in the alternator which simply cancel directly.

$$\frac{X_{\tau(j)} - X_{\tau(i)}}{X_j - X_i} = \frac{X_j - X_i}{X_j - X_i} = 1.$$

Finally there is one term involving both k and l.

$$\frac{X_{\tau(l)} - X_{\tau(k)}}{X_l - X_k} = \frac{X_k - X_l}{X_l - X_k} = -1.$$

The product of all terms is -1. □

By combining the two previous theorems we see that a permutation composed of an odd number of transpositions is odd and that a permutation composed of an even number of transpositions is even.

Computations with permutations are greatly facilitated by use of cyclic notation. We introduce this now, first with two examples and then with a definition.

Examples. (1 2 3) is a permutation σ of $\{1, 2, 3\}$ such that $\sigma(1) = 2, \sigma(2) = 3$, and $\sigma(3) = 1$. In other words, (1 2 3) sends 1 into 2, 2 into 3, and 3 into 1.

$$\begin{aligned} 1 &\overset{(1\ 2\ 3)}{\rightsquigarrow} 2 \\ 2 &\rightsquigarrow 3 \\ 3 &\rightsquigarrow 1. \end{aligned}$$

Secondly, (2 5)

$$\begin{aligned} 2 &\overset{(2\ 5)}{\rightsquigarrow} 5 \\ 5 &\rightsquigarrow 2. \end{aligned}$$

We now make a definition of *cycle*.

Definition. Let $i_1, i_2, \ldots, i_k$ all be distinct and belong to $\{1, 2, \ldots, n\}$. By $(i_1 i_2 \cdots i_k)$ we shall mean the permutation σ such that $\sigma(i_1) = i_2, \sigma(i_2) = i_3, \ldots, \sigma(i_{k-1}) = i_k, \sigma(i_k) = i_1$. σ holds all other members of $\{1, 2, \ldots, n\}$ fixt. We call the permutation $(i_1 i_2 i_3 \cdots i_k)$ a *cycle*.

The six members of $\mathfrak{S}_3$ which we tabulated in Section 1.8 are, in cyclic notation, $\sigma_1 = I, \sigma_2 = (2\,3), \sigma_3 = (1\,2\,3), \sigma_4 = (1\,3), \sigma_5 = (1\,3\,2), \sigma_6 = (1\,2)$. Products such as $\sigma_3 \circ \sigma_4 = (1\ 2\ 3) \circ (1\ 3)$ are easily computed by composition of the cycles. From right to left with the composition of the cycles (they are functions) and from left to right within the cycles we see

$$\begin{aligned} 1 &\overset{(1\ 3)}{\rightsquigarrow} 3 \overset{(1\ 2\ 3)}{\rightsquigarrow} 1 & \qquad 1 &\overset{(1\ 2\ 3)\circ(1\ 3)}{\rightsquigarrow} 1 \\ 2 &\rightsquigarrow 2 \rightsquigarrow 3 & 2 &\rightsquigarrow 3 \\ 3 &\rightsquigarrow 1 \rightsquigarrow 2 & 3 &\rightsquigarrow 2. \end{aligned}$$

Thus (1 2 3) ∘ (1 3) = (2 3). From now on we shall usually omit the composition sign between cycles. Another example of a composition of cycles is (1 3 4 2)(2 3 4)(1 3).

$$\begin{aligned} 1 &\rightsquigarrow 3 \rightsquigarrow 4 \rightsquigarrow 2 \\ 2 &\rightsquigarrow 2 \rightsquigarrow 3 \rightsquigarrow 4 \\ 4 &\rightsquigarrow 4 \rightsquigarrow 2 \rightsquigarrow 1 \\ 3 &\rightsquigarrow 1 \rightsquigarrow 1 \rightsquigarrow 3. \end{aligned}$$

We conclude that the composition is (1 2 4). We give one more example to assure the method is understood. Find (2 5 7)(2 5 6)(3 5 1).

$$\begin{array}{c} 1 \rightsquigarrow 3 \rightsquigarrow 3 \rightsquigarrow 3 \\ 3 \rightsquigarrow 5 \rightsquigarrow 6 \rightsquigarrow 6 \\ 6 \rightsquigarrow 6 \rightsquigarrow 2 \rightsquigarrow 5 \\ 5 \rightsquigarrow 1 \rightsquigarrow 1 \rightsquigarrow 1 \\ 2 \rightsquigarrow 2 \rightsquigarrow 5 \rightsquigarrow 7 \\ 7 \rightsquigarrow 7 \rightsquigarrow 7 \rightsquigarrow 2. \end{array}$$

The product is (2 7)(1 3 6 5).

Theorem. *Each composition of cycles or permutations can be written as a product of disjoint cycles*

PROOF. If σ is a permutation $\sigma(1) \in \{1, 2, \ldots, n\}$. If $\sigma(1) \neq 1$ we write $(1\sigma(1)$ which begins a cycle. If $\sigma(\sigma(1)) \neq 1$ we continue the cycle $(1\sigma(1)\sigma(\sigma(1)))$. Since $\{1, 2, \ldots, n\}$ is a finite set some application of σ must eventually produce a repetition. $\sigma^i(1) = \sigma^j(1)$ for some $i, j, i < j$. But permutations are invertible. We see $\sigma^{j-i}(1) = 1$ showing that 1 must be reached first. The cycle is then closed at this point $(1\sigma(1)\sigma^2(1) \cdots \sigma^{j-i}(1))$. One then chooses the first number in $\{1, 2, \ldots, n\}$ not mentioned in the cycle already produced and begins a second cycle $(m\sigma(m)\sigma^2(m) \cdots)$. In this manner one eventually places all members of $\{1, 2, \ldots, n\}$ in some cycle. One cannot produce in any cycle a number found in a preceding cycle for then one would have $\sigma^p(1) = \sigma^q(m)$ for some p, q indicating that m belongs to the preceding cycle or 1 to the following. If at any time one has $\sigma(1) = 1$ or $\sigma(m) = m$ then such a number is held fixt by σ and need not be mentioned in a cycle. If all members of $\{1, 2, \ldots, n\}$ are held fixt then $\sigma = I$. The produced cycle product is equal to σ. □

It is a simple matter to write down all the members of $\mathfrak{S}_1$, $\mathfrak{S}_2$, $\mathfrak{S}_3$, and $\mathfrak{S}_4$. Since $\mathfrak{S}_n$ has $n!$ members we know that these groups have 1, 2, 6, and 24 members respectively.

$\mathfrak{S}_1 = \{I\}$.
$\mathfrak{S}_2 = \{I, (1\ 2)\}$.
$\mathfrak{S}_3 = \{I, (1\ 2), (1\ 3), (2\ 3), (1\ 2\ 3), (1\ 3\ 2)\}$.
$\mathfrak{S}_4 = \{I, (1\ 2), (1\ 3), (1\ 4), (2\ 3), (2\ 4), (3\ 4),$
$(1\ 2\ 3), (1\ 2\ 4), (1\ 3\ 4), (2\ 3\ 4), (1\ 3\ 2), (1\ 4\ 2), (1\ 4\ 3), (2\ 4\ 3),$
$(1\ 2\ 3\ 4), (1\ 2\ 4\ 3), (1\ 3\ 2\ 4), (1\ 3\ 4\ 2), (1\ 4\ 2\ 3), (1\ 4\ 3\ 2),$
$(1\ 2)(3\ 4), (1\ 3)(2\ 4), (1\ 4)(2\ 3)\}$.

We now finish up this discussion of permutations with these two theorems.

Theorem. *Each permutation in $\mathfrak{S}_n$ can be written as a product of transpositions.*

PROOF. A cycle $(i_1 i_2 \cdots i_k)$ is equal to the composition $(i_1 i_k)(i_1 i_{k-1}) \cdots (i_1 i_3)(i_1 i_2)$ as may be readily checked. □

Theorem. *Each permutation in* $\mathfrak{S}_n$ *can be written as a product of the transpositions* $(1\ 2), (1\ 3), \ldots, (1\ n)$.

PROOF. Using the previous theorem and the factorization $(i_1 i_j) = (1 i_j)(1 i_1)$ the result immediately follows. □

Because of sign considerations it follows that every even permutation can be written as the product of an even number of transpositions.

We now begin our discussion of determinants by giving a definition of the determinant of a square matrix.

Definition. If A is an n by n matrix with entries in a commutative, unitary ring R then the *determinant* of A is

$$\det A = \sum_{\sigma \in \mathfrak{S}_n} \varepsilon(\sigma) A_{1\sigma(1)} A_{2\sigma(2)} \cdots A_{n\sigma(n)}.$$

The determinant, $\det A$, is often written $|A|$. From this value for $\det A$ there exists a mapping $\det : R^{n \times n} \to R$.

EXAMPLE. If A is a 2 by 2 matrix then

$$\det A = \sum_{\sigma \in \mathfrak{S}_2} \varepsilon(\sigma) A_{1\sigma(1)} A_{2\sigma(2)} = A_{11}A_{22} - A_{12}A_{21}.$$

$$\det \begin{pmatrix} 2 & -3 \\ 4 & -1 \end{pmatrix} = \begin{vmatrix} 2 & -3 \\ 4 & -1 \end{vmatrix} = (2)(-1) - (-3)(4) = 10.$$

If A is a 3 by 3 matrix $\det A = \sum_{\sigma \in \mathfrak{S}_3} \varepsilon(\sigma) A_{1\sigma(1)} A_{2\sigma(2)} A_{3\sigma(3)}$. There are six permutations in $\mathfrak{S}_3$, three with positive sign and three with negative sign. The sum for $\det A$ then has six terms. Det $A = A_{11}A_{22}A_{33} - A_{11}A_{23}A_{32} + A_{13}A_{21}A_{32} - A_{13}A_{22}A_{31} + A_{12}A_{23}A_{31} - A_{12}A_{21}A_{33}$. For a matrix with entries in $\mathbb{Z}$ such as

$$\begin{pmatrix} 2 & 3 & 5 \\ -1 & 0 & 2 \\ 4 & 1 & -6 \end{pmatrix}$$

the determinant is $(2)(0)(-6) - (2)(2)(1) + (5)(-1)(1) - (5)(0)(4) + (3)(2)(4) - (3)(-1)(-6) = -3$. There are in existence several schemes for remembering these six terms, one being the following illustration.

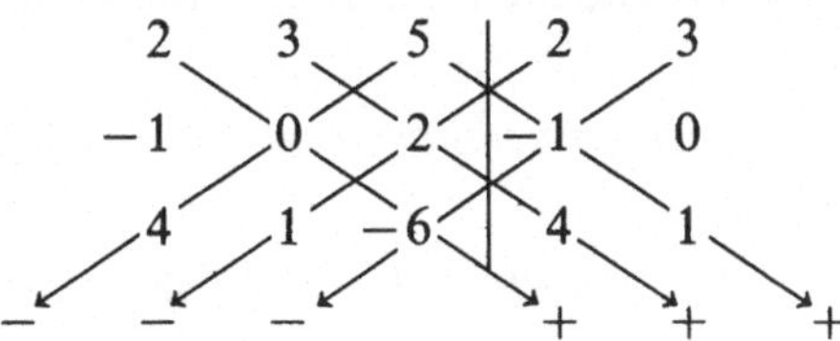

We now move on to develop some of the basic properties of determinants. We consider first some of the more manipulative aspects. These properties will facilitate calculation and prepare the ground for a vector space setting.

Theorem. *Let R be a commutative unitary ring and A be an n by n matrix with entries in R. Then* $\det A = \det A^*$.

PROOF. In the sum given in the definition of the determinant of A we rearrange each term so that the column numbers appear in increasing order instead of the row numbers. The term $A_{1\sigma(1)}A_{2\sigma(2)}\cdots A_{n\sigma(n)}$ is written $A_{\sigma^{-1}(1)1}A_{\sigma^{-1}(2)2}\cdots A_{\sigma^{-1}(n)n}$. This is possible because if the subscript 1, for example, is found coupled with k, $\sigma(k) = 1$, A_{k1}, then we know, because σ is a bijection that $\sigma^{-1}(1) = k$. A_{k1} becomes $A_{\sigma^{-1}(1)1}$. Furthermore, $\varepsilon(\sigma) = \varepsilon(\sigma^{-1})$.

$$\operatorname{Det} A = \sum_{\sigma\in\mathfrak{S}_n} \varepsilon(\sigma^{-1})A_{\sigma^{-1}(1)1}A_{\sigma^{-1}(2)2}\cdots A_{\sigma^{-1}(n)n}.$$

Now as σ runs through all permutations of $\mathfrak{S}_n$ so also does σ^{-1}.

$$\operatorname{Det} A = \sum_{\tau\in\mathfrak{S}_n} \varepsilon(\tau)A_{\tau(1)1}A_{\tau(2)2}\cdots A_{\tau(n)n}.$$

But this sum is the definition of the determinant of A^*. □

Before proving a most important theorem we demonstrate a lemma which gives the value of sums with duplications in the terms.

Lemma. *Let R be a commutative unitary ring and let A be an n by n matrix with values in R. Then if $k_i = k_j$ for some $i \neq j$*

$$\sum_{\sigma\in\mathfrak{S}_n} \varepsilon(\sigma)A_{k_1\sigma(1)}A_{k_2\sigma(2)}\cdots A_{k_n\sigma(n)} = \theta.$$

PROOF. Given any term $\varepsilon(\sigma)\, A_{k_1\sigma(1)}A_{k_2\sigma(2)}\cdots A_{k_n\sigma(n)}$ in the sum there is another term arising from the permutation $\sigma\circ\tau$ where τ is the transposition interchanging i and j and leaving the other integers fixt. This term is

$$\begin{aligned}\varepsilon(\sigma\circ\tau)&A_{k_1\sigma\circ\tau(1)}A_{k_2\sigma\circ\tau(2)}\cdots A_{k_i\sigma\circ\tau(i)}\cdots A_{k_j\sigma\circ\tau(j)}\cdots A_{k_n\sigma\circ\tau(n)}\\ &= -\varepsilon(\sigma)A_{k_1\sigma(1)}A_{k_2\sigma(2)}\cdots A_{k_i\sigma(j)}\cdots A_{k_j\sigma(i)}\cdots A_{k_n\sigma(n)}.\end{aligned}$$

The two terms cancel exactly since $k_i = k_j$. □

We now prove probably the most important theorem for determinants.

Theorem. *Let R be a commutative unitary ring. Let A and B be n by n matrices with entries in R. Then*

$$\det AB = \det A \det B.$$

Proof

Det AB

$$= \sum_{\sigma \in \mathfrak{S}_n} \varepsilon(\sigma)(AB)_{1\sigma(1)}(AB)_{2\sigma(2)} \cdots (AB)_{n\sigma(n)}$$

$$= \sum_{\sigma \in \mathfrak{S}_n} \varepsilon(\sigma)\left(\sum_{k_1=1}^{n} A_{1k_1}B_{k_1\sigma(1)}\right)\left(\sum_{k_2=1}^{n} A_{2k_2}B_{k_2\sigma(2)}\right)\cdots\left(\sum_{k_n=1}^{n} A_{nk_n}B_{k_n\sigma(n)}\right)$$

$$= \sum_{k_1=1}^{n}\sum_{k_2=1}^{n}\cdots\sum_{k_n=1}^{n} A_{1k_1}A_{2k_2}\cdots A_{nk_n}\sum_{\sigma \in \mathfrak{S}_n} \varepsilon(\sigma)B_{k_1\sigma(1)}B_{k_2\sigma(2)}\cdots B_{k_n\sigma(n)}$$

$$= \sum_{\tau \in \mathfrak{S}_n} A_{1\tau(1)}A_{2\tau(2)}\cdots A_{n\tau(n)}\sum_{\sigma \in \mathfrak{S}_n} \varepsilon(\sigma)B_{\tau(1)\sigma(1)}B_{\tau(2)\sigma(2)}\cdots B_{\tau(n)\sigma(n)}$$

$$= \sum_{\tau \in \mathfrak{S}_n} A_{1\tau(1)}A_{2\tau(2)}\cdots A_{n\tau(n)}\sum_{\sigma \in \mathfrak{S}_n} \varepsilon(\sigma)B_{1\sigma\tau^{-1}(1)}B_{2\sigma\tau^{-1}(2)}\cdots B_{n\sigma\tau^{-1}(n)}$$

$$= \sum_{\tau \in \mathfrak{S}_n} \varepsilon(\tau)A_{1\tau(1)}A_{2\tau(2)}\cdots A_{n\tau(n)}\sum_{\sigma\tau^{-1} \in \mathfrak{S}_n} \varepsilon(\sigma\tau^{-1})B_{1\sigma\tau^{-1}(1)}B_{2\sigma\tau^{-1}(2)}\cdots B_{n\sigma\tau^{-1}(n)}$$

$= \det A \det B$. □

To determine the effect of the elementary change of basis matrices upon the determinant of a matrix we first calculate the determinant of each of the three elementary change of basis matrices (see Section 7.4 for notation).

Theorem. *Let R be a commutative unitary ring. Then*

I. $\det E(p, q) = -v$

II. $\det E(r, q; p) = v, r \in R$

III. $\det E(s; p) = s$, *s a unit in R.*

Proof

Type I. Det $E(p, q) = \sum_{\sigma \in \mathfrak{S}_n} \varepsilon(\sigma)E_{1\sigma(1)}E_{2\sigma(2)}\cdots E_{p\sigma(p)}\cdots E_{q\sigma(q)}\cdots E_{n\sigma(n)}$. In this sum all terms are zero save for that one permutation σ such that $\sigma(1) = 1, \sigma(2) = 2, \ldots, \sigma(p) = q, \ldots, \sigma(q) = p, \ldots, \sigma(n) = n$. This permutation is a transposition and has sign -1. Det $E(p, q) = \varepsilon(\tau)v \cdot v \cdots v = -v$.

Type II. Det $E(r,q;p) = \sum_{\sigma \in \mathfrak{S}_n} \varepsilon(\sigma)E_{1\sigma(1)}E_{2\sigma(2)}\cdots E_{p\sigma(p)}\cdots E_{q\sigma(q)}\cdots E_{n\sigma(n)}$. In this sum all terms are zero except for those arising from some permutation such that $\sigma(1) = 1$, $\sigma(2) = 2, \ldots, \sigma(p) = p, \ldots, \sigma(q) = q, \ldots, \sigma(n) = n$ or $\sigma(1) = 1, \sigma(2) = 2, \ldots, \sigma(p) = q, \ldots, \sigma(n) = n$. The latter choice fails to be a permutation (is not actually a term) and the first is the identity.

$$\text{Det } E(r, q; p) = \varepsilon(I)v \cdot v \cdots v = v.$$

Type III. Det $E(s; p) = \sum_{\sigma \in \mathfrak{S}_n} \varepsilon(\sigma)E_{1\sigma(1)}E_{2\sigma(2)}\cdots E_{p\sigma(p)}\cdots E_{n\sigma(n)}$. In this sum all terms are zero except that arising from the identity permutation. Det $E(s; p) = \varepsilon(I)v \cdot v \cdots s \cdots v = s$. For all three types to see the proof it is helpful to write out fully the elementary matrix in block form. □

The coupling of the product theorem with the results just computed yields a corollary.

Corollary. *Let R be a commutative unitary ring. Let $r \in R$ and s be a unit in R. Then*

I. $\det E(p, q)A = \det AE(p, q) = -\det A$
II. $\det E(r, q; p)A = \det AE(r, q; p) = \det A$
III. $\det E(s; p)A = \det AE(s; p) = s \det A.$

In words, performing an elementary row (or column) operation of type I to a matrix changes the determinant by a factor $-\nu$. A type II operation leaves the determinant unchanged, and a type III operation multiplies the determinant by the factor involved.

EXAMPLE

$$\operatorname{Det}\begin{pmatrix}7 & 1 & 2\\ 2 & 4 & 6\\ 5 & -3 & 1\end{pmatrix} = -\det\begin{pmatrix}1 & 7 & 2\\ 4 & 2 & 6\\ -3 & 5 & 1\end{pmatrix} = 2\det\begin{pmatrix}1 & 7 & 2\\ 2 & 1 & 3\\ -3 & 5 & 1\end{pmatrix}$$

$$= 2\det\begin{pmatrix}1 & 7 & 2\\ 0 & -13 & -1\\ -3 & 5 & 1\end{pmatrix} = 2\det\begin{pmatrix}1 & 7 & 2\\ 0 & -13 & -1\\ 0 & 26 & 7\end{pmatrix}$$

$$= 2\det\begin{pmatrix}1 & 0 & 0\\ 0 & -13 & -1\\ 0 & 26 & 7\end{pmatrix} = 2\det\begin{pmatrix}1 & 0 & 0\\ 0 & -13 & -1\\ 0 & 0 & 5\end{pmatrix}$$

$$= 2\det\begin{pmatrix}1 & 0 & 0\\ 0 & -13 & 0\\ 0 & 0 & 5\end{pmatrix} = (2)(-13)(5)\det\begin{pmatrix}1 & 0 & 0\\ 0 & 1 & 0\\ 0 & 0 & 1\end{pmatrix}$$

$$= -130.$$

Theorem. *Let R be a commutative unitary ring. Then the determinant is a linear function of any row or column of a matrix. Typically, row p,*

$$\det\begin{pmatrix}A_{11} & A_{12} & \cdots & A_{1n}\\ \cdots & & & \\ rA_{p1} + sA'_{p1} & rA_{p2} + sA'_{p2} & \cdots & rA_{pn} + sA'_{pn}\\ \cdots & & & \\ A_{n1} & A_{n2} & \cdots & A_{nn}\end{pmatrix}$$

$$= r\det\begin{pmatrix}A_{11} & A_{12} & \cdots & A_{1n}\\ \cdots & & & \\ A_{p1} & A_{p2} & \cdots & A_{pn}\\ \cdots & & & \\ A_{n1} & A_{n2} & \cdots & A_{nn}\end{pmatrix} + s\det\begin{pmatrix}A_{11} & A_{12} & \cdots & A_{1n}\\ \cdots & & & \\ A'_{p1} & A'_{p2} & \cdots & A'_{pn}\\ \cdots & & & \\ A_{n1} & A_{n2} & \cdots & A_{nn}\end{pmatrix}.$$

PROOF. Because of our ability to interchange rows, interchange columns, transpose the matrix, it is sufficient to prove this result for row number one.

$$\begin{aligned}\sum_{\sigma\in\mathfrak{S}_n} \varepsilon(\sigma)(rA + sA')_{1\sigma(1)}A_{2\sigma(2)}\cdots A_{n\sigma(n)} \\ = \sum_{\sigma\in\mathfrak{S}_n} \varepsilon(\sigma)(rA)_{1\sigma(1)}A_{2\sigma(2)}\cdots A_{n\sigma(n)} + \sum_{\sigma\in\mathfrak{S}_n} \varepsilon(\sigma)(sA')_{1\sigma(1)}A_{2\sigma(2)}\cdots A_{n\sigma(n)} \\ = r\sum_{\sigma\in\mathfrak{S}_n} \varepsilon(\sigma)A_{1\sigma(1)}A_{2\sigma(2)}\cdots A_{n\sigma(n)} + s\sum_{\sigma\in\mathfrak{S}_n} \varepsilon(\sigma)A'_{1\sigma(1)}A_{2\sigma(2)}\cdots A_{n\sigma(n)}.\end{aligned}$$ □

In any treatment of determinants one finds evaluation of determinants by expansion of row or column. By this means one can express a determinant of an n by n matrix as a sum of determinants of $n-1$ by $n-1$ matrices. We first define these $n-1$ by $n-1$ matrices, or at least their determinants.

Definition. Let R be a commutative unitary ring. Let A be an n by n matrix with entries in R. For each i and j in $\{1, 2, \ldots, n\}$ the following is called a *cofactor* of A:

$$\hat{A}_{ij} = (-1)^{i+j}\det\begin{pmatrix} A_{11} & A_{12} & \cdots & A_{1j-1} & A_{1j+1} & \cdots & A_{1n} \\ A_{21} & A_{22} & & A_{2j-1} & A_{2j+1} & & A_{2n} \\ \cdots & & & & & & \\ A_{i-1\,1} & A_{i-1\,2} & & A_{i-1\,j-1} & A_{i-1\,j+1} & & A_{i-1\,n} \\ A_{i+1\,1} & A_{i+1\,2} & & A_{i+1\,j-1} & A_{i+1\,j+1} & & A_{i+1\,n} \\ \cdots & \cdots & & & & & \\ A_{n1} & A_{n2} & & A_{nj-1} & A_{nj+1} & & A_{nn} \end{pmatrix}.$$

We note that the cofactor is obtained from A by omitting the ith row and the jth column of A, taking the determinant of the resulting $n-1$ by $n-1$ matrix, and then affixing a sign. Without the affixed sign the determinant is called a *minor*.

Theorem. *Let R be a commutative unitary ring. Let A be an n by n matrix with entries in R. Then*

$$\begin{aligned} \det A &= A_{i1}\hat{A}_{i1} + A_{i2}\hat{A}_{i2} + \cdots + A_{in}\hat{A}_{in} \quad \textit{for any row } i \\ \det A &= A_{1j}\hat{A}_{1j} + A_{2j}\hat{A}_{2j} + \cdots + A_{nj}\hat{A}_{nj} \quad \textit{for any column } j. \end{aligned}$$

PROOF. Because of our ability to interchange rows, to interchange columns and to take the transpose it is sufficient to prove the theorem for an expansion by the first row.

$$\begin{aligned}
\operatorname{Det} A &= \sum_{\sigma\in\mathfrak{S}_n} \varepsilon(\sigma)A_{1\sigma(1)}A_{2\sigma(2)}\cdots A_{n\sigma(n)}\\
&= A_{11}\sum_{\substack{\sigma\in\mathfrak{S}_n\\ \sigma(1)=1}} \varepsilon(\sigma)A_{2\sigma(2)}\cdots A_{n\sigma(n)}\\
&\quad + A_{12}\sum_{\substack{\sigma\in\mathfrak{S}_n\\ \sigma(1)=2}} \varepsilon(\sigma)A_{2\sigma(2)}\cdots A_{n\sigma(n)} + \cdots\\
&\quad + A_{1n}\sum_{\substack{\sigma\in\mathfrak{S}_n\\ \sigma(1)=n}} \varepsilon(\sigma)A_{2\sigma(2)}\cdots A_{n\sigma(n)}\\
&= A_{11}\sum_{\tau\in\mathfrak{S}(2,\ldots,n)} \varepsilon(\tau)A_{2\tau(2)}A_{3\tau(3)}\cdots A_{n\tau(n)}\\
&\quad + (-1)A_{12}\sum_{\tau\in\mathfrak{S}(1,3,\ldots,n)} \varepsilon(\tau)A_{2\tau(2)}A_{3\tau(3)}\cdots A_{n\tau(n)}\\
&\quad + (-1)^2A_{13}\sum_{\tau\in\mathfrak{S}(1,2,4,\ldots,n)} \varepsilon(\tau)A_{2\tau(2)}A_{3\tau(3)}\cdots A_{n\tau(n)}\\
&\quad \cdots\\
&\quad + (-1)^{n-1}A_{1n}\sum_{\tau\in\mathfrak{S}(1,2,\ldots,n-1)} \varepsilon(\tau)A_{2\tau(2)}A_{3\tau(3)}\cdots A_{n\tau(n)}\\
&= A_{11}\hat{A}_{11} + A_{12}\hat{A}_{12} + \cdots + A_{1n}\hat{A}_{1n}. \qquad\square
\end{aligned}$$

EXAMPLES. An expansion by row number 1.

$$\operatorname{Det}\begin{pmatrix}3 & -1 & 4\\ 2 & 3 & 1\\ 7 & 1 & -2\end{pmatrix} = 3\det\begin{pmatrix}3 & 1\\ 1 & -2\end{pmatrix} - (-1)\det\begin{pmatrix}2 & 1\\ 7 & -2\end{pmatrix} + 4\det\begin{pmatrix}2 & 3\\ 7 & 1\end{pmatrix}$$
$$= -108.$$

An expansion of the same determinant by column number 2:

$$\begin{aligned}
\operatorname{Det}\begin{pmatrix}3 & -1 & 4\\ 2 & 3 & 1\\ 7 & 1 & -2\end{pmatrix} &= -(-1)\det\begin{pmatrix}2 & 1\\ 7 & -2\end{pmatrix} + (3)\det\begin{pmatrix}3 & 4\\ 7 & -2\end{pmatrix}\\
&\quad - (1)\det\begin{pmatrix}3 & 4\\ 2 & 1\end{pmatrix}\\
&= [(2)(-2) - (7)(1)] + 3[(3)(-2) - (7)(4)]\\
&\quad - [(3)(1) - (2)(4)]\\
&= -108.
\end{aligned}$$

Another example:

$$\begin{aligned}
\operatorname{Det}\begin{pmatrix}a & b & 0\\ 0 & a & b\\ b & 0 & a\end{pmatrix} &= \det\begin{pmatrix}a+b & b+a & b+a\\ 0 & a & b\\ b & 0 & a\end{pmatrix} = (a+b)\det\begin{pmatrix}1 & 1 & 1\\ 0 & a & b\\ b & 0 & a\end{pmatrix}\\
&= (a+b)\det\begin{pmatrix}1 & 0 & 0\\ 0 & a & b\\ b & -b & a-b\end{pmatrix}\\
&= (a+b)\det\begin{pmatrix}a & b\\ -b & a-b\end{pmatrix}\\
&= (a+b)(a^2 - ab + b^2).
\end{aligned}$$

In order to produce a relationship between determinants and the inverse of a matrix we now define the adjoint of a matrix.

Definition. Let B be a commutative unitary ring and let A be an n by n matrix with entries in R. We define the *adjoint* of A,

$$\text{adj } A = (\hat{A}_{ij} | i \in n, j \in n)^*.$$

In words, the adjoint of A is the transpose of the matrix of cofactors of A.

EXAMPLE. Let A be the matrix

$$\begin{pmatrix} 2 & -7 & 1 \\ 4 & 3 & 2 \\ -1 & 2 & 1 \end{pmatrix}.$$

$$\hat{A}_{11} = \begin{vmatrix} 3 & 2 \\ 2 & 1 \end{vmatrix} = -1. \quad \hat{A}_{12} = -\begin{vmatrix} 4 & 2 \\ -1 & 1 \end{vmatrix} = -6. \quad \hat{A}_{13} = \begin{vmatrix} 4 & 3 \\ -1 & 2 \end{vmatrix} = 11.$$

$$\hat{A}_{21} = -\begin{vmatrix} -7 & 1 \\ 2 & 1 \end{vmatrix}. \quad \hat{A}_{22} = \begin{vmatrix} 2 & 1 \\ -1 & 1 \end{vmatrix} = 3. \quad \hat{A}_{23} = -\begin{vmatrix} 2 & -7 \\ -1 & 2 \end{vmatrix} = 3.$$

$$\hat{A}_{31} = \begin{vmatrix} -7 & 1 \\ 3 & 2 \end{vmatrix} = -17. \quad \hat{A}_{32} = -\begin{vmatrix} 2 & 1 \\ 4 & 2 \end{vmatrix} = 0. \quad \hat{A}_{33} = \begin{vmatrix} 2 & -7 \\ 4 & 3 \end{vmatrix} = 34.$$

$$\text{Adj } A = \begin{pmatrix} -1 & -6 & 11 \\ 9 & 3 & 3 \\ -17 & 0 & 34 \end{pmatrix}^* = \begin{pmatrix} -1 & 9 & -17 \\ -6 & 3 & 0 \\ 11 & 3 & 34 \end{pmatrix}.$$

The product of this adjoint with the original matrix A encourages testing the general case.

$$\begin{pmatrix} 2 & -7 & 1 \\ 4 & 3 & 2 \\ -1 & 2 & 1 \end{pmatrix} \begin{pmatrix} -1 & 9 & -17 \\ -6 & 3 & 0 \\ 11 & 3 & 34 \end{pmatrix} = \begin{pmatrix} 51 & 0 & 0 \\ 0 & 51 & 0 \\ 0 & 0 & 51 \end{pmatrix}.$$

Theorem. *Let R be a commutative unitary ring. Let A be an n by n matrix with entries in R. Then A (adj A) = (adj A) A = (det A) δ.*

PROOF. In the product of A with the adjoint of A the entry in row i and column j is produced using row i of A and column j of adjoint A: $A_{i1}\hat{A}_{j1} + A_{i2}\hat{A}_{j2} + A_{in}\hat{A}_{jn}$. This sum, when $i = j$, is simply an expansion of det A by row number i. Alternatively, when $i \neq j$, the sum is an expansion of a determinant of a matrix different from A, one in which the jth row of A is replaced by the ith row. The value is then the determinant of a matrix with two rows alike, the ith row of A, and must therefore be zero. The product reversed has a similar proof. □

We can now calculate directly the inverse of a matrix in terms of determinants.

Theorem. *Let R be a commutative unitary ring. Let A be an n by n matrix with entries in R. Then A has an inverse if and only if* $\det A$ *is a unit in R. In case A^{-1} exists, $A^{-1} = (1/\det A)$* adj *A.*

PROOF. Suppose first that A has an inverse. $AA^{-1} = \delta$, the identity matrix. $\operatorname{Det}(AA^{-1}) = \det \delta$. Det $A \det A^{-1} = 1$. Both $\det A$ and $\det A^{-1}$ are members of R and therefore must be multiplicative inverses in R. Det A is a unit of R.

Now for the converse assume $\det A$ is a unit of R. $(\det A)^{-1}$ is also in R. $(A)[(\det A)^{-1} \operatorname{adj} A] = (\det A)^{-1} A \operatorname{adj} A = (\det A)^{-1}(\det A)\delta = \delta$. So also does $(\det A)^{-1} \operatorname{adj} A$ left multiply A to give δ and is therefore the inverse of A. □

If the given ring R is a field, then the only nonunit of R is θ. The condition that $\det A$ be a unit then simply means $\det A \neq \theta$.

EXAMPLE. For the matrix

$$\begin{pmatrix} 2 & -7 & 1 \\ 4 & 3 & 2 \\ -1 & 2 & 1 \end{pmatrix}$$

we previously used in computation

$$(\det A)^{-1} \operatorname{adj} A = \tfrac{1}{51}\begin{pmatrix} -1 & 9 & -17 \\ -6 & 3 & 0 \\ 11 & 3 & 34 \end{pmatrix} = \begin{pmatrix} -\frac{1}{51} & \frac{9}{51} & -\frac{17}{51} \\ -\frac{6}{51} & \frac{3}{51} & 0 \\ \frac{11}{51} & \frac{3}{51} & \frac{34}{51} \end{pmatrix}.$$

We now prove Cramer's Rule for solution of n equations in n unknowns by use of determinants. The proof given is interesting and simple.

Theorem. *Let A be an n by n matrix with entries in a commutative unitary ring R and let* $\det A$ *be a unit in R. Then the equations*

$$\begin{aligned} A_{11}X_1 + A_{12}X_2 + \cdots + A_{1n}X_n &= Y_1 \\ A_{21}X_1 + A_{22}X_2 + \cdots + A_{2n}X_n &= Y_2 \\ \cdots & \\ A_{n1}X_1 + A_{n2}X_2 + \cdots + A_{nn}X_n &= Y_n \end{aligned}$$

have unique solutions

$$X_1 = (1/\det A)\begin{vmatrix} Y_1 & A_{12} & \cdots & A_{1n} \\ Y_2 & A_{22} & \cdots & A_{2n} \\ \cdots & & & \\ Y_n & A_{n2} & \cdots & A_{nn} \end{vmatrix},$$

$$X_2 = (1/\det A)\begin{vmatrix} A_{11} & Y_1 & A_{13} & \cdots & A_{1n} \\ A_{21} & Y_2 & A_{23} & \cdots & A_{2n} \\ \cdots & & & & \\ A_{n1} & Y_n & A_{n3} & \cdots & A_{nn} \end{vmatrix}, \ldots,$$

$$X_n = (1/\det A)\begin{vmatrix} A_{11} & A_{12} & \cdots & A_{1\,n-1} & Y_1 \\ A_{21} & A_{22} & \cdots & A_{2\,n-1} & Y_2 \\ \cdots \\ A_{n1} & A_{n2} & \cdots & A_{n\,n-1} & Y_n \end{vmatrix}.$$

PROOF. Beginning with the equation $AX = Y$ with the usual matrix meaning we multiply on the left with adj A. (adj $A)AX =$ (adj $A)Y$. This yields (det A) $\delta X =$ (adj $A)Y$. Using the fact that det A is a unit we have the solution $X = (1/\det A)(\text{adj } A)Y$. (Adj $A)Y =$

$$\begin{pmatrix} \hat{A}_{11} & \hat{A}_{21} & \cdots & \hat{A}_{n1} \\ \hat{A}_{12} & \hat{A}_{22} & \cdots & \hat{A}_{n2} \\ \cdots & & & \cdots \\ \hat{A}_{1n} & \hat{A}_{2n} & \cdots & \hat{A}_{nn} \end{pmatrix}\begin{pmatrix} Y_1 \\ Y_2 \\ \\ Y_n \end{pmatrix} = \begin{pmatrix} \hat{A}_{11}Y_1 + \hat{A}_{21}Y_2 + \cdots + \hat{A}_{n1}Y_n \\ \hat{A}_{12}Y_1 + \hat{A}_{22}Y_2 + \cdots + \hat{A}_{n2}Y_n \\ \cdots \\ \hat{A}_{1n}Y_1 + \hat{A}_{2n}Y_2 + \cdots + \hat{A}_{nn}Y_n \end{pmatrix}.$$

Each entry in the column matrix is the numerator of the fraction given in the conclusion of the theorem. □

EXAMPLE. The standard solution in terms of determinants for two equations in two unknowns is given by the theorem. The equations

$$\begin{aligned} A_{11}X_1 + A_{12}X_2 &= Y_1 \\ A_{21}X_1 + A_{22}X_2 &= Y_2 \end{aligned}$$

have solutions

$$X_1 = \frac{\begin{vmatrix} Y_1 & A_{12} \\ Y_2 & A_{22} \end{vmatrix}}{\begin{vmatrix} A_{11} & A_{12} \\ A_{21} & A_{22} \end{vmatrix}} \qquad X_2 = \frac{\begin{vmatrix} A_{11} & Y_1 \\ A_{21} & Y_2 \end{vmatrix}}{\begin{vmatrix} A_{11} & A_{12} \\ A_{21} & A_{22} \end{vmatrix}}$$

whenever the determinant in the denominator is a unit. The equations

$$\begin{aligned} A_{11}X_1 + A_{12}X_2 + A_{13}X_3 &= Y_1 \\ A_{21}X_1 + A_{22}X_2 + A_{23}X_3 &= Y_2 \\ A_{31}X_1 + A_{32}X_2 + A_{33}X_3 &= Y_3 \end{aligned}$$

have the solution

$$X_1 = \frac{\begin{vmatrix} Y_1 & A_{12} & A_{13} \\ Y_2 & A_{22} & A_{23} \\ Y_3 & A_{32} & A_{33} \end{vmatrix}}{\begin{vmatrix} A_{11} & A_{12} & A_{13} \\ A_{21} & A_{22} & A_{23} \\ A_{31} & A_{32} & A_{33} \end{vmatrix}} \qquad X_2 = \frac{\begin{vmatrix} A_{11} & Y_1 & A_{13} \\ A_{21} & Y_2 & A_{23} \\ A_{31} & Y_3 & A_{33} \end{vmatrix}}{\begin{vmatrix} A_{11} & A_{12} & A_{13} \\ A_{21} & A_{22} & A_{23} \\ A_{31} & A_{32} & A_{33} \end{vmatrix}}$$

$$X_3 = \frac{\begin{vmatrix} A_{11} & A_{12} & Y_1 \\ A_{21} & A_{22} & Y_2 \\ A_{31} & A_{32} & Y_3 \end{vmatrix}}{\begin{vmatrix} A_{11} & A_{12} & A_{13} \\ A_{21} & A_{22} & A_{23} \\ A_{31} & A_{32} & A_{33} \end{vmatrix}}$$

when the determinant in the denominator is a unit.

We now connect to a modest extent (not nearly to the extent that is possible) some of our results on determinants with our module theory. Our n by n matrices arise, of course, also as matrices of endomorphisms of finite dimensional modules over commutative unitary rings. The determinant of the matrix is computed from the entries of the matrices. From each matrix of a given endomorphism can be computed a separate determinant. Our next theorem shows that regardless of which basis one chooses for the module, which matrix one chooses to represent the endomorphism, the determinant remains the same.

Theorem. *Let M be a finite dimensional free module over a commutative unitary ring R. Let A and B be any two matrices representing the endomorphism f of M. Then* $\det A = \det B$.

PROOF. The two matrices A and B are matrices of the endomorphism f with respect to two finite bases. The matrices are related by means of the formula $B = PAP^{-1}$ (cf. Section 7.4). Det $B = \det PAP^{-1} = \det P \det A \det P^{-1} = \det A \det P \det P^{-1} = \det A \det PP^{-1} = \det A$. □

Definition. If $f \in \mathscr{E}(M)$, the module of endomorphisms of a finite dimensional module M over a commutative unitary ring R, we define $\det f = \det A$, where A is any matrix of the endomorphism f.

Some quick results follow from connecting results on determinants to the definition.

Theorem. *Let M be a finite dimensional module over a commutative unitary ring R. Then* $\det : \mathscr{E}(M) \to R$ *preserves the composition of $\langle \mathscr{E}(M), \circ, I \rangle$ in the multiplication of R. f is an automorphism of M if and only if* $\det f$ *is a unit of R.*

PROOF. Det$(g \circ f) = \det \mu(g \circ f) = \det(\mu(g)\mu(f)) = \det \mu(g) \det \mu(f) = \det g \det f$. Det $\mu(f)$ is a unit if and only if $\mu(f)$ is invertible if and only if f is an automorphism. □

We remark again that if the ring is a field every nonzero element in the field is a unit. Thus if M is a vector space an endomorphism f is an automorphism if and only if $\det f \neq \theta$.

We now characterize matrices over fields which have possibly lesser rank.

Theorem. *Let K be a field and A an m by n (not necessarily square) matrix with entries in K.*

1. *If there exists a k by k submatrix of A with determinant value nonzero then* rank $A \geqslant k$.
2. *If all k by k submatrices of A have determinant values zero then* rank $A < k$.

PROOF. Assume first there exists a k by k submatrix of A with nonzero determinant. Using left and right multiplication on A by change of basis (invertible) matrices we can produce a matrix in which the k by k submatrix with nonzero determinant appears in rows $1, 2, \ldots, k$ and columns $1, 2, \ldots, k$. The rank of the resultant matrix is the same as the original matrix. The first k columns are linearly independent because if they were dependent then so also would be the first k columns terminating with k rows making the k by k determinant zero. This would contradict our hypothesis. Knowing that the first k columns are linearly independent tells us the rank of the matrix is at least k. So also has A rank at least k. To prove part 2 of the theorem we prove this equivalent statement: if the rank of A is at least k then there exists some k by k submatrix with determinant value nonzero. We begin with A having rank at least k. There exist k (at least) linearly independent columns of A. These k linearly independent columns make an m by k submatrix of A. k is certainly less than or equal to m since the rank of a matrix cannot exceed the number of rows or the number of columns. In this m by k submatrix of rank k there must exist k linearly independent rows. Extracting these k linearly independent rows we have a k by k submatrix of A with rank k. This k by k submatrix with rank k has nonzero determinant. □

EXAMPLE. The 3 by 4 matrix

$$\begin{pmatrix} 1 & 1 & 3 & 2 \\ 5 & 3 & 7 & 4 \\ 3 & 2 & 5 & 3 \end{pmatrix}$$

with entries in $\mathbb{Q}$ must have rank at least 2 because the submatrix

$$\begin{pmatrix} 1 & 1 \\ 5 & 3 \end{pmatrix}$$

has determinant -2. On the other hand it can be computed that every 3 by 3 submatrix has determinant 0. The rank of the matrix is 2.

QUESTIONS

1. Which of these statements are true?
   (A) The composition of even permutations is even.
   (B) The composition of odd permutations is odd.

(C) A permutation and its inverse have the same sign.
(D) The identity mapping is an odd permutation.
(E) None of the statements is true.

2. Which of the alternatives complete a true sentence? Composition of permutations
(A) is commutative
(B) is associative
(C) has a neutral element
(D) is such that every permutation has an inverse permutation.
(E) None of the alternatives completes a true sentence.

3. Which of these sentences are true?
(A) A transposition is an even permutation.
(B) An even permutation is the product of an even number of transpositions.
(C) An odd permutation cannot be the product of transpositions.
(D) The identity function is a transposition.
(E) None of the sentences is true.

4. The permutation $\sigma \in \mathfrak{S}_4$ such that $\sigma(1) = 4, \sigma(2) = 3, \sigma(3) = 1, \sigma(4) = 2$ expressed in cyclic notation is
(A) (1 4)(2 3)(3 1)(4 2)
(B) (1 4 2 3)
(C) (1 3)(2 4)(1 2)
(D) (4 3 1 2).
(E) None of the cyclic expressions is correct.

5. Which of the following statements are true?
(A) $(i_1 i_2 \cdots i_k) = (i_2 i_3 \cdots i_{k-1} i_k i_1)$.
(B) $(i_1 i_2 \cdots i_k)(j_1 j_2 \cdots j_l) = (j_1 j_2 \cdots j_l)(i_1 i_2 \cdots i_k)$ if all symbols $i_1, i_2, \ldots, i_k, j_1, j_2, \ldots, j_l$ are distinct.
(C) $(i_1 i_2 \cdots i_k)^k = I$.
(D) $(1\ 2\ 3)(2\ 3\ 4) \cdots (n-2\ n-1\ n) = I$.
(E) None of the statements is true.

6. Given the equations

$$\begin{aligned} 2X_1 + X_2 + X_3 &= 1 \\ X_1 - X_2 + X_3 &= 2 \\ 3X_1 \qquad + 2X_3 &= 0 \end{aligned}$$

with entries in $\mathbb{Q}$ which of the following statements are true?
(A) There are no solutions.
(B) There is a solution and it is unique.
(C) There are an infinite number of solutions; the set of solutions is of dimension 1.
(D) There are an infinite number of solutions; the set of solutions is of dimension 2.
(E) None of the alternatives is correct.

7. Which of these statements are correct for a 3 by 3 matrix with entries in $\mathbb{Q}$?
(A) $2 \det A = \det 8A$.
(B) $8 \det A = \det 2A$.
(C) $\det A = \frac{1}{2} \operatorname{adj} A$.

(D) $9 \det A = A^{-1}$.
(E) None of the alternatives is correct.

8. Let

$$A = \begin{pmatrix} 1 & 0 & 2 \\ 2 & 3 & 1 \\ -1 & 4 & -1 \end{pmatrix}.$$

Which of these statements are not correct?
(A) $\hat{A}_{21} = -8$.
(B) Det $A = 15$.
(C) $\hat{A}_{32} = 3$.
(D) $\hat{A}_{12} = 0$.
(E) All of the statements are correct.

9. Which of the following determinants are not equal to

$$\begin{vmatrix} 1 & 2 & 3 \\ 1 & 2 & 1 \\ 2 & 0 & 6 \end{vmatrix}?$$

(A) $\begin{vmatrix} 1 & 2 & 3 \\ 0 & 0 & -2 \\ 2 & 0 & 6 \end{vmatrix}$ (B) $\begin{vmatrix} 2 & 1 & 3 \\ 2 & 1 & 1 \\ 0 & 2 & 6 \end{vmatrix}$

(C) $\begin{vmatrix} 2 & 1 & 3 \\ 2 & 1 & 1 \\ 4 & 0 & 6 \end{vmatrix}$ (D) $4\begin{vmatrix} 1 & 1 & 3 \\ 1 & 1 & 1 \\ 1 & 0 & 3 \end{vmatrix}$.

(E) All of the determinants are equal to the given one.

10. A square matrix A with entries in a field K has an inverse
(A) if and only if det $A = \theta$
(B) only if the matrix equation $AX = B$ has a solution
(C) if A adj $A = (\det A)\,\delta$
(D) if A has all nonzero cofactors.
(E) None of the alternatives is true.

11. Given the matrices A, X with entries in a field K and the equation $AX = -(A + A)X$ which line contains the first error?
(A) $AX + (A + A)X = \theta$.
(B) $3AX = \theta$.
(C) $AX = \theta$.
(D) $X = \theta$.
(E) There is no error.

12. Which of the following are incorrect?
(A) A determinant of a square matrix A is zero if two rows of A are alike.
(B) A determinant of a square matrix A is zero if one column contains all zeros.
(C) A 3 by 3 determinant must be zero if 6 entries are zero.
(D) A determinant of a matrix with integer entries never has a proper fraction for its value.
(E) All statements are correct.

13. Let A be an n by n matrix with entries in a field K.
    (A) $AX = \theta$ has the trivial solution $X = \theta$; i.e., $X_1 = \theta, X_2 = \theta, \ldots, X_n = \theta$.
    (B) $AX = \theta$ has only the trivial solution $X = \theta$ if $\det A = \theta$.
    (C) $AX = B$ has a solution if $\det A = \theta$.
    (D) The solutions of $AX = \theta$ are arbitrary if $\det A = \theta$.
  (E) None of the four statements is correct.

EXERCISES

1. Write each of these permutations as the composition of disjoint cycles.
   (a) σ such that $\sigma(1) = 3, \sigma(2) = 4, \sigma(3) = 1, \sigma(4) = 2$.
   (b) τ such that $\tau(1) = 4, \tau(2) = 3, \tau(3) = 1, \tau(4) = 2$.
   (c) $(2\ 1\ 3\ 4)(3\ 5\ 1\ 2)(1\ 2)(4\ 6\ 1)$.
   (d) $(6\ 1\ 2\ 3)(6\ 1\ 2\ 4)(3\ 4\ 2)(5\ 1\ 2)(3\ 1)$.

2. Compute each of these compositions; write each as a product of disjoint cycles. Observe each example for the information it contains.
   (a) $(1\ 2)(1\ 2)$.
   (b) $(1\ 2)(2\ 1)$.
   (c) $(1\ 2\ 3)(3\ 2\ 1)$.
   (d) $(1\ 2\ 3)(1\ 3\ 2)$.
   (e) $(1\ 2\ 3\ 4)(4\ 3\ 2\ 1)$.
   (f) $(1\ 2\ 3)(1\ 2\ 3)$.
   (g) $(1\ 2\ 3)(1\ 2\ 3)(1\ 2\ 3)$.
   (h) $(1\ 2\ 3\ 4)(1\ 2\ 3\ 4)(1\ 2\ 3\ 4)(1\ 2\ 3\ 4)$.
   (i) $(i_1 i_2 \cdots i_k)^k$.
   (j) $(i_1 i_2 \cdots i_k)(i_k i_{k-1} \cdots i_2 i_1)$.

3. Verify each of these compositions. Observe each example.
   (a) $(1\ 2)(1\ 7) = (1\ 7\ 2)$.
   (b) $(1\ 7)(1\ 2) = (1\ 2\ 7)$.
   (c) $(3\ 1\ 2\ 4) = (3\ 4)(3\ 2)(3\ 1)$.
   (d) $(1\ 7)(1\ 3) = (1\ 7)(1\ 2)(1\ 2)(1\ 3) = (1\ 2\ 7)(1\ 3\ 2)$.
   (e) $(1\ 3\ 2) = (1\ 2\ 3)(1\ 2\ 3)$.
   (f) $(5\ 6) = (1\ 5)(1\ 6)(1\ 5) = (1\ 6)(1\ 5)(1\ 6)$.

4. Show that every permutation is the composition of 2-cycles. If a permutation is even then the number of 2-cycles is even. If a permutation is odd then the number of 2-cycles is odd. Is $(i_1 i_2 \cdots i_{2k})$ an even permutation or an odd permutation? What of the sign of the permutation $(i_1 i_2 \cdots i_{2k} i_{2k+1})$?

5. Show that every permutation in $\mathfrak{S}_n$ can be written as a product of $(1\ 2), (1\ 3), \ldots, (1 n)$. Repetitions are allowed.

6. Show that every even permutation in $\mathfrak{S}_n$ can be written as a product of $(1\ 2\ 3)$, $(1\ 2\ 4), \ldots, (1\ 2\ n)$, repetitions allowed. [*Hint*: Exercise 3.]

7. Show that the product of any two members of the subset $\{I, (1\ 2\ 3), (1\ 3\ 2)\}$ of $\mathfrak{S}_3$ is again a member of the subset.

8. Show that the product of any two members of the subset $\{I, (1\ 2)(3\ 4), (1\ 3)(2\ 4), (1\ 4)(2\ 3)\}$ of $\mathfrak{S}_4$ is again a member of the subset.

9. Fifteen men sit in a row and wish to reverse the order in which they sit. Can they do this switching seats in pairs? Can they do this switching seats in triples (3 at a time)?

10. Show that the matrix

$$\begin{pmatrix} 1 & a & a^2 \\ 1 & b & b^2 \\ 1 & c & c^2 \end{pmatrix}$$

has determinant $(a - b)(b - c)(c - a)$ by using the theorems on determinants of elementary change of basis matrices and $\det \delta = 1$.

11. For what values of a in $\mathbb{R}$ will

$$\begin{aligned} aX_1 \qquad\quad + aX_3 &= 4 \\ X_2 + 2X_3 &= -2 \\ aX_1 + 6X_2 + 3X_3 &= 1 \end{aligned}$$

have a unique solution?

12. Show that if the rows (or columns) of an n by n matrix are linearly dependent then the determinant is zero.

13. Show that if two rows (or columns) of an n by n matrix are alike then the determinant is zero.

14. What is the rank of the matrix

$$\begin{pmatrix} 2 & 0 & 1 & 2 \\ 1 & 2 & 1 & 4 \\ 0 & 4 & 2 & 8 \\ 3 & 6 & 4 & 14 \end{pmatrix}?$$

Is the family $((2, 0, 1, 2), (1, 2, 1, 4), (0, 4, 2, 8), (3, 6, 4, 14))$ linearly dependent in $\mathbb{R}^4$? What is the adjoint of the given matrix? What is the product of the given matrix with its adjoint? Does the matrix have an inverse?

15. Are there matrices A such that

$$\begin{pmatrix} 3 & 4 \\ 1 & -1 \end{pmatrix} A = \begin{pmatrix} 7 & -1 \\ 12 & 13 \end{pmatrix}?$$

If so, what are they?

16. Express the following system as a matrix equation and solve by multiplying by the matrix inverse.

$$\begin{aligned} X_1 + 2X_3 &= 1 \\ 3X_2 &= 0 \\ -2X_1 - \quad X_3 &= 0. \end{aligned}$$

17. Show that if $\operatorname{rank} (b_1 - a_1 \; b_2 - a_2) = 1$ then

$$\det \begin{pmatrix} X_1 & X_2 & 1 \\ a_1 & a_2 & 1 \\ b_1 & b_2 & 1 \end{pmatrix} = 0$$

is the equation of a line in $\mathbb{R}^2$ containing the noncollinear points (a_1, a_2) and (b_1, b_2).

18. Show that if

$$\operatorname{rank}\begin{pmatrix} b_1 - a_1 & b_2 - a_2 & b_3 - a_3 \\ c_1 - a_1 & c_2 - a_2 & c_3 - a_3 \end{pmatrix} = 2$$

then

$$\det\begin{pmatrix} X_1 & X_2 & X_3 & 1 \\ a_1 & a_2 & a_3 & 1 \\ b_1 & b_2 & b_3 & 1 \\ c_1 & c_2 & c_3 & 1 \end{pmatrix} = 0$$

is the equation of a plane in $\mathbb{R}^3$ containing the noncollinear points (a_1, a_2, a_3), (b_1, b_2, b_3), (c_1, c_2, c_3).

19. Let

$$A = \begin{pmatrix} 1 & 2 & 1 \\ 3 & 7 & 4 \\ 2 & -1 & 3 \end{pmatrix}.$$

Find A^{-1} by solving the equation $BA = \delta$ for B using linear equations and the row-reduced echelon matrix. Compare with the adjoint computation.

20. Show that

$$\det\begin{pmatrix} a & b & c & d \\ d & a & b & c \\ c & d & a & b \\ b & c & d & a \end{pmatrix}$$

has a factor $a + b + c + d$.

21. Let $A(x)$ be an n by n square matrix in which each entry $A_{ij}(x)$ is a real-valued differentiable function defined on the real interval $[a, b]$. Denote by $\tilde{A}^k$ the matrix obtained from A by replacing the kth row of functions $A_{k1}A_{k2} \cdots A_{kn}$ by their derivatives $DA_{k1}DA_{k2} \cdots DA_{kn}$. Show that $D \det A(x) = \sum_{k=1}^{n} \det \tilde{A}^k(x)$.

Abstract systems 8

This chapter is more abstract than the other chapters in this book because it discusses algebraic systems in general and not just particular systems such as vector space or ring. The aim of this chapter on structure is to reveal some of the organization given to the algebraic concepts we have been implicitly using in the rest of the book. It formalizes the material of the other chapters and is the organizational spirit of this book. Despite the fact that it is here that the organization is laid bare we do not feel a study of this chapter is necessary for a productive use of this text. Analogously, one can become a very good mathematician without being a professional logician even though logical thought is central to mathematics.

The general study of algebraic systems lies in a branch of mathematics called universal algebra. We will touch in this chapter on only *some* of the fundamentals of the subject. Our purpose is not a study of universal algebra per se but rather to use some of its elementary principles as a basis on which to organize our study of elementary algebra.

For intuitive purposes we split our study of algebraic systems into operational systems and relational systems. We study independence (so important in modules) and closure classes. We define morphisms in terms of the operations that they preserve and not in terms of the axioms satisfied by the domain or codomain. Some concise formulations are given as necessary and sufficient conditions for a function to be a morphism. We give a presentation of quotient systems that is independent of normal subsystem and then in turn define normal subsystem from quotient system. We speak of kernels and then close with products and sums.

8.1 Algebraic systems

In this section we lay down the definitions of algebraic systems, operational and relational, and give examples of algebraic systems.

The common property of systems such as groups, rings, vector spaces, and natural numbers is a set with operations. We will call a set with operations an algebraic system. Because we wish also to be able to speak of ordered integral domains and the like, we include the possibility of having relations given on the set as well as operations.

Definition. An *algebraic system* $\langle M, (\beta_i | i \in I); (\sigma_j | j \in J)\rangle$ is a set M together with a family of operations $(\beta_i | i \in I)$ and a family of relations $(\sigma_j | j \in J)$ on the set M. The set M together with the family of operations is called an *operational system* while the set M together with the family of relations is called a *relational system.*

Definitions of operations, relations, and closure can be found in Section 2.1. For definiteness we repeat some definitions here. An n-ary operation β on the set M is a function $\beta: M^n \to M$. This includes the special case of a nullary operation $v: \{0\} \to M$. We speak of an n-ary operation as having *size n*. A subset S of M is closed under the n-ary operation β if $(x_1, x_2, \ldots, x_n) \in S^n$ implies $\beta(x_1, x_2, \ldots, x_n) \in S$. An n-ary relation on M is a subset of M^n. Orders and equivalence relations on M are binary relations on M. An n-ary relation has size n.

Definition. The *type* of an algebraic system is a listing of the sizes of the operations and relations on the set. Algebraic systems of the same type are called *similar*.

In the type we shall count the set itself as having size 1; this is reasonable because the set can be thought of as the identity operation, a unary operation.

We now list examples of algebraic systems. Most of these systems we have already studied.

Examples

1. A group $\langle G, \beta, v\rangle$ is a set M, a binary operation β, a nullary operation v such that β is associative, v is a neutral element for β, and for each x in G there exists $y \in G$ such that $\beta(x, y) = \beta(y, x) = v$. The group has type $(1, 2, 0)$.
2. An ordered group $\langle G, \beta, v; \sigma\rangle$ is a set M, a binary operation β, a nullary operation v, a binary relation σ such that $\langle G, \beta, v\rangle$ is a group and σ is an order compatible with the binary operation β, ($(a, b) \in \sigma$ implies $(\beta(a, x), \beta(b, x)) \in \sigma$ and $(\beta(x, a), \beta(x, b)) \in \sigma$ for all a, b, x in G). The ordered group has type $(1, 2, 0; 2)$.
3. A monoid $\langle M, \beta\rangle$ is a set M, a binary operation β which is associative. The monoid has type $(1, 2)$.

4. A unitary monoid $\langle M, \beta, \nu \rangle$ is a set M, a binary operation β, a nullary operation ν such that β is associative and ν is the neutral element for β. The unitary monoid has type (1, 2, 0).
5. A cancellative unitary monoid $\langle M, \beta, \nu \rangle$ is a set M, a binary operation β, a nullary operation ν such that $\langle M, \beta, \nu \rangle$ is a unitary monoid and $\beta(x, a) = \beta(x, b)$ or $\beta(a, x) = \beta(b, x)$ implies $a = b$. The cancellative unitary monoid has type (1, 2, 0).

It should be noted that unitary monoids, cancellative unitary monoids, and groups are algebraic systems with the same type (1, 2, 0). There are many possible algebraic systems with the type (1, 2, 0). Only those satisfying the axioms for a group are groups. We will also take note of the fact that there are different equivalent axiomatizations of a group. Several alternatives to the one given are

1B. A group $\langle G, \beta \rangle$ is a set G, a binary operation β such that β is associative and the equations $\beta(a, x) = b$ and $\beta(y, c) = d$ are always uniquely solvable for x and y in G. The type here for the group is (1, 2).

1C. A group $\langle G, \beta, \nu, \gamma \rangle$ is a set G, a binary operation β, a nullary operation ν, a unary operation γ such that β is associative, ν is a neutral element for β, $\gamma(x)$ is the β-inverse of x in G. The type here for the group is (1, 2, 0, 1). In Section 8.2 we shall see that the operations in (1C) more completely characterize the group in that every algebraic subsystem will be a subgroup.

6. A ring $\langle R, +, \cdot, \theta \rangle$ is a set R, a binary operation $+$, a binary operation $\cdot$, a nullary operation θ such that $\langle R, +, \theta \rangle$ is a group, $+$ is commutative, and $\cdot$ is distributive with respect to $+$. The ring has type (1, 2, 2, 0).
7. A unitary ring $\langle R, +, \cdot, \theta, \nu \rangle$ is a set R, a binary operation $+$, a binary operation $\cdot$, a nullary operation θ, a nullary operation ν, such that $\langle R, +, \cdot, \theta \rangle$ is a ring and ν is a neutral element for $\cdot$. The unitary ring has type (1, 2, 2, 0, 0).
8. An integral domain $\langle R, +, \cdot, \theta, \nu \rangle$ is a set R, a binary operation $+$, a binary operation $\cdot$, a nullary operation θ, a nullary operation ν such that $\langle R, +, \cdot, \theta, \nu \rangle$ is a unitary ring, $\cdot$ is commutative, $\theta \neq \nu$, $xa = ya$ and $a \neq \theta$ imply $x = y$. The integral domain has type (1, 2, 2, 0, 0).
9. A field $\langle R, +, \cdot, \theta, \nu \rangle$ is a set R, a binary operation $+$, a binary operation $\cdot$, a nullary operation θ, a nullary operation ν such that $\langle R, +, \cdot, \theta, \nu \rangle$ is an integral domain and for each $x \in R$, $x \neq \theta$ there exists $y \in R$ such that $xy = \nu$. The field has type (1, 2, 2, 0, 0).

In order to define a module or a vector space and fit within the definition of algebraic system we reinterpret the scalar multiplication $R \times M \to M$ with value rx for each r in R, $x \in M$, as a family of unary operations on M, one for each $r \in R$. For each $r \in M$, define $\gamma_r : M \to M$ such that $\gamma_r(x) = rx$.

10. A module $\langle M, +, \zeta, (\gamma_r | r \in R) \rangle$ over a ring R is a set M, a binary operation $+$, a nullary operation ζ, a family of unary operations $(\gamma_r | r \in R)$ such that

$\langle M, +, \zeta\rangle$ is a commutative group, γ_r is additive for each r in R, $\gamma_r \circ \gamma_s = \gamma_{rs}$, $\gamma_{r+s} = \gamma_r + \gamma_s$, $\gamma_1 = I$. The module over a ring R has type $(1, 2, 0, (1|r \in R))$.

11. An ordered set $\langle S; \sigma\rangle$ is a set S with binary relation σ on S which is an order. The ordered set has type $(1; 2)$.
12. An ordered integral domain $\langle R, +, \cdot, \theta, v; \sigma\rangle$ is a set R, a binary operation $+$, a binary operation $\cdot$, a nullary operation θ, a nullary operation v; a binary relation σ such that $\langle R, +, \cdot, \theta, v\rangle$ is an integral domain, σ is a total order, $(x, y) \in \sigma$ implies $(x + a, y + a) \in \sigma$, $(0, a) \in \sigma$ and $(x, y) \in \sigma$ imply $(xa, ya) \in \sigma$. The ordered integral domain has type $(1, 2, 2, 0, 0; 2)$.
13. A vector space $\langle M, +, \zeta, (\gamma_r|r \in K)\rangle$ over a field K is a set M, a binary operation $+$, a nullary operation ζ, a family $(\gamma_r|r \in K)$ of unary operations such that $\langle M, +, \zeta, (\gamma_r|r \in K)\rangle$ is a module. The type of the vector space is $(1, 2, 0, (1|r \in K))$.
14. A Boolean algebra $\langle S, \vee, \wedge, ', 0, 1; \sigma\rangle$ is a set S, a binary operation $\vee$, a binary operation $\wedge$, a unary operation $'$ (complement), a nullary operation 0, a nullary operation 1, a binary relation σ such that $\vee$ and $\wedge$ are associative, commutative, and distributive each to the other, $a \vee a = a$, $a \wedge a = a, 0 \wedge a = 0, 0 \vee a = a, 1 \wedge a = a, 1 \vee a = 1$, σ is an order on S such that $(a, b) \in \sigma$ if and only if $a \wedge b = a$ if and only if $a \vee b = b$. Also $(a \vee b)' = a' \wedge b', (a \wedge b)' = a' \vee b', a \wedge a' = 0, a \vee a' = 1, a'' = a$. The Boolean algebra has type $(1, 2, 2, 1, 0, 0; 2)$.

Questions

1. Let S represent the binary operation of subtraction and $\div$ represent division. Which of the following are operational systems?
   (A) $\langle \mathbb{Z}, +, 0, S\rangle$
   (B) $\langle \mathbb{Z}, +, 0, 1\rangle$
   (C) $\langle \mathbb{Z}, +, \cdot, 0, 1, \div\rangle$
   (D) $\langle \mathbb{Q}, \cdot, 1, \div\rangle$.
   (E) None of the options is an operational system.

2. Which of these statements are true?
   (A) There can be two groups with the same type.
   (B) Similar algebraic systems can have different types.
   (C) An integral domain has the same type as a field.
   (D) Similar groups are isomorphic.
   (E) None of the statements is true.

3. Which of these statements are true?
   (A) The natural numbers $\langle \mathbb{N}, +, \cdot, 0, 1\rangle$ have type $(1, 2, 2, 0, 0)$.
   (B) A unitary ring has type $(1, 2, 2, 0, 0)$.
   (C) The natural number system is a unitary ring.
   (D) The natural number system and a unitary ring are similar operational systems.
   (E) None of the statements is true.

EXERCISES

1. Prove that Definitions 1 and 1C of a group are equivalent.

2. Let $\langle G, +, 0\rangle$ be a commutative group. Let S be the binary operation of subtraction on $G: S(a, b) = a - b = a + (-b)$. Formulate a definition of the group using the operational system $\langle G, S, 0\rangle$.

3. Give a definition of the natural numbers as an operational system using the Peano axioms.

8.2 Algebraic subsystems

In this section we discuss algebraic subsystems, closure systems, generation of subsystems, and independent subsystems.

We give first the very obvious definition of a subsystem.

Definition. A subset S of an operational system $\langle M, (\beta_i|i \in I)\rangle$ is a *operational subsystem* if and only if each operation β_i, $i \in I$, is closed on S; that is, $\beta_i(x_1, x_2, \ldots, x_{n_i}) \in S$ whenever $(x_1, x_2, \ldots, x_{n_i}) \in S^{n_i}$ for all $i \in I$.

We make the observation that the subsystem requires only closure under the operations and not satisfaction of any further conditions. Under the definition of group given in Example 1, Section 8.1, a subset may well be closed under the operations β and ν without being a subgroup, although the subset will be a subsystem. Such a subsystem could fail to contain some inverses. The set of all subgroups of a given group $\langle G, \beta, \nu\rangle$ is, in general, a proper subset of the set of all operational subsystems of the operational system $\langle G, \beta, \nu\rangle$. The class of all groups is a proper subclass of the class of all algebraic systems of type (1, 2, 0). The class of all groups consists precisely of those operational systems of type (1, 2, 0) which also satisfy the conditions for a group.

With the goal of treating the disparities just alluded to, we make the following definition.

Definition. A collection $\mathscr{C}$ of subsets of M is called a *closure class* if and only if given $\mathscr{D} \subseteq \mathscr{C}$ and $\mathscr{D} \neq \varnothing$ it follows that $\bigcap \mathscr{D} \in \mathscr{C}$ and $M \in \mathscr{C}$.

EXAMPLE. The set of all subgroups of a given group is a closure class. This follows from the intersection of a collection of subgroups' being a subgroup.

Theorem. *The set of all operational subsystems of a given operational system* $\langle M, (\beta_i|i \in I)\rangle$ *is a closure class.*

PROOF. Let $\mathscr{C}$ be the set of all operational subsystems of $\langle M, (\beta_i|i \in I)\rangle$. Let $\mathscr{D} \subseteq \mathscr{C}$. Let β_i be an operation on M and suppose β_i is an n_i-ary operation. To avoid excessive use of subscripts we omit the subscript when it is unessential to the argument. Consider $x_1, x_2, \ldots, x_n$ in $\bigcap \mathscr{D}$. $\beta(x_1, x_2, \ldots, x_n) \in D$ for every D in $\mathscr{D}$. Therefore, $\beta(x_1, x_2, \ldots, x_n) \in \bigcap \mathscr{D}$ and $\bigcap \mathscr{D}$ is closed under the operation β. This is true for all β_i, $i \in I$. $\bigcap \mathscr{D}$ is therefore an operational subsystem of M. □

We now treat relational subsystems.

Definition. $\langle N; (\tau_j | j \in J)\rangle$ is a *relational subsystem* of the relational system $\langle M; (\sigma_j | j \in J)\rangle$ if and only if $N \subseteq M$ and $\tau_j \subseteq \sigma_j$ for all $j \in J$.

Given any relational system $\langle M; (\sigma_j | j \in J)\rangle$ and any subset N of M the restrictions of the relations σ_j to N always produce a relational subsystem on N: $\langle N; (\sigma_j \cap N^{n_j} | j \in J)\rangle$. This is the largest relational system possible on N which is a relational subsystem of $\langle M; (\sigma_j | j \in J)\rangle$. This relational subsystem on N is called the *full* relational subsystem of $\langle M; (\sigma_j | j \in J)\rangle$ on the subset N.

EXAMPLE. Given the integers with their usual order, $\langle \mathbb{Z}; \leqslant\rangle$, the set of even integers, $2\mathbb{Z}$, can be given two different orders. The first $\langle 2\mathbb{Z}; \rho\rangle$ is the full relational subsystem in which even integers are ordered just as they are in $\mathbb{Z}$. In a second relational subsystem $\langle 2\mathbb{Z}; \tau\rangle$ we order only the multiples of 4: $\tau = \{(4m, 4n) | m \leqslant n\}$. $\langle 2\mathbb{Z}; \tau\rangle$ is a relational subsystem of $\langle \mathbb{Z}; \leqslant\rangle$, but not a full relational subsystem.

We now discuss the interaction of relations and operations of an algebraic system. An ordered group $\langle G, \cdot, v; \leqslant\rangle$ is a group $\langle G, \cdot, v\rangle$ and a totally ordered set $\langle G; \leqslant\rangle$ such that $x_1 \leqslant y_1$ implies $ax_1 \leqslant ay_1$ and $x_1 a \leqslant y_1 a$ for all a in G. These conditions involving both the order and the multiplication are equivalent to the single condition $x_1 \leqslant y_1$ and $x_2 \leqslant y_2$ imply $x_1 x_2 \leqslant y_1 y_2$ as one can easily verify. We restate this condition for an arbitrary binary relation σ and G. $(x_1, y_1) \in \sigma$ and $(x_2, y_2) \in \sigma$ imply $(x_1 x_2, y_1 y_2) \in \sigma$. We now frame a definition for an n-ary operation.

Definition. An r-ary relation σ is *compatible* with an n-ary operation β if and only if $(x_{11}, x_{12}, \ldots, x_{1r}) \in \sigma$, $(x_{21}, x_{22}, \ldots, x_{2r}) \in \sigma, \ldots, (x_{n1}, x_{n2}, \ldots, x_{nr}) \in \sigma$ imply $(\beta(x_{11}, x_{21}, \ldots, x_{n1}), \ldots, \beta(x_{1r}, x_{2r}, \ldots, x_{nr})) \in \sigma$.

The notation can be made more compact for this definition. We define the power of a relation σ as follows: $((x_{11}, x_{21}, \ldots, x_{n1}), (x_{12}, x_{22}, \ldots, x_{n2}), \ldots, (x_{1r}, x_{2r}, \ldots, x_{nr})) \in \sigma^n$ if and only if $(x_{11}, x_{12}, \ldots, x_{1r}) \in \sigma$ and $(x_{21}, x_{22}, \ldots, x_{2r}) \in \sigma$ and $\cdots$ and $(x_{n1}, x_{n2}, \ldots, x_{nr}) \in \sigma$. This allows restating the condition of compatibility: $((x_{11}, x_{21}, \ldots, x_{n1}), \ldots, (x_{1r}, x_{2r}, \ldots, x_{nr})) \in \sigma^n$ implies $(\beta(x_{11}, x_{21}, \ldots, x_{n1}), \ldots, \beta(x_{1r}, x_{2r}, \ldots, x_{nr})) \in \sigma$. We also write $(\beta(x_{11}, x_{21}, \ldots, x_{n1}), \ldots, \beta(x_{1r}, x_{2r}, \ldots, x_{nr}))$ as $\beta^r((x_{11}, x_{21}, \ldots, x_{n1}), \ldots, (x_{1r}, x_{2r}, \ldots, x_{nr}))$. Again we restate the compatibility condition: $((x_{11}, x_{21}, \ldots, x_{n1}), \ldots, (x_{1r}, x_{2r}, \ldots, x_{nr})) \in \sigma^n$ implies $\beta^r((x_{11}, x_{21}, \ldots, x_{n1}), \ldots, (x_{1r}, x_{2r}, \ldots, x_{nr})) \in \sigma$. Finally, we state the most compact formulation as a definition.

Definition. An r-ary relation σ is *compatible* with an n-ary operation β if and only if $\beta^r(\sigma^n) \subseteq \sigma$.

In this presentation of algebraic systems we have treated operations and relations separately. They are not actually independent of each other. An n-ary operation $\beta: M^n \to M$ expressed as a set of ordered pairs, $\{((x_1, x_2, \ldots, x_n), \beta(x_1, x_2, \ldots, x_n)) | (x_1, x_2, \ldots, x_n) \in M^n\}$ is a subset of $M^n \times M$ or M^{n+1}. β is then an $(n + 1)$-ary relation on M. It is sufficient then to consider only relational systems expressing all operational concepts in terms of relations. This would, however, be more artificial and nonintuitive. Nevertheless, it is an interesting observation that it can be done. In the reverse direction it is possible to reformulate relations in terms of operations if we give up the requirement that the operation be defined over all of M; that is, we permit partial operations. For an n-ary relation $\sigma = \{(x_1, x_2, \ldots, x_n) | (x_1, x_2, \ldots, x_n) \in \sigma\}$ we define $\hat{\sigma}(x_1, x_2, \ldots, x_n) = x_1$ for each $(x_1, x_2, \ldots, x_n) \in \sigma$. $\hat{\sigma}$ is then an n-ary operation defined on a subset σ of M^n which takes values in M. The relation σ is recoverable from $\hat{\sigma}$ as the domain of $\hat{\sigma}$. The study of partial algebras considers algebraic systems constructed from partial operations.

Our next topic for this section is the finding of a general context for the idea of algebraic generation; for example, the subspace of a vector space generated by a given subset of the vector space. We return to the closure class.

Theorem. *Let $\langle M, (\beta_i | i \in I)\rangle$ be an operational system and $\mathscr{C}$ a closure class of M. Then given any subset S of M there exists a smallest member $[S]$ of $\mathscr{C}$ which includes S.*

PROOF. Let $\mathscr{D}$ be the set of all members of the closure class $\mathscr{C}$ which have S as a subset. $S \subseteq \bigcap \mathscr{D}$. Moreover, $\bigcap \mathscr{D} \in \mathscr{C}$. We denote $\bigcap \mathscr{D}$ as $[S]$. Then if $D \in \mathscr{D}$ we have $[S] = \bigcap \mathscr{D} \subseteq D$ proving $[S]$ is the smallest subset of $\mathscr{C}$ having S as a subset. Note $\mathscr{D} \neq \varnothing$ because $M \in \mathscr{D}$. □

The previous theorem suggests a definition.

Definition. Given an operational system $\langle M, (\beta_i | i \in I)\rangle$ and a closure class $\mathscr{C}$, for any subset S of M we define $[S]$ to be the member of the closure class *generated* by S.

EXAMPLE. The set of all subgroups $\mathscr{C}$ of a given group $\langle G, \beta, v\rangle$ is a closure class. Given any subset S of G, $[S]$ denotes the smallest member of the closure class containing S, the smallest subgroup of G containing S. We call this the subgroup generated by S. We note that in looking for the subgroup generated by S we choose the smallest member of the closure class (subgroups) containing S and not the smallest operational subsystem containing S.

In terms of the concept of generation we make the following definition of an algebraic closure class.

Definition. A closure class $\mathscr{C}$ is an *algebraic closure class* if and only if $[A] = \bigcup\{[F] | F \text{ is a finite subset of } A\}$ for all sets A.

This condition means that if $x \in [A]$ then $x \in [a_1, a_2, \ldots, a_n]$ for some $a_1, a_2, \ldots, a_n$ in A; this to say, each element of a closure is finitely generated.

EXAMPLE. The set of all subgroups of a given group is an algebraic closure class. We demonstrate this fact. Let A be an arbitrary subset of the group G. We show that $\bigcup\{[F] | F \text{ is a finite subset of } A\}$ is a subgroup of G. Let $x, y \in \bigcup\{[F] | F \text{ is a finite subset of } A\}$. $x \in [a_1, a_2, \ldots, a_m]$, $a_1, a_2, \ldots, a_m \in A$. $y \in [a'_1, a'_2, \ldots, a'_n]$, $a'_1, a'_2, \ldots, a'_n \in A$. Then x, y, and xy belong to $[a_1, a_2, \ldots, a_m, a'_1, a'_2, \ldots, a'_n] \subseteq \bigcup\{[F] | F \text{ is a finite subset of } A\}$. The union is also closed under the neutral element and inverses. Since the union contains every element of A it must contain the smallest subgroup containing A, namely, $[A]$. For the reverse inclusion we notice $[F] \subseteq [A]$ for every finite subset of A so that the union is also a subset of $[A]$. $[A] = \bigcup\{[F] | F \text{ is a finite subset of } A\}$.

Theorem. *The set of all operational subsystems of a given operational system* $\langle M, (\beta_i | i \in I)\rangle$ *is an algebraic closure class.*

PROOF. We have earlier proven the collection to be a closure class. To prove the closure class algebraic we need only repeat the argument given in the previous example. We consider the operational system $[A]$ generated by the arbitrary set A. We wish to show $U = \bigcup\{[F] | F \text{ is a finite subset of } A\}$ is closed under all operations. Let β be any n-ary operation of the system and $x_1, x_2, \ldots, x_n \in U$. $x_1 \in [F_1]$, $x_2 \in [F_2], \ldots, x_n \in [F_n]$ for some finite subsets $F_1, F_2, \ldots, F_n$ of A. Then $x_1, x_2, \ldots, x_n \in [F_1 \cup F_2 \cup \cdots \cup F_n]$. $\beta(x_1, x_2, \ldots, x_n) \in [F_1 \cup F_2 \cup \cdots \cup F_n] \subseteq U$. U is an operational subsystem. $U = [A]$ by the same argument given given in the example. □

If we now consider a group $\langle G, \beta, v, \gamma\rangle$ according to the definition of group located in Example 1C of Section 8.1, an operational system of type (1, 2, 0, 1), we arrive at the interesting situation where the set of all subgroups and the set of all operational subsystems coincide. The closure class of subgroups and the closure class of all operational subsystems are the same. To verify this situation we have only to show that every operational subsystem is a subgroup; for, certainly every subgroup is an operational subsystem. Let H be a subset of G which is closed under β, v, and γ. β is associative on H, v is a neutral element for H and a member of H, and $\gamma(x)$ is the β-inverse of x in H. H is a subgroup.

We now raise the following question. Is it always possible to define operations on a set in such a manner that the operational subsystems coincide with some given subclass of the system? An answer lies in the following theorem.

Theorem. *Let M be a set. Let $\mathscr{C}$ be an algebraic closure class on M. Then there exist operations $(\beta_i|i \in I)$ on M such that the set of all operational subsystems of $\langle M, (\beta_i|i \in I)\rangle$ coincide with the given subclass $\mathscr{C}$.*

PROOF. The proof consists in defining operations for M and in this sense is constructive. Each subset S of M generates a member of the closure class denoted by $[S]$.

For each $a \in [\varnothing]$ define $\beta_a:\{0\} \to M$ such that $\beta_a(0) = a$. This gives a set of nullary operations, one for each member of the closure class generated by the empty set. Let n be a positive integer. For each $(a_1, a_2, \ldots, a_n) \in M^n$ and $b \in [a_1, a_2, \ldots, a_n]$ define

$$\beta_b^{(a_1, a_2, \ldots, a_n)}(x_1, x_2, \ldots, x_n) = \begin{cases} b & \text{if } (x_1, x_2, \ldots, x_n) \neq (a_1, a_2, \ldots, a_n), \\ x_1 & \text{if } (x_1, x_2, \ldots, x_n) = (a_1, a_2, \ldots, a_n). \end{cases}$$

We consider the operational system

$$\langle M, (\beta_a|a \in [\varnothing]), (\beta_b^{(a_1, \ldots, a_n)}|(a_1, \ldots, a_n) \in M^n, b \in [a_1, \ldots, a_n], n \in \mathbb{N}^+\rangle.$$

We must now show that the set of all operational subsystems of this operational system coincides with the given algebraic closure system $\mathscr{C}$.

We begin by showing every member of the closure class is an algebraic subsystem of M. Let $C \in \mathscr{C}$. If $C = \varnothing$ then there are no nullary operations. If $C \neq \varnothing$ then because $\varnothing \subseteq C$ we have $[\varnothing] \subseteq [C] = C$. If $a \in [\varnothing]$ then $a \in C$ and $\beta_a(0) = a \in C$. Thus every nullary operation is closed in C. Now let $\beta_b^{(a_1, \ldots, a_n)}$ be an n-ary operation on M. For simplicity abbreviate the operation with β alone. Let $c_1, c_2, \ldots, c_n \in C$. The value of β is given in two cases. If $(c_1, \ldots, c_n) \neq (a_1, \ldots, a_n)$ then $\beta(c_1, \ldots, c_n) = c_1$ which belongs to C. If $(c_1, \ldots, c_n) = (a_1, \ldots, a_n)$ then $\beta(c_1, \ldots, c_n) = b \in [a_1, \ldots, a_n] = [c_1, \ldots, c_n] \subseteq C$. In either case the value of β lies in C. C is closed under all the constructed operations and is therefore an operational subsystem.

Secondly, we wish to show that any operational subsystem is a member of the given algebraic closure class $\mathscr{C}$. Let B be an operational subsystem of M according to the defined operations. We wish to show now that B is a member of the closure class, i.e., $[B] = B$. Let $b \in [B]$. Then $b \in [a_1, a_2, \ldots, a_n]$ for some $a_1, a_2, \ldots, a_n \in B$. Range $\beta_b^{(a_1, a_2, \ldots, a_n)}$ includes b. Since $a_1, a_2, \ldots, a_n \in B$ and B is closed under all operations, $b \in B$. Thus, $[B] \subseteq B$. Clearly, $B \subseteq [B]$. $B = [B]$. □

We define next a concept, with respect to a given closure class, called independence. It is a generalization of the linear independence found in vector spaces.

Definition. Let M be a set and $\mathscr{C}$ be a closure class on M. A subset A of M is *independent* if and only if $[A - \{x\}] \neq [A]$ for every $x \in A$.

In other words, we can say that an independent set is a minimal member among those sets which generate a given member of the closure class.

EXAMPLE. As we mentioned, any linearly independent set of a vector space is an example of an independent set. In the group $\mathfrak{S}_3$ of permutations the sets $\{(1\ 2\ 3)\}$, $\{(1\ 2)\}$, $\{(1\ 2\ 3), (1\ 2)\}$ of cycles are independent sets. On the other hand, the set $\{(1\ 2\ 3), (1\ 2), (1\ 3)\}$ fails to be independent.

We now prove a number of results that show some of the properties of independent sets.

Theorem. *Let M be a set and $\mathscr{C}$ be a closure class on M. Then A is an independent set if and only if $\{x\} \cap [A - \{x\}] = \varnothing$ for every $x \in A$.*

PROOF. If there exists $y \in \{x\} \cap [A - \{x\}]$ then $x \in [A - \{x\}]$ for some $x \in A$. $A - \{x\} \subseteq [A - \{x\}]$ and $x \in [A - \{x\}]$ imply $A \subseteq [A - \{x\}]$. $[A] \subseteq [A - \{x\}]$. $[A] = [A - \{x\}]$. Conversely, if $[A] = [A - \{x\}]$ for some $x \in A$, then $x \in [A]$. $x \in [A - \{x\}]$. $\{x\} \cap [A - \{x\}] \neq \varnothing$. This proves the result. Some easy properties of $[A]$ are found in Exercise 2. Some of them are used in this proof. □

Theorem. *Let M be a set and $\mathscr{C}$ a closure class on M.*

1. *If A is an independent set and $B \subseteq A$ then B is an independent set.*
2. *If A is an independent set and $B \subset A$ then $[B] \subset [A]$.*
3. *If B is an independent set then no proper subset of B generates $[B]$.*
4. *If B is a dependent set then some proper subset of B generates $[B]$.*

PROOF. $\{x\} \cap [B - \{x\}] \subseteq \{x\} \cap [A - \{x\}] = \varnothing$ for each $x \in B$ proves part 1. For part 2, let $a \in A - B$. $\{a\} \cap [A - \{a\}] = \varnothing$ because A is assumed to be independent. Suppose $[B] = [A]$. $\{a\} \cap [A - \{a\}] \supseteq \{a\} \cap [B] = \{a\} \cap [A] = \{a\} \neq \varnothing$. This contradicts the independence of A. $[A] \neq [B]$. For part 3, let C be any proper subset of B, an independent set. Let $x \in B - C$. $[C] \subseteq [B - \{x\}] \subset [B]$ if B is independent. For part 4, if B is dependent then there exists x in B such that $[B - \{x\}] = [B]$. $B - \{x\}$ is a proper subset of B. □

QUESTIONS

1. Which of the following are closure classes?
   (A) The set of all subsets of an operational system $\langle S, \gamma \rangle$ of type $(1, 0)$
   (B) The set of all subrings of the ring $\langle R, +, \cdot, \theta \rangle$, an operational system of type $(1, 2, 2, 0)$

(C) The set of all subsets of a given set S containing some one fixt subset A of S
(D) The set of all commutative subgroups of a given group.
(E) None of the collections is a closure class.

2. Which of the following statements are true?
(A) A binary relation ρ is compatible with a unary operation γ if and only if $(x, y) \in \rho$ implies $(\gamma(x), \gamma(y)) \in \rho$.
(B) A binary relation ρ is compatible with a binary operation β if and only if $\beta^2(\rho) \subseteq \rho$.
(C) A binary relation ρ is compatible with a unary operation γ if and only if $x\gamma y$ implies $\gamma(x)\rho\gamma(y)$.
(D) A binary relation ρ is compatible with a unary operation γ if and only if $\gamma^2(\rho) \subseteq \rho$.
(E) None of the statements is correct.

3. Let M have a closure class $\mathscr{C}$. Then $\bigcup\{[F] | F \subseteq A, F \text{ is finite}\}$
(A) is a member of the closure class
(B) is a subset of $[A]$
(C) equals $[A]$ if $\mathscr{C}$ is an algebraic closure class
(D) includes $[A]$.
(E) None of the alternatives completes a true sentence.

4. Which of the following statements are true?
(A) If A is independent then so also is $[A]$.
(B) If A is independent then so also is $A - \{x\}$ for some x in A.
(C) If A is independent then $A \cup \{x\}$ is not independent for some x in A.
(D) If A and B are independent then so also is $A \cup B$.
(E) None of the statements is true.

Exercises

1. If $\langle M, (\beta_i | i \in I)\rangle$ is an operational system then there is at least one equivalence relation which is compatible with every operation.

2. Prove the following three elementary properties of any closure class on a set M.
(a) For any subset S, $S \subseteq [S]$.
(b) $S \subseteq T$ implies $[S] \subseteq [T]$ for any subsets S, T.
(c) $[[S]] = [S]$ for any subset S.

3. Show that $\mathbb{N}$ is an operational subsystem of $\langle \mathbb{Z}, +, 0\rangle$, yet $\mathbb{N}$ is not a subgroup of $\mathbb{Z}$.

4. Show that the set of all subrings of a given ring $\langle R, +, \cdot, \theta\rangle$ is an algebraic closure class.

5. Show that the set of all ideals of a given commutative unitary ring $\langle R, +, \cdot, \theta, v\rangle$ is an algebraic closure class.

6. Show that the set of all closed intervals of the real numbers (and include $\varnothing$, $\mathbb{R}$, and single points) is a closure class. Show that it is not algebraic.

7. For any closure class prove: $[C \cup D] = [[C] \cup [D]]$, $[\bigcup_{i \in I} A_i] = [\bigcup_{i \in I} [A_i]]$, $[\bigcap_{i \in I} A_i] \subseteq \bigcap_{i \in I} [A_i]$.

8. Show that in a group $\langle G, \cdot, \nu \rangle$ in which the elements are totally ordered (every pair of elements are comparable) the following two conditions are equivalent:

$$x_1 \leqslant y_1 \text{ implies } ax_1 \leqslant ay_1 \text{ and } x_1 a \leqslant y_1 a,$$
$$x_1 \leqslant y_1 \text{ and } x_2 \leqslant y_2 \text{ imply } x_1 x_2 \leqslant y_1 y_2.$$

9. An r-ary relation σ is defined on $\mathbb{Q}$ as follows: $(x_1, x_2, \ldots, x_r) \in \sigma$ if and only if $x_1 x_2 \cdots x_r = 1$. Is σ compatible with multiplication?

10. Show that if every nonempty subclass $\mathscr{D}$ of an algebraic closure class $\mathscr{C}$ has a maximal element then every member of $\mathscr{C}$ is finitely generated.

11. Find an operational system for a ring R for which the set of operational subsystems and the set of subrings is the same.

12. Consider the closure class of all operational subsystems of an operational system containing at least one nullary operation. Show that no subsystem can be independent.

8.3 Morphisms

In this section we discuss morphisms of operational and relational systems.

Definition. Let β be an n-ary operation on a set M and β' be an n-ary operation on a set M'. We say that a function $f: M \to M'$ *preserves* the operation β in the operation β' if and only if $f(\beta(x_1, x_2, \ldots, x_n)) = \beta'(f(x_1), f(x_2), \ldots, f(x_n))$ for all $x_1, x_2, \ldots, x_n$ in M.

Instances of this property are such equations as $f(x_1 + x_2) = f(x_1) +' f(x_2)$, $f(x_1 x_2) = f(x_1)f(x_2)$, $e^{x_1 + x_2} = e^{x_1}e^{x_2}$, $\log x_1 x_2 = \log x_1 + \log x_2$, $f(\theta) = \theta'$, $f(-x) = -f(x)$, $f(x^-) = f(x)^-$.

If we denote the mapping $(x_1, x_2, \ldots, x_n) \mapsto (f(x_1), f(x_2), \ldots, f(x_n))$ by $f^n: M^n \to M'^n$ then the equation given in the definition above can be written as $f(\beta(x_1, x_2, \ldots, x_n)) = \beta'(f^n(x_1, x_2, \ldots, x_n))$ for all $(x_1, x_2, \ldots, x_n) \in M^n$. This equation asserts the equality of the two functions $f \circ \beta$ and $\beta' \circ f^n$ illustrated in this diagram.

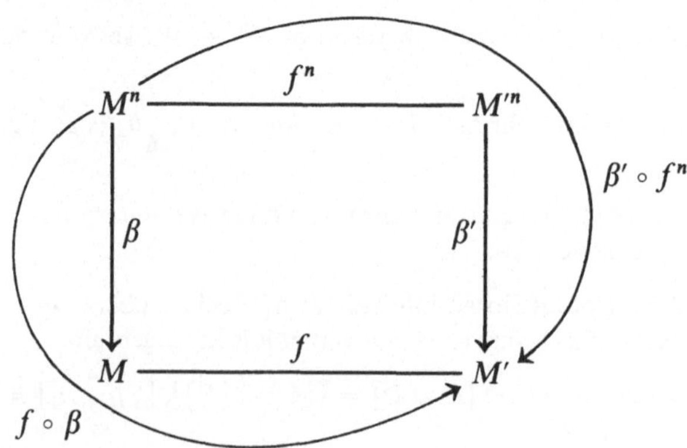

The function $f:M \to M'$ preserves the operation β in the operation β' if and only if $f \circ \beta = \beta' \circ f^n$.

With this brief preparation we define morphism.

Definition. Let $\langle M, (\beta_i|i \in I)\rangle$ and $\langle M', (\beta'_j|j \in J\rangle$ be operational systems. Let β_i and β'_i both be operations of the same type n_i for each $i \in K$, some subset of both I and J. We say that $f:M \to M'$ is a *type* $(1, (n_i|i \in K))$ *morphism* if and only if f preserves β_i in β'_i for each i in K.

EXAMPLE. The mapping $\mathbb{N} \to R$ such that $f(m) = a^m$ for some nonzero $a \in R$ is a type (1, 2) morphism preserving the operation + of the natural numbers $\langle \mathbb{N}, +, \cdot, 0, 1\rangle$ in the operation $\cdot$ of the ring $\langle R, +, \cdot, \theta\rangle$. $f(m + n) = f(m)f(n)$.

It is to be noticed that the quantities determining a morphism are the operations and their sizes, not which axioms are satisfied by the domain and codomain. In many cases of interest the two operational systems and the morphism all have the same type. One then has $f:M \to M'$ preserving β_i in β'_i for each $i \in I$ giving a morphism from the system $\langle M, (\beta_i|i \in I)\rangle$ to the system $\langle M, (\beta'_i|i \in I)\rangle$.

EXAMPLE. A mapping $f:R \to R'$ such that $f(x_1 + x_2) = f(x_1) +' f(x_2)$, $f(x_1x_2) = f(x_1)f(x_2)$, $f(\theta) = \theta'$, is a type (1, 2, 2, 0) morphism from the ring $\langle R, +, \cdot, \theta\rangle$ to the ring $\langle R', +', \cdot', \theta'\rangle$. That the domain and codomain are rings is not necessary for f to be a morphism. The operational type determines the type of the morphism.

EXAMPLE. A mapping $f:R \to R'$ such that $f(x_1 + x_2) = f(x_1) +' f(x_2)$, $f(x_1x_2) = f(x_1)f(x_2)$, $f(\theta) = \theta'$, $f(v) = v'$, is a type (1, 2, 2, 0, 0) morphism from the unitary ring $\langle R, +, \cdot, \theta, v\rangle$ to the unitary ring $\langle R', +', \cdot', \theta', v'\rangle$. Notice that the zero mapping is a type (1, 2, 2, 0) morphism, but not a type (1, 2, 2, 0, 0) morphism.

With our definitions well established we can now prove the expected theorem on images and preimages of subsystems.

Theorem. *If $f:M \to M'$ is a type $(1, (n_i|i \in I))$ morphism of the operational system $\langle M, (\beta_i|i \in I)\rangle$ into the operational system $\langle M', (\beta'_i|i \in I)\rangle$ then*

N is a subsystem of M implies $f(N)$ is a subsystem of M',
N' is a subsystem of M' implies $f^{-1}(N')$ is a subsystem of M.

PROOF. Let β be any operation on M and β' be the corresponding operation on M'; we omit the subscript for brevity. Let $x'_1, x'_2, \ldots, x'_n \in f(N)$. There exist $x_1, x_2, \ldots, x_n$ in N such that $f(x_1) = x'_1, f(x_2) = x'_2, \ldots, f(x_n) = x'_n$. $\beta(x_1, x_2, \ldots, x_n) \in N$. $f(\beta(x_1, x_2, \ldots, x_n)) \in f(N)$. But $f \circ \beta = \beta' \circ f^n$. $\beta' \circ$

$f^n(x_1, x_2, \ldots, x_n) \in f(N)$. $\beta'(f(x_1), f(x_2), \ldots, f(x_n)) \in f(N)$. $\beta'(x_1', x_2', \ldots, x_n') \in f(N)$. As β stood for any operation, $f(N)$ is closed under all operations and is therefore a subsystem.

For the second part we again use a typical operation β and the corresponding β'. Let $x_1, x_2, \ldots, x_n \in f^{-1}(N')$. $f(x_1), f(x_2), \ldots, f(x_n) \in N'$. $\beta'(f(x_1), f(x_2), \ldots, f(x_n)) \in N'$. $\beta' \circ f^n(x_1, x_2, \ldots, x_n) \in N'$. $f \circ \beta(x_1, x_2, \ldots, x_n) \in N'$. $\beta(x_1, x_2, \ldots, x_n) \in f^{-1}(N')$. □

We also have a theorem on morphisms and closure systems.

Theorem. *Let $f: M \to M'$ be a $(1, (n_i|i \in I))$ type morphism of the operational system $\langle M, (\beta_i|i \in I)\rangle$ into the operational system $\langle M, (\beta_i'|i \in I)\rangle$. Let $\mathscr{C}'$ be a closure system of M'. Then $\mathscr{C} = \{f^{-1}(C')|C' \in \mathscr{C}'\}$ is a closure system of M.*

PROOF. The sizes of the morphism and the operations are really irrelevant. This is really a theorem about functions. Let $\mathscr{B}$ be a collection of sets in $\mathscr{C}$. Then $\mathscr{B} = \{f^{-1}(B')|B' \in \mathscr{B}'\}$ for some subcollection $\mathscr{B}'$ of $\mathscr{C}'$. $\bigcap\mathscr{B} = \bigcap\{B|B \in \mathscr{B}\} = \bigcap\{f^{-1}(B')|B' \in \mathscr{B}'\} = f^{-1}(\bigcap\mathscr{B}') \in \mathscr{C}$ since $\bigcap\mathscr{B}' \in \mathscr{C}'$. □

We now turn to the interaction of functions with relational systems.

Definition. A function $f: M \to M'$ *preserves* an r-ary relation σ on M in an r-ary relation σ' on M' if and only if $(x_1, x_2, \ldots, x_r) \in \sigma$ implies $(f(x_1), f(x_2), \ldots, f(x_r)) \in \sigma'$. Or more briefly, $f^r(\sigma) \subseteq \sigma'$.

If $f^r(\sigma) = \sigma'$, i.e., $(x_1, x_2, \ldots, x_r) \in \sigma$ if and only if $(f(x_1), f(x_2), \ldots, fx_r)) \in \sigma'$, then we say f *fully preserves* σ in σ'.

A function $f: M \to M'$ of relational systems $\langle M; (\sigma_i|i \in I)\rangle$ and $\langle M'; (\sigma_j|j \in J)\rangle$ is a *type* $(r_k|k \in K)$ *relation preserving morphism* if and only if f preserves relation σ_k in σ_k' for all $k \in K$, a subset of J and K. If, moreover, f fully preserves relation σ_k in relation σ_k' for each $k \in K$ then f is called a *relation fully preserving morphism.*

$$\begin{array}{ccc} M^r & \xrightarrow{\quad f^r \quad} & (M')^r \\ & & \cup\!\backslash \\ \cup\!| & & \sigma \\ & & \cup\!/ \\ \sigma & \xrightarrow{\quad f^r \quad} & f^r(\sigma) \end{array}$$

Theorem. *Let $\langle M; (\sigma_j|j \in J)\rangle$ and $\langle M'; (\sigma_j'|j \in J)\rangle$ be similar relational systems. Let $f: M \to M'$ be a relation preserving morphism.*

1. *If $\langle N; (\tau_j|j \in J)\rangle$ is a relational subsystem of M then $\langle f(N); (f^r(\tau_j)|j \in J)\rangle$ is a subsystem of M'.*

2. *If* $\langle N; (\tau_j | j \in J)\rangle$ *is a full relational subsystem of* M *and* f *is a full relation preserving morphism then* $\langle f(N); (f^r(\tau_j)|j \in J)\rangle$ *is a full subsystem of* M'.
3. *If* $\langle N'; (\tau_j | j \in J)\rangle$ *is a subsystem of* M' *and* f *is a full relation preserving morphism then* $\langle f^{-1}(N'); ((f^r)^{-1}(\tau_j')|j \in J)\rangle$ *is a relational subsystem of* M.

PROOF

1. $N \subseteq M$ and $\tau \subseteq \sigma$ imply $f(N) \subseteq f(M) \subseteq M'$ and $f^r(\tau) \subseteq f^r(\sigma) \subseteq \sigma'$. Hence $\langle f(N); (f^r(\tau_j)|j \in J)\rangle$ is a relational subsystem of M'.

2. For $(x_1, x_2, \ldots, x_r) \in N^r$ we have $(x_1, x_2, \ldots, x_r) \in \tau$ if and only if $(x_1, x_2, \ldots, x_r) \in \sigma$ if and only if $(f(x_1), f(x_2), \ldots, f(x_r)) \in \sigma'$. Therefore, $f^r(\tau) = \sigma' \cap f(N)^r$, proving $\langle f(N); (f^r(\tau_j)|j \in J)\rangle$ to be a full subsystem of M'.

3. $(f^r)^{-1}(\tau_j) = \{(x_1, x_2, \ldots, x_r)|(x_1, x_2, \ldots, x_r) \in M^r$ and $(f(x_1), f(x_2), \ldots, f(x_r)) \in \tau'\}$. Since $\tau' \subseteq \sigma'$ and $(x_1, x_2, \ldots, x_r) \in \sigma$ if and only if $(f(x_1), f(x_2), \ldots, f(x_r)) \in \sigma'$. $(f^r)^{-1}(\tau') \subseteq \sigma$ proving $\langle f^{-1}(N'); (f^r)^{-1}(\tau_j')|j \in J\rangle$ to be a subsystem of M. □

EXAMPLE. A full order preserving of ordered integral domains $\langle R, +, \cdot, \theta, v; \leqslant\rangle$ and $\langle R', +', \cdot', \theta', v'; \leqslant'\rangle$ preserves the algebraic operations and order relation so that $f(x_1 + x_2) = f(x_1) +' f(x_2)$, $f(x_1 x_2) = f(x_1)f(x_2)$, $f(\theta) = \theta'$, $f(v) = v'$; $x_1 \leqslant x_2$ if and only if $f(x_1) \leqslant' f(x_2)$. In terms of the notation of this section we have $f \circ (+) = (+') \circ f^2$, $f \circ (\cdot) = (\cdot') \circ f^2$, $f \circ \theta = \theta' \circ f^0, f \circ v = v' \circ f^0; f^2(\leqslant) = (\leqslant')$.

QUESTIONS

1. Which of these statements are true?
   (A) $f \circ \beta^n = \beta' \circ f$ implies f preserves β in β'.
   (B) $\sigma \cap S \times S$ is an equivalence relation on a subset S of M if σ is an equivalence relation on M.
   (C) $f^r(\sigma) = \sigma'$ implies the r-ary relation σ is preserved by f in the r-ary relation σ'.
   (D) $f: \mathbb{N} \to \mathbb{N}$ such that $f(x) = x^p$ is a type (1, 2) morphism.
   (E) None of the statements is correct.

2. Which of these statements are true?
   (A) $\langle \mathbb{R}^+, \cdot, 1\rangle$ and $\langle \mathbb{R}, +, 0\rangle$ are commutative groups.
   (B) $\log: \mathbb{R}^+ \to \mathbb{R}$ preserves $\cdot$ in $+$.
   (C) $\log: \mathbb{R}^+ \to \mathbb{R}$ is a type (1, 2, 0) morphism.
   (D) $\log a^{-1} = -\log a$ for all $a \in \mathbb{R}^+$.
   (E) None of the statements is correct.

3. Let γ stand for the unary operation of taking reciprocals and n for the operation of taking negatives. Which of these statements are true?
   (A) $\mathbb{Q}^+$ is a subgroup of $\langle \mathbb{R}^+, \cdot, 1, \gamma\rangle$.
   (B) $\log \mathbb{Q}^+$ is a subgroup of $\langle \mathbb{R}, +, 0, n\rangle$.

(C) $\log(n(1)) = 0$.
(D) $\log^{-1} \mathbb{Q}$ is a subgroup of $\langle \mathbb{R}^+, \cdot, 1, \gamma \rangle$.
(E) None of the statements is true.

4. Which of these statements are true?
(A) $\log: \mathbb{R}^+ \to \mathbb{R}$ fully preserves $\leqslant$ in $\leqslant$.
(B) $\log: \mathbb{R}^+ \to \mathbb{R}$ preserves $\leqslant$ in $\leqslant$.
(C) $\leqslant$ is compatible with $\cdot$ of $\langle \mathbb{R}^+, \cdot, 1 \rangle$.
(D) $\leqslant$ is compatible with $+$ of $\langle \mathbb{R}, +, 0 \rangle$.
(E) None of the statements is true.

EXERCISES

1. Show that there are an infinite number of morphisms of $\mathbb{Z}$ into $\mathbb{Z}$ preserving addition and zero, two morphisms preserving addition and zero and multiplication, and only one morphism preserving addition, zero, multiplication, and the unity.

8.4 Congruences and quotient systems

In this section we have congruences, quotient systems, and isomorphism theorems.

From our earlier definition of the compatibility of an r-ary relation with an n-ary operation (Section 8.2) we specialize now for the binary relation.

Definition. A binary relation ρ on a set M is *compatible* with an n-ary operation β if and only if $x_1\rho y_1, x_2\rho y_2, \ldots, x_n\rho y_n$ imply $\beta(x_1, x_2, \ldots, x_n)\rho\beta(y_1, y_2, \ldots, y_n)$.

In more compact form (Section 8.2), a relation ρ on a set M is compatible with an operation β if and only if $\beta^2(\rho^n) \subseteq \rho$. We now give equivalence relations which are compatible with all operations of an operational system a special name.

Definition. A relation ρ on a set M of an operational system $\langle M, (\beta_i i \in I)\rangle$ is a *congruence* if and only if ρ is an equivalence relation compatible with every β_i, $i \in I$.

EXAMPLE. $\langle \mathbb{Z}, +, 0 \rangle$ has an equivalence relation ρ such that $x\rho y$ if and only if $x - y \in 3\mathbb{Z}$. Since $x_1\rho y_1$ and $x_2\rho y_2$ imply $(x_1 + x_2)\rho(y_1 + y_2)$, the equivalence relation ρ is compatible with the binary operation $+$. ρ is compatible with any nullary operation. ρ is therefore a congruence.

We now have the theorem showing that any morphism introduces a congruence upon its domain.

Theorem. *Let $f: M \to M'$ be a morphism of the operational systems $\langle M, (\beta_i | i \in I)\rangle$, $\langle M', (\beta'_i | i \in I)\rangle$. The relation $\rho = \{(x, y) | (x, y) \in M^2$ and $f(x) = f(y)\}$ is a congruence of M.*

Proof. That ρ is an equivalence relation was proved in Section 1.7. Now let β be any operation on M. Let $(\beta(x_1, x_2, \ldots, x_n), \beta(y_1, y_2, \ldots, y_n)) \in \beta^2(\rho^n)$. $((x_1, x_2, \ldots, x_n), (y_1, y_2, \ldots, y_n)) \in \rho^n$. $(f(x_1), f(x_2), \ldots, f(x_n)) = (f(y_1), f(x_2), \ldots, f(y_n))$. $\beta'(f(x_1), f(x_2), \ldots, f(x_n)) = \beta'(f(y_1), f(y_2), \ldots, f(y_n))$. $(\beta' \circ f^n)(x_1, x_2, \ldots, x_n) = (\beta' \circ f^n)(y_1, y_2, \ldots, y_n)$. $(f \circ \beta)(x_1, x_2, \ldots, x_n) = (f \circ \beta)(y_1, y_2, \ldots, y_n)$. $(\beta(x_1, x_2, \ldots, x_n), \beta(y_1, y_2, \ldots, y_n)) \in \rho$. $\beta^2(\rho^n) \subseteq \rho$. ρ is therefore a congruence. □

Associated with every congruence on an operational system there is a quotient system, a partition of the original set with well-defined operations derived representatively from the original operations. As throughout this book, monomorphisms are injective morphisms and epimorphisms are surjective morphisms.

Theorem. *Let $\langle M, (\beta_i | i \in I)\rangle$ be an operational system and ρ a congruence on M. Then $\langle M/\rho, (\bar{\beta}_i | i \in I)\rangle$ is an operational system where $\bar{\beta}_i(x_1/\rho, x_2/\rho, \ldots, x_n/\rho) = \beta_i(x_1, x_2, \ldots, x_n)/\rho$ for each $i \in I$.*

Proof. An equivalence relation ρ defined a partition $\{x/\rho | x \in M\}$ of M. That each β_i is compatible with this equivalence relation will mean that the operation $\bar{\beta}_i$ is well defined, independent of the representative used in the definition. Suppose $(x_1/\rho, x_2/\rho, \ldots, x_n/\rho) = (y_1/\rho, y_2/\rho, \ldots, y_n/\rho)$. $((x_1, x_2, \ldots, x_n), (y_1, y_2, \ldots, y_n)) \in \rho^n$. $(\beta(x_1, x_2, \ldots, x_n), \beta(y_1, y_2, \ldots, y_n)) \in \rho$. $\beta(x_1, x_2, \ldots, x_n)/\rho = \beta(y_1, y_2, \ldots, y_n)/\rho$. $\bar{\beta}(x_1/\rho, x_2/\rho, \ldots, x_n/\rho) = \bar{\beta}(y_1/\rho, y_2/\rho, \ldots, y_n/\rho)$. □

Corollary. *If ρ is a congruence on the operational system $\langle M, (\beta_i | i \in I)\rangle$ then the quotient map $\varphi: M \to M/\rho$ such that $\varphi(x) = x/\rho$ is an epimorphism.*

We now prove our fundamental morphism theorem on the factorization of a morphism into an epimorphism and a monomorphism.

Theorem. *Given a morphism $f: M \to M'$ of the operational systems $\langle M, (\beta_i | i \in I)\rangle$ and $\langle M', (\beta'_i | i \in I)\rangle$ there exist a congruence ρ on M, an epimorphism $\varphi: M \to M/\rho$ and a monomorphism $f': M/\rho \to M'$ such that $f' \circ \varphi = f$.*

Proof. The morphism f defines the congruence ρ and the epimorphism as already proved. The mapping f' is defined by $f'(x/\rho) = f(x)$ which was shown in Section 1.7 to be well defined as a function. That f' is a morphism is verified as follows. $f'(\overline{\beta}(x_1/\rho, x_2/\rho, \ldots, x_n/\rho)) = f'(\beta(x_1, x_2, \ldots, x_n)/\rho) = f(\beta(x_1, x_2, \ldots, x_n)) = \beta' \circ f^n(x_1, x_2, \ldots, x_n) = \beta'(f(x_1), f(x_2), \ldots, f(x_n)) = \beta'(f'(x_1/\rho), f'(x_2/\rho), \ldots, f'(x_n/\rho)) = \beta' \circ (f')^n(x_1/\rho, x_2/\rho, \ldots, x_n/\rho)$. Therefore, $f' \circ \overline{\beta} = \beta' \circ (f')^n$. This shows that f' preserves $\overline{\beta}$ in β' for every operation $\overline{\beta}$ of the operational system. □

We have considered the compatibility of operations with an equivalence relation in order to construct quotient operational systems. We now consider compatibility of r-ary relations with an equivalence relation in order to construct quotient relational systems.

Definition. An r-ary relation σ is *compatible* with an equivalence relation ρ if and only if $x_1 \rho y_1, x_2 \rho y_2, \ldots, x_r \rho y_r$ and $(x_1, x_2, \ldots, x_r) \in \sigma$ imply $(y_1, y_2, \ldots, y_r) \in \sigma$.

In essence, this definition means that whenever an r-ple is related by σ then all equivalent r-ples are also related by σ.

Definition. Let $\langle M; (\sigma_j | j \in J)\rangle$ be a relational system. Let ρ be an equivalence relation on M and let every σ_j, $j \in J$, be compatible with ρ. Then we call ρ a *relational congruence* for the relational system.

After these preliminary definitions we prove a theorem which shows the existence of the quotient relational system.

Theorem. *Let $\langle M; (\sigma_j | j \in J)\rangle$ be a relational system and a relational congruence for the relational system. Then $\langle M/\rho; (\overline{\sigma}_j | j \in J)\rangle$ is a relational system where $(x_1/\rho, x_2/\rho, \ldots, x_r/\rho) \in \overline{\sigma}_j$ if and only if $(x_1, x_2, \ldots, x_r) \in \sigma_j$ for all $j \in J$. We call $\langle M/\rho; (\sigma_j | j \in J)\rangle$ the quotient relational system.*

Proof. That ρ is a congruence ensures that the new quotient relation σ is well defined (for every σ_j; we drop the subscript for brevity). $(x_1/\rho, x_2/\rho, \ldots, x_r/\rho) = (y_1/\rho, y_2/\rho, \ldots, y_r/\rho)$ if and only if $(x_1, x_2, \ldots, x_r)\rho^r(y_1, y_2, \ldots, y_r)$. $(x_1/\rho, x_2/\rho, \ldots, x_r/\rho) \in \overline{\sigma}$ if and only if $(x_1, x_2, \ldots, x_r) \in \sigma$ if and only if $(y_1, y_2, \ldots, y_r) \in \sigma$ if and only if $(y_1/\rho, y_2/\rho, \ldots, y_r/\rho) \in \overline{\sigma}$. □

Theorem. *The quotient function $\varphi: M \to M/\rho$ such that $\varphi(x) = x/\rho$ mapping the relational system $\langle M; (\sigma_j | j \in J)\rangle$ into the quotient relational system (defined by a congruence ρ) $\langle M/\rho; (\overline{\sigma}_j | j \in J)\rangle$ is a full relation preserving morphism.*

Proof. $\varphi^r(\sigma) = \{(\varphi(x_1), \varphi(x_2), \ldots, \varphi(x_r)) | (x_1, x_2, \ldots, x_r) \in \sigma\} = \{(x_1/\rho, x_2/\rho, \ldots, x_r/\rho) | (x_1, x_2, \ldots, x_r) \in \sigma\} = \overline{\sigma}$ for each relation σ of the relational system. □

The apparent need for separate theorems for operations and relations is not totally real. The important question is whether an operation or relation is totally or partially defined. In operations we are treating the totally defined case and in relations the partially defined case.

We now offer a few examples to show some possibilities for interaction between subsystems and quotient systems. In the first example every equivalence class or coset in the quotient system is itself a subsystem. Let $\langle \mathbb{N}, \beta \rangle$ be such that β is a unary operation on $\mathbb{N}: \beta(2n) = 2n + 1$, $\beta(2n + 1) = 2n$. Define ρ, a relation on $\mathbb{N}$, such that $(2n + 1, 2n)$ and $(2n, 2n + 1)$ belong to ρ for every n in $\mathbb{N}$ and also include (m, m) for every $m \in N$ so that ρ is reflexive. This relation is an equivalence relation on $\mathbb{N}$ which produces a partition $\{\{0, 1\}, \{2, 3\}, \{4, 5\}, \ldots\}$ of $\mathbb{N}$. ρ is a congruence and N/ρ is a quotient system because $x\rho y$ implies $\beta(x)\rho\beta(y)$. It is easily verified that each coset of the partition is a subsystem of $\langle \mathbb{N}, \beta \rangle$.

In our second example we shall construct a quotient system in which no coset is an operational subsystem. Let $\langle \mathbb{N}, \gamma \rangle$ be such that γ is a unary operation on $\mathbb{N}: \gamma(n) = n + 2$. Let ρ be defined such that $(2n + 1, 2n)$, $(2n, 2n + 1), (m, m) \in \rho$ for every $m, n \in \mathbb{N}$. $\mathbb{N}/\rho$ is the same set as in example 1. Again $x\rho y$ implies $\gamma(x)\rho\gamma(y)$. ρ is a congruence. In this example no equivalence class is closed under γ and is therefore not an operational subsystem.

In our third example let $\langle \mathbb{N}, \delta \rangle$ be such that $\delta(0) = 1$, $\delta(1) = 0$, $\delta(m) = m + 2$ for $m \neq 0$, $m \neq 1$. We introduce two different equivalence relations for this operational system. We let ρ be as in the two previous examples leading to the partition $\mathbb{N}/\rho = \{\{0, 1\}, \{2, 3\}, \{4, 5\}, \ldots\}$. We let σ be $\{(n, n) | n \in \mathbb{N}\} \cup \{(0, 1), (1, 0)\}$ leading to the partition $\mathbb{N}/\sigma = \{\{0, 1\}, \{2\}, \{3\}, \{4\}, \ldots\}$. Both ρ and σ are congruences and both $\mathbb{N}/\rho$ and $\mathbb{N}/\sigma$ contain the operational subsystem $\{0, 1\}$ as a member. The moral of this third example is that one cannot always construct the quotient system by knowing that some subsystem is a member of the quotient operational system. For some special operational systems such a procedure is possible; for example, from a given ideal of a ring one can construct the quotient ring.

The following theorem is a version of what is sometimes known as the second isomorphism theorem.

Theorem. *Let ρ and σ be congruences of the operational system $\langle M, (\beta_i | i \in I) \rangle$ such that $\rho \subseteq \sigma$. Then there exists a congruence τ of M/ρ such that $(M/\rho)/\tau$ is isomorphic to M/σ.*

PROOF. Define $f: M/\rho \to M/\sigma$ such that $f(x/\rho) = x/\sigma$. $x/\rho = y/\rho$ implies $(x, y) \in \rho$ which implies $(x, y) \in \sigma$, which implies $x/\sigma = y/\sigma$, which in turn implies $f(x/\rho) = f(y/\rho)$. Note in particular that $x/\rho \subseteq x/\sigma$. We now verify that f is a morphism. $\overline{\overline{\beta}} \circ f^n(x_1/\rho, x_2/\rho, \ldots, x_n/\rho) = \overline{\overline{\beta}}(f(x_1/\rho), f(x_2/\rho), \ldots, f(x_n/\rho)) = \overline{\overline{\beta}}(x_1/\sigma, x_2/\sigma, \ldots, x_n/\sigma) = \beta(x_1, x_2, \ldots, x_n)/\sigma = f(\beta(x_1, x_2, \ldots, x_n)/\rho) = f(\overline{\beta}(x_1/\rho, x_2/\rho, \ldots, x_n/\rho))$. By the fundamental morphism theorem there exist a τ and mappings φ and f' such that $f': (M/\rho)/\tau \to M/\sigma$ is an isomorphism. □

In the usual notation of rings and groups, vector spaces, this theorem is not stated in terms of congruences but rather in terms of ideals, normal subgroups, and subspaces which define congruences. We now make a generalization of these theorems in the latter form for operational systems.

Definition. A subsystem N of an operational system $\langle M, (\beta_i|i \in I)\rangle$ is called a *normal subsystem* if and only if N is a member of M/ρ for some congruence ρ.

From the third example we just gave before the previous theorem we see that the normal operational subsystem does not, in general, completely determine the congruence. Nevertheless, we shall normally denote M/ρ by M/N; we shall leave to context the full meaning of M/N. We give now the altered form of the theorem.

Theorem. *Let K and L be normal operational subsystems of the operational system $\langle M, (\beta_i|i \in I)\rangle$ such that $K \subseteq L$. Then $(M/K)/(L/K)$ is isomorphic with M/L.*

PROOF. That K and L are normal subsystems means there exist congruences ρ and σ of M such that $K \in M/\rho$ and $L \in M/\sigma$. From the previous theorem we have that there exists an isomorphism $f':(M/K)/\tau \to M/L$. We complete the proof by showing that $L/K \in (M/K)/\tau$. Let $x/K \in L/K$ and $y/K \in M/K$. $x/K\tau y/K$ if and only if $f(x/K) = f(y/K)$ if and only if $x/L = y/L$ if and only if $x\sigma y$ if and only if $y \in x/\sigma = L$. $\{y/K|y/K\tau x/K\} = \{y/K|y \in L\} = L/K$. Therefore, $L/K \in (M/K)/\tau$ and we can denote $(M/K)/\tau$ by $(M/K)/(L/K)$. □

We prove now one theorem typical of a number of possibilities which discuss closure classes of congruences, normal subsystems and generation.

Theorem. *Any nonempty subset of an operational system $\langle M, (\beta_i|i \in I)\rangle$ is contained in some smallest normal operational subsystem N.*

PROOF. We obtain a smallest normal subsystem by choosing or constructing the smallest possible congruence in the class of all congruences which have the given property. This is equivalent to our problem because the fewer elements congruent to a given element then the fewer the members of the coset containing that member.

Let $P = \{\rho|\text{there exists } x \in M \text{ such that } S \subseteq x/\rho \text{ and } x/\rho \text{ is an operational subsystem of } M\}$. P is nonempty because $M \times M$ is a congruence yielding the partition $\{M\}$; M is a subsystem containing S. The remainder of the proof consists in showing that $\bigcap P$ has all the desired properties. It is obviously the smallest member of P if it belongs to P. It is clear how to show that $\bigcap P$ is an equivalence relation. We continue the proof by showing it to be a congruence. Let $((x_1, x_2, \ldots, x_n), (y_1, y_2, \ldots, y_n)) \in (\bigcap P)^n$. Then $((x_1, x_2, \ldots, x_n), (y_1, y_2, \ldots, y_n)) \in \rho^n$ for all ρ in P. $(\beta(x_1, x_2, \ldots, x_n), \beta(y_1, y_2, \ldots, y_n)) \in \rho$ for all ρ in P. $(\beta(x_1, x_2, \ldots, x_n), \beta(y_1, y_2, \ldots, y_n)) \in \bigcap P$.

We now show that S is a subset of some coset of $M/\bigcap P$. By the definition of P we have $S \subseteq x/\rho$ for some $x \in M$ for every ρ in P. Let $a \in S$. Then $S \subseteq a/\rho$ for every $\rho \in P$. We wish to show $S \subseteq a/\bigcap P$. Let $y \in S$. $y \in a/\rho$ for every ρ in P. $(y, a) \in \rho$ for each ρ in P. $(y, a) \in \bigcap P$. $y \in a/\bigcap P$.

Finally, we show that $a/\bigcap P$ is an operational subsystem. Let $x_1, x_2, \ldots, x_n \in a/\bigcap P$. $(x_1, a), (x_2, a), \ldots, (x_n, a) \in \bigcap P$. $(x_1, a), (x_2, a), \ldots, (x_n, a) \in \rho$ for each ρ in P. $x_1, x_2, \ldots, x_n \in a/\rho$ for each ρ in P. $\beta(x_1, x_2, \ldots, x_n) \in a/\rho$ for each ρ in P because a/ρ is an operational subsystem. $(\beta(x_1, x_2, \ldots, x_n), a)$ for each ρ in P. $(\beta(x_1, x_2, \ldots, x_n), a) \in \bigcap P$. $\beta(x_1, x_2, \ldots, x_n) \in a/\bigcap P$. $a/\bigcap P$ is the desired normal operational subsystem with S as subset. □

We now introduce the necessary definitions in order to be able to have the kernel of a morphism.

Definition. An operational system $\langle M, (\beta_i | i \in I)\rangle$ is a *zero system* if and only if the operational system has a nullary operation θ such that $\{\theta(0)\}$ is an operational subsystem of M.

We note that in a zero system we have $[\varnothing] = \{\theta(0)\}$.

EXAMPLE. In the ring $\langle R, +, \cdot, \theta\rangle$ we have $\{\theta\}$ is closed under $+$ and $\cdot$ and therefore $\langle R, +, \cdot, \theta\rangle$ is a zero system.

Definition. For any morphism between zero systems $\langle M, (\beta_j | j \in J), \theta\rangle$ and $\langle M', (\beta'_j | j \in J), \theta'\rangle$, $f: M \to M'$, we define the *kernel* of f to be the set $f^{-1}(\{\theta'(0)\})$.

Theorem. *Let $f: M \to M'$ be a morphism between zero systems $\langle M, (\beta_j | j \in J), \theta\rangle$ and $\langle M', (\beta'_j | j \in J), \theta'\rangle$. Then* kernel f *is a normal subsystem of M.*

PROOF. Since the preimage of an operational subsystem is an operational subsystem we know kernel f is an operational subsystem. The relation ρ is defined so that $x\rho y$ if and only if $f(x) = f(y)$ and is a congruence. $x \in$ kernel f if and only if $f(x) = \theta'(0)$ if and only if $f(x) = f(\theta(0))$ if and only if $x\rho\theta(0)$ if and only if $x \in \theta(0)/\rho$. Kernel $f \in M/\rho$ proves kernel f is normal. We then write the quotient system as $M/\ker f$. □

Corollary. *$M/\ker f$ is isomorphic to $f(M)$.*

We note that a unitary ring is not a zero system because $\{\theta\}$ is not closed under the nullary operation ν and is therefore not an operational subsystem.

We do not define kernel f for a type (1, 2, 2, 0, 0) morphism between unitary rings $\langle R, +, \cdot, \theta, v\rangle$ and $\langle R', +', \cdot', \theta', v'\rangle$. If we work with a kernel in this case we work with the unitary ring merely as a ring and consider type (1, 2, 2, 0) morphisms.

Questions

1. Let $\langle M, (\beta_i | i \in I)\rangle$ be an operational system. Which of these statements are true?
   (A) An equivalence relation is compatible with every operation of M.
   (B) A relation ρ is a congruence if and only if ρ is compatible with every operation of M.
   (C) A quotient system of M is a subsystem which is normal.
   (D) There exists at least one quotient system of M.
   (E) None of the four statements is true.

2. Let $\langle M, (\beta_i | i \in I)\rangle$ be an operational system. Which of these statements are true?
   (A) A partition of M may fail to be a quotient system.
   (B) A quotient system of M must contain some subsystem of M.
   (C) Any subsystem of M is a member of some quotient system of M.
   (D) Any morphism of M into M has a kernel.
   (E) None of the statements is true.

3. Define $x\rho y$ if and only if $x - y \in \mathbb{Q}$ on the operational system $\langle \mathbb{R}, +, \cdot, 0, n\rangle$ where n stands for negation. Which of these statements are true?
   (A) ρ is a congruence.
   (B) $\mathbb{Q}$ is a normal subsystem of $\mathbb{R}$.
   (C) $\langle \mathbb{R}, +, \cdot, 0, n\rangle$ is a zero system.
   (D) $(x, y) \in \rho$ implies $(n(x), n(y)) \in \rho$.
   (E) None of the statements is true.

4. Let A and B be ideals of the ring $\langle R, +, \cdot, \theta, n\rangle$, an operational system, such that $A \subseteq B$ and n stands for negation. Which of the following statements are true?
   (A) $B/A \subseteq R/A$.
   (B) $R/B \approx (R/A)/(B/A)$.
   (C) $R/B \subseteq R/A$.
   (D) $R/B \underline{\in} R/A$.
   (E) None of the statements is true.

Exercises

1. Let $\langle R, +, \cdot, \theta, n\rangle$ be a ring in which n stands for negation. Show that every operational subsystem is a subring and vice versa. Show that any subset A of R is a normal operational subsystem if and only if A is an ideal of R as a ring.

2. Let $\langle M, (\beta_i | i \in I); \sigma\rangle$ be an algebraic system. Let the relation σ be compatible with every operation β_i, $i \in I$. Let ρ be an equivalence relation compatible with every operation β_i, $i \in I$, and also compatible with the relation σ. Show that $\langle M/\rho, (\bar{\beta}_i | i \in I); \bar{\sigma}\rangle$ is an algebraic system in which $\bar{\sigma}$ is compatible with every $\bar{\beta}_i$, $i \in I$, and therefore is a quotient algebraic system.

8.5 Products and sums

In this section we place constructions of Cartesian products, weak Cartesian products, and powers.

From two given operational systems of the same type we can construct a new operational system of the same type including the given operational systems.

Definition. Let $\langle M, (\beta_i | i \in I)\rangle$ and $\langle N, (\gamma_i | i \in I)\rangle$ be similar operational systems. For each i in I we define an n-ary operation $\beta_i \times \gamma_i$ on $M \times N$ such that $(\beta_i \times \gamma_i)((x_1, y_1), (x_2, y_2), \ldots, (x_n, y_n)) = (\beta_i(x_1, x_2, \ldots, x_n), \gamma_i(y_1, y_2, \ldots, y_n))$. The operational system $\langle M \times N, (\beta_i \times \gamma_i | i \in I)\rangle$ is defined to be the *Cartesian product* and has the same type as both the given systems.

We note that the mapping or operation $\beta \times \gamma : (M \times N)^n \to M \times N$ arises from $\beta : M^n \to M$ and $\gamma : N^n \to N$ which yield, strictly speaking, $\beta \times \gamma : M^n \times N^n \to M \times N$ from pairing. We choose to identify $(M \times N)^n$ and $M^n \times N^n$ by their obvious isomorphism.

The original operational systems are included in the Cartesian product system in a sense illustrated by this diagram and lemma.

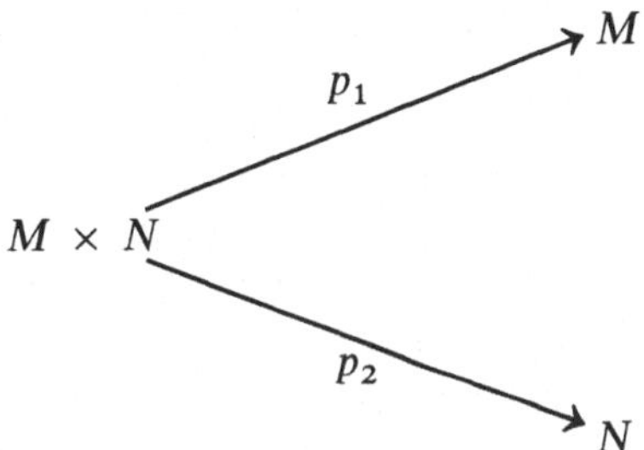

Lemma. *The projections $p_1 : M \times N \to M$ and $p_2 : M \times N \to N$ of the Cartesian product into its components are epimorphisms.*

Proof. We assume both M and N are nonempty, for otherwise, the statement may be false. Let $b \in N$. Then for any $x \in M$ we have $p_1(x, b) = x$ proving p_1 to be a surjection. To show that p_1 is a morphism we show that it preserves the operation $\beta \times \gamma$ of $M \times N$ in the corresponding operation β of M.

$$\begin{aligned}
(p_1 \circ (\beta \times \gamma))((x_1, y_1), (x_2, y_2), \ldots, (x_n, y_n)) & \\
&= p_1(\beta(x_1, x_2, \ldots, x_n), \gamma(y_1, y_2, \ldots, y_n)) \\
&= \beta(x_1, x_2, \ldots, x_n) \\
&= \beta(p_1(x_1, y_1), p_1(x_2, y_2), \ldots, p_1(x_n, y_n)) \\
&= (\beta \circ p_1^n)((x_1, y_1), (x_2, y_2), \ldots, (x_n, y_n)).
\end{aligned}$$

$$p_1 \circ (\beta \times \gamma) = \beta \circ p_1^n.$$

Likewise p_2 is an epimorphism. □

The following theorem shows how it is possible to combine two separate morphisms through the Cartesian product.

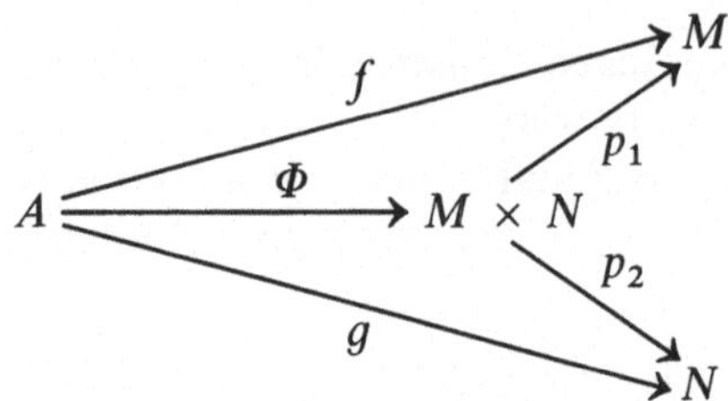

Theorem. *Let $\langle M, (\beta_i|i \in I)\rangle$ and $\langle N, (\gamma_i|i \in I)\rangle$ be similar operational systems. Then for any operational system $\langle A, (\alpha_i|i \in I)\rangle$ of the same type and morphisms $f: A \to M, g: A \to N$ there exists a unique morphism $\Phi: A \to M \times N$ such that $f = p_1 \circ \Phi$ and $g = p_2 \circ \Phi$.*

PROOF Define $\Phi(u) = (f(u), g(u))$. Then $\Phi(\alpha(x_1, x_2, \ldots, x_n) = (f(\alpha(x_1, x_2, \ldots, x_n), g(\alpha(x_1, x_2, \ldots, x_n))) = (\beta \circ f^n(x_1, x_2, \ldots, x_n), \gamma \circ g^n(x_1, x_2, \ldots, x_n)) = (\beta \times \gamma)(f^n(x_1, x_2, \ldots, x_n), g^n(x_1, x_2, \ldots, x_n)) = (\beta \times \gamma) \circ \Phi^n(x_1, x_2, \ldots, x_n)$. □

EXAMPLE. The rings $\langle \mathbb{Z}_2, +, \cdot, \bar{0}\rangle$ and $\langle \mathbb{Z}_3, +, \cdot, \bar{0}\rangle$ have a Cartesian product which is the ring $\langle \mathbb{Z}_2 \times \mathbb{Z}_3, + \times +, \cdot \times \cdot, (\bar{0}, \bar{0})\rangle$.

The definition given for the Cartesian product of two systems can be extended to any number of operational systems. Let $(\langle M_j, (\beta_{ij}|i \in I)\rangle | j \in J)$ be a family of similar operational systems, indexed by the set J. The *Cartesian product* $\times_{j \in J} M_j$ of the sets consists of all families of the form $(x_j|j \in J)$ in which $x_j \in M_j$ for each $j \in J$. An operation $\times_{j \in J} \beta_{ij}$ on the Cartesian product is defined from the operations $(\beta_{ij}|j \in J)$ so that $(\times_{j \in J} \beta_{ij})((x_{1j}|j \in J), (x_{1j}|j \in J), \ldots, (x_{nj}|j \in J)) = (\beta_{ij}(x_{1j}, x_{2j}, \ldots, x_{nj})|j \in J)$. This gives us the operational system $\langle \times_{j \in J} M_j, (\times_{j \in J} \beta_{ij}|i \in I)\rangle$.

If all the component systems are identical, say $M_j = M$ for all $j \in J$ and $\beta_{ij} = \beta_i$ for all $j \in J$, then we call the Cartesian product a *power* and write the system $\langle M^J, (\beta_i^J|i \in I)\rangle$. We have associated with each Cartesian product a collection of projections which are epimorphisms. $p_k: \times_{j \in J} M_j \to M_k$ such that $p_k(x_j|j \in J) = x_k$.

For the case when J is infinite we can define a proper subsystem of the Cartesian product called the *weak Cartesian product*. We briefly discuss this only for the power case. Let a be some fixt element of M (for example, it could be the neutral element of a group). We define $(M^J)^w = \{(x_j|j \in J)|(x_j|j \in J) \in M^J$ and $x_j = a$ for all but a finite number of $j\}$.

EXAMPLE. We form of $\langle \mathbb{R}, +, 0\rangle$ the power $\mathbb{R}^{\mathbb{N}}$. $\mathbb{R}^{\mathbb{N}}$ consists of all infinite sequences. Addition in $\mathbb{R}^{\mathbb{N}}$ is coordinatewise, $(x_i|i \in \mathbb{N}) + (y_i|i \in \mathbb{N}) = (x_i + y_i|i \in \mathbb{N})$, and the zero element is $(0|i \in \mathbb{N})$. The weak Cartesian power is the subset $(\mathbb{R}^{\mathbb{N}})^w$ of all sequences which have all save a finite number of entries equal to zero.

Relational systems also admit a natural Cartesian product construction not unlike the operational systems.

Definition. If $\langle M; (\sigma_j | j \in J)\rangle$ and $\langle N; (\tau_j | j \in J)\rangle$ are relational systems then $\langle M \times N; (\sigma_j \times \tau_j | j \in J)\rangle$ is a relational system and is called the *Cartesian product*. $\sigma_j \times \tau_j = \{((x_1, y_1), (x_2, y_2), \ldots, (x_n, y_n)) | (x_1, x_2, \ldots, x_n) \in \sigma_j$ and $(y_1, y_2, \ldots, y_n) \in \tau_j\}$.

This relational product $\sigma_j \times \tau_j$ is an abuse of notation as a rearrangement of ordered pairs must take place.

Theorem. *Let $\langle M; (\sigma_j | j \in J)\rangle$ and $\langle N; (\tau_j | j \in J)\rangle$ be relational systems. Then the projections $p_1 : M \times N \to M$ and $p_2 : M \times N \to N$ are relation preserving morphisms from the Cartesian product relational system to the component relational systems.*

PROOF $((x_1, y_1), (x_2, y_2), \ldots, (x_n, y_n)) \in \sigma \times \tau$ if and only if $(x_1, x_2, \ldots, x_n) \in \sigma$ and $(y_1, y_2, \ldots, y_n) \in \tau$ if and only if $(p_1(x_1, y_1), p_1(x_2, y_2), \ldots, p_1(x_n, y_n)) \in \sigma$ and $(p_2(x_1, y_1), p_2(x_2, y_2), \ldots, p_2(x_n, y_n)) \in \tau$ if and only if $p_1^n((x_1, y_1), (x_2, y_2), \ldots, (x_n, y_n)) \in \sigma$ and $p_2^n((x_1, y_1), (x_2, y_2), \ldots, (x_n, y_n)) \in \tau$. Thus $p_1^n(\sigma \times \tau) \subseteq \sigma$ and $p_2^n(\sigma \times \tau) \subseteq \tau$. □

For the case of an arbitrary number of relational systems $(\langle M_k : (\sigma_{jk} | j \in J)\rangle | k \in K)$ has a Cartesian product relational system $\langle \bigtimes_{k \in K} M_k ; (\bigtimes_{k \in K} \sigma_{jk} | j \in J)\rangle$. The Cartesian product relation is defined by this statement: $((x_{1k} | k \in K), (x_{2k} | k \in K), \ldots, (x_{nk} | k \in K)) \in \bigtimes_{k \in K} \sigma_{jk}$ if and only if $(x_{1k}, x_{2k}, \ldots, x_{nk}) \in \sigma_{jk}$ for all $k \in K$.

By analogy with the theorem characterizing the Cartesian product operational system in terms of morphisms we use the dual characterization to define the coproduct of two operational systems.

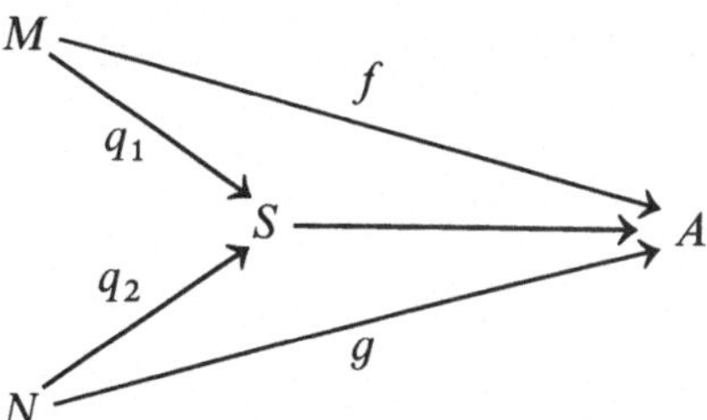

Definition. Let $\langle M, (\beta_i | i \in I)\rangle$ and $\langle N, (\gamma_i | i \in I)\rangle$ be similar operational systems. If there exists a similar operation system $\langle S, (\delta_i | i \in I)\rangle$ and monomorphisms $q_1 : M \to S$, $q_2 : M \to S$ such that if $\langle A, (\varepsilon_i | i \in I)\rangle$ is any similar operational system and $f : M \to A$ and $g : N \to A$ are morphisms implies there exists a unique morphism $\Psi: S \to A$ such that $\Psi \circ q_1 = f$ and $\Psi \circ q_2 = g$ then we call the operational system $\langle S, (\delta_i | i \in I)\rangle$ a *coproduct* of M and N and we write $S = M \vee N$.

EXAMPLE. Let $\langle M, \theta\rangle$ and $\langle N, \eta\rangle$ be operational systems with one nullary operation. The type is (1, 0). The coproduct exists for these systems and is the disjoint union with identified special points. In detail, $S = \{(x, 0) | x \in M - \{\theta(0)\}\} \cup \{(x, 1) | x \in N - \{\eta(0)\}\} \cup \{P\}$ with nullary operation ζ on S such that $\zeta(0) = P$. $q_1: M \to S$ such that $q_1(x) = x$ for $x \neq \theta(0)$, $q_1(x) = P$ for $x = \theta(0)$. $q_2: N \to S$ such that $q_2(x) = x$ for $x \neq \eta(0)$, $q_2(x) = P$ for $x = \eta(0)$. Define for $f: M \to A$ and $g: N \to A$ the function Ψ so that $\Psi(y) = f(x)$ if $y = q_1(x)$ for some $x \in M$ and $\Psi(y) = g(x)$ if $y = q_2(x)$ for some $x \in N$. Then $\Psi(q_1(x)) = f(x)$ for $x \in M$ and $\Psi(q_2(x)) = g(x)$ for $x \in N$. $\Psi(P) = f(\theta(0)) = g(\eta(0)) = \zeta(0)$.

EXAMPLE. The commutative groups $\langle M, \cdot, \nu, \gamma\rangle$ and $\langle M', \cdot', \nu', \gamma'\rangle$ have a coproduct $\langle M \times M', \cdot \times \cdot', \nu \times \nu', \gamma \times \gamma'\rangle$. In this instance, the coproduct is also the Cartesian product, an unusual event. We briefly illustrate how the Cartesian product is also the coproduct. $M \times M' = \{(x, x') | x \in M, x' \in M'\}$. Define $q_1: M \to M \times M'$ such that $q_1(x) = (x, \nu'(0))$ and $q_2: M' \to M \times M'$ such that $q_2(x') = (\nu(0), x')$. If $f: M \to A$ and $g: M' \to A$ are morphisms into some commutative group A then we define $\Psi: M \times M' \to A$ such that $\Psi(x, x') = f(q_1^{-1}(x))g(q_2^{-1}(x'))$.

QUESTIONS

1. Which of the following complete a true sentence? The Cartesian product of the integers, $\langle \mathbb{Z}, +, \cdot, 0, n\rangle$, with themselves is
   (A) an integral domain
   (B) a ring
   (C) a field
   (D) not defined.
   (E) None of the alternatives completes a true sentence.

2. The weak power, $(\mathbb{Z}^{\mathbb{N}})^w$, of the integers, $\langle \mathbb{Z}, +, \cdot, 0, n\rangle$
   (A) is a ring
   (B) is an integral domain
   (C) contains sequences of integers
   (D) contains a neutral element of multiplication.
   (E) None of the choices completes a true sentence.

3. In the Cartesian product $\times_{n=2,3,\ldots} \mathbb{Z}_n$ of rings $\langle \mathbb{Z}_n, +, \cdot, \bar{0}, n\rangle$
   (A) there are an infinite number of elements
   (B) some elements have reciprocals
   (C) there are subfields isomorphic to $\mathbb{Z}_p$ for every prime p
   (D) each element is a sequence.
   (E) None of the choices completes a true sentence.

Monoids and groups 9

This chapter is devoted to algebraic systems of one binary operation. Groups have been used many times before in the text, as permutations, as modules (a commutative group with an exterior ring multiplication), and elsewhere, but here we organize the systems of one binary operation into one chapter and give some additional specialized results. The most general algebraic system we consider in this chapter is the monoid, a set with one associative binary operation. Some special monoids discussed are unitary monoids, cancellative monoids, and groups. We discuss subsystems, quotient systems, morphisms, and the fundamental morphism theorem. In Sections 9.4 and 9.5 we study more specialized results available for groups alone. We study cyclic groups and connect the order of an element with this concept. Several topics such as center, normalizer, conjugacy classes are all organized around inner automorphisms ($\varphi_a(x) = a^{-}xa$). We then apply these results to answer some questions about elements of prime order. We offer some standard theorems relating direct products and Cartesian products of groups. We define simple groups and solvable groups. We give an inductive proof of the fundamental theory of Abelian groups which is subsumed in Chapter 10 by the direct sum resolution of a finitely generated module.

9.1 Monoids, unitary monoids, cancellative monoids, and groups

In this section we give the basic definitions for algebraic systems having one binary operation. Although some of the systems have been previously defined we will repeat the definitions here.

Definitions. A *monoid* $\langle M, \cdot\rangle$ is a set M together with an associative binary operation $\cdot$.

A *unitary monoid* $\langle M, \cdot, v\rangle$ is a set M together with an associative binary operation $\cdot$ and a nullary operation v such that $v(0)$ is a neutral element for the binary operation.

A *cancellative monoid* $\langle M, \cdot\rangle$ is a set together with an associative binary operation $\cdot$ for which the cancellative properties holds: $xy = xz$ implies $y = z$ and $yx = zx$ implies $y = z$. One might also consider monoids with only left cancellation or only right cancellation.

A *group* $\langle G, \cdot, v\rangle$ is a set G, an associative binary operation $\cdot$, and a nullary operation v such that $v(0)$ is a neutral element for $\cdot$ and every x in G has an inverse with respect to the binary operation. An equivalent definition is that a group $\langle G, \cdot, v, \gamma\rangle$ is a set G, an associative binary operation $\cdot$, a nullary operation v, a unary operation γ such that $v(0)$ is the neutral element for multiplication, and $\gamma(x)$ is the multiplicative inverse of x.

If the binary operation of a monoid is commutative then the monoid is called a commutative monoid. A commutative group is frequently called an Abelian group (from Niels Abel, a principal worker in early group theory).

EXAMPLES. $\langle \mathbb{N}, +, 0\rangle$ is a unitary commutative cancellation monoid. $\langle \mathbb{N}, \cdot, 1\rangle$ is a unitary commutative monoid. $\langle \mathbb{N}^+, \cdot, 1\rangle$ is a unitary commutative cancellation monoid. If $\mathbb{Z}^* = \mathbb{Z} - \{0\}$ then $\langle \mathbb{Z}^*, \cdot, 1\rangle$ is a unitary commutative cancellation monoid. $\langle \mathbb{Z}, +, 0\rangle$ is a commutative group. $\langle \mathbb{Q}^*, \cdot, 1\rangle$ is a commutative group. $\langle \mathfrak{S}(X), \circ, I\rangle$, the set of all bijections on a set X together with composition and the identity function, is a group. The group is noncommutative if X contains more than two members. $\langle \mathbb{Z}_5, +, \bar{0}\rangle$ is a commutative group. $\langle \mathbb{Z}_4, \cdot, \bar{1}\rangle$ is a unitary commutative monoid. $\langle 2\mathbb{N}, +, 0\rangle$ is a unitary commutative monoid. For any set X, $\langle X^X, \circ, I\rangle$ is a unitary monoid. If $\mathrm{Sur}(X)$ stands for the set of all surjections on a set X then $\langle \mathrm{Sur}(X), \circ, I\rangle$ is a unitary right cancellative monoid.

We now simply list, without proof, necessary and sufficient conditions for a subset of an algebraic system to be a subsystem.

Theorem. *F is a submonoid of a monoid $\langle E, \cdot\rangle$ if and only if $x, y \in F$ imply $xy \in F$. F is a unitary submonoid of a unitary monoid $\langle E, \cdot, v\rangle$ if and only if $v \in F$ and $x, y \in F$ imply $xy \in F$. F is a subgroup of a group $\langle G, \cdot, v\rangle$ if and only if $x, y \in F$ imply $xy \in F$, $x \in F$ implies $x^- \in F$, and $v \in F$ (or $F \neq \varnothing$).*

EXAMPLES. $2\mathbb{N}$ is a submonoid of the monoid $\langle \mathbb{N}, \cdot\rangle$ but not a unitary submonoid of the unitary monoid $\langle \mathbb{N}, \cdot, 1\rangle$. $\mathbb{Z}^*$ is a unitary submonoid of the unitary monoid $\langle \mathbb{Q}^*, \cdot, 1\rangle$ but not a subgroup of the group $\langle \mathbb{Q}^*, \cdot, 1\rangle$. We also recall a result from the chapter on integers: the only subgroups of $\langle \mathbb{Z}, +, 0\rangle$ are of the form $n\mathbb{Z}$ for some $n \in \mathbb{N}$.

On the generation of subsystems we have these results.

Theorem. *If $\langle M, \cdot\rangle$ is a monoid and S is a subset of M then there exists a smallest submonoid $[S]$ of M including S. If $\langle M, \cdot, v\rangle$ is a unitary monoid and S is a subset of M then there exists a smallest unitary submonoid $[S]$ of M including S. If $\langle M, \cdot, v\rangle$ is a unitary cancellative monoid and S is a subset of M then there exists a smallest unitary cancellative submonoid $[S]$ of M including S. If $\langle G, \cdot, v\rangle$ is a group and S is a subset of G then there exists a smallest subgroup $[S]$ of G including S.*

PROOF. In each case from the set $\mathscr{C}$ of all subsystems which include the given set S, prove that the class is nonempty, and show that the intersection of the class is itself a member of the class. □

Definition. If there exists a finite subset S of M such that $[S] = M$ then we say that the monoid $\langle M, \cdot\rangle$ is *finitely generated.*

EXAMPLE. $\langle \mathbb{N}, +\rangle$ is generated by $\{0, 1\}$ and is therefore a finitely generated monoid. The group $\langle \mathbb{Z}, +, 0\rangle$ is generated by the subset $\{1\}$. The monoid $\langle \mathbb{N}, \cdot\rangle$ is not finitely generated. The monoid $\langle \mathbb{N}, \cdot\rangle$ is generated by the subset consisting of 0, 1, and all prime numbers.

Definition. A subset S of a monoid $\langle M, \cdot\rangle$ is *independent* if and only if no proper subset of S generates $[S]$. An independent generating subset of a monoid is called a basis.

EXAMPLE. $\{0, 1\}$ is a basis for the monoid $\langle \mathbb{N}, +\rangle$. $\{0, 1, 2\}$ generates the monoid $\langle \mathbb{N}, +\rangle$ but fails to be an independent set. $\{2, 3\}$ and $\{1\}$ are both bases for the group $\langle \mathbb{Z}, +, 0\rangle$.

QUESTIONS

1. If $\langle M, \cdot\rangle$ is a monoid then
   (A) there cannot be a neutral element for M
   (B) there cannot be two distinct neutral elements for M
   (C) there cannot be two distinct inverses for any element of M
   (D) a cancellation law cannot hold.
   (E) None of the alternatives completes a satisfactory sentence.

2. A unitary monoid $\langle M, \cdot, v\rangle$ which fails to be a group
   (A) cannot contain a subgroup other than $\{v\}$
   (B) cannot be a cancellative monoid
   (C) contains at least two elements
   (D) must be infinite.
   (E) None of the alternatives completes a satisfactory sentence.

3. Which of the following statements are correct?
   (A) The natural numbers, $\langle \mathbb{N}, +, \cdot, 0, 1\rangle$, do not make a ring.
   (B) $\langle \mathbb{N}, +, 0\rangle$ is a commutative cancellative unitary monoid.

(C) $\langle \mathbb{N}, \cdot, 1 \rangle$ is a commutative cancellative unitary monoid.
(D) The monoid $\langle \mathbb{N}, + \rangle$ has an infinite number of submonoids.
(E) None of the statements is true.

4. Let X be a given set. Which of these statements are true?
(A) The set $\mathrm{Sur}(X)$ of all surjections of X onto X is a right cancellative unitary monoid under composition.
(B) The set $\mathrm{Inj}(X)$ of all injections of X into X is a left cancellative unitary monoid under composition.
(C) The set $\mathfrak{S}(X)$ of all bijections of X onto X is a unitary submonoid of both $\mathrm{Sur}(X)$ and $\mathrm{Inj}(X)$.
(D) Some member of $\mathrm{Inj}(X)$ can have two distinct left inverses in the monoid $\mathrm{Inj}(X)$.
(E) None of the sentences is true.

5. Which of the following statements are true?
(A) $\{1\}$ generates the monoid $\langle \mathbb{N}, + \rangle$.
(B) The set of all prime numbers generates $\langle \mathbb{N}, \cdot \rangle$.
(C) The set of all prime numbers generates the unitary monoid $\langle \mathbb{N}^+, \cdot, 1 \rangle$.
(D) $\{p/q | p$ and q are prime numbers$\}$ generates the unitary monoid $\langle \mathbb{Q}^*, \cdot, 1 \rangle$.
(E) None of the statements is true.

6. Which of the following statements are true?
(A) $\mathbb{Q}$ generates the group $\langle \mathbb{R}, +, 0 \rangle$.
(B) $\{2, 5\}$ generates the group $\langle \mathbb{Z}, +, 0 \rangle$.
(C) $\{2, 5\}$ generates the monoid $\langle \mathbb{N}, + \rangle$.
(D) $\{0, 1, 2\}$ generates the monoid $\langle \mathbb{N}, + \rangle$.
(E) None of the statements is true.

7. The elements of the group $\langle \mathfrak{S}_3, \circ, I \rangle$ in this question are written in cyclic notation (cf. Section 7.7). Which of the following sets generate $\mathfrak{S}_3$?
(A) $\{(1\ 2), (1\ 3)\}$
(B) $\{(1\ 2\ 3)\}$
(C) $\{(1\ 2\ 3), (1\ 3\ 2)\}$
(D) $\{(1\ 2\ 3), (1\ 2)\}$.
(E) None of the sets generates $\mathfrak{S}_3$.

8. Which of the following statements are true?
(A) $\{(1\ 2), (1\ 3)\}$ is a basis for $\langle \mathfrak{S}_3, \circ, I \rangle$.
(B) $\{\bar{2}, \bar{3}\}$ is a basis for $\langle \mathbb{Z}_4, +, \bar{0} \rangle$.
(C) $\{-1\}$ is a basis for $\langle \mathbb{Z}, +, 0 \rangle$.
(D) $\{2, 7\}$ is a basis for $\langle \mathbb{Z}, +, 0 \rangle$.
(E) None of the statements is true.

Exercises

1. Construct a multiplication table for a group with elements v, a, b, c such that $ab = c$, $ca = b$, $bc = a$, $aa = v$, $bb = v$, $cc = v$, assuming the group to be commutative. What are the subgroups of this group? This group of four elements is called Klein's four group.

2. Construct the multiplication table for the monoid $\langle \mathbb{Z}_4, +\rangle$. Is there a neutral element? Is the monoid a group? Are there subgroups?

3. Verify that $\langle X^X, \circ, I\rangle$ is a unitary monoid where X is a given set.

4. Let U be the subset of X^X of all functions which leave all but a finite number of members of X fixt. A member c of X is left fixt by a function f if and only if $f(c) = c$. Prove U is a unitary submonoid of X^X.

5. Let V be the subset of X^X of all bijections of X which leave all but a finite number of elements fixt. Prove that V is a submonoid of both U (of Exercise 4) and $\mathfrak{S}(X)$. Prove V is a subgroup of $\mathfrak{S}(X)$.

6. Let T be the subset of V (of Exercise 5) of all transpositions. $T = \{f \mid f \in V$ and $f(x) = y, f(y) = x$ for some $x, y \in X, x \neq y$ and otherwise f is the identity on $X\}$. Show that the subgroup of $\mathfrak{S}(X)$ generated by T is V. Show furthermore that T is a minimal generating set for V.

7. Find all groups of cardinality 1, 2, 3, 4, 5, and 6. *Hint*: Construct all possible multiplication tables keeping in mind the definition of a group.

8. Prove there is at least one group of cardinality n for every positive integer n.

9. For the group $\mathfrak{S}_4$ find the subgroup generated by $\{(1\ 2), (1\ 3), (1\ 4)\}$. Find also the subgroup generated by $\{(1\ 2)(3\ 4), (1\ 3)(2\ 4), (1\ 4)(2\ 3)\}$. These elements are written in cyclic notation.

10. Show that $\mathfrak{S}_n$ is not commutative if $n \geqslant 3$.

11. Let $\{a_1, a_2, \ldots, a_n\}$ be a set of n ($\geqslant 1$) distinct symbols. Let M be the set of all strings of symbols obtained by placing any finite number of symbols adjacent to each other. Any symbol can be repeated any finite number of times in a string. For example, $a_1a_1a_3a_2a_4a_4a_4$, $a_2a_3a_5a_3a_3a_1$, and a_2a_2 are all three strings, members of M. We introduce an operation on M which is simply the juxtaposition of two strings to form one new string. $a_2a_3a_5a_3a_3a_1$ is juxtaposed with a_2a_2 to form $a_2a_3a_5a_3a_3a_1a_2a_2$. Show that M is a monoid with cancellation. If we include in M an empty string then M is a cancellative unitary monoid. Show that $\{1, a_1, a_2, \ldots, a_n\}$ is a basis for M.

12. We let $\{a_1, a_2, \ldots, a_n, a_1^-, a_2^-, \ldots, a_n^-\}$ be a set of $2n$ distinct symbols and form G, the set of all finite strings, and include the empty string. We further stipulate that a_j juxtaposed with a_j^- in either order contracts to the empty string, $j = 1, 2, \ldots, n$. Prove that G is a group.

13. Give an example of a group: (a) with exactly two subgroups; (b) with exactly three subgroups; (c) with exactly four subgroups.

14. If $\langle G, \cdot, v\rangle$ is a group prove that the equations $ax = b$ and $yc = d$ have solutions in G for x and y and that the solutions are unique (given $a, b, c, d \in G$).

15. Let $\langle R, +, \cdot, \theta\rangle$ be a ring. Verify that $\langle R, \cdot\rangle$ is a monoid. If $\langle R, +, \cdot, \theta, v\rangle$ is an integral domain what kind of monoid is $\langle R^*, \cdot, v\rangle$? If $\langle R, +, \cdot, \theta, v\rangle$ is a field then what kind of monoid is $\langle R^*, \cdot, v\rangle$?

16. Let $\langle G, +, \theta\rangle$ be a commutative group and define another operation on G, a multiplication, such that $xy = \theta$ for all $x, y \in G$. Is $\langle G, +, \cdot, \theta\rangle$ a ring? Is there a unity in the ring?

17. Let $\langle G, \cdot, v\rangle$ be a group. Prove that G is commutative if and only if $(xy)^2 = x^2y^2$ for all $x, y \in G$.

18. Let $\langle G, \cdot, v\rangle$ be a group. Prove that S, a nonempty subset of G, is a subgroup if and only if $x, y \in S$ imply $xy^{-} \in S$.

19. Let $\langle G, \cdot, v\rangle$ be a group in which every element is its own inverse. Prove that G is commutative.

20. Let $\langle G, \cdot, v\rangle$ be a unitary cancellative monoid with only a finite number of elements. Prove that G is a group. Do not assume G to be commutative.

21. By example show that the union of two subgroups of a given group is not necessarily a subgroup.

22. Prove that if H and K are subgroups of a given group $\langle G, \cdot, v\rangle$ and $H \cup K$ is also a subgroup of G then $H \subseteq K$ or $K \subseteq H$.

23. Let $\langle G, \cdot\rangle$ be a monoid. Suppose further that the equations $ax = b$ and $yc = d$ always have unique solutions for x and y in G. Prove that G is a group. [*Hint*: This problem must be done in several steps. Begin by proving the existence of left and right neutral elements for G. Do not assume that "neutral behavior" for one element is "neutral behavior" for all elements.]

24. Let $\langle G, \cdot\rangle$ be a monoid. Let v be a right neutral element for G. Suppose also that for each x in G there exists y in G so that $xy = n$ for *some* right neutral element n in G (Do not assume $n = v$). Prove G is a group. *Hint*: $x \in G$ and $xx = x$ imply $x = v$. $n \in G$ and n is a right neutral element imply $n = v$. If $uv = v$ then $vu = v$ also. $vx = x$ for all x in G. If $x \in G$ and $y, z \in G$ such that $xy = v$ and $xz = v$ then $y = z$.

9.2 Congruences and quotient systems

In this section we discuss equivalence relations and quotient systems for monoids.

We remind ourselves that an equivalence relation $\sim$ is compatible with a binary relation $\cdot$ if and only if $x_1 \sim y_1$ and $x_2 \sim y_2$ imply $x_1x_2 \sim y_1y_2$. An equivalence relation $\sim$ is compatible with a unary operation γ if and only if $x \sim y$ implies $\gamma(x) \sim \gamma(y)$. An equivalence relation is always compatible with a nullary relation. A congruence is, by definition, an equivalence relation compatible with all the operations of a system.

Theorem. *Let $\langle M, \cdot\rangle$ be a monoid and $\sim$ a congruence. Then $\bar{\cdot}$ such that $x/\!\sim \bar{\cdot}\; y/\!\sim\; = xy/\!\sim$ is a well-defined binary operation on the quotient set $M/\!\sim$ and $\langle M/\!\sim, \bar{\cdot}\rangle$ is a monoid (the quotient monoid of M).*

Proof. The compatibility of the congruence ensures that the operation on the quotient set is well-defined. □

For groups a little compatibility leads to a lot of compatibility.

Theorem. *Let $\langle G, \cdot, \nu, \gamma\rangle$ be a group. Let $\sim$ be an equivalence relation on G compatible with the binary operation of multiplication. Then $\sim$ is a congruence for the group defining a quotient group $\langle G/\sim, \bar{\cdot}, \nu/\sim, \bar{\gamma}\rangle$.*

PROOF. We show that an equivalence compatible with multiplication must also be compatible with the neutral element and with the inverse. An equivalence relation is compatible with any nullary operation vacuously. Denote $\gamma(x)$ with x^-. Suppose $x \sim y$. Then $x^-xy = x^-yy^-$ yielding $y^- \sim x^-$. The equivalence relation is therefore also compatible with the taking of inverses. $\sim$ is a congruence for the group. The neutral element of $G/\sim$ must be $\nu/\sim$ and the inverse of $x/\sim$ must be $x^-/\sim$. □

For both unitary monoids and groups the neutral element in the quotient system is a subsystem.

Theorem. *Let $\langle M, \cdot, \nu\rangle$ be a unitary monoid (group) and $\sim$ a congruence. Then $\nu/\sim$ is a unitary submonoid (group).*

PROOF. Let x and y be in $\nu/\sim$. $x \sim \nu$ and $y \sim \nu$. $xy \sim \nu$. $xy \in \nu/\sim$. $\nu/\sim$ is a unitary submonoid.

If, moreover, M is a group then let $x \in \nu/\sim$. $x \sim \nu$. $x^-x \sim x^-\nu$. $\nu \sim x^-$. $x^- \in \nu/\sim$. $\nu/\sim$ is a subgroup of M. □

As a partial converse we now prove that the only coset of the quotient unitary monoid which is a subsystem is the coset containing the neutral element.

Theorem. *Let $\langle M, \cdot, \nu\rangle$ be a unitary monoid (group) and N a unitary submonoid (subgroup) of M which belongs to $M/\sim$ for some congruence $\sim$. Then $N = \nu/\sim$.*

PROOF. If $N \in M/\sim$ then $N = a/\sim$ for some a in M. But if N is a unitary submonoid then $\nu \in N$. $\nu \sim a$. $\nu/\sim = a/\sim = N$. □

The combination of the two theorems tells us that a subgroup is a member of a quotient group if and only if it is the coset defined by the neutral element.

Definition. We define a submonoid, unitary monoid, or subgroup to be *normal* if and only if it is a member of the appropriate quotient system.

This definition is actually given in Chapter 8 for algebraic systems in general. It is here specialized to systems with one binary operation. For groups, we immediately derive the following equivalent condition often used as a definition for a normal or invariant subgroup.

Theorem. *If N is a subgroup of a group $\langle G, \cdot, \nu\rangle$ then N is normal if and only if $x^-Nx \subseteq N$ for all $x \in G$.*

PROOF. Suppose first N to be a normal subgroup of G according to our definition. Then $N = v/\sim$ for some congruence $\sim$. Let $y \in x^- Nx$. $y = x^- nx$ for some n in N. $n \sim v$. $x^- nx \sim x^- vx = v$. $y \sim v$. $y \in N$. Thus $x^- Nx \subseteq N$.

Now suppose N to be a subgroup of G such that $x^- Nx \subseteq N$ for all x in G. We must define a congruence on G so that N is a member of the corresponding quotient group. We define $x \sim y$ if and only if $xy^- \in N$. It follows from the subgroup properties of N that $\sim$ is an equivalence relation on G. Let $x_1 \sim y_1$ and $x_2 \sim y_2$. $x_1 y_1^- \in N$. $x_2 y_2^- \in N$. $(x_1 x_2)(y_1 y_2)^- = x_1 x_2 y_2^- y_1^- = x_1 y_1^- y_1 x_2 y_2^- y_1^- = (x_1 y_1^-) y_1 (x_2 y_2^-) y_1^- = (x_1 y_1^-)(y_1^-)^-(x_2 y_2^-) \cdot (y_1^-) \in N$. Thus $x_1 x_2 \sim y_1 y_2$ proving $\sim$ to be a congruence. Letting $y = v$ in the definition of the relation $\sim$ we see that $x \sim v$ if and only if $x \in N$ proving $N = v/\sim$. N is normal according to our definition. □

Consistent with the use of notation in earlier chapters we denote the quotient group or quotient unitary monoid in terms of its associated normal subsystem which we have proved unique. If N is the normal subgroup associated with the quotient group $\langle G/\sim, \bar{\cdot}, v/\sim \rangle$ then we write $\langle G/N, \cdot, N \rangle$ to denote the quotient group.

It is interesting to look at the equivalence relation defined by a subgroup S of a group $\langle G, \cdot, v \rangle$ even if the subgroup is not normal.

Theorem. *For any subgroup S of a group $\langle G, \cdot, v \rangle$ there exist two equivalence relations ρ and σ on G as follows: $x\rho y$ if and only if $xy^- \in S$, $x\sigma y$ if and only if $x^- y \in S$. Furthermore, $G/\rho = \{Sa | a \in G\}$ and $G/\sigma = \{aS | a \in G\}$.*

PROOF. It is routine to verify that ρ and σ are both equivalence relations on G. Furthermore, $Sa = \{sa | s \in S\} = \{x | x = sa,\ s \in S\} = \{x | xa^- = s, s \in S\} = \{x | xa^- \in S\} = \{x | x\rho a\} = a/\rho$. $a/\rho \in G/\rho$. $aS = \{x | x = as, s \in S\} = \{x | a^- x \in S\} = \{x | a\sigma x\} = \{x | x\sigma a\} = a/\sigma$. $a/\sigma \in G/\sigma$. □

We have therefore defined two partitions of G by means of the subgroups S: the set of right cosets of S, $\{Sa | a \in G\}$, and the set of left cosets of S, $\{aS | a \in G\}$. That the right coset Sa be equal to the left coset aS for every a in G, $Sa = aS$ for every a in G, is equivalent to the condition that S be a normal subgroup of G. The argument in detail is as follows. Suppose $Sa = aS$ for every a in G. Consider $x^- Sx$ for x in G. The set Sx equals xS. Hence $x^- Sx = x^- xS \subseteq S$. This proves S is normal. Now assume $x^- Sx \subseteq S$ for every x in S. Let $y \in Sa$. $y = sa$ for some s in S. $y = aa^- sa = as' \in aS$. Similarly, $aS \subseteq Sa$. $aS = Sa$. Thus we have seen that if the subgroup S is normal then the two partitions of left cosets and right cosets coincide and make a quotient group. If S is not normal then neither equivalence relation is a congruence and there is no quotient group for that subgroup.

We have, even in the case that there is no quotient group, the very important Lagrange theorem showing that the cosets of the quotient set are equal in size.

Theorem. *In the partitions of a group $\langle G, \cdot, v\rangle$ produced by a subgroup S each coset, left or right, has the same number of members as S itself.* Crd G = crd $S \cdot$ crd G/S.

PROOF. Let $\varphi: S \to aS$ such that $\varphi(x) = ax$. We show φ to be a bijection. Let $\varphi(x) = \varphi(y)$. $ax = ay$. $a^{-}ax = a^{-}ay$. $x = y$. For any y in aS, $\varphi(a^{-}y) = a(a^{-}y) = y$. This bijection's existence shows that every left coset has the same number of elements as does S. The total number of elements in the group is the sum of the number of elements in all the cosets of G/σ. The number of elements in the group equals then the number of elements in S times the number of cosets in G/σ. The result naturally holds also for right cosets. □

There will be later several important applications of this theorem. For the time being, however, we are content to note that the number of elements in any subgroup is a factor of the number of elements in the entire group. Concerning the two partitions defined by a subgroup S which is not a normal subgroup of G, we will occasionally denote both of them by G/S leaving it to context to determine whether we are writing of left or right cosets.

QUESTIONS

1. Which of the following relations are congruences?
   (A) $x \sim y$ if and only if $x - y \in 4\mathbb{Z}$ on the group $\langle \mathbb{Z}, +, 0\rangle$.
   (B) $x \sim y$ if and only if $x = 2y$ or $y = 2x$ or $x = y$ on the monoid $\langle \mathbb{N}, \cdot\rangle$.
   (C) $x \sim y$ if and only if $x^2 = y^2$ on the group $\langle \mathbb{Z}, +, 0\rangle$.
   (D) $x \sim y$ if and only if $x - y \in \mathbb{Q}$ on the group $\langle \mathbb{R}, +, 0\rangle$.
   (E) None of the relations is a congruence.

2. Which of the following statements are true?
   (A) Some equivalence relations are not congruences.
   (B) Every congruence is an equivalence relation.
   (C) Not every equivalence relation is not a congruence.
   (D) Some relations are not equivalence relations.
   (E) None of the statements is true.

3. Define a relation on $\mathfrak{S}_3$ such that $x \sim y$ if and only if $x \circ y^{-1} \in \{(1), (1\ 2)\}$. Which of the following are true?
   (A) $\sim$ is a congruence on $\mathfrak{S}_3$.
   (B) $\{(1), (1\ 2)\}$ is a subgroup of $\mathfrak{S}_3$.
   (C) $\{(1), (1\ 2)\}$ generates $\mathfrak{S}_3$.
   (D) $\sim$ is an equivalence relation on $\mathfrak{S}_3$.
   (E) None of the statements is true.

4. Which of these statements are true?
   (A) On a commutative group every equivalence relation is a congruence.
   (B) If $aS = Sa$ for every a in G, a group, then the subgroup S is a normal subgroup.

(C) If the number of elements in each coset aS of a subgroup S is the same as the number of elements in S then S is a normal subgroup.
(D) If σ is a congruence for the monoid $\langle M, \cdot\rangle$ then σ is also a congruence for any unitary monoid $\langle M, \cdot, v\rangle$ with the same set M and binary operation $\cdot$.
(E) None of the statements is true.

5. The set of all functions f such that $f(x) = ax + b$, $a \neq 0$, $a, b \in \mathbb{Q}$, is a group under composition. Which of these statements are true?
(A) The set of all f with $a = 1$ is a subgroup.
(B) The set of all f with $a = 1$ is a normal subgroup.
(C) "$(f(x) = ax + b) \sim (g(x) = cx + d)$ if and only if $a = c$" is a congruence.
(D) "$(f(x) = ax + b) \sim (g(x) = cx + d)$ if and only if $ad - bc = 0$" is a congruence.
(E) None of the statements is true.

6. The number of cosets defined by the subgroup $\{I, (1\ 2)\}$ of $\mathfrak{S}_3$ is
(A) 12
(B) 1
(C) 3
(D) 2
(E) 6.

7. The number of elements in the quotient group $\mathfrak{S}_4/V_4$ is (V_4 is the normal subgroup $\{(1), (1\ 2)(3\ 4), (1\ 3)(2\ 4), (1\ 4)(2\ 3)\}$ of $\mathfrak{S}_4$)
(A) 12
(B) 1
(C) 3
(D) 2
(E) 6.

8. The theorem on groups proving that the number of elements in each coset is the same as the number of elements in the defining subgroup is named for
(A) Bolzano
(B) Weierstrass
(C) Mozart
(D) Lagrange
(E) Euclid.

Exercises

1. Give examples whenever possible of a group G and nontrivial normal subgroup N such that
(a) G is commutative and G/N is commutative
(b) G is commutative and G/N is noncommutative
(c) G is noncommutative and G/N is commutative
(d) G is noncommutative and G/N is noncommutative.

2. Give an example of a group and a subgroup which defines a left coset different from its right coset: find G, H, $a \in G$ such that $aH \neq Ha$.

3. Prove that if a group has $2m$ elements (some even number) then any subgroup of order m is a normal subgroup.

4. Prove that if M and N are normal subgroups of a group $\langle G, \cdot, \nu\rangle$ then both MN and $M \cap N$ are normal subgroups of G.

5. Which subgroups of $\mathfrak{S}_3$ are normal and which are not normal?

6. A group is defined to be simple if it has no normal subgroups except G itself and the singleton neutral element. Give examples of a commutative and a noncommutative simple group. Give an example of a nonsimple group.

7. Find all normal subgroups of V_4, Klein's four group, and find their quotient groups.

8. Let $\langle G, \cdot, \nu\rangle$ be a group. The set $\mathrm{Com}(G) = \{xyx^{-}y^{-}|x, y \in G\}$ is called the set of commutators of the group G. Prove that the set of commutators generates a normal subgroup of G: i.e., $[\mathrm{Com}(G)]$ is a normal subgroup of G. [*Hint*: $a^{-}(xyx^{-}y^{-})a = (a^{-}xa)(a^{-}ya)(a^{-}xa)^{-}(a^{-}ya)^{-}$.]

9. Prove that a group G is a commutative group if and only if $\mathrm{Com}(G) = \{\nu\}$. Prove that if H is a subgroup of G and $\mathrm{Com}(G) \subseteq H$ then H is a normal subgroup of G.

10. For any group G prove that $G/[\mathrm{Com}(G)]$ is a commutative group. Show that if N is a normal subgroup such that G/N is commutative then $\mathrm{Com}(G) \subseteq N$.

9.3 Morphisms

In this section we list theorems on morphisms of monoids, prove the Cayley representation theorem for groups, and investigate some function spaces with values in monoid and groups.

A *morphism* is a function that preserves operations from one algebraic system to another. In our study of monoids, unitary monoids and groups the relevant operations are the binary one called multiplication, the nullary one which chooses the neutral element and the unary one which denotes inverses. We now list some easy results, but do not prove them. The reader should consider the proofs to be exercises.

Theorem. *Let $\langle M, \cdot\rangle$ and $\langle M', \cdot'\rangle$ be monoids and $f: M \to M'$ be a morphism (preserving $\cdot$ in $\cdot'$). If A is a submonoid of M then $f(A)$ is a submonoid of M' and if B is a submonoid of M' then $f^{-1}(B)$ is a submonoid of M.*

Let $\langle M, \cdot, \nu, \gamma\rangle$ and $\langle M', \cdot', \nu', \gamma'\rangle$ be groups and $f: M \to M'$ be a morphism (preserving $\cdot$, ν, γ in $\cdot'$, ν', γ' respectively). If A is a subgroup of M then $f(A)$ is a subgroup of M' and if B is a subgroup of M' then $f^{-1}(B)$ is a subgroup of M.

The preservations of the three operations of a group by a function are not independent of one another.

Theorem. *Let $f: M \to M'$ preserve the binary operation of multiplication. If $\langle M, \cdot, \nu\rangle$ and $\langle M', \cdot', \nu'\rangle$ are unitary cancellative monoids then f is a morphism. If $\langle M, \cdot, \nu, \gamma\rangle$ and $\langle M', \cdot', \nu', \gamma'\rangle$ are groups then f is a morphism.*

PROOF. $f(v) = f(vv) = f(v)f(v)$. $f(v)v' = f(v)f(v)$. Cancelling from both sides the element $f(v)$ we have $v' = f(v)$. $f(x)f(x^-) = f(xx^-) = f(v) = v'$. Since inverses are unique $f(x^-) = (f(x))^-$. □

The fundamental morphism theorem appears just as with sets and rings.

Theorem. *Let $\langle M, \cdot \rangle$ and $\langle M', \cdot' \rangle$ be monoids and $f: M \to M'$ a morphism. The relation $\sim$ on M defined by ($x \sim y$ if and only if $f(x) = f(y)$) is a congruence. The quotient mapping $\varphi: M \to M/\sim$ is an epimorphism. There exists a monomorphism $f': M/\sim \to M'$ such that $f = f' \circ \varphi$.*

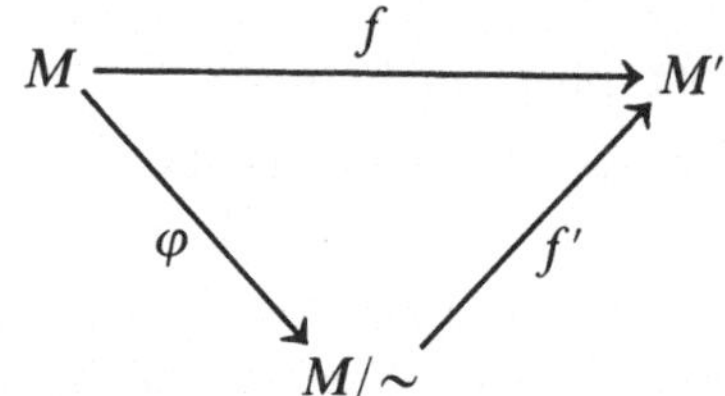

We again leave the details of the proof to the reader. The theorem also holds true for unitary monoids $\langle M, \cdot, v \rangle$ and $\langle M', \cdot', v' \rangle$ and for groups $\langle M, \cdot, v, \gamma \rangle$ and $\langle M', \cdot', v', \gamma' \rangle$. In the case of the unitary monoid and group the subset $\{v'\}$ of M' is a unitary submonoid or subgroup implying $f^{-1}(v')$ is a unitary submonoid or subgroup of M. It is furthermore normal and by definition the kernel of f. In this case $M/\sim$ can be written $M/\ker f$.

We now follow with some special isomorphism theorems for groups.

Theorem. *Let K and L be normal subgroups of a group $\langle G, \cdot, v, \gamma \rangle$ and let $K \subseteq L$. Then there is an isomorphism from $(G/K)/(L/K)$ to G/L.*

PROOF. This theorem is proved more generally for operational systems in Section 8.4 but we summarize a proof here. From G/K to G/L we define a function f taking each coset of G/K into the coset of G/L which contains it: $f(xK) = xL$. This function $f: G/K \to G/L$ is an epimorphism. The kernel of f is $\{xK \mid xL = L\}$ which is $\{xK \mid x \in L\} = L/K$. There exists an isomorphism $f': (G/K)/(L/K) \to G/L$. □

Theorem. *Let H and L be subgroups of a group $\langle G, \cdot, v, \gamma \rangle$ with N a normal subgroup. Then there exists an isomorphism from $H/(H \cap N)$ to HN/N.*

PROOF. Define $f: H \to HN/N$ such that $f(h) = hN$. $hN \in HN/N$ for each h in H and furthermore the mapping is surjective. It is not difficult to see that N is a normal subgroup of the subgroup HN of G. $f(h_1h_2) = h_1h_2N = h_1Nh_2N = f(h_1)f(h_2)$. Kernel $f = \{h \mid h \in H$ and $hN = N\} = \{h \mid h \in H$ and $h \in N\} = H \cap N$. There exists an isomorphism $f': H/(H \cap N) \to HN/N$. □

Our final isomorphism theorem travels under the name of Zassenhaus.

Theorem. *Let H_1, H_2, K_1, K_2 be subgroups of a group $\langle G, \cdot, v, \gamma\rangle$. Let K_1 be a normal subgroup of H_1, K_2 a normal subgroup of H_2. Then*

$$\frac{(H_1 \cap H_2)K_1}{(H_1 \cap K_2)K_1} \quad \textit{is isomorphic to} \quad \frac{H_1 \cap H_2}{(H_1 \cap K_2)(K_1 \cap H_2)}.$$

Proof. The second isomorphism theorem we have just proved establishes the isomorphism between

$$\frac{H_1 \cap H_2}{[(H_1 \cap K_2)K_1] \cap (H_1 \cap H_2)} \quad \text{and} \quad \frac{(H_1 \cap H_2)(H_1 \cap K_2)K_1}{(H_1 \cap K_2)K_1}$$

and the theorem is proved after we have established the following three statements:

1. $(H_1 \cap K_2)K_1$ is a normal subgroup of $H_1 \cap H_2$
2. $(H_1 \cap H_2)(H_1 \cap K_2)K_1 = (H_1 \cap H_2)K_1$
3. $[(H_1 \cap K_2)K_1] \cap (H_1 \cap H_2) = (H_1 \cap K_2)(K_1 \cap H_2)$.

Each of these three statements is left as an exercise. □

We include now a theorem due to Arthur Cayley. This theorem is a spectacular result although it does not solve all problems of group theory. In the first chapter we introduced the group through permutations. We found that for a given set S, the set of all bijections of that set, together with the operation of composition and the identity function constitute a group, which we called the symmetric group $\langle \mathfrak{S}(S), \circ, I\rangle$. The following theorem demonstrates that not only are symmetric groups excellent examples of groups, they typify groups. Every group whatsoever is isomorphic to some subgroup of some symmetric group.

Lemma. *If $\langle G, \cdot, v\rangle$ is a group then left (or right) multiplication by a fixt element of the group is a bijection of G.*

Proof. Let $f_a : G \to G$ such that $f_a(x) = ax$. The cancellative property of a in G proves f_a to be an injection. That a has an inverse in G proves f_a to be a surjection. $f_a \in \mathfrak{S}(G)$. □

We denote $\{f_a | a \in G\}$ by $\mathscr{T}(G)$, meaning the set of left translations of G.

Theorem. *If $\langle G, \cdot, v\rangle$ is a group then G is isomorphic to some subgroup of $\langle \mathfrak{S}(G), \circ, I\rangle$.*

Proof. We define $\Phi : G \to \mathfrak{S}(G)$ such that $\Phi(a) = f_a$. $\Phi(ab) = f_{ab} = f_a f_b = \Phi(a)\Phi(b)$. The interior equation is checked by elements as follows: $f_{ab}(x) = (ab)x = a(bx) = f_a(bx) = f_a(f_b(x)) = (f_a \circ f_b)(x)$. Φ is a morphism.

Let $\Phi(a) = \Phi(b)$. $f_a = f_b$. $f_a(x) = f_b(x)$ for all x in G. $ax = bx$ for all x in G, including v. $av = bv$. $a = b$. Φ is a monomorphism and G and $\Phi(G)$ are isomorphic groups. $\Phi(G) = \{f_a | a \in G\} = \mathscr{T}(G)$. □

After this taste of functional representation of G by $\mathscr{T}(G)$ we move more deeply into a discussion of functional representation of structures. The structure of the monoid carries over to functions having their values in the monoid.

Theorem. *If $\langle M, \cdot\rangle$ is a given monoid and S is a nonempty set then $\langle M^S, \cdot\rangle$ is a monoid. Moreover, if $\langle M, \cdot, v\rangle$ is a unitary monoid so also is $\langle M^S, \cdot, u\rangle$. If $\langle M, \cdot, v\rangle$ is a group so also is $\langle M^S, \cdot, u\rangle$ a group. u denotes the unity function.*

Proof. Similar results have been treated in earlier chapters. Given $f:S \to M$ and $g:S \to M$ we define $(f \cdot g):S \to M$ such that $(f \cdot g)(x) = f(x)g(x)$. Let v be the neutral element of M. Define $u:S \to M$ such that $u(x) = v$ for all x in S. For the function f in M^S define $f^-:S \to M$ such that $f^{-1}(x) = (f(x))^{-1}$. Then $(f \cdot f^-)(x) = f(x)(f(x))^{-1} = v$. $f \cdot f^- = u$. Also $f^- \cdot f = u$. Note carefully that f^- in this situation is not the compositional inverse of f. □

The original monoid can be embedded within the function space.

Theorem. *$\langle M, \cdot\rangle$ is isomorphic to the submonoid of $\langle M^S, \cdot\rangle$ consisting of the constant functions.*

Proof. Define $\chi:M \to M^S$ such that $\chi(a)$ has constant value a. $\chi(ab)(x) = ab = \chi(a)(x)\chi(b)(x) = [\chi(a)\chi(b)](x)$. Let $\chi(a) = \chi(b)$. Then $a = b$. χ is a monomorphism. □

By restricting attention to those functions which are morphisms between two monoids we obtain a submonoid of the entire function space.

Theorem. Mor(M, M'), *the set of morphisms between two monoids $\langle M, \cdot\rangle$ and $\langle M', \cdot'\rangle$, is a submonoid of $\langle (M')^M, \cdot\rangle$ when M' is commutative.* Mor(M, M') *is also commutative. If M' is a commutative group then so also is* Mor(M, M').

Proof. We show that the product of two morphisms is again a morphism. $(fg)(xy) = f(xy)g(xy) = f(x)f(y)g(x)g(y) = f(x)g(x)f(y)g(y) = (fg)(x) \cdot (fg)(y)$. $(fu)(x) = f(x)u(x) = f(x)v = f(x)$. $fu = f$. $u(xy) = v = vv = u(x)u(y)$. $f^-(xy) = f(xy)^{-1} = f(y)^{-1}f(x)^{-1} = f^-(y)f^-(x) = f^-(x)f^-(y)$. Again here f^- represents the inverse with respect to the binary operation of multiplication and not the compositional inverse. □

For the case of a morphism of a given monoid into itself, we use, as usual, the name endomorphism and abbreviate Mor(M, M) by $\mathscr{E}(M)$. We know by our previous theorems that if $\langle M, \cdot\rangle$ is a commutative monoid then $\langle \mathscr{E}(M), \cdot\rangle$ is a commutative monoid. We now move into the composition structure made possible by the identification of domain and codomain.

Theorem. *For any set M, $\langle M^M, \circ, I\rangle$ is a unitary monoid. If $\langle M, +, \theta\rangle$ is a commutative monoid then $\langle \mathscr{E}(M), \circ, I\rangle$ is a unitary submonoid of $\langle M^M, \circ, I\rangle$.*

PROOF. Given $f:M \to M$ and $g:M \to M$ then $g \circ f:M \to M$. $f \circ I = I \circ f = f$ for any $f:M \to M$. That $\mathscr{E}(M)$ is a submonoid is ensured by the fact that the composition of two endomorphisms is another endomorphism. The identity function is obviously an endomorphism. $(g \circ f)(x + y) = g(f(x + y)) = g(f(x) + f(y)) = g(f(x)) + g(f(y)) = (g \circ f)(x) + (g \circ f)(y)$. $I(x + y) = x + y = I(x) + I(y)$. □

We note that even if $\langle M, \cdot, v\rangle$ is a group $\langle \mathscr{E}(M), \circ, I\rangle$ will not usually be a group. The composition structure is quite different from the original binary operation. Inverses in the composition structure are found in bijections. As usual we call an endomorphism which is also a bijection an automorphism.

Theorem. *If $\langle M, \cdot\rangle$ is a monoid then $\langle \mathscr{A}(M), \circ, I\rangle$, the set of all automorphisms of $\langle M, \cdot\rangle$, is a group and a unitary submonoid of $\langle \mathscr{E}(M), \circ, I\rangle$ and $\langle M^M, \circ, I\rangle$.*

PROOF. The composition of two bijections is a bijection and the compositional inverse of an automorphism is also an automorphism. □

EXAMPLE. The unitary monoid $\langle \mathbb{N}, +, 0\rangle$ generates the functional monoid of sequences $\langle \mathbb{N}^{\mathbb{N}}, +, z\rangle$. The neutral element is $(0, 0, 0, \ldots)$. The endomorphisms of $\langle \mathbb{N}, +, 0\rangle$ are all of the form $f_k:\mathbb{N} \to \mathbb{N}$ such that $f_k(n) = kn$ for some k in $\mathbb{N}$. Thus $\mathscr{E}(\mathbb{N}) = \{f_k | k \in \mathbb{N}\}$. $\langle \mathscr{E}(\mathbb{N}), \circ, I\rangle$ is a monoid under composition. The only automorphism in $\mathscr{E}(\mathbb{N})$ is $I = f_1$. $\mathscr{A}(\mathbb{N}) = \{I\}$ is the trivial group.

Of particular interest among the automorphisms of a group $\langle G, \cdot, v\rangle$ are the *inner automorphisms* $\varphi_a:G \to G$ such that $\varphi_a(x) = a^- xa$. For a commutative group they are all the identity. We denote the set of all inner automorphisms of G by $\mathscr{I}(G)$.

Theorem. *Let $\langle G, \cdot, v\rangle$ be a group. Then the inner automorphisms, $\mathscr{I}(G)$, form a normal subgroup of $\langle \mathscr{A}(G), \circ, I\rangle$.*

PROOF. It can be verified that $\varphi_b \circ \varphi_a = \varphi_{ba}$ and $(\varphi_a)^{-1} = \varphi_{a^{-1}}$ by evaluating the functions. Also $(g \circ \varphi_a \circ g^{-1})(x) = g(\varphi_a(g^{-1}(x))) = g(a^- g^{-1}(x)a) = (g(a))^{-1} x g(a) = \varphi_{g(a)}(x)$. □

$\mathscr{A}(M)$, the group of automorphisms of a monoid $\langle M, \cdot\rangle$ is a subgroup of $\mathfrak{S}(M)$, the symmetric group for the set M. A member of $\mathfrak{S}(M)$ is merely a bijection of M and not in general a morphism of the monoid structure. $\mathscr{A}(M)$ is therefore, in general, a proper subgroup of $\mathfrak{S}(M)$. The before mentioned left translations $\mathscr{T}(M)$, are also bijections of interest when M is a group.

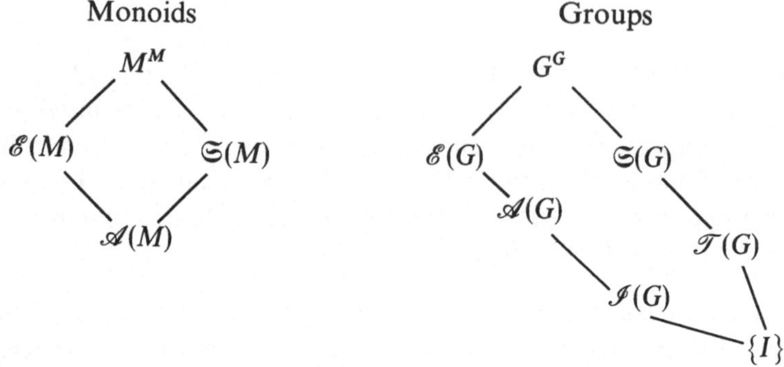

For the case of the commutative group $\langle G, +, \theta\rangle$ we can combine the group's binary operation with the functional composition to produce *the ring of endomorphisms.*

Theorem. *If $\langle G, +, \theta\rangle$ is a commutative group then $\langle \mathscr{E}(G), +, \circ, z, I\rangle$ is a unitary ring and the invertible elements of $\mathscr{E}(G)$ are precisely the members of $\mathscr{A}(G)$.*

Proof. $\langle \mathscr{E}(G), +, z\rangle$ is a commutative group with operations based upon those of $\langle G, +, \theta\rangle$. $\langle \mathscr{E}(G), \circ, I\rangle$ is a unitary monoid. We have left to verify the composition distributivity with respect to the addition. $[f \circ (g + h)](x) = f((g + h)(x)) = f(g(x) + h(x)) = f(g(x)) + f(h(x)) = (f \circ g)(x) + (f \circ h)(x) = ((f \circ g) + (f \circ h)(x)$. The morphism property of f has been used for this left distributivity. $((g + h) \circ f)(x) = (g + h)(f(x)) = g(f(x)) + h(f(x)) = (g \circ f)(x) + (h \circ f)(x) = ((g \circ f) + (h \circ f))(x)$. □

Example. The endomorphisms of $\langle \mathbb{Z}, +, \theta\rangle$ are precisely $\{f_k | k \in \mathbb{Z}\}$ where $f_k(x) = kx$. Two of these endomorphisms are automorphisms: $\mathscr{A}(\mathbb{Z}) = \{f_1, f_{-1}\}$. The addition of the endomorphisms obeys the rule $f_m + f_n = f_{m+n}$. Composition obeys the rule $(f_m \circ f_n)(x) = f_m(f_n(x)) = mnx = f_{mn}(x)$.

Questions

1. Let $\langle G, \cdot, v\rangle$ and $\langle G', \cdot', v'\rangle$ be groups and $f: G \to G'$ be a function preserving the binary operation of multiplication. Which of these statements are true?
   (A) $f(v)$ is a neutral element for $f(G)$, the range of f.
   (B) $f(v)$ is a neutral element for G', the codomain of f.
   (C) $f(v) = v'$.
   (D) $f^{-1}(v') = v$.
   (E) None of the statements is true.

2. Let $f: M \to M'$ be a morphism of the monoids $\langle M, \cdot\rangle$ and $\langle M', \cdot'\rangle$. Which of the following statements are true?
   (A) If A is a submonoid of M and is also a group, then $f(A)$ is a submonoid of M' and is a group.

(B) If v' is a neutral element for $\langle M', \cdot' \rangle$ then $f^{-1}(v')$ is a submonoid of M.
(C) If $\langle M', \cdot' \rangle$ is a cancellative monoid and f is an injection then $\langle M, \cdot \rangle$ is also a cancellative monoid.
(D) If $\langle M, \cdot \rangle$ is a commutative monoid then $\langle M', \cdot' \rangle$ is also a commutative monoid.
(E) None of the statements is true.

3. Let $f: \mathbb{N} \to \{0, 1\}$ be defined such that $f(n) = 0$ if n is even and $f(n) = 1$ if n is odd. Which of these statements are true?
(A) f is a morphism of the unitary monoids $\langle \mathbb{N}, +, 0 \rangle$ and $\langle \{0, 1\}, \cdot, 1 \rangle$.
(B) Kernel $f = 2\mathbb{N}$.
(C) $\mathbb{N}/\ker f = 2\mathbb{N}$.
(D) $\varphi(\mathbb{N}) = \{2\mathbb{N}, 2\mathbb{N} + 1\}$ where φ is the quotient mapping.
(E) None of the statements is true.

4. Which of these statements are true?
(A) $\mathbb{Z}_{2n}/(n\mathbb{Z}/2n\mathbb{Z})$ is isomorphic to $\mathbb{Z}_n$.
(B) $\mathbb{Z}_n/\mathbb{Z}_2$ is isomorphic to $(2n)\mathbb{Z}$.
(C) $n\mathbb{Z}/\mathbb{Z}_2$ is isomorphic to $\mathbb{Z}$.
(D) $\mathbb{Z}_{n^2}/(n\mathbb{Z}/n^2\mathbb{Z})$ is isomorphic to $\mathbb{Z}_n$.
(E) None of the statements is true.

5. The monoid of functions $\langle M^M, \cdot, u \rangle$ constructed from a given unitary monoid $\langle M, \cdot, v \rangle$
(A) is a group when $\langle M, \cdot, v \rangle$ is a group
(B) contains fewer members than $\langle M, \cdot, v \rangle$
(C) contains more members than $\langle M, \cdot, v \rangle$ if M is not trivial
(D) cannot be a group if M is noncommutative.
(E) None of the choices completes a satisfacotry sentence.

6. For monoids $\langle M, \cdot \rangle$ and $\langle M', \cdot' \rangle$ we see that $\langle \mathrm{Mor}(M, M'), \cdot \rangle$
(A) is a subgroup of $(M')^M$ if M is a group
(B) is closed under products if $\langle M', \cdot' \rangle$ is commutative
(C) fails to have a unity if M fails to have a unity
(D) is a commutative group when $\langle M', \cdot' \rangle$ is a commutative group.
(E) None of the choices completes a true sentence.

7. Which of the following statements are true?
(A) $[3]/[6] = [3]/([3] \cap [2])$ is isomorphic to $([3] + [2])/[2] = \mathbb{Z}/[2] = \mathbb{Z}_2$.
(B) $\mathcal{T}(G)$ is isomorphic to G for any group $\langle G, \cdot, v \rangle$.
(C) $G \subseteq \mathcal{T}(G)$ for any group $\langle G, \cdot, v \rangle$.
(D) $\mathcal{T}(G) \neq \mathfrak{S}(G)$ for all nontrivial groups G.
(E) None of the statements is true.

8. Which of the following statements are true?
(A) The mapping $\varphi_a: \mathbb{N} \to M$ such that $\varphi_a(n) = a^n$ from the natural numbers $\langle \mathbb{N}, +, 0 \rangle$ to any unitary monoid $\langle M, \cdot, v \rangle$ is a morphism.
(B) $\langle \mathbb{N}, \cdot \rangle$ and $\langle \mathbb{Q}^+, \cdot \rangle$ are isomorphic monoids.
(C) All automorphisms are endomorphisms.
(D) All automorphisms are isomorphisms.
(E) None of the statements is true.

EXERCISES

1. Let $\mathbb{C}^* = \mathbb{C} - \{0\}$. Define $f:\mathbb{Z} \to \mathbb{C}^*$ such that $f(n) = i^n$, $i = \sqrt{-1}$. Show that f is a morphism from the additive group of integers to the multiplicative group of nonzero complex numbers. Find kernel f and range f. Are domain f and range f isomorphic?

2. Find an isomorphism $\mathbb{Z}_3 \to \mathbb{Z}_3$ not the identity isomorphism of the group $\langle \mathbb{Z}_3, +, \bar{0} \rangle$.

3. Show that $\langle \mathbb{Z}_2, +, \bar{0} \rangle$ and $\langle \mathfrak{S}_2, \circ, I \rangle$ are isomorphic.

4. Show that the groups $\langle \mathbb{Z}_4, +, \bar{0} \rangle$ and $\langle V_4, \circ, I \rangle$ are not isomorphic.

5. Show that $\{\sigma | \sigma \in \mathfrak{S}_{n+1}$ and $\sigma(n + 1) = n + 1\}$ is a subgroup of $\mathfrak{S}_{n+1}$. Show that this subgroup of $\mathfrak{S}_{n+1}$ is isomorphic with $\mathfrak{S}_n$. Is this subgroup a normal subgroup of $\mathfrak{S}_{n+1}$?

6. If $f:M \to M$ and $g:N \to P$ are monomorphisms of unitary monoids show that $g \circ f$ is also a monomorphism. If f and g are epimorphisms show that $g \circ f$ is an epimorphism.

7. For the group $\langle \mathbb{Z}_3, +, \bar{0} \rangle$ find $\mathscr{I}(\mathbb{Z}_3)$, $\mathscr{A}(\mathbb{Z}_3)$, $\mathscr{T}(\mathbb{Z}_3)$, $\mathfrak{S}(\mathbb{Z}_3)$.

8. Prove that the subgroups $\mathscr{T}(G)$ and $\mathscr{A}(G)$ of $\langle \mathfrak{S}(G), \circ, I \rangle$ have only I in common.

9. For the group V_4 find $\mathscr{I}(V_4)$, $\mathscr{A}(V_4)$, $\mathscr{T}(V_4)$, $\mathfrak{S}(V_4)$.

10. Let $\langle M, \cdot, v \rangle$ be a commutative cancellative unitary monoid. Prove there exists a group $\langle \bar{M}, \cdot, v \rangle$ and a monomorphism $\varphi:M \to \bar{M}$ such that for any $y \in \bar{M}$ there exist x_1, x_2 in M such that $y = \varphi(x_1)\varphi(x_2)^-$. The gist of this result is that every commutative cancellative unitary monoid can be embedded in a group. Consult the theorems on rings of fractions and fashion a similar proof.

11. Apply the result of Exercise 10 to the monoid $\langle \mathbb{N}, +, 0 \rangle$. The existence of what familiar set is proved by this application?

12. Show that $\mathfrak{S}(G)$ is isomorphic with $\{f_a \circ \mathscr{A}(G) | f_a \in \mathscr{T}(G)\}$ for any group $\langle G, \cdot, v \rangle$. [*Hint*: Use one of the isomorphism theorems.]

13. Let $\langle G, \cdot, v \rangle$ be a group. Show that $f:G \to G$ such that $f(x) = x^-$ is a group automorphism if and only if G is commutative.

14. Find the endomorphisms of $\langle \mathbb{Z}^*, \cdot, 1 \rangle$.

15. Let $\langle G, \cdot, v \rangle$ be a group and N a normal subgroup of G. Denote by $\mathscr{S}(N)$ the set of all subgroups of G containing N and by $\mathscr{Q}(N)$ the set of all quotient groups $\{S/N | S \in \mathscr{S}(N)\}$. Prove that the mapping $\Phi:\mathscr{S}(N) \to \mathscr{Q}(N)$ such that $\Phi(S) = S/N$ is a bijection.

9.4 Cyclic groups and order

We discuss in this section cyclic subgroups, the order of an element, the center and normalizer, and conjugacy classes. While some of the material of this section can be adapted to monoids, the section deals principally with groups.

Definition. A monoid $\langle G, \cdot\rangle$ is *cyclic* or *simply generated* if and only if G is generated by a singleton subset: $G = [a]$ for some a in G.

EXAMPLES. $\langle \mathbb{Z}_6, +, \bar{0}\rangle$ is generated by $\{\bar{1}\}$. $\langle \mathbb{Z}, +, 0\rangle$ is generated by $\{1\}$. Both groups are cyclic.

Definition. The *order* of an element of a monoid is the cardinality of the submonoid generated by the element.

EXAMPLES. In $\langle \mathbb{Z}_6, +, \bar{0}\rangle$ the order of $\bar{1}$ is 6, the order of $\bar{2}$ is 3. In $\langle \mathbb{Z}, +, 0\rangle$ the order of 0 is 1 and the order of every other element is infinite.

In the next theorem the reader will recognize the techniques utilized earlier in the section on the characteristic of a ring.

Theorem. *Let $\langle G, \cdot, v\rangle$ be a group. The mapping $\psi_a:\mathbb{Z} \to G$ such that $\psi_a(n) = a^n$ is a morphism*; kernel $\psi_a = m\mathbb{Z}$ *for some nonnegative integer m. The order of the element a is m when m is a positive integer and the order of a is infinite when $m = 0$.*

PROOF. Kernel $\psi_a = \{n | n \in \mathbb{Z}$ and $\psi_a(n) = v\}$ is a normal subgroup of the additive group of $\mathbb{Z}$. This subgroup must be $m\mathbb{Z}$ for some nonnegative integer m. This includes the possibility of $m = 0$. By the fundamental morphism theorem the range of ψ_a, $\psi_a(\mathbb{Z})$, is isomorphic with $\mathbb{Z}/m\mathbb{Z} = \mathbb{Z}_m$ for some nonnegative integer m. $m = 0$ corresponds to the case of the trivial kernel and ψ_a being a monomorphism. $\psi_a(\mathbb{Z})$ is a subgroup of which a is a member and therefore includes $[a]$. On the other hand, $\psi_a(\mathbb{Z})$, consisting as it does simply of powers of a, must be included within $[a]$. $\psi_a(\mathbb{Z}) = [a]$. Thus $[a]$ is isomorphic to $\mathbb{Z}_m$ for some nonnegative integer m. If $m > 0$ then $[a]$ has a finite number of elements, namely, m. If $m = 0$ then $[a]$ is isomorphic with $\mathbb{Z}$ itself and has an infinite number of elements. Order a is then infinite. □

EXAMPLE. In the symmetric group $\langle \mathfrak{S}_3, \circ, I\rangle$ the permutation (1 2) is of order 2. In the additive group $\langle \mathbb{Z}, +, 0\rangle$ all elements are of infinite order except 0 which has order 1. In V_4 all elements except the identity have order 2.

We now use Lagrange's theorem to relate the order of an element of a group with the size of the group.

Theorem. *If G is group and a is an element of G then* crd G *is a multiple of* ord a; crd G *is a prime number implies G is a cyclic group.*

PROOF. By Lagrange's theorem the cardinality of a group is a multiple of the size of any subgroup. In particular, crd G is a multiple of crd$[a]$ for any a in G. Ord a = crd$[a]$. For the second part of the theorem assume crd G

is prime. Then crd $G \geqslant 2$ and there must be an element a in G different from the neutral element. Crd a must be a factor of crd G, a prime element. Hence crd$[a]$ = crd G. $[a]$ is a subgroup of G with the same number of elements as G. $[a] = G$. G is a cyclic group. □

The theorem tells us that every group of prime cardinality is cyclic. The converse statement is not true. $\mathbb{Z}_4$ and $\mathbb{Z}$, for example, are both additive cyclic groups. In fact, there are cyclic groups for every positive integer m, namely, $\langle \mathbb{Z}_m, +, \bar{0} \rangle$. The theorem also tells us that if a group contains an element of infinite order then the group has infinite cardinality. We now show that if a group is cyclic and infinite then it cannot have any elements of finite order except the neutral element.

Theorem. *Let $\langle G, \cdot, v \rangle$ be an infinite cyclic group. Then every element except v has infinite order.*

PROOF. There exists a function $\psi_a : \mathbb{Z} \to G$ such that $\psi_a(n) = a^n$, an isomorphism for some $a \in G$. Let x be in G. $x = \psi_a(m)$ for some m in $\mathbb{Z}$. $x = a^m$. $[x] = \{x^k | k \in \mathbb{Z}\} = \{a^{mk} | k \in \mathbb{Z}\}$. Suppose $x^k = x^l$ for some $k \neq l$. $a^{mk} = a^{ml}$, $mk \neq ml$. This contradicts ψ_a being an injection. Therefore, each x^k, k in $\mathbb{Z}$, is distinct and the subgroup $[x]$ is infinite. Ord x is infinite. □

EXAMPLE. In the additive group of integers the subgroups [2], [3], . . . , are all of infinite order, yet are proper subgroups of $\mathbb{Z}$.

We relate the inner automorphism φ_a to the center of the group.

Theorem. *Let $\langle G, \cdot, v \rangle$ be a group. An element a in G commutes with every element in G if and only if $\varphi_a = I$.*

PROOF. An element a commutes with every element of G if and only if $ax = xa$ for all x in G if and only if $x = a^{-}xa$ for all x in G if and only if $\varphi_a(x) = x$ for all x in G if and only if $\varphi_a = I$. □

The elements of a group which commute with every other element in the group are exactly those elements which define trivial inner automorphisms. The collection of these elements is called the center of the group.

Definition. Let $\langle G, \cdot, v \rangle$ be a group. The *center* of the group G is the set $Z(G) = \{a | ax = xa \text{ for all } x \text{ in } G\}$.

Theorem. *The center, $Z(G)$, of a group, $\langle G, \cdot, v \rangle$, is a normal subgroup of G and $G/Z(G)$ is isomorphic to $\mathscr{I}(G)$.*

PROOF. $\Phi : G \to \mathscr{I}(G)$ such that $\Phi(a) = \varphi_a$ is an epimorphism of G onto $\mathscr{I}(G)$. There exists an isomorphism $\Phi' : G/\ker \Phi \to \mathscr{I}(G)$. Kernel $\Phi = \{a | \varphi_a = I\} = Z(G)$. The kernel is always a normal subgroup. □

We now introduce another subgroup of G associated with the inner automorphism.

Definition. By the *normalizer* of a subset S of a group $\langle G, \cdot, \rangle$ we mean the set $\{a | a \in G \text{ and } a^{-}Sa = S\}$. We denote the normalizer of S by $\mathcal{N}(S)$.

The normalizer of a set S is the set of all a in G for which the associated inner automorphism φ_a leaves S invariant.

Theorem. *The normalizer $\mathcal{N}(S)$ of a subset S of a group $\langle G, \cdot, v\rangle$ is a subgroup of G.*

Proof. Let a and b belong to $\mathcal{N}(S)$. $a^{-}Sa = S$ and $b^{-}Sb = S$. If $s \in S$ then $(ab)^{-}sab = b^{-}a^{-}sab = b^{-}s_1b = s_2$ for some s_1, s_2 in S. Thus $(ab)^{-}Sab \subseteq S$. Now let $s_3 \in S$. $s_3 = b^{-}s_4b$ for some $s_4 \in S$. $s_4 = a^{-}s_5a$ for some $a_5 \in S$. $s_3 = (ab)^{-}s_5ab$. $S \subseteq (ab)^{-}Sab$. This proves $(ab)^{-}Sab = S$. $ab \in \mathcal{N}(S)$.

Now let $a \in \mathcal{N}(S)$. We wish to show that a^{-} belongs to $\mathcal{N}(S)$ also. This requires demonstrating that $(a^{-})^{-}Sa^{-} = S$, or alternatively, $aSa^{-} = S$. First consider asa^{-} in aSa^{-}. $asa^{-} = t$ commutes to $a^{-}ta = s$. But from what is given $a^{-}ta \in S$ implies $t \in S$. Therefore, $asa^{-} \in S$ and we have proved $aSa^{-} \subseteq S$. Now consider s in S. From what is given $a^{-}sa = t$ also belongs to S. Solving we have $s = ata^{-} \in aSa^{-}$. $S \subseteq aSa^{-}$. Finally, $\mathcal{N}(S)$ contains v since $v^{-}Sv = S$. □

We remark that the normalizer of a set is a group while the original set may not be a group. The normalizer of a set may not even include the original set. However, if H is a subgroup then $\mathcal{N}(H)$ does include H. The normalizer $\mathcal{N}(H)$ of a subgroup H is the largest subgroup of G of which H is a normal subgroup.

Since both the center and the normalizer are defined by inner automorphisms we have this theorem relating the concepts.

Theorem. *Let $\langle G, \cdot, v\rangle$ be a group and x be in G. $\mathcal{N}(x) = G$ if and only if $x \in Z(G)$.*

Proof. $x \in Z(G)$ if and only if $\varphi_x = I$ if and only if $x^{-}yx = y$ for all y in G if and only if $yx = xy$ for all y in G if and only if $y^{-}\{x\}y = \{x\}$ for all y in G if and only if $\{y | y^{-}\{x\}y = \{x\}\} = G$ if and only if $\mathcal{N}(x) = G$. It is to be understood here that the normalizer of an element x is the normalizer of the subset $\{x\}$. □

We now go further in exploring the relationship between the center and the normalizer.

Theorem. *Let $\langle G, \cdot, v\rangle$ be a group. $Z(G) \subseteq \mathcal{N}(a)$ for any a in G. $Z(G) = \bigcap_{a \in G} \mathcal{N}(a)$.*

PROOF. Suppose $y \in Z(G)$. $\varphi_y = I$. $y^{-}\{a\}y = a$ for all a in G. $y \in \mathcal{N}(a)$ for each a in G. $y \in \bigcap_{a\in G} \mathcal{N}(a)$. $Z(G) \subseteq \bigcap_{a\in G} \mathcal{N}(a)$.

Now let $x \in \mathcal{N}(a)$ for every a in G. $x^{-}ax = a$ for every a in G. $\varphi_x = I$. $x \in Z(G)$. $Z(G) = \bigcap_{a\in G} \mathcal{N}(a)$. □

We now introduce the third principal concept associated with the inner automorphism, that of the *conjugacy relation.*

Definition. Let $\langle G, \cdot, v\rangle$ be a group and $x, y \in G$. x and y are *conjugate* if and only if there exists a φ_a in $\mathcal{I}(G)$ such that $y = \varphi_a(x)$. We use the notation $x \backsim y$ to denote x and y are conjugate.

According to the definition $x \backsim y$ if and only if there exists an a in G such that $y = a^{-}xa$.

Theorem. *Conjugacy is an equivalence relation on the group G.*

PROOF. $x \backsim x$ because $x = \varphi_v(x)$. If $x \backsim y$ then $y = a^{-}xa$ for some a in G. Then $x = (a^{-})^{-}ya^{-}$ with a^{-} in G. $x = \varphi_{a^{-}}(y)$. $y \backsim x$. If $x \backsim y$ and $y \backsim z$ we have $y = \varphi_a(x)$, $z = \varphi_b(y)$ for some $a, b \in G$. Then $z = \varphi_b(\varphi_a(x)) = \varphi_{ab}(x)$ with ab in G. $x \backsim z$. □

The equivalence relation of conjugacy partitions the group G into equivalence classes called conjugacy classes. The equivalence relation of conjugacy is not a congruence; it is not compatible with the binary operation of the group. The set of equivalence classes is not a quotient group and the equivalence relation does not arise from a subgroup. Let us denote the *conjugacy class* containing an element x of G by ccl(x). Ccl(x) $= \{y|y \in G$ and $y \backsim x\}$. The partition defined by conjugacy is the set of conjugacy classes $\{\text{ccl}(x)|x \in G\}$. Since conjugacy is not defined from a subgroup we do not have the benefit of Lagrange's theorem yielding equivalence classes of the same size. The classes do, in fact, vary in size as we shall see. The following two conditions will prove useful in the theorems that follow. Ccl(x) = ccl(y) if and only if $x \backsim y$ if and only if $y = \varphi_a(x)$ for some a in G if and only if $y = a^{-}xa$ for some a in G. Ccl(x) = ccl(y) if and only if there exist b, c in G such that $b^{-}xb = c^{-}yc$.

We continue the discussion of conjugacy classes by looking at the center of the group. Each element of the center is a class by itself.

Theorem. *Let $\langle G, \cdot, v\rangle$ be a group and let x be in G. $x \in Z(G)$ if and only if* ccl(x) $= \{x\}$.

PROOF. $x \in Z(G)$ if and only if $\varphi_x = I$ if and only if $x^{-}ax = a$ for all a in G if and only if $a^{-}xa = x$ for all a in G if and only if $\varphi_a(x) = x$ for all a in G if and only if ccl(x) $= \{x\}$. □

The only conjugacy classes containing exactly one element are those classes containing an element from the center of G. This count result can be extended to give a count for each conjugacy class: $\text{crd}(\text{ccl}(a)) = \text{crd } G/\mathcal{N}(a)$. This extended result agrees with the earlier result when $a \in Z(G)$ for then $\mathcal{N}(a) = G$ yielding $\text{crd } G/G = 1$. Since $\mathcal{N}(a)$ is not, in general, a normal subgroup of G, $G/\mathcal{N}(a)$ does not indicate a quotient group, but merely the Lagrange partition of G defined by the subgroup $\mathcal{N}(a)$. We will in the next theorem use right cosets of $\mathcal{N}(a)$.

Theorem. *Let $\langle G, \cdot, v\rangle$ be a group. Then $\text{crd}(\text{ccl}(a)) = \text{crd } G/\mathcal{N}(a)$ where $\text{ccl}(a)$ is the conjugacy class containing a and $G/\mathcal{N}(a)$ is the partition of G consisting of right cosets defined by the normalizer subgroup $\mathcal{N}(a)$.*

PROOF. $\text{Ccl}(a) = \{b \mid b \in G \text{ and } b \backsim a\} = \{y^{-}ay \mid y \in G\}$. We define a function $\chi:\text{ccl}(a) \to G/\mathcal{N}(a)$ such that $\chi(y^{-}ay) = \mathcal{N}(a)y$. We now establish simultaneously that the map is well defined and is injective. $x^{-}ax = y^{-}ay$ if and only if $ayx^{-} = yx^{-}a$ if and only if $(yx^{-})^{-}a(yx^{-}) = a$ if and only if $yx^{-} \in \mathcal{N}(a)$ if and only if $\mathcal{N}(a) = \mathcal{N}(a)yx^{-}$ if and only if $\mathcal{N}(a)x = \mathcal{N}(a)y$. Clearly, χ is surjective because given any $\mathcal{N}(a)y$ in $G/\mathcal{N}(a)$ there exists an element $y^{-}ay$ in $\text{ccl}(a)$ such that $\chi(y^{-}ay) = \mathcal{N}(a)y$. □

Having established this means of counting the number of members of any conjugacy class in terms of the normalizer we derive the class equation: a total count of the members of the conjugacy classes.

Theorem. *Let $\langle G, \cdot, v\rangle$ be a group. Then*

$$\text{crd } G = \text{crd } Z(G) + \sum_{\substack{a' \in R \\ a' \notin Z(G)}} \text{crd } G/\mathcal{N}(a')$$

where R is a set of representative elements from the partition $\{\text{ccl}(a) \mid a \in G\}$.

PROOF. Let R be a subset of G containing one element from each conjugacy class $\text{ccl}(a)$; in other words, let R be a complete set of representatives from the partition $\{\text{ccl}(a) \mid a \in G\}$.

$$\begin{aligned} \text{Crd } G &= \sum_{a' \in R} \text{crd ccl}(a') = \sum_{a' \in R} \text{crd } G/\mathcal{N}(a') \\ &= \sum_{\substack{a' \in R \\ a' \in Z(G)}} \text{crd } G/\mathcal{N}(a') + \sum_{\substack{a' \in R \\ a' \notin Z(G)}} \text{crd } G/\mathcal{N}(a') \\ &= \sum_{a' \in Z(G)} 1 + \sum_{\substack{a' \in R \\ a' \notin Z(G)}} \text{crd } G/\mathcal{N}(a') \\ &= \text{crd } Z(G) + \sum_{\substack{a' \in R \\ a' \notin Z(G)}} \text{crd } G/\mathcal{N}(a'). \end{aligned}$$ □

The conjugacy relation is now carried over to subgroups.

Definition. Let $\langle G, \cdot, v \rangle$ be a group and H and K subgroups. H and K are *conjugate* subgroups if and only if there exists an element x in G such that $x^{-}Hx = K$.

We use the same symbol for *conjugacy of subgroups* as we did for elements: $H \backsim K$. Conjugacy of subgroups is, of course, a relation on the set of all subgroups of G and not a relation on G.

Theorem. *Conjugacy* ($H \backsim K$) *is an equivalence relation on the set of all subgroups of a group* $\langle G, \cdot, v \rangle$.

PROOF. $H \backsim H$ because $v^{-}Hv = H$. $x^{-}Hx = K$ implies $(x^{-})^{-}Kx^{-} = H$. If $x^{-}Hx = K$ and $y^{-}Ky = L$ then $(xy)^{-}Hxy = L$. □

A normal subgroup of G satisfies the equation $x^{-}Hx = H$ for all x in G. This shows that a normal subgroup is conjugate only to itself; it is a singleton conjugacy class. For this reason normal subgroups are often called self-conjugate subgroups.

The counting theorem for the normalizer of a subgroup also generalizes.

Theorem. *Let* $\langle G, \cdot, v \rangle$ *be a group. The number of subgroups of* G *conjugate to a given subgroup* H *equals* crd $G/\mathcal{N}(H)$.

PROOF. Define, as before, a mapping $\psi : \text{ccl}(H) \to \{\mathcal{N}(H)y \mid y \in G\}$, such that $\psi(x^{-}Hx) = \mathcal{N}(H)x$. Ccl($H$) stands for the collection of all subgroups of G that are conjugate to H (ccl(H) is *not* a subset of G). Since $\mathcal{N}(H)$ is a subgroup of G then $\{\mathcal{N}(H)y \mid y \in G\}$ is a partition of G. Now $x^{-}Hx = y^{-}Hy$ if and only if $(xy^{-})^{-}Hxy^{-} = H$ if and only if $xy^{-} \in \mathcal{N}(H)$ if and only if $\mathcal{N}(H)y = \mathcal{N}(H)x$ proving ψ to be a well-defined injection. Clearly, it is also a surjection. □

We now propose to use the accumulated results of this section to produce some facts about finite groups.

Theorem. *If* p *is a prime number and* $\langle G, \cdot, v \rangle$ *is a group with* crd $G = p^n$, $n \in \mathbb{N}$, *then* $Z(G) \neq \{v\}$.

PROOF. We remark that in case $n = 1$ then crd $G = p$ and the group is a cyclic one as we have seen earlier. A cyclic group is commutative and its center is the entire group.

In beginning the proof of the theorem we notice $x \in Z(G)$ if and only if $x^{-}ax = a$ for all a in G if and only if $\mathcal{N}(x) = G$ if and only if crd $\mathcal{N}(x) = p^n$. We consider now the class equation

$$\text{crd } G = \text{crd } Z(G) + \sum_{\substack{a' \in R \\ a' \notin Z(G)}} \text{crd } G/\mathcal{N}(a').$$

Each $a' \notin Z(G)$ means $\mathcal{N}(a') \neq G$ and crd $\mathcal{N}(a')$ is a power of p strictly smaller than n. p then divides crd $G/\mathcal{N}(a')$ for each $a' \notin Z(G)$. p divides crd G on the left side of the equation by hypothesis. p must divide crd $Z(G)$. $Z(G)$ must have at least p elements. □

Theorem. *Let p be a prime number and let $\langle G, \cdot, v\rangle$ be a group with* crd $G = p^2$. *Then G is a commutative group.*

Proof. By the previous theorem crd $Z(G) = p$ or crd $Z(G) = p^2$. We proceed to eliminate the first alternative. If crd $Z(G) = p$ then there exists an $x \notin Z(G)$. $Z(G) \subseteq \mathcal{N}(x)$ but $Z(G) \neq \mathcal{N}(x)$ because $x \in \mathcal{N}(x)$. We then have $Z(G) \subset \mathcal{N}(x)$. This requires that $\mathcal{N}(x)$ have p^2 elements because any subgroup of G has 1, p, or p^2 elements. Then $\mathcal{N}(x) = G$, itself. $x \in Z(G)$, a contradiction. □

Theorem. *Let $\langle G, \cdot, v\rangle$ be a finite commutative group and let p be a prime number dividing* crd G. *Then G contains an element of order p.*

Proof. We have earlier seen that Lagrange's theorem requires the order of any element or subgroup to divide the cardinality of the group. We know the converse to be false ($\mathfrak{A}_4$ has no element of order 6). This theorem is a partial converse showing that, at least, for primes the converse is true for commutative groups. The proof is by induction on $n =$ crd G. Assume the conclusion true for all groups with cardinality strictly smaller than n. If G has no nontrivial subgroups then G is generated by a single element of prime order and is cyclic. Crd $G = p$, since p divides n.

Now suppose G does have a nontrivial subgroup H. If p divides crd H then crd $H < n$. H is commutative and has therefore an element of order p. The element of order p belongs to H and therefore to G.

Now suppose p does not divide crd H. H is a normal subgroup of commutative G and crd $G/H < n$. Since crd $G =$ crd(G/H) crd H and p does not divide crd H, p must divide crd G/H. There exists an element aH of G/H of order p by the inductive hypothesis. $(aH)^p = H$ and $a^m H \neq H$ for $m = 1, 2, \ldots, p - 1$. Then $(a^p)^{\text{crd } H} = v$ and furthermore $(a^{\text{crd } H})^p = v$. Since p is prime and the order of $a^{\text{crd } H}$ in G must divide p we need only show that $a^{\text{crd } H} \neq v$ in order to demonstrate that ord$(a^{\text{crd } H}) = p$. Suppose for the sake of argument that $a^{\text{crd } H} = v$. $a^{\text{crd } H}H = H$. $(aH)^{\text{crd } H} = H$. But $(aH)^p = H$. Thus p divides crd H yielding a contradiction for this part of the proof. $a^{\text{crd } H} \in G$ and has order p. □

We now extend the result to groups that are not necessarily commutative by use of the class equation.

Theorem. *Let p be a prime natural number and $\langle G, \cdot, v\rangle$ be a finite group such that p divides* crd G. *Then G contains an element of order p.*

PROOF. We note that we have proved the theorem for the case when G is given to be commutative. Assume that the result is true for all groups (not necessarily commutative) of cardinality $<n$. Suppose G has a proper subgroup H which has its cardinality divisible by p. Since crd $H < n$, H contains by the inductive assumption an element of order p. This element of H of order p is an element of G of order p. In this case the theorem is trivial. Now assume G has no proper subgroups which have their cardinality divisible by p. If H is any subgroup of G then the equation of Lagrange, crd $G =$ crd G/H crd H, tells us that p divides crd G/H for every proper subgroup H (under the assumption that p does not divide crd H). We look at the class equation

$$\text{crd } G = \text{crd } Z(G) + \sum_{\substack{a' \in R \\ a' \notin Z(G)}} \text{crd } G/\mathcal{N}(a').$$

Each $\mathcal{N}(a')$ is a subgroup of G means p divides each term of the sum, $\sum_{a' \in R, a' \notin Z(G)} \text{crd } G/\mathcal{N}(a')$, as well as crd G. p then divides crd $Z(G)$. But $Z(G)$ is a commutative subgroup of G, and must by the previous theorem have an element of order p. This element is then an element of G of order p. □

QUESTIONS

1. Which of the following statements about cyclic groups are true?
   (A) No cyclic group can have more than a finite number of subgroups.
   (B) Every element in a cyclic group except the unity generates the group.
   (C) Every subgroup of a cyclic group is also cyclic.
   (D) Every element in a cyclic group generates a cyclic subgroup.
   (E) All cyclic groups of the same cardinality are isomorphic.

2. Which of the following statements are true?
   (A) Every infinite cyclic group is isomorphic to $\langle \mathbb{Z}, +, 0 \rangle$.
   (B) There are precisely 16 nonisomorphic cyclic groups of cardinality less than or equal to 16.
   (C) A cyclic group of order mn has elements of order m and elements of order n even if m and n fail to be relatively prime.
   (D) A cyclic group cannot have cardinality n^2 if n is an odd number.
   (E) None of the statements is true.

3. Which of these statements are true?
   (A) A Lagrange partition of an infinite group by a finite subgroup must have an infinite number of distinct cosets.
   (B) A group of prime order cannot have any proper subgroups.
   (C) If a group contains an element of infinite order then the group must be cyclic.
   (D) If a group has a minimal generating set containing two elements then the group cannot be cyclic.
   (E) None of the statements is true.

4. Which of these statements are true?
   (A) If the center of a group is not the entire group then the group cannot be cyclic.

(B) Every commutative group is cyclic.
(C) The number of elements in any group is always a multiple of the number of elements in the center of the group.
(D) Any commutative subgroup of a group must contain the center.
(E) None of the statements is true.

5. Let $\langle G, \cdot, v\rangle$ be a group and φ_a be the inner automorphism of G associated with the element a. Which of these statements are true?
(A) If the group is noncommutative then there must be at least two elements not in the center.
(B) $\varphi_a = \varphi_b$ if and only if $a = b$.
(C) $a \in Z(G)$ implies $\varphi_a = \varphi_v$.
(D) $\varphi_a = \varphi_b$ if and only if $ab^- \in Z(G)$.
(E) None of the statements is true.

6. Which of the following statements are true for the group $\langle G, \cdot, v\rangle$?
(A) The normalizer of G, $\mathcal{N}(G)$, is a subgroup of $\mathcal{A}(G)$.
(B) $\{a | a \in G \text{ and } \varphi_a(S) \subseteq S\}$ is called the normalizer of the set S.
(C) If G is a commutative group then G is the normalizer of every nonempty subset of G.
(D) $\{a | a \in G \text{ and } \varphi_a(S) = S\}$ is the normalizer of the set S.
(E) None of the statements is true.

7. For the group $\langle \mathfrak{S}_3, \circ, I\rangle$ which of the following statements are true?
(A) $(1\ 2) \backsim (2\ 3)$.
(B) $(1\ 2\ 3) \backsim (1\ 3\ 2)$.
(C) $I \backsim (1\ 2\ 3)$.
(D) $I \backsim (1\ 2)$.
(E) None of the statements is true.

8. For the group $\langle \mathfrak{S}_3, \circ, I\rangle$ which of the following statements are true?
(A) $\mathfrak{S}_3$ has three elements of order 3.
(B) $\mathfrak{S}_3$ has three elements of order 2.
(C) There is an element of order 2 in $\mathfrak{S}_3$ and an element of order 3 in $\mathfrak{S}_3$ for which the product has order 6.
(D) $\mathfrak{S}_3$ is a cyclic group.
(E) None of the statements is true.

Exercises

1. Show that every cyclic group is commutative.

2. Find all the generating subsets of $\langle \mathbb{Z}_4, +, \bar{0}\rangle$ and $\langle \mathbb{Z}_6, +, \bar{0}\rangle$. Find the orders of each of the elements of both groups.

3. Show that every noncommutative group has a nontrivial, proper, commutative subgroup.

4. Show that every nontrivial group has a nontrivial cyclic subgroup.

5. Let $\langle G, \cdot, v\rangle$ be a group and a be in G such that $a^n = v$ for some positive integer n. Prove ord a divides n.

6. Let $\langle G, \cdot, v\rangle$ be a commutative group and $a, b \in G$ such that ord $a = m$ and ord $b = n$. Let m, n be relatively prime. Prove ord $ab = mn$.

7. Prove that the only element is a group with order 1 is the neutral element.

8. Show that in the symmetric group $\langle \mathfrak{S}_n, \circ, I\rangle$ there are elements of each order $1, 2, 3, \ldots, n$.

9. Let $\langle G, \cdot, v\rangle$ be a group with pq elements where p and q are primes. Show that every proper subgroup is cyclic.

10. Let $\langle G, \cdot, v\rangle$ be a commutative group and F be the subset of G of all elements with finite order. Show that F is a normal subgroup of G. Show that the only element of G/F with finite order is the neutral element F.

11. Show that if a group has no subgroups other than itself and the trivial one then it is a finite group.

12. Let $\langle G, \cdot, v\rangle$ be a group. Show that for each a in G both a and a^- have the same order. Show also that ab and ba have the same order for any a, b in G.

13. Show that every finite cyclic group is isomorphic to $\langle \mathbb{Z}_n, +, \bar{0}\rangle$ for some $n \in \mathbb{N}^+$.

14. Show that every infinite cyclic group has precisely two generators.

15. Prove that every subgroup of an infinite cyclic group is either trivial or isomorphic to the entire group.

16. Let $\langle G, \cdot, v\rangle$ be a group in which every element has finite order. Prove that if S is a nonempty subset of G closed under products then S is a subgroup.

17. Let $\langle G, \cdot, v\rangle$ and $\langle G', \cdot', v'\rangle$ be groups and $f: G \to G'$ a morphism. Prove that ord x is a multiple of ord $f(x)$.

18. Prove that neither $\langle \mathbb{Q}, +, 0\rangle$ nor $\langle \mathbb{Q}^*, \cdot, 1\rangle$ is a cyclic group.

19. Let $\langle G, \cdot, v\rangle$ and $\langle G', \cdot', v'\rangle$ be groups. Let $f: G \to G'$ be a morphism. Prove that if G is a cyclic group then so also is $f(G)$.

20. Show that every subgroup of a cyclic group is cyclic. Show also that any quotient group of a cyclic group is cyclic.

21. Show that two cyclic groups with the same number of elements are isomorphic.

22. Can $\langle \mathbb{Q}, +, 0\rangle$ be isomorphic to $\langle \mathbb{Q}^*, \cdot, 1\rangle$ or a subgroup of $\langle \mathbb{Q}^*, \cdot, 1\rangle$?

23. Can $\langle \mathbb{Z}, +, 0\rangle$ be isomorphic to $\langle \mathbb{Q}^*, \cdot, 1\rangle$ or a subgroup of $\langle \mathbb{Q}^*, \cdot, 1\rangle$?

24. Let $\langle G, \cdot, v\rangle$ and $\langle H, \cdot, v\rangle$ be groups and $f: G \to H$ be an isomorphism. Prove that if an element a generates G then $f(a)$ generates H. Prove that H cyclic implies G cyclic. Prove that if $\{a, b\}$ generates G then $\{f(a), f(b)\}$ generates H.

25. Let $\langle G, \cdot, v\rangle$ and $\langle H, \cdot, v\rangle$ be groups and $f: G \to H$ be a monomorphism. Prove ord $a =$ ord $f(a)$.

26. How many generators does a cyclic group with n (finite) elements have?

27. How many isomorphisms are there between a cyclic group with m elements and a cyclic group with n elements?

28. What familiar function of calculus is an isomorphism from $\mathbb{R}$ to $\mathbb{R}^+$ of groups $\langle \mathbb{R}, +, 0\rangle$ and $\langle \mathbb{R}^+, \cdot, 1\rangle$? What is its inverse?

29. Prove that an infinite cyclic group can form only finite quotient groups.

30. Let $\langle G, \cdot, v\rangle$ be a commutative group and $g_a: G \to G$ be the mapping $g_a(x) = a^2x$. Show that $\{g_a | a \in G\} \subseteq \mathfrak{S}(G)$. Define $\Phi: G \to \mathfrak{S}(G)$ such that $\Phi(a) = g_a$. Prove that Φ is a morphism with kernel consisting exactly of elements of order 2 and the neutral element. Can an element of $G/\ker \Phi$ have order 2?

31. Let $\langle M, \cdot, v\rangle$ be a unitary monoid. For each invertible element a of M define the map $\varphi_a: M \to M$ such that $\varphi_a(x) = a^{-}xa$. Show that $\{\varphi_a | a$ is an invertible element of $M\}$ is a subgroup of $\mathscr{A}(M)$.

32. By example prove there exist groups G such that $\mathscr{A}(G)$ is not a normal subgroup of $\mathfrak{S}(G)$.

33. Show that $\langle \mathbb{C}^*, \cdot, 1\rangle$ contains elements of infinite order and also every finite order.

34. Let $\langle K, +, \cdot, \theta, v\rangle$ be a field with n elements. Prove that $x^{n-1} = v$ for every nonzero x in K. Prove $x^n = x$ for every x in K.

35. Prove Fermat's theorem: $x^p = x$ modulo p for p a prime integer.

36. Let $\langle K, +, \cdot, \theta, v\rangle$ be a field with n elements. Prove that every x in K is a root of the polynomial $X^n - X$. Represent the n elements of K by $a_1, a_2, \ldots, a_n$. Prove $X^n - X = (X - a_1)(X - a_2)\cdots(X - a_n)$. Prove also $X^{n-1} - v = \prod_{a \in K^*} (X - a)$. Prove $-v = \prod_{a \in K^*} a$ for $n \neq 2$.

37. Prove Wilson's theorem: $-1 = (p - 1)!$ modulo p, a prime integer.

38. Prove $\mathscr{N}(\varnothing) = G$ and $\mathscr{N}(v) = G$ for any group $\langle G, \cdot, v\rangle$.

39. Find for the group $\langle \mathfrak{S}_3, \circ, I\rangle$ the normalizers $\mathscr{N}((1\ 2))$, $\mathscr{N}(I, (1\ 2))$, $\mathscr{N}(\mathfrak{A}_3)$.

40. The following group is noncommutative and has 27 elements. It has three generators each with order three which we name a, b, and c. $a^3 = v$, $b^3 = v$, $c^3 = v$. We further stipulate that $ca = ac$, $cb = bc$, and $bac = ab$; two of the three generators commute with one another but the third does not. Show that $\{a^ib^jc^k | i, j, k = 0, 1, 2\}$ constitutes a complete representation of the group; every product can be reduced to the form $a^ib^jc^k$ with $i, j, k \in \{0, 1, 2\}$. Making a complete multiplication table for the group is unreasonably tedious. What is the inverse of $a^ib^jc^k$?

41. The multiplicative group generated by the two matrices $\begin{pmatrix} 0 & 1 \\ -1 & 0 \end{pmatrix}$ and $\begin{pmatrix} 0 & i \\ i & 0 \end{pmatrix}$ is a noncommutative group of order 8. Make a multiplication table (a long project) for the group. Find the order of each element and find all subgroups. This group is known as the quaternion group.

42. Show that the two matrices $\begin{pmatrix} 0 & 1 \\ -1 & 0 \end{pmatrix}$ and $\begin{pmatrix} 0 & 1 \\ 1 & 0 \end{pmatrix}$ also generate a multiplicative group of order 8. Is this group isomorphic to the quaternion group?

43. Each $\mathfrak{S}_n$ is a subgroup of $\mathfrak{S}_{n+1}$ in the sense that a cycle such as $(1\ 2 \cdots n) \in \mathfrak{S}_{n+1}$ as well as $\mathfrak{S}_n$. Show that $\mathfrak{S} = \bigcup_{n \in \mathbb{N}^+} \mathfrak{S}_n$ is a group and contains all $\mathfrak{S}_n$ as subgroups. Show that this group contains precisely one normal subgroup $\mathfrak{A}$ which is not $\mathfrak{S}$ nor $\{I\}$.

44. Show that any subgroup S of $\mathfrak{S}_n$ containing at least one odd permutation has a normal subgroup N such that crd $S/N = 2$. [*Hint*: Use an isomorphism theorem.]

9.5 Products

In this section we present product constructions for monoids. We also discuss simple groups and solvable groups and finish with a product theorem for commutative groups.

The Cartesian product of two monoids, when endowed with the product operations, yields a monoid which is called the product monoid.

Theorem. *Given monoids $\langle M, \cdot_1\rangle$ and $\langle M_2, \cdot_2\rangle$, the Cartesian product $\langle M_1 \times M_2, \cdot_1 \times \cdot_2\rangle$ is a monoid. If $\langle M_1, \cdot_1, v_1\rangle$ and $\langle M_2, \cdot_2, v_2\rangle$ are unitary monoids then $\langle M_1 \times M_2, \cdot_1 \times \cdot_2, (v_1, v_2)\rangle$ is a unitary monoid. If both unitary monoids are groups then the product is a group.*

PROOF. $(x_1, x_2)\cdot_1 \times \cdot_2 (y_1, y_2) = (x_1 \cdot_1 y_1, x_2 \cdot_2 y_2)$. □

EXAMPLE. $\mathbb{Z}_2 \times \mathbb{Z}_3 = \{(\bar{0}, \bar{0}), (\bar{0}, \bar{1}), (\bar{0}, \bar{2}), (\bar{1}, \bar{0}), (\bar{1}, \bar{1}), (\bar{1}, \bar{2})\}$ is the Cartesian product of $\langle \mathbb{Z}_2, +, \bar{0}\rangle$ and $\langle \mathbb{Z}_3, +, \bar{0}\rangle$ and is a group. This product group is cyclic with generator $(\bar{1}, \bar{1})$. $0(\bar{1}, \bar{1}) = (\bar{0}, \bar{0})$. $1(\bar{1}, \bar{1}) = (\bar{1}, \bar{1})$. $2(\bar{1}, \bar{1}) = (\bar{0}, \bar{2})$. $3(\bar{1}, \bar{1}) = (\bar{1}, \bar{0})$. $4(\bar{1}, \bar{1}) = (\bar{0}, \bar{1})$. $5(\bar{1}, \bar{1}) = (\bar{1}, \bar{2})$.

EXAMPLE. $\mathfrak{S}_3 \times \mathfrak{S}_3$ with composition in each component is the Cartesian product of the group $\langle \mathfrak{S}_3, \circ, I\rangle$ with itself. It is a noncommutative group of order 36. A sample calculation, using cycles in each component, is ((1 2 3), (2 3))((1 3), (1 3 2)) = ((1 2 3)(1 3), (2 3)(1 3 2)) = ((2 3), (1 2)).

The Cartesian product of a finite number of monoids is an easy generalization from two monoids. $\langle \times_{j \in n} M_j, \cdot\rangle$ has the binary operation defined coordinatewise as with two monoids. $(x_1, x_2, \ldots, x_n)(y_1, y_2, \ldots, y_n) = (x_1 \cdot_1 y_1, x_2 \cdot_2 y_2, \ldots, x_n \cdot_n y_n)$, or alternatively expressed $(x_i | i \in n)(y_i | i \in n) = (x_i \cdot_i y_i | i \in n)$. $\times_{j \in n} M_j = \{(x_j | j \in n) | x_j \in M_j\}$. We also write $\times_{j \in n} M_j$ as $M_1 \times M_2 \times \cdots \times M_n$. The notation $M_1 \times M_2 \times \cdots \times M_n$ is perhaps easier to comprehend but adapts poorly to the infinite case. We now turn our attention briefly to this infinite situation. We assume we have given a family of monoids $(\langle M_j, \cdot_j\rangle | j \in J)$, one for each member of the infinite set J. The infinite Cartesian product $\times_{j \in J} M_j = \{(x_j | j \in J) | x_j \in M_j\}$, consists of all possible families of elements in which the coordinate x_j is chosen from the monoid M_j of corresponding index. The operation of the product is simply multiplication in each coordinate j using the given multiplication in M_j. $(x_j | j \in J)(y_j | j \in J) = (x_j \cdot_j y_j | j \in J)$.

A simple example of such an infinite product is to use the rational numbers, $\mathbb{Q}$, a countable number of times. $\times_{j\in\mathbb{N}} \mathbb{Q} = \{(x_j | j \in \mathbb{N}) | x_j \in \mathbb{Q}\}$. This set is precisely the set of all possible sequences of rational numbers. Each component $\langle \mathbb{Q}, \cdot \rangle$ is a monoid and so also is the Cartesian product $\times_{j\in\mathbb{N}} \mathbb{Q}$. $\langle \mathbb{Q}, \cdot \rangle$ has a unity 1 and the Cartesian product has the unity $(1, 1, 1, \ldots)$.

For unitary monoids we have another infinite product which is a proper subset of the full Cartesian product. This is the weak Cartesian product which consists of all families which have all but a finite number of coordinate entries equal to the unity. Given a family of unitary monoids $(\langle M_j, \cdot, v_j \rangle | j \in J)$ the weak Cartesian product is $\times^w_{j\in J} M_j = \{(x_j | j \in J) | x_j \in M_j$ and $x_j = v_j$ for all but a finite number of j in $J\}$. It is easily verifiable that $\times^w_{j\in J} M_j$ is a proper unitary submonoid of $\times_{j\in J} M_j$. Furthermore, if each M_j is a group then $\times^w_{j\in J} M_j$ is a normal subgroup of the group $\times_{j\in J} M_j$.

Associated with any Cartesian product are the projection mappings of the product back into one of the component monoids.

Definition. The mapping $p_j : \times_{i\in I} M_i \to M_j$ is called the jth coordinate projection.

More simply, for two monoids M_1 and M_2 we have the first projection $p_1 : M_1 \times M_2 \to M_1$ such that $p_1(x_1, x_2) = x_1$ and the second projection $p_2 : M_1 \times M_2 \to M_2$ such that $p_2(x_1, x_2) = x_2$. It is easy to verify this theorem.

Theorem. *The projection mappings are epimorphisms.*

Analogously, we also have the insertion (or embedding) mappings $q_j : M_j \to \times_{i\in I} M_i$ such that $q_j(x) = (y_i | i \in J$ and $y_i = v_i$ for all i except j and $y_j = x)$. We must use here the unity elements for the monoids. For two unitary monoids M_1 and M_2, we have $q_1 : M_1 \to M_1 \times M_2$ such that $q_1(x) = (x, v_2)$ and $q_2(x) = (v_1, x)$. Range $q_1 = M_1 \times \{v_2\}$ and range $q_2 = \{v_1\} \times M_2$. Each component of a product of unitary monoids is isomorphic to a unitary submonoid of the Cartesian product.

In the case of groups we have this theorem.

Theorem. *The range of each insertion map $q_j : M_j \to \times_{i\in I} M_i$ of a group into the product group is a normal subgroup of the product. Furthermore, if $x \in q_j(M_j)$ and $y \in q_k(M_k)$ and $j \neq k$ then x and y commute and $q_j(M_j) \cap q_k(M_k) = (v_i | i \in I)$ for all $j \neq k$. $\times_{i\in I} M_i = \prod_{i\in I} q_i(M_i)$.*

Example. Three groups G_1, G_2, G_3 have a Cartesian product $G_1 \times G_2 \times G_3$. The three insertion mappings are $q_1 : G_1 \to G_1 \times G_2 \times G_3$ with range $G_1 \times \{v_2\} \times \{v_3\}$, $q_2 : G_2 \to G_1 \times G_2 \times G_3$ with range $\{v_1\} \times G_2 \times \{v_3\}$, $q_3 : G_3 \to G_1 \times G_2 \times G_3$ with range $\{v_1\} \times \{v_2\} \times G_3$. These ranges are all normal subgroups of $G_1 \times G_2 \times G_3$. Elements from these separate groups commute with each other even though the original groups G_1, G_2, G_3 are not necessarily commutative. The ranges have only the unity of $G_1 \times G_2 \times G_3$ in common. $(x_1, v_2, v_3)(v_1, x_2, v_3) = (x_1, x_2, v_3) = (v_1, x_2, v_3)(x_1, v_2, v_3)$. If (x_1, x_2, x_3) belongs to $G_1 \times G_2 \times G_3$ then (x_1, x_2, x_3) is equal to the product $(x_1, v_2, v_3)(v_1, x_2, v_3)(v_1, v_2, x_3)$ showing $G_1 \times G_2 \times G_3 = q_1(G_1)q_2(G_2)q_3(G_3)$.

On the basis of the properties exhibited by these normal subgroups defined by the insertion maps we frame a definition of a direct product (or direct sum if additive notation is employed).

Definition. A group $\langle G, \cdot, v\rangle$ is the *direct product* of normal subgroups H_1, $H_2, \ldots, H_n$ if and only if

1. $H_1, H_2, \ldots, H_n$ are nontrivial proper normal subgroups of G,
2. $G = H_1H_2 \cdots H_n$,
3. $H_i \cap H_1H_2 \cdots H_{i-1}H_{i+1} \cdots H_n = \{v\}$,
4. $x \in H_i$ and $y \in H_j$ and $i \neq j$ imply $xy = yx$.

For the direct product we write $H_1 \odot H_2 \odot \cdots \odot H_n$. If the operation is denoted additively then we will write $H_1 \oplus H_2 \oplus \cdots \oplus H_n$ and call it the direct sum. We also denote the direct product with $\bigodot_{i=1}^n H_i$ and the direct sum with $\bigoplus_{i=1}^n H_i$.

EXAMPLE. The cyclic group $\{1, a, a^2, a^3, a^4, a^5\}$ of six elements generated by a is a direct product of the two normal subgroups $\{1, a^3\}$ and $\{1, a^2, a^4\}$. $\{1, a, a^2, a^3, a^4, a^5\} = \{1, a^3\} \odot \{1, a^2, a^4\}$. The same example in additive notation is $\mathbb{Z}_6 = \{\overline{0}, \overline{3}\} \oplus \{\overline{0}, \overline{2}, \overline{4}\}$.

We have used the previous examples of Cartesian products to help motivate our definition of direct product. We now show that the two concepts are isomorphic.

Theorem. *Let $\langle G, \cdot, v\rangle$ be a group and $H_1, H_2, \ldots, H_n$ be normal subgroups, nontrivial and proper. If $G = H_1 \odot H_2 \odot \cdots \odot H_n$ then G is isomorphic to $H_1 \times H_2 \times \cdots \times H_n$, the Cartesian product.*

PROOF. We define a mapping $F: H_1 \times H_2 \times \cdots \times H_n \to G$ such that $F(x_1, x_2, \ldots, x_n) = x_1x_2 \cdots x_n$. F is easily seen to be a morphism. We must, however, use the fact that members of different normal subgroups of the direct product commute with one another. $F((x_1, x_2, \ldots, x_n)(y_1, y_2, \ldots, y_n)) = F(x_1y_1, x_2y_2, \ldots, x_ny_n) = x_1y_1x_2y_2 \cdots x_ny_n = x_1x_2 \cdots x_ny_1y_2 \cdots y_n = F(x_1, x_2, \ldots, x_n)\,F(y_1, y_2, \ldots, y_n)$. Ker $F = \{(x_1, x_2, \ldots, x_n) | x_1x_2 \cdots x_n = v\}$. Since $x_j = x_1^{-}x_2^{-} \cdots x_{j-1}^{-}x_{j+1}^{-} \cdots x_n^{-} \in H_1H_2 \cdots H_{j-1}H_{j+1} \cdots H_n$ and $x_j \in H_j$ we have $x_j = v$. This shows F to be a monomorphism. Finally, since $G = H_1H_2 \cdots H_n$ we have x in G implies $x = x_1x_2 \cdots x_n = F(x_1, x_2, \ldots, x_n)$, showing F to be a surjection. □

Example. $\mathbb{Z}_6 = \{\overline{0}, \overline{3}\} \oplus \{\overline{0}, \overline{2}, \overline{4}\}$. $\mathbb{Z}_6 \approx \{\overline{0}, \overline{3}\} \times \{\overline{0}, \overline{2}, \overline{4}\}$. Since $\{\overline{0}, \overline{3}\} \approx \mathbb{Z}_2$ and $\{\overline{0}, \overline{2}, \overline{4}\} \approx \mathbb{Z}_3$ we can write $\mathbb{Z}_6 \approx \mathbb{Z}_2 \times \mathbb{Z}_3$ using the isomorphism result that $A \approx C$ and $B \approx D$ imply $A \times B \approx C \times D$.

The direct product can also be defined for an infinite number of normal subgroups and the isomorphism between the direct product and the weak Cartesian product can be established. Several exercises are devoted to this end.

For groups disassembling into direct products we can use the fundamental isomorphism theorem to produce this theorem.

Theorem. *Let $\langle G, \cdot, v\rangle$ be a group. If $G = H_1 \odot H_2$ then $G/H_1 \approx H_2$. If $G = H_1 \odot H_2 \odot \cdots \odot H_n$ then $G/(H_1H_2 \cdots H_{j-1}H_{j+1} \cdots H_n) \approx H_j$.*

Proof. The projection $p_j : H_1 \times H_2 \times \cdots \times H_n \to H_j$ has kernel equal to $H_1 \times \cdots \times H_{j-1} \times \{v\} \times H_{j+1} \times \cdots \times H_n$. $G/\ker p_j \approx H_j$. □

Groups or monoids representable as direct products are, in a sense, decomposed into constituent components. This is a method for analyzing a group. We continue our study of normal subgroups.

Definition. A group $\langle G, \cdot, v\rangle$ is called *simple* if and only if the only normal subgroups of G are $\{v\}$ and G itself.

Example. Any cyclic group of prime order is simple. It has been shown that all other simple groups have even order. The problem of cataloging simple groups is a difficult one and remains quite imcomplete. We will show later that $\mathfrak{A}_5$, the even permutation subgroup of $\mathfrak{S}_5$, is a simple group. Meanwhile, we have this useful result.

Theorem. *Let $\langle G, \cdot, v\rangle$ be a group. N is a maximal proper normal subgroup of G if and only if G/N is a nontrivial simple group.*

Proof. We first establish that for N and H normal subgroups of G that $N \subseteq H$ implies H/N is a normal subgroup of G/N. Suppose there exists a normal subgroup H such that $N \subseteq H \subseteq G$. Let xN and yN belong to H/N. $xN\ yN = xyN \in H/N$. $(xN)^- = x^-N \in H/N$. For $zN \in G/N$ we have $(zN)^-xN(zN) = z^-xzN \in H/N$ proving H/N is normal in G/N.

Now suppose G/N is a simple group and $N \subseteq H$, also a normal subgroup. $N \subseteq H$ implies H/N is normal in G/N. But G/N is simple. $H/N = \{N\}$ or $H/N = G/N$. If $H/N = \{N\}$ then $H = N$. If $H/N = G/N$ then $H = G$. Hence N is a maximal normal subgroup.

For the converse, assume G/N has some nontrivial proper normal subgroup M. M consists of some, but not all, of the cosets of G/N. $\{xN | xN \in M\} = M$. Let S be the set of all members of G which belong to some coset in M, i.e., $S = \bigcup M$. S is a normal subgroup of G. If $x, y \in S$ then $xN, yN \in M$. $xN(yN)^- \in M$. $xy^-N \in M$. $xy^- \in S$. If $z \in G$ then $(zN)^-xN\ zN = z^-xzN \in M$.

$z^{-}xz \in S$. $M \neq \{N\}$ implies there exists an x in S, $x \notin N$. M is proper means there exists $xN \in G/N$ such that $xN \notin M$. Therefore, $x \in G$ and $x \notin S$. $N \subset S \subset G$. N is not maximal. □

We now show one of the ways normal subgroups are used to analyze the structure of groups.

Definition. Let $G_0, G_1, \ldots, G_{n+1}$ be a finite number of subgroups of a group $\langle G, \cdot, v\rangle$ such that $G_0 = G$, $G_{n+1} = \{v\}$ and G_{j+1} is a maximal proper normal subgroup of G_j, $j = 0, 1, \ldots, n$. $G_0, G_1, \ldots, G_{n+1}$ is then called a *composition series for the group G*. $G/G_1, G_1/G_2, \ldots, G_n/\{v\}$ are the *factor groups* or *quotient groups of the composition series*. The cardinality of these quotient groups are called the *factors of the composition series* for G.

EXAMPLE. For the additive group $\mathbb{Z}_{12}$ two different composition series are $\mathbb{Z}_{12}$, $\{\bar{0}, \bar{2}, \bar{4}, \bar{6}, \bar{8}, \overline{10}\}$, $\{\bar{0}, \bar{4}, \bar{8}\}$, $\{\bar{0}\}$ and $\mathbb{Z}_{12}$, $\{\bar{0}, \bar{3}, \bar{6}, \bar{9}\}$, $\{\bar{0}, \bar{6}\}$, $\{\bar{0}\}$. The quotient groups for the two composition series are respectively $\mathbb{Z}_{12}/\{\bar{0}, \bar{2}, \bar{4}, \bar{6}, \bar{8}, \overline{10}\}$, $\{\bar{0}, \bar{2}, \bar{4}, \bar{6}, \bar{8}, \overline{10}\}/\{\bar{0}, \bar{4}, \bar{8}\}$, $\{\bar{0}, \bar{4}, \bar{8}\}/\{\bar{0}\}$ and $\mathbb{Z}_{12}/\{\bar{0}, \bar{3}, \bar{6}, \bar{9}\}$, $\{\bar{0}, \bar{3}, \bar{6}, \bar{9}\}/\{\bar{0}, \bar{6}\}$, $\{\bar{0}, \bar{6}\}/\{\bar{0}\}$. The factors for the two composition series are 2, 2, 3 and 3, 2, 2.

Theorem. *Any two composition series of a finite group $\langle G, \cdot, v\rangle$ have their quotient groups isomorphic in some order.*

PROOF. If the group has no nontrivial normal subgroups (is simple) then there is but one composition series possible: G, $\{v\}$. We proceed by induction on the number of prime factors in the integer n which denotes the cardinality of G. If n is prime, has but one prime factor, then G is cyclic and has no nontrivial subgroups. The theorem in this case is true. Now suppose for induction that the theorem is true if crd G has fewer than k prime factors. Let crd G now have k prime factors. Let two composition series for G be $G'_0, G'_1, \ldots, G'_{l+1}$ and $G''_0, G''_1, \ldots, G''_{m+1}$. By the definition of composition series $G'_0 = G''_0 = G$ and $G'_{l+1} = G'_{m+1} = \{v\}$. If in addition $G'_1 = G''_1$ denote both by G_1. Then $G'_1, G'_2, \ldots, G'_{l+1}$ and $G''_1, G''_2, \ldots, G''_{m+1}$ are composition series for G_1. Since G_1 is a proper normal subgroup of G we must have crd $G_1 <$ crd G and crd G_1 a factor of crd G. Crd G_1 must have fewer than k prime factors. By the inductive assumption $G'_1, G'_2, \ldots, G'_{l+1}$ and $G''_1, G''_2, \ldots, G''_{m+1}$ have their quotient groups isomorphic in some order: $G'_1/G'_2 \approx G''_{i_1}/G''_{i_2}$, $G'_2/G'_3 \approx G''_{i_2}/G''_{i_3}, \ldots, G'_l/G^i_{l+1} \approx G''_{i_l}/G''_{i_{l+1}}$. Thus $l = m$. Furthermore, $G'_0/G'_1 \approx G''_0/G''_1$ so that the original series are the same length and isomorphic in some order.

If it is not the case that $G'_1 = G''_1$ then a more elaborate argument is necessary. Suppose $G'_1 \neq G''_1$ and let $G'_1 \cap G''_1 = H_1$. By one of the isomorphism theorems of Section 9.3, $(G''_1G'_1)/G'_1 \approx G''_1/(G'_1 \cap G''_1)$. $G''_1G'_1 = G$ because the product is a normal subgroup of G properly larger than G'_1.

Thus $G/G_1' \approx G_1''/(G_1' \cap G_1'')$. By the same reasoning $G/G_1'' \approx G_1'/(G_1'' \cap G_1')$. G_0'/G_1' and G_0''/G_1'' are both simple groups since the subgroups defining the quotient groups are maximal normal subgroups. As isomorphic groups $G_1'/(G_1' \cap G_1'')$ and also $G_1''/(G_1' \cap G_1'')$ must be simple groups too. This shows $(G_1' \cap G_1'')$ is a maximal normal subgroup of both G_1' and of G_1''. Using $H_1 = G_1' \cap G_1''$, H_1 is maximal in G_1' and in G_1''. Let $H_1, H_2, \ldots, H_{r+1}$ be a composition series for H_1. Then G_1 has two composition series: G_1', $G_2', \ldots, G_{l+1}'$ and $G_1', H_1, H_2, \ldots, H_{r+1}$. In some order their quotient groups must be isomorphic. G_1'/H_1, $H_1/H_2, \ldots, H_r/H_{r+1}$ and G_1'/G_2', $G_2'/G_3', \ldots, G_l'/G_{l+1}'$ are isomorphic in some order. So also are the extended series $G_0'/G_1', G_1'/H_1, H_1/H_2, \ldots, H_r/H_{r+1}$ and $G_0'/G_1', G_1'/G_2', G_2'/G_3', \ldots,$ G_l'/G_{l+1}' isomorphic in some order. By a symmetrical argument G_0''/G_1'', $G_1''/G_2'', G_2''/G_3'', \ldots, G_m''/G_{m+1}''$ are also isomorphic in some order to the quotient groups $G_0'/H_1, H_1/H_2, \ldots, H_r/H_{r+1}$. Thus $l = r = m$ and the two given composition series have their quotient groups isomorphic in some order. □

EXAMPLE. In the previous example of $\mathbb{Z}_{12}$ there are two composition series. The isomorphisms are $\mathbb{Z}_{12}/\{\bar{0}, \bar{2}, \bar{4}, \bar{6}, \bar{8}, \overline{10}\} \approx \{\bar{0}, \bar{3}, \bar{6}, \bar{9}\}/\{\bar{0}, \bar{6}\}$, $\{\bar{0}, \bar{2}, \bar{4}, \bar{6}, \bar{8}, \overline{10}\}/\{\bar{0}, \bar{4}, \bar{8}\} \approx \{\bar{0}, \bar{6}\}/\{\bar{0}\}$, $\{\bar{0}, \bar{4}, \bar{8}\}/\{\bar{0}\} \approx \mathbb{Z}_{12}/\{\bar{0}, \bar{3}, \bar{6}, \bar{9}\}$.

Definition. A finite group is *solvable* if and only if the quotient groups of the composition series are commutative.

Theorem. *A finite group is solvable if and only if the quotient groups of the composition series have prime cardinality and are cyclic.*

PROOF. This theorem amounts to proving that if N is a maximal normal subgroup of a finite group G then G/N is commutative if and only if G/N has prime cardinality and is cyclic. Certainly, every cyclic group with the number of elements prime is commutative. Now assume that G/N is a commutative group. Since N is a maximal normal subgroup we have G/N both simple and commutative. If G/N is not cyclic of prime size then some element of G/N would generate a normal subgroup which would be proper. But G/N is simple. G/N must be cyclic and have a prime number of elements. □

EXAMPLE. Both of these are composition series. $\mathfrak{S}_3, \mathfrak{A}_3, \{I\}$. $\mathfrak{S}_4, \mathfrak{A}_4, V_4, \mathfrak{S}_2, \{I\}$.

EXAMPLE. $\mathfrak{S}_1, \mathfrak{S}_2, \mathfrak{S}_3, \mathfrak{S}_4$ are all solvable groups.

Lemma. *Any normal subgroup N of $\mathfrak{A}_n$, $n = 4, 5, \ldots$, containing a three cycle must be $\mathfrak{A}_n$ itself.*

PROOF. If $(1\ 2\ 3) \in N$ then $(1\ 2\ k) \in N$ because $[(1\ 2)(3\ k)]^{-}(1\ 2\ 3)^2[(1\ 2) \cdot (3\ k)] = (1\ 2)(3\ k)(1\ 3\ 2)(1\ 2)(3\ k) = (1\ 2\ k)$. From this we can conclude that if (1 2 3) belongs to N then (1 2 4), (1 2 5), ..., (1 2 n) all belong to N. From Section 7.7 we know (1 2 3), (1 2 4), ..., (1 2 n) generate $\mathfrak{A}_n$. Thus if $(1\ 2\ 3) \in N$ then $N = \mathfrak{A}_n$.

If the particular given three cycle of N is not (1 2 3) then we can by appropriate compositions produce (1 2 3) as a member of N. Let $(i_1\ i_2\ i_3)$ belong to N and consider the following cases.

1. $(j\ k\ l) \in N$ implies $[(k\ l)(j\ 1)]^{-}(j\ k\ l)\ [(k\ l)(j\ 1)] = (1\ l\ k) \in N$. $(1\ l\ k) \in N$ implies $(1\ l\ k)^2 = (1\ k\ l) \in N$.
2. $(1\ l\ 2) \in N$ implies $(1\ l\ 2)^2 = (1\ 2\ l) \in N$.
3. $(1\ k\ l) \in N$ implies $[(1\ k)(2\ l)]^{-}(1\ k\ l)\ [(1\ k)(2\ l)] = (1\ 2\ k) \in N$.
4. $(1\ 2\ l) \in N$ implies $[(1\ 2)(3\ l)]^{-}(1\ 2\ l)^2[(1\ 2)(3\ l)] = (1\ 2\ 3) \in N$. □

Theorem. *$\mathfrak{A}_n$ is simple for $n = 5, 6, 7, \ldots$.*

PROOF. Let N be a normal subgroup of $\mathfrak{A}_n$ not $\{I\}$. We wish to show that N is equal to $\mathfrak{A}_n$ by showing N must contain some permutation which is a three cycle.

Suppose N, a nontrivial normal subgroup, fails to contain a three cycle from $\mathfrak{A}_n$, $n \geqslant 5$. Then let σ be the nonidentical permutation in N which disturbs (does not leave fixt) the fewest members of $\{1, 2, \ldots, n\}$. If σ disturbs 4 members it cannot be in the form (1 2 3 4) which is an odd permutation, but must be of the form (1 2)(3 4). Then $[(3\ 4\ 5)^{-}\sigma(3\ 4\ 5)]\sigma^{-} = (3\ 5\ 4) \cdot (1\ 2)(3\ 4)(3\ 4\ 5)(1\ 2)(3\ 4) = (3\ 4\ 5)$ belongs to N also. But (3 4 5) disturbs only 3 members of $\{1, 2, \ldots, n\}$, a contradiction. σ must then disturb at least 5 members of $\{1, 2, \ldots, n\}$. σ will then be of the form (1 2 3)(4 5 6) ... or (1 2)(3 4)(5 6)(7 8) ... or (1 2 3 4)(5 6) .... σ must involve at least 5 digits which we have for convenience chosen to call 1, 2, 3, 4, 5. If other numbers are involved one can change the argument accordingly. One can now see that $\tau = [(3\ 4\ 5)^{-}\sigma(3\ 4\ 5)]\sigma^{-}$ can disturb no member of $\{1, 2, \ldots, n\}$ not already disturbed by σ. The number 2, however is held fixt by τ. τ will then disturb fewer members of $\{1, 2, \ldots, n\}$ than does σ. This contradicts σ disturbing the fewest members. Thus N must contain a three cycle and by the lemma is equal to $\mathfrak{A}_n$. □

We finish this section with results on direct product resolutions of commutative groups. Since the groups are commutative we use additive notation and speak of direct sums.

Theorem. *Let $\langle G, +, \theta\rangle$ be a commutative group with cardinality $p_1^{\alpha_1}p_2^{\alpha_2}\cdots p_n^{\alpha_n}$ with each p_j a prime natural number, all distinct. Then $G = \bigoplus_{j=1}^{n} P_j$ such that order of each element in P_j is a power of p_j and* crd $P_j = p_j^{\alpha_j}$.

PROOF. Let x, y both be elements of order some power of p. $p^i x = \theta$ and $p^j y = \theta$ for appropriate integers i and j. Then $p^{i+j}(x + y) = p^{i+j}x + p^{i+j}y = p^j(p^i x) + p^i(p^j y) = \theta + \theta = \theta$. $p^i(-x) = -p^i x = -\theta = \theta$. Thus the elements of order some power of p form a subgroup of G. Call this subgroup P.

In order to distinguish between the several prime components we define P_j to be the subgroup of G of all elements having order some power of the prime p_j. We wish to demonstrate that $G = P_1 \oplus P_2 \oplus \cdots \oplus P_n$. Because the given group is commutative all pairs of elements commute and in particular, pairs from different P_j. Because only zero can have order $1 = p^0$, only zero can belong to more than one of the P_j. $P_i \cap P_j = \{\theta\}$ for $i \neq j$. Now let y belong to G and have some order $p_1^{\beta_1} p_2^{\beta_2} \cdots p_n^{\beta_n}$. We prove by induction on n that $G = P_1 + P_2 + \cdots + P_n$. If $x \in G$ and the order of x has two prime components, ord $x = p_1^{\beta_1} p_2^{\beta_2}$, $\gcd(p_1^{\beta_1}, p_2^{\beta_2}) = 1$. There exist integers m', n' such that $m' p_1^{\beta_1} + n' p_2^{\beta_2} = 1$. We consider then the two elements $m' p_1^{\beta_1} x$ and $n' p_2^{\beta_2} x$. $x = 1x = (m' p_1^{\beta_1})x + (n' p_2^{\beta_2})x$. Ord $m' p_1^{\beta_1} x = p_2^{\beta_2}$ and ord $n' p_2^{\beta_2} x = p_1^{\beta_1}$. Thus an element of order $p_1^{\beta_1} p_2^{\beta_2}$ is the sum of an element of order $p_2^{\beta_2}$ and an element of order $p_1^{\beta_1}$. For the induction step let x have order $p_1^{\beta_1} p_2^{\beta_2} \cdots p_k^{\beta_k}$. $\operatorname{Gcd}(p_i^{\beta_i}, p_j^{\beta_j}) = 1$ for $i \neq j$. There exist m', n' such that $m' p_1^{\beta_1} p_2^{\beta_2} \cdots p_{k-1}^{\beta_{k-1}} + n' p_k^{\beta_k} = 1$. Therefore, $x = m' p_1^{\beta_1} p_2^{\beta_2} \cdots p_{k-1}^{\beta_{k-1}} x + n' p_k^{\beta_k} x$ belongs to $B + P_k$ with order of $m' p_1^{\beta_1} p_2^{\beta_2} \cdots p_{k-1}^{\beta_{k-1}} x = p_k^{\beta_k}$ and order of $n' p_k^{\beta_k} x = p_1^{\beta_1} p_2^{\beta_2} \cdots p_{k-1}^{\beta_{k-1}}$. Hence by induction $G = P_1 + P_2 + \cdots + P_n$. □

Having proved that every commutative group can be resolved into a direct sum of subgroups in which each subgroup contains elements with order equal to powers of some prime we now continue the resolution of these prime subgroups.

Theorem. *Let P be a finite commutative group with every element having order some power of the prime p. The P is the direct sum of cyclic subgroups.*

Proof. Let P have cardinality p^α. Let p^β be the order of an element in P of greatest order. We make an induction on β. If $\beta = 1$ then every non-neutral element has order p. Let $a_1 \in P$. If $P = [a_1]$ then the theorem is proved. If not let $a_2 \in P - [a_1]$. The set $\{m_1 a_1 + m_2 a_2 | m_1, m_2 = 0, 1, \ldots, p - 1\}$ has p^2 members because if $m_1 a_1 + m_2 a_2 = n_1 a_1 + n_2 a_2$ then $(m_1 - n_1)a_1 = (n_2 - m_2)a_2$ showing $a_2 \in [a_1]$ unless $m_1 = n_1$, $m_2 = n_2$. If $P = [a_1, a_2]$ then the theorem is done. Otherwise continue the argument. Eventually, $[a_1, a_2, \ldots, a_t] = P$ for some elements and P has a generating set. Each element is of order p and P is the direct sum $[a_1] \oplus [a_2] \oplus \cdots \oplus [a_t]$ of t cyclic subgroups.

Now suppose the highest order element appearing in P has order p^k and the theorem is true for $\beta < k$. Construct the set $\hat{Q} = \{pa | a \in P\}$. $\hat{Q}$ is a group in which the maximum order element appearing has order p^{k-1}. By the inductive assumption $\hat{Q} = [\hat{a}_1] \oplus [\hat{a}_2] \oplus \cdots \oplus [\hat{a}_u]$, the direct sum of cyclic subgroups of order, respectively, $p^{n_1}, p^{n_2}, \ldots, p^{n_u}$; $1 \leqslant n_1, \ldots, n_u \leqslant k - 1$. Let $Q = [a_1] \oplus [a_2] \oplus \cdots \oplus [a_u]$, where $pa_j = \hat{a}_j$. Q is a subgroup of P. The order of a_{j_1} is $p^{n_j + 1}$. The subgroup Q of P has $p^{n_1+1} \cdot p^{n_2+1} \cdots p^{n_u+1}$ members. $Q = \{m_1 a_1 + m_2 a_2 + \cdots + m_u a_u | m_1 = 0, \ldots,$

$p^{n_1+1} - 1; m_2 = 0, \ldots, p^{n_2+1} - 1; m_u = 0, \ldots, p^{n_u+1} - 1\}$. The order of each element of Q is strictly greater than p. If this is not all of P choose a nonzero element b in $P - Q$. $pb \in \hat{Q}$. $-pb$ also is in $\hat{Q}$. Suppose $-pb = l_1\hat{a}_1 + \cdots + l_u\hat{a}_u$. $-pb = l_1pa_1 + \cdots + l_upa_u = p(l_1a_1 + \cdots + l_ua_u)$. Let $a_{u+1} = b + l_1a_1 + \cdots + l_ua_u$. Since $pa_{u+1} = p(b + l_1a_1 + \cdots + l_ua_u) = pb - pb = \theta$ we see a_{u+1} has order p. $b \in [a_1, a_2, \ldots, a_u, a_{u+1}] \subseteq P$. Also $[a_1, a_2, \ldots, a_u] \cap [a_{u+1}] = \{\theta\}$ because if $r_1a_1 + r_2a_2 + \cdots + r_ua_u = r_{u+1}a_{u+1}$ then $p(r_1a_1 + \cdots + r_ua_u) = \theta$ which implies $r_1a_1 + \cdots + r_ua_u = \theta = r_{u+1}a_{u+1}$. Hence, $[a_1] \oplus [a_2] \oplus \cdots \oplus [a_u] \oplus [a_{u+1}]$ is a direct sum of cyclic subgroups of P. If P is not exhausted, then continue the process. Eventually, $P = [a_1] \oplus [a_2] \oplus \cdots \oplus [a_u] \oplus [a_{u+1}] \oplus \cdots \oplus [a_v]$, where $[a_{u+1}], \ldots, [a_v]$ are all subgroups of order p. □

Questions

1. Of the Cartesian product of the two groups $\langle \mathbb{Z}_3, +, \bar{0} \rangle$ and $\langle \mathbb{Z}_5, +, \bar{0} \rangle$ which of these statements are true?
   (A) The group has 8 elements.
   (B) There are elements of order 3.
   (C) There are elements of order 6.
   (D) The product is $\mathbb{Z}_8$.
   (E) None of the statements is true.

2. Which of these statements about groups are true?
   (A) $\mathbb{Z}_{12} \approx \mathbb{Z}_6 \times \mathbb{Z}_2$. ($\approx$ means *is isomorphic to.*)
   (B) $\mathbb{Z}_{12} \approx \mathbb{Z}_3 \times \mathbb{Z}_4$.
   (C) $\mathbb{Z}_{12} \approx \mathbb{Z}_2 \times \mathbb{Z}_2 \times \mathbb{Z}_3$.
   (D) $\mathbb{Z}_{12} = \{\bar{0}, \bar{6}\} \oplus \{\bar{0}, \bar{2}, \bar{4}, \bar{6}, \bar{8}, \overline{10}\}$.
   (E) None of the statements is true.

3. Which of these are composition series?
   (A) $\mathbb{Z}_{12}, \{\bar{0}, \bar{2}, \bar{4}, \bar{6}, \bar{8}, \overline{10}\}, \{\bar{0}, \bar{4}, \bar{8}\}$.
   (B) $\mathfrak{S}_4, \mathfrak{A}_4, \{I\}$.
   (C) $\mathfrak{S}_5, \mathfrak{A}_5, \{I\}$.
   (D) $\mathfrak{S}_3, \mathfrak{A}_3, \{I\}$.
   (E) None is a composition series.

4. Which of these groups are solvable?
   (A) $\mathbb{Z}_{12}$
   (B) $\mathfrak{S}_4$
   (C) $\mathfrak{S}_5$
   (D) $\mathfrak{S}_3$.
   (E) None of the groups is solvable.

5. The weak Cartesian product $\bigtimes^{w}_{2 \leqslant j} \mathbb{Z}_j$ of the family of additive groups
   (A) has only elements of finite order
   (B) has subgroups of every finite order
   (C) has an infinite number of elements
   (D) is commutative.
   (E) None of the statements completes a true sentence.

Exercises

1. Let $\langle M_1, \cdot\rangle, \langle M_2, \cdot\rangle, \langle M'_1, \cdot\rangle, \langle M'_2, \cdot\rangle$ be monoids and $f_1: M_1 \to M'_1, f_2: M_2 \to M'_2$ be morphisms. By $f_1 \times f_2: M_1 \times M_2 \to M'_1 \times M'_2$ we denote the mapping such that $(f_1 \times f_2)(x_1, x_2) = (f_1(x_1), f_2(x_2))$. Show that $f_1 \times f_2$ is also a morphism. Show that if f_1 and f_2 are monomorphisms then so also is $f_1 \times f_2$. Show that if f_1 and f_2 are epimorphisms then $f_1 \times f_2$ is an epimorphism.

2. Prove that $\mathfrak{S}_3$ is not the direct product of $\mathfrak{A}_3$ and $\{I, (1\ 2)\}$.

3. Show that the group $\mathfrak{A}_3 \times \{I, (1\ 2)\}$ is isomorphic to $\mathbb{Z}_3 \times \mathbb{Z}_2$.

4. Show that if $\langle G_1, \cdot, v_1\rangle$ and $\langle G_2, \cdot, v_2\rangle$ are groups then $G_1 \times G_2$ is isomorphic to the group $G_2 \times G_1$ (Do not assume the groups G_1 and G_2 are commutative.)

5. The converse of the theorem of this section, $G = H_1 \odot H_2$ implies $G/H_1 \approx H_2$, is not true. Demonstrate this with an example.

6. Show the following isomorphism result for groups: $A \approx C$ and $B \approx D$ imply $A \times B \approx C \times D$.

7. If G_1 and G_2 are isomorphic groups and H_1 is a normal subgroup of G_1 and H_2 is a normal subgroup of G_2 and $H_1 \approx H_2$ then prove $G_1/H_1 \approx G_2/H_2$.

8. We say that a group $\langle G, \cdot, v\rangle$ is the direct product of a family $(H_j | j \in J)$ of non-trivial proper normal subgroups of G if and only if for each x in G, x is the finite sum of elements from H_j, $j \in J$, and $H_j \cap [\prod_{i \neq j} H_i] = \{v\}$ and $x \in H_i$, $y \in H_j$, $i \neq j$ imply $xy = yx$. Show that if $G = \bigodot_{j \in J} H_j$ then $G \approx \mathsf{X}^{w}_{j \in J} H_j$.

9. Show that $\bigoplus_{j \in N} R$ is isomorphic with the additive group of polynomials $\langle R[X], +, \theta\rangle$.

10. Show that the group $\mathsf{X}^{w}_{n \in N^+} \mathfrak{S}_n$ has elements of every finite order, has no elements of infinite order, and is an infinite group. Show that the group $\mathsf{X}_{n \in N^+} \mathfrak{S}_n$ has elements of every possible order.

11. Show that the group $\mathsf{X}^{w}_{n \in N^+} \mathfrak{S}_n$ has a normal subgroup isomorphic to $\mathfrak{A}_j$ for every $j = 1, 2, \ldots$.

12. Show that $\mathbb{Z}_{12} \approx \mathbb{Z}_4 \oplus \mathbb{Z}_3$ (additive groups). Is $\mathbb{Z}_{12}/\{\bar{0}, \bar{4}, \bar{8}\} \approx \mathbb{Z}_4$?

13. Give an example of a simple group of even cardinality not 2.

14. Make two possible composition series for $\mathbb{Z}_{24}$ and show that the quotient groups are isomorphic in some order.

15. Make all possible composition series for $\mathbb{Z}_{60}$.

16. Find all possible commutative groups of cardinality 24 by expressing as direct products (sums).

17. Let a group $\langle G, \cdot, v\rangle$ be given with a subgroup H. Define a map $f_x: G/H \to G/H$ such that $f_x(aH) = xaH$. Here G/H denotes the set of left cosets of H, a quotient set, but not a quotient group because H is not necessarily normal. Show that f_x is a bijection of G/H. Let aH be a given left coset of G/H. Show that the set of all x such that f_x leaves aH fixt, $\{x | x \in G \text{ and } f_x(aH) = aH\}$, is exactly aHa^{-}. Show

that aHa^{-} is a subgroup of G. Define a mapping $F: G \to \mathfrak{S}(G/H)$ such that $F(x) = f_x$. Show that F is a morphism. Show that $\ker F = \{x | F(x) = I\} = \{x | f_x = I\} = \bigcap_{a \in G} aHa^{-}$. Show that the normal subgroup $\bigcap_{a \in G} aHa^{-}$ is a subgroup of H and is the largest normal subgroup of G contained in H.

18. Let G be a given group with a subgroup H such that G/H has n members (finite). Show that there exists a normal subgroup K of H so that $\operatorname{crd}(G/K)$ divides $n!$.

19. Show that no subgroup H of $\mathfrak{S}_n$, $n \geqslant 5$, $H \neq \mathfrak{A}_n$ or $\mathfrak{S}_n$, can generate fewer than n cosets in $\mathfrak{S}_n$; i.e., there is no subgroup H of $\mathfrak{S}_n$ such that $\operatorname{crd}(\mathfrak{S}_n/H) < n$.

Linear algebra: Modules over principal domains and similarity 10

This chapter treats further properties of modules over rings which are not fields. We have a discussion of the order of module elements not unlike the order of an element of a group or the characteristic of a ring. We resolve elements into sums of elements of elements of relatively prime orders. We introduce determinant divisors invariant under matrix equivalence. Using the theory developed for modules over principal ideal domains we find the invariant factor matrix: a canonical form for the equivalence of matrices over a principal domain. We solve linear equations with coefficients in a principal domain. We give a direct sum resolution of a finitely generated module over a principal domain. We consider the relation of similarity of matrices with entries in a field and apply the theorems of this chapter to yield several canonical forms. The technique is to construct, from a given vector space and endomorphism, a new module, resolve this module into a direct sum of cyclic submodules, and use this resolution to produce a basis for the vector space which yields for the endomorphism an especially simple matrix: the canonical form. We close with a study of the characteristic equation and characteristic values.

10.1 Cyclic modules

In this section we study cyclic modules in terms of annihilating ideals from the ring of the module. To preserve the line of thought we will repeat some definitions from Chapters 6 and 7.

Definitions. A module M over a ring R is *finitely generated* if and only if M is generated by some finite subset of M. A module M that is generated by a singleton subset ($M = [x]$ for some x in M) is called a *cyclic* module.

Theorem. *If M is a cyclic module over a unitary ring R with generating subset $\{x\}$ then $M = Rx$.*

PROOF. $Rx = \{rx|r \in R\}$ is a submodule of M which contains x. On the other hand, any submodule of M containing x must contain Rx. □

EXAMPLES. $\mathbb{Z}_5$ is a cyclic module over the ring $\mathbb{Z}$. A generator of the cyclic module is the coset $1 + \langle 5 \rangle = \bar{1}$. We can then write $\mathbb{Z}_5 = [\bar{1}]$. By the theorem just proved $\mathbb{Z}_5 = \mathbb{Z}\bar{1}$. Similarly, $\mathbb{Z}_6$ is a cyclic module over the ring $\mathbb{Z}$. $\mathbb{Z}_6 = [\bar{1}] = [\bar{5}]$. $\mathbb{Z}_6 = \mathbb{Z}\bar{1} = \mathbb{Z}\bar{5}$. None of the elements $\bar{0}, \bar{2}, \bar{3}, \bar{4}$ will be generators of $\mathbb{Z}_6$. While $\mathbb{Z}[X]$, the polynomial ring of $\mathbb{Z}$, is a $\mathbb{Z}$-module it is not a cyclic module. $\mathbb{Z}[X]$ is generated by the set $\{1, X, X^2, \ldots\}$. $\mathbb{Z}[X]$ is not generated by any finite set.

Definition. Let M be an R-module. Let S be a subset of M. A ring element a of R is an *annihilator* of S if and only if $ax = \zeta$ for all x in S.

We also shall speak of an annihilator of an element x of M and by this we mean simply an annihilator of the singleton set $\{x\}$.

EXAMPLE. 4 annihilates the entire $\mathbb{Z}$-module $\mathbb{Z}_4$. 2 annihilates the subset $\{\bar{0}, \bar{2}\}$ of $\mathbb{Z}_4$. The polynomial $X + 1$ of the $\mathbb{Z}$-module $\mathbb{Z}[X]$ is annihilated only by the integer 0.

Theorem. *Let M be a module over a commutative unitary ring R. The set of all annihilators of a given subset S of M is an ideal of R.*

PROOF. If $ax = \zeta$ and $bx = \zeta$ for any x in S then $(a - b)x = ax - bx = \zeta - \zeta = \zeta$. If $ax = \zeta$ for any x in S then $(ra)x = r(ax) = r\zeta = \zeta$. There is always at least one annihilator of S, namely θ. □

Definition. The set of all annihilators of a given subset S of a module over a commutative ring R is called the *annihilating ideal* of S.

In case the ring R is a principal domain then every such annihilating ideal is generated by a single ring element. Such a generator of the annihilating ideal is called an *order* of the set S. We sometimes call the annihilating ideal an order ideal of S. Because an ideal may have several generators an order is not unique. However, we do observe that any generator of the annihilating ideal of a subset S is an associate (unit multiple) of any other generator of the same ideal. We include this observation in the exercises of this section.

EXAMPLE. We return to the $\mathbb{Z}$-module $\mathbb{Z}_4$ and list each vector in $\mathbb{Z}_4$ with its annihilating ideal: $\bar{0}$, $\mathbb{Z}$; $\bar{1}$, $\langle 4 \rangle$; $\bar{2}$, $\langle 2 \rangle$; $\bar{3}$, $\langle 4 \rangle$. The ideal $\langle 4 \rangle$ could also be written $\langle -4 \rangle$; both 4 and -4 are orders of $\bar{1}$. An order of $\bar{0}$ is 1.

EXAMPLE. Since every commutative group G can be considered to be a $\mathbb{Z}$-module the order of an element of the group G in the group theoretic sense will be an order in the $\mathbb{Z}$-module sense.

EXAMPLE. Modules can range from two extremes in which every vector has nontrivial annihilating ideal to the case when only the vector ζ has nontrivial annihilating ideal. Every element of the $\mathbb{Z}$-module $\mathbb{Z}_4$ has nontrivial annihilating ideal while only 0 has nontrivial annihilating ideal in the $\mathbb{Z}$-module $\mathbb{Z}$. If we consider the product $\mathbb{Z}$-module $\mathbb{Z} \times \mathbb{Z}_4$ then some elements in addition to $(0, \overline{0}) = \zeta$ have nontrivial annihilating ideal and some elements have trivial annihilating ideal.

EXAMPLE. A module may have trivial annihilating ideal for every single nonzero element yet may fail to be a free module (have a basis). For example, the $\mathbb{Z}$-module $\mathbb{Q}$ has $\langle 0 \rangle$ as the annihilating ideal for each of its elements (save zero, of course) yet is not a free module.

The following result demonstrates how the annihilating ideal of a submodule can be constructed in terms of the annihilating ideals of the individual elements of the submodule.

Theorem. *Let M be a module over a principal domain R. Let A_x be the annihilating ideal of x for each x in N, a submodule of M. Then $\bigcap\{A_x | x \in N\}$ is the annihilating ideal of N.*

PROOF. If $a \in R$ annihilates every element of N then $a \in A_x$ for each x in N. Then $a \in \bigcap\{A_x | x \in N\}$. The argument is reversible. □

We can analyze the character of the cyclic submodule by studying the character of the annihilating ideal.

Theorem. *Let M be a module over a principal domain R. Let $x \in M$. Then $[x]$ is isomorphic to the R-module $R/\langle a \rangle$ where $\langle a \rangle$ is the annihilating ideal of x.*

PROOF. $x \in M$ is a generator of the cyclic submodule $[x]$. The annihilating ideals of $\{x\}$ and $[x]$ are the same ideal of R. Because R is a principal domain this annihilating ideal is generated by some single element a of R. We define $\omega_x : R \to [x]$ such that $\omega_x(r) = rx$. ω_x is an R-module morphism, preserving both vector addition and the R-exterior multiplication. The range of ω_x is Rx or $[x]$ making ω_x an epimorphism. There exists an isomorphism $\omega'_x : R/\ker \omega_x \to [x]$. Ker $\omega_x = \{r | r \in R \text{ and } rx = \zeta\} = \langle a \rangle$. Thus $R/\langle a \rangle$ is an R-module isomorphic with the R-module $[x]$. □

One must carefully distinguish between vectors which are generators of submodules of M and ring elements which are generators of ideals of R.

We can decompose cyclic modules with annihilating ideals which have reducible orders.

Theorem. *Let R be a principal domain. Let $c, a, b \in R$ with a and b relatively prime, $c = ab$. Then $R/\langle c \rangle = P_1 \oplus P_2$ with $P_1 \approx R/\langle b \rangle$ and $P_2 \approx R/\langle a \rangle$.*

PROOF. $R/\langle c \rangle$ is a cyclic R-module with generator $v + \langle c \rangle$. Because a and b are relatively prime there exist α and β in R such that $\alpha a + \beta b = v$. If $d + \langle c \rangle \in R/\langle c \rangle$ then $d + \langle c \rangle = d(\alpha a + \beta b) + \langle c \rangle = d\alpha(a + \langle c \rangle) + d\beta(b + \langle c \rangle) \in [a + \langle c \rangle] + [b + \langle c \rangle]$. $[a + \langle c \rangle] \cap [b + \langle c \rangle]$ is annihilated by both a and b and therefore by $v = \alpha a + \beta b$. $[a + \langle c \rangle] \cap [b + \langle c \rangle] = \{\langle c \rangle\}$. Thus $R/\langle c \rangle = [a + \langle c \rangle] \oplus [b + \langle c \rangle]$, a direct sum.

The annihilating ideal of $a + \langle c \rangle$ is calculated as follows: $k(a + \langle c \rangle) = \langle c \rangle$ if and only if $ka \in \langle c \rangle$ if and only if $c(= ab)$ is a factor of ka if and only if b is a factor of k if and only if $k \in \langle b \rangle$. $\langle b \rangle$ is therefore the annihilating ideal of $a + \langle c \rangle$. $R/\langle b \rangle$ is then an R-module isomorphic to $[a + \langle c \rangle]$. We prove the result for the other component similarly. □

This just established result can be extended by induction to any finite number of components.

Corollary. *If $c = up_1^{\alpha_1}p_2^{\alpha_2} \cdots p_k^{\alpha_k}$ is an irreducible factorization of c, u is a unit, then $R/\langle c \rangle = \bigoplus_{i=1}^k P_i$ with P_i isomorphic to $R/\langle p_i^{\alpha_i} \rangle$, $i = 1, 2, \ldots, k$.*

PROOF. We leave this extension to the reader. □

Finally from this section we show how it is possible to select from the module vectors with specified orders.

Theorem. *Let M be a module over a principal domain R. Let x_1 and x_2 be in M with orders a_1 and a_2, nonzero and relatively prime. Then $x_1 + x_2$ has order $a_1 a_2$.*

PROOF. $a_1 a_2$ is in the annihilating ideal of $x_1 + x_2$ because $a_1 a_2(x_1 + x_2) = a_2(a_1 x) + a_1(a_2 x) = \zeta$. Now let b be any member of the annihilating ideal of $x_1 + x_2$. $b(x_1 + x_2) = \zeta$. Since a_1 and a_2 are relatively prime there exist α_1, α_2 in R such that $\alpha_1 a_1 + \alpha_2 a_2 = v$. $\alpha_2 a_2(x_1 + x_2) = \alpha_2 a_2 x_1 + \alpha_2 a_2 x_2 = \alpha_2 a_2 x_1 = (v - \alpha_1 a_1)x_1 = x_1$. x_1 is a multiple of $x_1 + x_2$. Any annihilator of $x_1 + x_2$, namely b, must annihilate x_1. $b \in \langle a_1 \rangle$. a_1 is a factor of b. By a symmetrical argument a_2 is also a factor of b. Because a_1 and a_2 are relatively prime the product $a_1 a_2$ divides b. Thus any member b of the annihilating ideal of $x_1 + x_2$ is a multiple of $a_1 a_2$ and therefore $a_1 a_2$ is a generator of that ideal.

Theorem. *Let M be a module over a principal domain with a nontrivial annihilating ideal $\langle a \rangle$. Then M has an element with order equal to a.*

PROOF. If a is a unit of R then $\langle a \rangle = R$. The annihilating ideal of M is R itself. An element of M annihilated by R is the zero vector ζ. The given unit a is an order of ζ.

Now suppose that a is not a unit. $a = up_1^{\alpha_1}p_2^{\alpha_2} \cdots p_k^{\alpha_k}$ with all p_i distinct irreducibles, relatively prime in pairs and u is a unit. Each a/p_i, $i = 1, 2, \ldots, k$, is not a multiple of a. Since a is a minimal annihilator of M then there exists x_i in M such that $(a/p_i)x_i \neq \zeta$. We now consider the element $(p_1^{\alpha_1}p_2^{\alpha_2} \cdots p_k^{\alpha_k}/p_i^{\alpha_i})x_i$ of M. $p_i^{\alpha_i}(p_1^{\alpha_1}p_2^{\alpha_2} \cdots p_k^{\alpha_k}/p_i^{\alpha_i})x_i = \zeta$. $p_i^{\alpha_i - 1}(p_1^{\alpha_1}p_2^{\alpha_2} \cdots p_k^{\alpha_k}/p_i^{\alpha_i})x_i \neq \zeta$. An order of the vector is $p_i^{\alpha_i}$. By the previous theorem an order of $\sum_{i=1}^{k} (p_1^{\alpha_1}p_2^{\alpha_2} \cdots p_k^{\alpha_k}/p_i^{\alpha_i})x_i$ is the product $p_1^{\alpha_1}p_2^{\alpha_2} \cdots p_k^{\alpha_k}$. $a = up_1^{\alpha_1}p_2^{\alpha_2} \cdots p_k^{\alpha_k}$ is also an order. □

QUESTIONS

1. Which of the following alternatives complete a true sentence? A module M over a ring R has a finite basis
   (A) implies M is finitely generated
   (B) implies M has nonzero elements with nonzero order
   (C) if every nonzero element has annihilating ideal $\langle \theta \rangle$
   (D) provided the ring R is a field.
   (E) None of the alternatives completes a true statement.

2. Let M be a module over a principal domain R. Which of these statements are true?
   (A) θ generates the annihilating ideal of ζ.
   (B) Every submodule of M is isomorphic to some module $R/\langle a \rangle$, $a \in R$.
   (C) $R/\langle ab \rangle \approx R/\langle a \rangle + R/\langle b \rangle$ for all $a, b \in R$.
   (D) $[x] + [y] = [x + y]$ for all $x, y \in M$.
   (E) None of the statements is true.

3. Which of these statements are true?
   (A) $\mathbb{Z}/\langle 2 \rangle \times \mathbb{Z}/\langle 4 \rangle \approx \mathbb{Z}/\langle 8 \rangle$.
   (B) $\mathbb{Z}/\langle 2 \rangle \times \mathbb{Z}/\langle 3 \rangle \approx \mathbb{Z}/\langle 6 \rangle$.
   (C) $\mathbb{Z}/\langle 1 \rangle \times \mathbb{Z}/\langle -1 \rangle \approx \mathbb{Z}/\langle 1 \rangle$.
   (D) $\mathbb{Z}/\langle 2 \rangle \times \mathbb{Z}/\langle 2 \rangle \approx \mathbb{Z}/\langle 4 \rangle$.
   (E) None of the statements is true.

4. Which of these statements are true?
   (A) $\mathbb{Z}[X]$ is a finitely generated $\mathbb{Z}$-module generated by $\{X\}$.
   (B) $\mathbb{Q}$ is a $\mathbb{Z}$-module generated by $\{1, -1\}$.
   (C) $\mathbb{Z} \times \mathbb{Z}$ is a finitely generated $\mathbb{Z}$-module.
   (D) If S is a subring of a principal ring R then R is an S-module.
   (E) None of the statements is true.

EXERCISES

1. Show that the annihilating ideal of a cyclic submodule $[x]$ does not depend upon which generator of $[x]$ is chosen; that is, show that if $[x] = [y]$ then the annihilating ideals of x and y are equal.

2. If M is an R-module and $aM = \{\zeta\}$ (a in R annihilates M), then with the proper definition of $R/\langle a\rangle$-exterior multiplication, M may be regarded an $R/\langle a\rangle$-module.
3. Suppose M is an $R/\langle a\rangle$-module. Then M is an R-module and $aM = \{\zeta\}$.
4. Let R be a principal domain and P a cyclic R-module of order $a \neq \theta$. Let $x \in P$ such that $bx = \zeta$ with $a = bc$. Then there exists y in P such that $x = cy$.
5. Let M be a module over a principal domain R. Let M be generated by a submodule N and one additional element w. Then M/N is cyclic with generator $w + N$.
6. Verify that $\mathbb{Q}$ as a $\mathbb{Z}$-module has every nonzero element of zero order yet is not a free module.
7. Prove that a free module over a principal domain has every element except ζ of zero order.
8. Prove that the annihilating ideal of an element is unique.

10.2 Invariant factors

In this section we discuss determinant divisors and invariant factors and produce a canonical form under equivalence for a matrix with coefficients in a principal domain.

Definition. Let A be a matrix with entries in a principal domain R. A *k-minor* of A is the determinant of a k by k submatrix preserving row and column order.

EXAMPLE. Given the 4 by 3 matrix

$$A = \begin{pmatrix} 4 & -2 & 1 \\ 3 & 2 & 5 \\ 1 & -1 & 0 \\ 3 & 2 & 0 \end{pmatrix}$$

with entries in the principal domain $\mathbb{Z}$. Some 2-minors are

$$\begin{vmatrix} 4 & -2 \\ 3 & 2 \end{vmatrix} = 16 \quad \text{and} \quad \begin{vmatrix} -1 & 0 \\ 2 & 0 \end{vmatrix} = 0.$$

A 1-minor is -2 or 5.

Definition. A kth *determinant divisor* of a matrix A is a greatest common divisor of all the k-minors of A. We use the notation $\mathscr{D}_k(A)$ for the kth determinant divisor of A.

EXAMPLES. For the matrix A in the previous example we have $\mathscr{D}_1(A) = 1$, $\mathscr{D}_2(A) = 1$, $\mathscr{D}_3(A) = 5$.

We remind the reader again of our convention choosing any associate for a greatest common divisor when the principal domain has more than one unit. We also remind the reader that the greatest common divisor of a set of ring elements with at least one nonzero member can be computed by neglecting any zero members since all ring elements are divisors of zero ($\theta \cdot r = \theta$ for all r in R). In case all k-minors of a matrix are zero then we adopt the convention that $\mathscr{D}_k(A) = \theta$.

EXAMPLE. For the matrix

$$\begin{pmatrix} 6 & 2 & 4 \\ 6 & 2 & 4 \end{pmatrix}$$

we have $\mathscr{D}_1 = 2$ and $\mathscr{D}_2 = 0$.

We now show that the determinant divisors are invariant under equivalence of matrices.

Theorem. *Let A, P, Q be matrices with entries in a principal domain R. Let P and Q be invertible matrices. Then $\mathscr{D}_k(QAP^{-1}) = \mathscr{D}_k(A)$.*

PROOF. It is to be understood that the equality $\mathscr{D}_k(QAP^{-1}) = \mathscr{D}_k(A)$ is modulo any unit of R, that the two determinant divisors are associates.

Given two matrices B, C we show that any k-minor of BC is a linear combination of k-minors of C. Denoting the product BC by F we have typically

$$\det \begin{pmatrix} F_{11} & F_{12} & \cdots & F_{1k} \\ F_{21} & F_{22} & \cdots & F_{2k} \\ \cdots & & & \\ F_{k1} & F_{k2} & \cdots & F_{kk} \end{pmatrix}$$

$$\begin{aligned} &= \sum_{\sigma \in \mathfrak{S}_k} \varepsilon(\sigma) F_{1\sigma(1)} F_{2\sigma(2)} \cdots F_{k\sigma(k)} \\ &= \sum_{\sigma \in \mathfrak{S}_k} \varepsilon(\sigma) \left[\sum_{j_1=1}^{n} B_{1j_1} C_{j_1\sigma(1)} \right] \cdots \left[\sum_{j_k=1}^{n} B_{1j_k} C_{j_k\sigma(k)} \right] \\ &= \sum_{j_1=1}^{n} \cdots \sum_{j_k=1}^{n} B_{1j_1} \cdots B_{kj_k} \left[\sum_{\sigma \in \mathfrak{S}_k} \varepsilon(\sigma) C_{j_1\sigma(1)} \cdots C_{j_k\sigma(k)} \right], \end{aligned}$$

which is a linear combination of the k-minors of C (and zeros). Similarly by running the permutations on the rows we can show k-minors of BC are linear combinations of the k-minors of B.

A common divisor of the k-minors of A will be a common divisor of the k-minors of $QAP^{-1} = D$. Conversely, a common divisor of the k-minors of D will be a common divisor of the k-minors of $Q^{-1}DP = A$. Thus A and D have the same or associated greatest common divisors of k-minors. □

We propose to set a canonical form for equivalence of matrices with entries in a principal domain. We recall that given a matrix A with entries in a field, A is equivalent to a matrix in which the number of unities on the diagonal equals the rank of A. This diagonal matrix is the canonical form for equivalence of matrices with entries in a field.

$$\begin{pmatrix} v & \theta & \theta & \cdots & \theta & \theta & \cdots & \theta \\ \theta & v & \theta & \cdots & \theta & \theta & \cdots & \theta \\ \theta & \theta & v & \cdots & \theta & \theta & \cdots & \theta \\ \cdots & & & & & & & \\ \theta & \theta & \theta & \cdots & v & \theta & \cdots & \theta \\ \theta & \theta & \theta & \cdots & \theta & \theta & \cdots & \theta \\ \theta & \theta & \theta & \cdots & \theta & \theta & \cdots & \theta \end{pmatrix} \begin{matrix} \\ \\ \\ \\ \leftarrow \text{row } r \\ \\ \\ \end{matrix}$$

$\uparrow$
column r

Because of the absence of inverses in a principal domain it is not possible, in general, to obtain a canonical form with only unity on the diagonal. The canonical form to be described in the theorem following which can be obtained for the principal domain will, however, still be a diagonal matrix.

The reader will recall our frequent earlier use of elementary transformation or change of basis matrices described in Section 7.4. These were $E(p, q)$, $E(r, q; p)$ and $E(s; p)$ which are invertible matrices with determinants $-v$, v and s, respectively. We shall also employ a fourth transformation matrix which will be described in the theorem when we first use it. The proof of the theorem and the finding of the canonical matrix in practice will utilize the transformation matrices step by step to alter the given matrix to the desired form.

Theorem. *Let A be an m by n matrix with entries in a principal domain R. Then there exist matrices P, Q, D such that*

1. *P and Q are invertible*
2. *$QAP^{-1} = D$*
3. *All entries of D except diagonal entries are zero*
4. *If we denote D_{ii} by d_i then there exists a natural number r, $0 \leqslant r \leqslant \min(m, n)$ such that $j > r$ implies $d_j = \theta$ and $j \leqslant r$ implies $d_j \neq \theta$ and $1 \leqslant j < r$ implies d_j divides d_{j+1}.*

PROOF. In order to proceed with the step by step reduction to D we must have a gauge of the size of the entries in the matrix. This gauge will be used much as absolute value for the integers and degree for polynomials. Such a gauge was necessary, for example, with the division algorithm in order to state that the remainder is smaller than the divisor. We have available in the principal domain R the irreducible factorization of the elements. For a nonzero element $a = up_1^{\alpha_1}p_2^{\alpha_2} \cdots p_s^{\alpha_s}$ we define the length of a, $l(a)$, to be $\alpha_1 + \alpha_2 + \cdots + \alpha_s$, the number of irreducible factors of a. This irreducible factorization is unique up to unit multiples so that $l(a)$ is well-defined. It is also clear that if b is a proper factor of a then $l(b) < l(a)$.

We can at any point in our reduction process by interchange of rows (left multiplication by $E(p, q)$) and by interchange of columns (right multiplication by $E(p, q)$) bring to position (1, 1) any entry of the given matrix. In particular, we can interchange until the entry in position (1, 1) has minimum length among all the nonzero entries. If there is an entry B_{1j} in the first row which *is a multiple* of the entry B_{11} in position (1, 1) we can add $-B_{1j}/B_{11}$ (an element of R) times the first column to the jth column to produce a zero in position $(1, j)$. This transformation is achieved by right multiplication by $E(-B_{1j}/B_{11}; 1, j)$. For all other entries in row 1 which are multiples of B_{11} we also add suitable multiples of column 1 to produce zeros in those positions in row 1. For all entries in column 1 which are multiples of B_{11} and not in position (1, 1) we can also produce zeros.

For any entry B_{1k} in row 1 which is not a multiple of B_{11} we use right multiplication by an invertible matrix to produce a new matrix with $\gcd(B_{11}, B_{1k})$ in position (1, 1). Since B_{1k} is not a multiple of B_{11} this $\gcd(B_{11}, B_{1k})$ has strictly smaller length than does B_{11}. The invertible matrix by which we multiply to produce this greatest common divisor in position (1, 1) is determined as follows. Suppose $d = \gcd(B_{11}, B_{1k})$. There exist α, β in R such that $\alpha B_{11} + \beta B_{1k} = d$. Then right multiplication by the illustrated matrix produces a matrix with d in position (1, 1) and θ in position $(1, k)$.

$$\begin{pmatrix}
\alpha & \theta & \theta & \cdots & -B_{1k}/d & \cdots & \theta & \theta \\
\theta & \nu & \theta & \cdots & \theta & \cdots & \theta & \theta \\
\theta & \theta & \nu & \cdots & \theta & \cdots & \theta & \theta \\
\cdots & & & & & & & \\
\beta & \theta & \theta & \cdots & B_{11}/d & \cdots & \theta & \theta \\
\cdots & & & & & & & \\
\theta & \theta & \theta & \cdots & \theta & \cdots & \nu & \theta \\
\theta & \theta & \theta & \cdots & \theta & \cdots & \theta & \nu
\end{pmatrix} \leftarrow \text{row } k.$$

$$\uparrow$$
$$\text{column } k$$

This matrix is invertible because its determinant is ν.

We are now at the point where we have produced a new matrix with the entry d in position (1, 1) and d has smaller length than the previous entry B_{11} in the position (1, 1). We now return to our beginning steps to replace with zeros any entries of the first row or first column which are multiples of the entry d in position (1, 1). Then any entries of the first row or first column which are not zeros and are not multiples of the entry in position (1, 1) are replaced with zeros putting a greatest common divisor in position (1, 1). After a finite number of steps we obtain a matrix which has in position (1, 1) a nonzero entry of smallest length and all other entries of row one and column one are zero. Suppose that this matrix equivalent to A is

$$\begin{pmatrix} C_{11} & \theta & \theta & \cdots & \theta \\ \theta & C_{22} & C_{23} & \cdots & C_{2n} \\ \cdots & & & & \\ \theta & C_{m2} & C_{m3} & \cdots & C_{mn} \end{pmatrix}.$$

If C_{11} fails to be a factor of every nonzero entry of the matrix say, C_{11} fails to be a factor of C_{ij} then row i is added to row 1 to produce

$$\begin{pmatrix} C_{11} & C_{i2} & C_{i3} & \cdots & C_{ij} & \cdots & C_{in} \\ \theta & C_{22} & C_{23} & \cdots & C_{2j} & \cdots & C_{2n} \\ \cdots & & & & & & \\ \theta & C_{i2} & C_{i3} & \cdots & C_{ij} & \cdots & C_{in} \\ \cdots & & & & & & \\ \theta & C_{m2} & C_{m3} & \cdots & C_{mj} & \cdots & C_{mn} \end{pmatrix} \begin{matrix} \\ \\ \\ \leftarrow \text{row } i. \\ \\ \\ \end{matrix}$$

We now have a matrix in which there is in the first row in position $(1, j)$ an entry C_{ij} which is not a multiple of C_{11}, the entry in position (1, 1). We can then produce a new matrix equivalent to this matrix which has $\gcd(C_{11}, C_{ij})$ in position (1, 1) and a zero in position $(1, j)$. The new entry in position (1, 1) has smaller length than both C_{11} and C_{ij}.

We repeat all procedures until eventually after a finite number of transformations we have a matrix

$$\begin{pmatrix} d_1 & \theta & \theta & \cdots & \theta \\ \theta & G_{22} & G_{23} & \cdots & G_{2n} \\ \cdots & & & & \\ \theta & G_{m2} & G_{m3} & \cdots & G_{mn} \end{pmatrix}$$

in which d_1 has smaller length than any other entry in the matrix and d_1 is a factor of every nonzero entry in the matrix. This matrix is equivalent to A.

In the same manner that we have treated the matrix A we now treat the submatrix

$$\begin{pmatrix} G_{22} & G_{23} & \cdots & G_{2n} \\ G_{32} & G_{33} & \cdots & G_{3n} \\ \cdots & & & \\ G_{m2} & G_{m3} & \cdots & G_{mn} \end{pmatrix}.$$

to obtain an entry d_2 in position (2, 2) which has smaller length than any other entry and is a factor of every nonzero entry of the submatrix. The rest of the entries of row 2 and column 2 are zero. We actually work with the entire matrix

$$\begin{pmatrix} d_1 & \theta & \theta & \cdots & \theta \\ \theta & G_{22} & G_{23} & \cdots & G_{2n} \\ \cdots \\ \theta & G_{m2} & G_{m3} & \cdots & G_{mn} \end{pmatrix}$$

but use transformations which leave the first row and the first column undisturbed. Since d_1 is a factor of every entry of the submatrix then d_1 will be a factor of d_2, a linear combination of entries in the submatrix.

The process of finding diagonal entries $d_1, d_2, \ldots$ will terminate whenever the subscript exceeds $\min(m, n)$, or when after obtaining, say d_r, we discover that all remaining entries of the obtained matrix are zero. The matrix D finally obtained has the form

$$\begin{pmatrix} d_1 & \theta & \cdots & \theta & \theta & \cdots & \theta \\ \theta & d_2 & \cdots & \theta & \theta & \cdots & \theta \\ \cdots \\ \theta & \theta & \cdots & d_r & \theta & \cdots & \theta \\ \theta & \theta & \cdots & \theta & \theta & \cdots & \theta \\ \cdots \\ \theta & \theta & \cdots & \theta & \theta & \cdots & \theta \end{pmatrix}.$$

The matrix Q is the composite of the left multiplied transformation matrices and the matrix P^{-1} is the composite of the right multiplied transformation matrices. Both are invertible because they are products of invertible transformation matrices. Actually, in the procedures of this theorem we have used only transformation matrices with determinant $\pm v$ and therefore P^{-1} and Q are matrices with determinants equal to v or $-v$. We have called the matrix on the right P^{-1} instead of simply P in order to fit earlier patterns on change of basis we used. □

Example. In order to illustrate the constructure procedures and results of this theorem we take a simple example and perform the reductions step by step. We take the matrix

$$\begin{pmatrix} 2 & 3 & 2 \\ 1 & 6 & 4 \\ 3 & -2 & 4 \end{pmatrix}$$

with entries in the principal domain $\mathbb{Z}$ and reduce it to canonical form. We do the work in tabular form keeping the products of row transformations on the left and the products of column transformations on the right. The center column begins with A and terminates with D, the canonical form.

1	0	0	2	3	2	1	0	0
0	1	0	1	6	4	0	1	0
0	0	1	3	−2	4	0	0	1
0	1	0	1	6	4	1	0	0
1	0	0	2	3	2	0	1	0
0	0	1	3	−2	4	0	0	1
0	1	0	1	6	4	1	0	0
1	−2	0	0	−9	−6	0	1	0
0	0	1	3	−2	4	0	0	1
0	1	0	1	6	4	1	0	0
1	−2	0	0	−9	−6	0	1	0
0	−3	1	0	−20	−8	0	0	1
0	1	0	1	0	4	1	−6	0
1	−2	0	0	−9	−6	0	1	0
0	−3	1	0	−20	−8	0	0	1
0	1	0	1	0	0	1	−6	−4
1	−2	0	0	−9	−6	0	1	0
0	−3	1	0	−20	−8	0	0	1
0	1	0	1	0	0	1	−4	−6
1	−2	0	0	−6	−9	0	0	1
0	−3	1	0	−8	−20	0	1	0
0	1	0	1	0	0	1	−4	−6
1	1	−1	0	2	11	0	0	1
0	−3	1	0	−8	−20	0	1	0
0	1	0	1	0	0	1	−4	14
1	1	−1	0	2	1	0	0	1
0	−3	1	0	−8	20	0	1	−5
0	1	0	1	0	0	1	14	−4
1	1	−1	0	1	2	0	1	0
0	−3	1	0	20	−8	0	−5	1
0	1	0	1	0	0	1	14	−32
1	1	−1	0	1	0	0	1	−2
0	−3	1	0	20	−48	0	−5	11
0	1	0	1	0	0	1	14	−32
1	1	−1	0	1	0	0	1	−2
−20	−23	21	0	0	−48	0	−5	11

From the above procedures we conclude

$$\begin{pmatrix} 0 & 1 & 0 \\ 1 & 1 & -1 \\ -20 & -23 & 21 \end{pmatrix}\begin{pmatrix} 2 & 3 & 2 \\ 1 & 6 & 4 \\ 3 & -2 & 4 \end{pmatrix}\begin{pmatrix} 1 & 14 & -32 \\ 0 & 1 & -2 \\ 0 & -5 & 11 \end{pmatrix} = \begin{pmatrix} 1 & 0 & 0 \\ 0 & 1 & 0 \\ 0 & 0 & -48 \end{pmatrix}.$$

The matrix

$$\begin{pmatrix} 1 & 0 & 0 \\ 0 & 1 & 0 \\ 0 & 0 & -48 \end{pmatrix}$$

is in canonical form.

Definition. Let A, a matrix with entries in a principal domain R, be equivalent to a matrix

$$D = \begin{pmatrix} d_1 & \theta & \cdots & \theta & \theta & \cdots & \theta \\ \theta & d_2 & \cdots & \theta & \theta & \cdots & \theta \\ \cdots \\ \theta & \theta & \cdots & d_r & \theta & \cdots & \theta \\ \theta & \theta & \cdots & \theta & \theta & \cdots & \theta \\ \cdots \\ \theta & \theta & \cdots & \theta & \theta & \cdots & \theta \end{pmatrix} \leftarrow \text{row } r$$

$$\uparrow \text{ column } r$$

in which each of the nonzero $d_1, d_2, \ldots, d_r$ is a factor of the successive one. We then call D the *invariant factor matrix* of A and $d_1, d_2, \ldots, d_r$ the *invariant factors* of A.

We would like now to show that the invariant factor matrix for A and the invariant factors are essentially unique. This is to say that the values of $d_1, d_2, \ldots, d_r$ do not depend upon the particular procedures chosen to calculate them.

Theorem. *If A is a matrix with entries in a principal domain R then the invariant factors are unique (modulo unit multiples).*

PROOF. We assume that D and D' are both invariant factor matrices of A. Since D is equivalent to A and D' is equivalent to A we have D' equivalent to D. Equivalent matrices have determinant divisors which are associates; $\mathscr{D}_k(D)$ is a unit multiple of $\mathscr{D}_k(D')$ for each $k: 1 \leqslant k \leqslant \min(m, n)$. $\mathscr{D}_k(D) = d_1 d_2 \cdots d_k$ for $k \leqslant r$ and $\mathscr{D}_k(D) = \theta$ for $k > r$. $\mathscr{D}_k(D') = d'_1 d'_2 \cdots d'_k$ for $k \leqslant r'$ and $\mathscr{D}_k(D') = \theta$ for $k > r'$. All entries $d_1, d_2, \ldots, d_r, d'_1, d'_2, \ldots, d'_{r'}$ are nonzero. We can now conclude $r = r'$ and $d_1 = u_1 d'_1, d_1 d_2 = u_2 d'_1 d'_2, \ldots,$ $d_1 d_2 \cdots d_r = u_r d'_1 d'_2 \cdots d'_{r'}$ for some units $u_1, u_2, \ldots, u_r$ in R. By successive substitution and cancellation we have $d_1 = u_1 d'_1$, $d_2 = u_1^{-1} u_2 d'_2$, $d_3 = u_2^{-1} u_3 d'_3, \ldots, d_r = u_{r-1}^{-1} u_r d'_r$. This shows that d_k is a unit multiple of d'_k, $1 \leqslant k \leqslant r$; d_k and d'_k are associates in R. This is the qualified uniqueness of the invariant factors. □

QUESTIONS

1. Let the matrix A with entries in $\mathbb{Z}$ be

$$\begin{pmatrix} 2 & 0 & 0 & 0 \\ 0 & 2 & 2 & 2 \\ 0 & 2 & 2 & 0 \\ 2 & 0 & 0 & 2 \end{pmatrix}.$$

Which of the following statements are true?
(A) $\mathscr{D}_4(A) = 16$.
(B) $\mathscr{D}_3(A) = 8$.
(C) $\mathscr{D}_2(A) = 4$.
(D) $\mathscr{D}_1(A) = 2$.
(E) None of the statements is true.

2. Which of the following matrices are invariant factor matrices with entries in $\mathbb{Z}$?

(A) $\begin{pmatrix} -1 & 0 & 0 \\ 0 & 2 & 0 \\ 0 & 0 & 0 \end{pmatrix}$ (B) $\begin{pmatrix} 2 & 0 & 0 \\ 0 & 16 & 0 \\ 0 & 0 & 8 \end{pmatrix}$

(C) $\begin{pmatrix} 2 & 0 & 0 \\ 0 & 2 & 0 \\ 0 & 0 & -2 \end{pmatrix}$ (D) $\begin{pmatrix} 0 & 0 & 0 \\ 0 & 2 & 4 \\ 0 & 0 & 4 \end{pmatrix}$.

(E) None of the matrices is an invariant factor matrix.

3. Which of the following matrices are invariant factor matrices with entries in $\mathbb{Q}$ (a principal domain).

(A) $\begin{pmatrix} -1 & 0 & 0 \\ 0 & 2 & 0 \\ 0 & 0 & 0 \end{pmatrix}$ (B) $\begin{pmatrix} 2 & 0 & 0 \\ 0 & 16 & 0 \\ 0 & 0 & 8 \end{pmatrix}$

(C) $\begin{pmatrix} 2 & 0 & 0 \\ 0 & 2 & 0 \\ 0 & 0 & -2 \end{pmatrix}$ (D) $\begin{pmatrix} 0 & 0 & 0 \\ 0 & 2 & 0 \\ 0 & 0 & 4 \end{pmatrix}$.

(E) None of the matrices is an invariant factor matrix.

4. If the entries are taken from the ring of polynomials $\mathbb{Q}[X]$ which of these matrices are invariant factor matrices?

(A) $\begin{pmatrix} X-1 & 0 & 0 \\ 0 & X^2-1 & 0 \\ 0 & 0 & 0 \end{pmatrix}$ (B) $\begin{pmatrix} 2X-2 & 0 & 0 \\ 0 & X^2+X-2 & 0 \\ 0 & 0 & 0 \end{pmatrix}$

(C) $\begin{pmatrix} X^2+2 & 0 & 0 \\ 0 & X^2+2 & 0 \\ 0 & 0 & X^2+2 \end{pmatrix}$ (D) $\begin{pmatrix} X+1 & 0 & 0 \\ 0 & X^2-1 & 0 \\ 0 & 0 & X^3-1 \end{pmatrix}$.

(E) None of the matrices is an invariant factor matrix.

5. Two m by n matrices A and B with entries in a principal domain R are equivalent if and only if

(A) they have the same rank

(B) they have the same or associated invariant factors

(C) they are matrices of the same morphism $R^n \to R^m$

(D) they are both triangular ($A_{ij} = \theta$, $B_{ij} = \theta$ for $j < i$).

(E) None of the choices completes a true sentence.

10.3 Linear equations in a principal domain

In this section we discuss the solution of linear equations with coefficients in a principal domain.

We have earlier discussed the rank of a morphism and the rank of a matrix which represents the morphism. The rank of a morphism is the dimension of the range of the morphism. The matrix of a morphism is defined with respect to some choice of basis in the domain and codomain of the morphism. Given any m by n matrix A with coefficients in a principal domain R then we may consider the matrix to be the matrix of a morphism f from the free module R^n to the free module R^m. Range f is a submodule of the free module R^m and is itself therefore a free module with dimension $\leqslant m$. Range f has a basis of r linearly independent generating vectors, where r stands for the rank of f. Range f is also generated by the columns of the matrix of f. The maximum number of linearly independent columns of the matrix cannot differ from r for otherwise range f would have a basis with dimension differing from r.

We now establish other criteria for the rank of a matrix with entries in a principal domain.

Theorem. *Let D be an invariant factor matrix with nonzero entries $d_1, d_2, \ldots, d_r$ in a principal domain R. Then* rank $D = r$.

PROOF. The first r columns of D are linearly independent and r is the maximum number of linearly independent columns of D. □

Theorem. *Let A be an m by n matrix with entries in a principal domain R. Then* rank $A = r$ *if and only if $\mathscr{D}_k(A) = \theta$ for $k > r$ and $\mathscr{D}_k(A) \neq \theta$ for $k \leqslant r$.*

PROOF. Let D be an invariant factor matrix for A. $\mathscr{D}_k(A)$ and $\mathscr{D}_k(D)$ are associates in R for $k = 1, 2, \ldots, \min(m, n)$. Suppose rank $A = r$. D then has nonzero entries $d_1, \ldots, d_r$. $\mathscr{D}_k(D) = \theta$ for $k > r$ and $\mathscr{D}_k(D) = d_1 d_2 \cdots d_k \neq \theta$ for $k \leqslant r$. $\mathscr{D}_k(A) = \theta$ for $k > r$ and $\mathscr{D}_k(A) \neq \theta$ for $k \leqslant r$. The converse is argued in the reverse order. □

Theorem. *Let A be an m by n matrix with entries in a principal domain R. Then* rank $A = r$ *if and only if all k-minors of A with $k > r$ are zero and some r-minor is nonzero.*

PROOF. $\mathscr{D}_k(A) = \theta$ if and only if all k-minors are zero. □

We now see that the row rank and the column rank of a matrix are the same; the maximum number of linearly independent rows equals the maximum number of linearly independent columns.

Theorem. *Let A be an m by n matrix with entries in a principal domain R. The rank of A is equal to the maximum number of linearly independent rows of A.*

PROOF. If we consider instead of A the transpose of A and remember that $\det B = \det B^*$ for any square submatrix B then the criterion of the previous theorem shows that the number of linearly independent rows is also equal to r, the rank of A. □

We now discuss the kernel of a morphism and matrix.

Theorem. *Let D be an invariant factor matrix with nonzero entries $d_1, d_2, \ldots, d_r$ in a principal domain R. Then* $\dim\{\hat{X} | D\hat{X} = \theta\} = n - r$.

PROOF. The matrix equation is equivalent to the linear equations $d_1\hat{X}_1 = \theta$, $d_2\hat{X}_2 = \theta, \ldots, d_r\hat{X}_r = \theta$, which have set of solutions $\{(\theta, \theta, \ldots, \theta, t_{r+1}, \ldots, t_n) | t_{r+1}, \ldots, t_n \in R\}$. This is a submodule of R^n with basis $e_{r+1}, \ldots, e_n$ ($e_1, e_2, \ldots, e_n$ denote the standard basis for R^n) and thus has dimension $n - r$. □

We now show that the kernel of a matrix mapping must have a basis with dimension $n - r$ so that dim ker $A\cdot$ + dim range $A\cdot$ = dim domain $A\cdot$. However, we do caution that a proper submodule and the containing module can have the same dimension.

Theorem. *Let A be an m by n matrix of* rank *r with entries in a principal domain R. Then* $\dim\{X | AX = \theta\} = n - r$.

PROOF. Let D be an invariant factor matrix of A. Then $D = QAP^{-1}$ for some invertible matrices Q and P with determinant ν or $-\nu$. $\{X | AX = \theta\} = \{X | Q^{-1}DPX = \theta\} = \{X | DPX = \theta\}$. Because $P\cdot$ is an isomorphism we have $\dim\{X | DPX = \theta\} = \dim\{\hat{X} | D\hat{X} = \theta\} = n - r$. Since $(e_{r+1}, \ldots, e_n)$

is a basis for $\{\hat{X}|D\hat{X} = \theta\} = \{PX|DPX = \theta\}$, $(P^{-1}e_{r+1}, \ldots, P^{-1}e_n)$ must be a basis for $\{X|DPX = \theta\} = \{X|AX = \theta\}$. □

Now we turn to the solution of linear equations with coefficients in a principal domain.

Theorem. *The linear equations*

$$\begin{aligned} A_{11}X_1 + A_{12}X_2 + \cdots + A_{1n}X_n &= Y_1 \\ A_{21}X_1 + A_{22}X_2 + \cdots + A_{2n}X_n &= Y_2 \\ \ldots \\ A_{m1}X_1 + A_{m2}X_2 + \cdots + A_{mn}X_n &= Y_m \end{aligned}$$

with coefficients $A_{11}, \ldots, A_{mn}, Y_1, \ldots, Y_m$ *in a principal domain* R *have a solution in* R *if and only if* rank A = rank $A:Y$ *and if* rank $A = r$ *then* $\mathscr{D}_r(A) = \mathscr{D}_r(A:Y)$.

Proof. We recall the solution of linear equations in which the coefficients lie in a field (cf. Section 7.6). $A:Y$ is the matrix A augmented by the extra column Y. We, as before, interpret the solution of linear equations as a problem of whether or not the vector y with coordinates Y belongs to the range of the morphism f which has matrix A. We choose M, M' to be free R-modules with bases $(e_1, e_2, \ldots, e_n)$ and $(e'_1, e'_2, \ldots, e'_m)$ and define the morphism f to be the morphism with matrix A with respect to these bases. By a change of basis in M and a change of basis in M' we can obtain the invariant factor matrix D for the morphism f. D is equivalent to A. We denote the new basis in M by $(u_1, u_2, \ldots, u_n)$ and the new basis in M' by $(v_1, v_2, \ldots, v_m)$ and the change of basis matrices by P and Q respectively. To orient the reader we offer again a diagram.

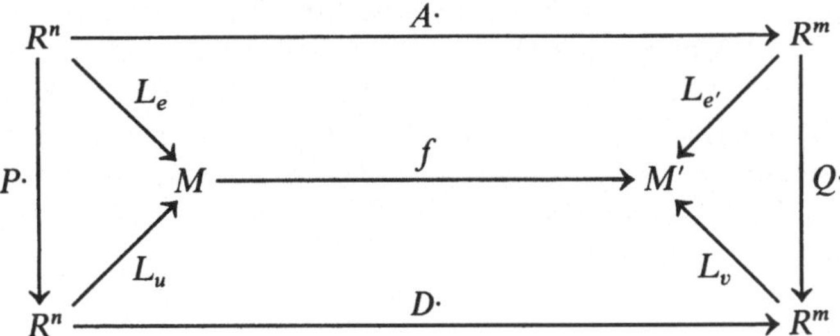

As with our discussion of linear equations with coefficients in a field the equation $AX = Y$ has a solution if and only if there exists an X such that $AX = Y$ if and only if Y belongs to the range of $A\cdot$. In terms of M, M', and f, an equivalent condition is that y belongs to the range of f. Alternatively, we shift to u and v coordinates and use U for the u coordinates of the vector x and V for the v coordinates of the vector y. Then we ask whether or not there exists a U such that $DU = V$ or whether V belongs to the range of $D\cdot$.

The equation $DU = V$ is equivalent to the system of linear equations

$$\begin{aligned} d_1U_1 \qquad\qquad\qquad &= V_1 \\ d_2U_2 \qquad\qquad &= V_2 \\ \cdots \qquad\qquad & \\ d_rU_r &= V_r \\ \theta &= V_{r+1} \\ \cdots & \\ \theta &= V_m. \end{aligned}$$

A necessary and sufficient condition that a solution exist for these equations is $V_{r+1} = \theta, V_{r+2} = \theta, \ldots, V_m = \theta$ and d_1 is a factor of V_1, d_2 is a factor of $V_2, \ldots, d_r$ is a factor of V_r. The solution set will be $\{(V_1/d_1, V_2/d_2, \ldots, V_r/d_r, t_{r+1}, t_{r+2}, \ldots, t_n) | t_{r+1}, \ldots, t_n \in R\}$, a coset of R^n of dimension $n - r$. This is the set of solutions for U which can be converted to a set of solutions for X by left multiplication by P^{-1}, the inverse of the change of basis matrix P. A complete set of solutions for $AX = Y$ is

$$\left\{ P^{-1}\begin{pmatrix} V_1/d_1 \\ \cdots \\ V_r/d_r \\ t_{r+1} \\ \cdots \\ t_n \end{pmatrix} \middle| \; t_{r+1}, \ldots, t_n \in R \right\}.$$

We recall that the kth determinant divisor of a matrix B is a greatest common divisor of all the k-minors of the matrix B. In terms of determinant divisors we rewrite the condition d_1 divides V_1, d_2 divides $V_2, \ldots, d_r$ divides V_r and $V_{r+1} = \theta, \ldots, V_m = \theta$ to the condition $\mathscr{D}_r(D) = \mathscr{D}_r(D\colon V)$ and rank $D = \text{rank}\,(D\colon V)$. However, $\mathscr{D}_r(D\colon V) = \mathscr{D}_r((DP)\colon V) = \mathscr{D}_r(Q^{-1}((DP)\colon V)) = \mathscr{D}_r(Q^{-1}(DP)\colon(Q^{-1}V)) = \mathscr{D}_r(A\colon Y)$. Couple this with $\mathscr{D}_r(D) = \mathscr{D}_r(A)$ and we can conclude $\mathscr{D}_r(A) = \mathscr{D}_r(A\colon Y)$ is an equivalent condition. The condition for rank involves some r-minor's being nonzero which is given by the previous condition and all k-minors, $k > r$, being zero which follows from the invariancy of $\mathscr{D}_k$. □

EXAMPLES. The equations

$$\begin{aligned} 2X_1 + 3X_2 &= 8 \\ 2X_1 + X_2 &= 4 \end{aligned}$$

with coefficients in $\mathbb{Z}$ have rank $A = \text{rank}\ A\colon Y = 2$. $\mathscr{D}_2(A) = -4$ and $\mathscr{D}_2(A\colon Y) = \gcd\{-4, -8, 4\} = 4$. The equations therefore have solutions. We carry out the reduction of A to the invariant factor matrix D in a tabular form.

1	0	2	3	1	0
0	1	2	1	0	1
0	1	2	1	1	0
1	0	2	3	0	1
0	1	1	2	0	1
1	0	3	2	1	0
0	1	1	2	0	1
1	−3	0	−4	1	0
0	1	1	0	0	1
1	−3	0	−4	1	−2

The equation $D\hat{X} = QY$ is

$$\begin{pmatrix} 1 & 0 \\ 0 & -4 \end{pmatrix}\begin{pmatrix} \hat{X}_1 \\ \hat{X}_2 \end{pmatrix} = \begin{pmatrix} 4 \\ -4 \end{pmatrix}.$$

This gives solutions $\hat{X} = \begin{pmatrix} 4 \\ 1 \end{pmatrix}$. Then

$$X = P^{-1}\hat{X} = \begin{pmatrix} 0 & 1 \\ 1 & -2 \end{pmatrix}\begin{pmatrix} 4 \\ 1 \end{pmatrix} = \begin{pmatrix} 1 \\ 2 \end{pmatrix}.$$

On the other hand if we consider the equations

$$\begin{aligned} 2X_1 + X_2 &= 4 \\ 2X_1 + 3X_2 &= 10 \end{aligned}$$

with coefficients in $\mathbb{Z}$ we have rank A = rank $A:Y = 2$, but we have $\mathscr{D}_2(A) = 4$ and $\mathscr{D}_2(A:Y) = 1$. These numbers are not associates in $\mathbb{Z}$ and therefore there are no solutions in $\mathbb{Z}$. If, however, we ask about solutions in $\mathbb{Q}$ then 4 and 1 are associates (both units) and therefore there are solutions in $\mathbb{Q}$.

QUESTIONS

1. m linear equations in n unknowns, $AX = Y$, with coefficients in $\mathbb{Z}$
- (A) have a solution in $\mathbb{Q}$ whenever they have a solution in $\mathbb{Z}$
- (B) have a solution in $\mathbb{Z}$ whenever they have a solution in $\mathbb{Q}$
- (C) have a solution if $m = n$ and det A is a unit
- (D) and with $m = n$ and having a solution must have det A equal to a unit.
- (E) None of the alternatives completes a true sentence.

2. The rank of an m by n matrix A with coefficients in a principal domain R
- (A) is $<k$ if $\mathscr{D}_k(A) = \theta$
- (B) is $\geqslant k$ if there exists a nonzero k-minor
- (C) is $\leqslant k$ if there exists a k-minor with value zero

(D) can equal $m + n$.
(E) None of the possibilities completes a true sentence.

3. m homogeneous linear equations in n unknowns, $AX = \theta$, with coefficients in a principal domain R
(A) always have a solution
(B) have no solution if $\mathscr{D}_k(A) \neq \theta$, $k = \min(m, n)$
(C) have only zero solutions if $m = n$ and $\mathscr{D}_m(A) \neq \theta$
(D) have nonzero solutions if $m \neq n$.
(E) None of the alternatives is satisfactory.

4. Let there be given m linear equations in n unknowns, $AX = Y$, with coefficients in a principal domain R. Which of these statements are true?
(A) If $m = n$ and $\mathscr{D}_m(A) \neq \theta$ then $A^{-1}Y$ is a unique solution of $AX = Y$.
(B) If X and X' are solutions of $AX = Y$ then $X - X'$ is also a solution.
(C) If X is a solution of $AX = Y$ then $X + \ker A\cdot$ is a complete set of solutions of $AX = Y$.
(D) If $m = n$ and solutions exist for $AX = Y$ then $\mathscr{D}_m(A)$ is a unit.
(E) None of the statements is true.

Exercises

1. Let A be an m by n matrix with entries in a principal domain R. Show that if $\mathscr{D}_k(A) = \theta$ then $\mathscr{D}_{k+1}(A) = \theta$ for any $k = 0, 1, \ldots, \min(m, n) - 1$.

2. Determine whether or not the following equations with coefficients in $\mathbb{Z}$ have solutions in $\mathbb{Z}$. If they do have solutions find them.

$$\begin{aligned} 2X_1 + X_2 + X_3 &= 8 \\ X_1 - X_2 + 2X_3 &= 1. \end{aligned}$$

3. For what integral values of a do the equations

$$\begin{aligned} aX_1 + X_2 &= 1 \\ X_1 + X_2 &= a \end{aligned}$$

have integral solutions? What are the solutions?

4. Using the techniques of this section show that the equation $aX_1 + bX_2 = c$ with a, b, c in $\mathbb{Z}$ and a, b nonzero has a solution if and only if $\gcd(a, b)$ divides c. Find the solution when it exists.

5. An egg farmer needs, per year, 25 sacks of grain to feed 6 hens. From each hen he receives 230 eggs per year. If this farmer refuses to deal in fractions how many hens must he own to receive at least 20,000 eggs per year?

10.4 A direct sum resolution of a finitely generated module

In this section we prove a generalization of the fundamental theorem of Abelian groups, a resolution of a finitely generated module over a principal domain into the direct sum of cyclic submodules.

We have earlier seen that any commutative group may be regarded as a module over the principal domain $\mathbb{Z}$ with the integral multiples of a group element treated as $\mathbb{Z}$-exterior multiplication. We have given in Section 9.5 theorems yielding direct sum resolutions of finite Abelian groups. We now treat the more general case of a module over a principal domain, not necessarily $\mathbb{Z}$. We first show what the invariant factor theorem means in terms of bases.

Theorem. *Let M be a finite dimensional free module over a principal domain R. Let N be a submodule of M. Then there exist bases $(u_1, u_2, \ldots, u_m)$ of M and $(v_1, v_2, \ldots, v_n)$ of N, $n \leqslant m$, such that $v_j = d_j u_j$, $j = 1, 2, \ldots, n$, d_j is a divisor of d_{j+1}, $j = 1, 2, \ldots, n-1$, for some $d_1, d_2, \ldots, d_n$ in R.*

PROOF. M is a free module and so also must N be a free module with dimension of $N \leqslant$ dimension of M. M has some basis $(x_1, x_2, \ldots, x_m)$ and N has some basis $(y_1, y_2, \ldots, y_n)$ and $n \leqslant m$. Each element of N is also an element of M and therefore each y_j can be expressed as a linear combination of $(x_1, x_2, \ldots, x_m)$: $y_j = \sum_{i=1}^m A_{ij} x_i$, $j = 1, 2, \ldots, n$. By means of row and column transformations (multiplication by change of basis matrices), the invariant factor theorem, the matrix A can be reduced to the equivalent invariant factor matrix D. $D = QAP^{-1}$, where the Q and P are invertible change of basis matrices (with determinant v or $-v$) for N and M respectively. From the matrices P and Q we can obtain the bases $(v_1, v_2, \ldots, v_n)$ and $(u_1, u_2, \ldots, u_m)$.

$$v_l = \sum_{j=1}^{n} P_{jl}^{-1} y_j, \quad l = 1, 2, \ldots, n.$$

$$u_k = \sum_{i=1}^{m} Q_{ik}^{-1} x_i, \quad k = 1, 2, \ldots, m.$$

The bases $(v_1, v_2, \ldots, v_n)$ and $(u_1, u_2, \ldots, u_m)$ are related by the matrix D.

$$v_j = \sum_{i=1}^{m} D_{ij} u_i = d_j u_j, \quad j = 1, 2, \ldots, n.$$

The invariant factors $d_1, d_2, \ldots, d_n$ have the property d_j is a divisor of d_{j+1}, $j = 1, 2, \ldots, n-1$. □

We now apply the previous theorem to the finding of the direct sum resolution.

Theorem. *Let M be a finitely generated module over a principal domain R. Then M is the direct sum of cyclic submodules $[\Phi(u_1)]$, $[\Phi(u_2)], \ldots, [\Phi(u_m)]$ with annihilating ideals $\langle d_1 \rangle, \langle d_2 \rangle, \ldots, \langle d_m \rangle$. Each d_j is a factor of d_{j+1}, $j = 1, \ldots, m-1$.*

PROOF. Let M be generated by the finite set $\{x_1, x_2, \ldots, x_m\}$. The mapping $\Phi: R^m \to M$ such that $\Phi(r_1, r_2, \ldots, r_m) = r_1x_1 + r_2x_2 + \cdots + r_mx_m$ is an epimorphism. We denote by Φ' the associated isomorphism $R/N \to M$ where N denotes kernel $\Phi = \{(r_1, r_2, \ldots, r_m)|r_1x_1 + r_2x_2 + \cdots + r_mx_m = \zeta\}$. Let us denote the dimension of the free submodule N of the free module R^m by n. The dimension of R^m is, of course, m and we are using the standard basis for R^m. We now use the preceding theorem and choose bases $(u_1, u_2, \ldots, u_m)$ for R^m and $(z_1, z_2, \ldots, z_n)$ for N, $n \leqslant m$, such that $z_j = d_ju_j$, $j = 1, 2, \ldots, n$, and each d_j is a factor of $d_{j+1}, j = 1, 2, \ldots, n - 1$.

An element x of R^m belongs to N if and only if $x = s_1z_1 + s_2z_2 + \cdots + s_nz_n$ for some $s_1, s_2, \ldots, s_n$ in R if and only if $x = (s_1d_1)u_1 + (s_2d_2)u_2 + \cdots + (s_nd_n)u_n$ for some $s_1, s_2, \ldots, s_n$ in R. Hence an element $r_1u_1 + r_2u_2 + \cdots + r_mu_m$ belongs to N if and only if d_1 is a factor of r_1, d_2 is a factor of $r_2, \ldots$, and d_n is a factor of r_n and $r_{n+1} = \theta, \ldots$, and $r_m = \theta$.

Corresponding to each u_i, $i = 1, 2, \ldots, m$, there is an element $\Phi(u_i)$ in M. The range of Φ, namely M, is generated by the set $\{\Phi(u_1), \Phi(u_2), \ldots, \Phi(u_m)\}$. An element $r_1\Phi(u_1) + r_2\Phi(u_2) + \cdots + r_m\Phi(u_m)$ is zero if and only if $\Phi(r_1u_1 + r_2u_2 + \cdots + r_mu_m) = \zeta$ if and only if $r_1u_1 + r_2u_2 + \cdots + r_mu_m$ belongs to N if and only if d_j divides $r_j, j = 1, 2, \ldots, m$, and $r_{n+1} = \cdots = r_m = \theta$. The annihilating ideals of the various submodules $[\Phi(u_1)]$, $[\Phi(u_2)], \ldots, [\Phi(u_m)]$, are $\langle d_1\rangle, \langle d_2\rangle, \ldots, \langle d_n\rangle, \langle\theta\rangle, \ldots, \langle\theta\rangle$.

That M is the sum of the submodules, $[\Phi(u_1)] + [\Phi(u_2)] + \cdots + [\Phi(u_m)]$, is immediate from $\{\Phi(u_1), \Phi(u_2), \ldots, \Phi(u_m)\}$ generating M. If $y \in [\Phi(u_i)] \cap [\Phi(u_j)]$ then $y = r\Phi(u_i)$ and $y = s\Phi(u_j)$ for some r, s in R. $r\Phi(u_i) = s\Phi(u_j)$ implies $\Phi(ru_i - su_j) = \zeta$. $ru_i - su_j \in N$. d_i divides r and d_j divides s or one or both of r and s are zero. In all cases r annihilates u_i and s annihilates u_j. y is therefore zero. $M = [\Phi(u_1)] \oplus [\Phi(u_2)] \oplus \cdots \oplus [\Phi(u_m)]$. If any of the annihilating ideals are generated by units then the ideal itself is the entire ring R. If, say, $d_1 = v$, then $\langle d_1\rangle = R$. In this case $[\Phi(u_1)] = \{\zeta\}$. □

EXAMPLE. The purpose of this transparent example is to illustrate the application of the preceding theorem. We take the commutative group ($\mathbb{Z}$-module) $M = \mathbb{Z}_3 \times \mathbb{Z}_4 \times \mathbb{Z}_2$ which obviously is generated by the set $\{(\bar{1}, \bar{0}, \bar{0}), (\bar{0}, \bar{1}, \bar{0}), (\bar{0}, \bar{0}, \bar{1})\}$. We define the epimorphism $\Phi: \mathbb{Z} \times \mathbb{Z} \times \mathbb{Z} \to M$ such that $\Phi(r_1, r_2, r_3) = (\bar{r}_1, \bar{r}_2, \bar{r}_3)$. For $\mathbb{Z}$ we use the standard basis and remember that $\mathbb{Z}^3$ is a free module. We denote kernel Φ by N and see that $N = 3\mathbb{Z} \times 4\mathbb{Z} \times 2\mathbb{Z}$. N is a free submodule of $\mathbb{Z}^3$ with basis $((3, 0, 0), (0, 4, 0), (0, 0, 2))$. The matrix expressing the N basis in terms of the $\mathbb{Z}^3$ basis is

$$\begin{pmatrix} 3 & 0 & 0 \\ 0 & 4 & 0 \\ 0 & 0 & 2 \end{pmatrix}.$$

When we reduce this matrix to the invariant factor matrix we find invertible P, Q such that $D = QAP^{-1}$.

$$\begin{pmatrix} 1 & 0 & 0 \\ 0 & 2 & 0 \\ 0 & 0 & -12 \end{pmatrix} = \begin{pmatrix} 1 & 1 & 0 \\ 0 & 0 & 1 \\ -4 & -3 & 0 \end{pmatrix} \begin{pmatrix} 3 & 0 & 0 \\ 0 & 4 & 0 \\ 0 & 0 & 2 \end{pmatrix} \begin{pmatrix} -1 & 0 & 4 \\ 1 & 0 & -3 \\ 0 & 1 & 0 \end{pmatrix}.$$

We also find Q^{-1} to be the matrix

$$\begin{pmatrix} -3 & 0 & -1 \\ 4 & 0 & 1 \\ 0 & 1 & 0 \end{pmatrix}.$$

This yields a new basis $(u_1, u_2, u_3) = ((-3, 4, 0), (0, 0, 1), (-1, 1, 0))$ for $\mathbb{Z}^3$. Thus $M = [\Phi(u_1)] \oplus [\Phi(u_2)] \oplus [\Phi(u_3)] = [(-\bar{3}, \bar{4}, \bar{0})] \oplus [(\bar{0}, \bar{0}, \bar{1})] \oplus [(-\bar{1}, \bar{1}, \bar{0})] = [(\bar{0}, \bar{0}, \bar{0})] \oplus [(\bar{0}, \bar{0}, \bar{1})] \oplus [(-\bar{1}, \bar{1}, \bar{0})] = [(\bar{0}, \bar{0}, \bar{1})] \oplus [(-\bar{1}, \bar{1}, \bar{0})]$. $[(0, 0, 1)] \approx \mathbb{Z}/\langle 2\rangle = \mathbb{Z}_2$. $[(-1, 1, 0)] \approx \mathbb{Z}/\langle -12\rangle = \mathbb{Z}_{12}$. This gives the expected result that $\mathbb{Z}_3 \times \mathbb{Z}_4 \times \mathbb{Z}_2 \approx \mathbb{Z}_2 \times \mathbb{Z}_{12}$.

EXAMPLE. We use the theorem to list all commutative groups of a given cardinality. Knowing that each invariant factor must be a divisor of the succeeding one allows us quickly to fashion all alternatives. All possible commutative groups of 24 elements are $\mathbb{Z}_2 \times \mathbb{Z}_2 \times \mathbb{Z}_6$, $\mathbb{Z}_2 \times \mathbb{Z}_{12}$, and $\mathbb{Z}_{24}$ (or isomorphic images). We can further use the theorem of Section 10.1 to split factors which are relatively prime to obtain $\mathbb{Z}_2 \times \mathbb{Z}_2 \times \mathbb{Z}_2 \times \mathbb{Z}_3$, $\mathbb{Z}_2 \times \mathbb{Z}_4 \times \mathbb{Z}_3$ and $\mathbb{Z}_8 \times \mathbb{Z}_3$. All commutative groups of cardinality 60 are $\mathbb{Z}_2 \times \mathbb{Z}_{30} \approx \mathbb{Z}_2 \times \mathbb{Z}_2 \times \mathbb{Z}_3 \times \mathbb{Z}_5$ and $\mathbb{Z}_{60} \approx \mathbb{Z}_4 \times \mathbb{Z}_3 \times \mathbb{Z}_5$. All commutative groups of cardinality 72 are $\mathbb{Z}_{72} \approx \mathbb{Z}_8 \times \mathbb{Z}_9$, $\mathbb{Z}_2 \times \mathbb{Z}_{36} \approx \mathbb{Z}_2 \times \mathbb{Z}_4 \times \mathbb{Z}_9$, $\mathbb{Z}_2 \times \mathbb{Z}_2 \times \mathbb{Z}_{18} \approx \mathbb{Z}_2 \times \mathbb{Z}_2 \times \mathbb{Z}_2 \times \mathbb{Z}_9$, $\mathbb{Z}_2 \times \mathbb{Z}_6 \times \mathbb{Z}_6 \approx \mathbb{Z}_2 \times \mathbb{Z}_2 \times \mathbb{Z}_3 \times \mathbb{Z}_2 \times \mathbb{Z}_3$, $\mathbb{Z}_3 \times \mathbb{Z}_{24} \approx \mathbb{Z}_3 \times \mathbb{Z}_3 \times \mathbb{Z}_8$, $\mathbb{Z}_6 \times \mathbb{Z}_{12} \approx \mathbb{Z}_2 \times \mathbb{Z}_3 \times \mathbb{Z}_4 \times \mathbb{Z}_3$.

QUESTIONS

1. Which of the following groups are not isomorphic to $\mathbb{Z}_{24}$?
   (A) $\mathbb{Z}_4 \times \mathbb{Z}_6$
   (B) $\mathbb{Z}_2 \times \mathbb{Z}_{12}$
   (C) $\mathbb{Z}_2 \times \mathbb{Z}_2 \times \mathbb{Z}_6$
   (D) $\mathbb{Z}_3 \times \mathbb{Z}_8$.
   (E) All of the listed groups are isomorphic to $\mathbb{Z}_{24}$.

2. How many different (nonisomorphic) commutative groups of 12 elements are there?
   (A) 1
   (B) 2
   (C) 3
   (D) 4
   (E) 0.

3. Which of these statements are true?
   (A) Every commutative group of prime cardinality is cyclic.
   (B) Every finitely generated commutative group is the direct sum of cyclic subgroups.

(C) Every commutative group with cardinality the power of some prime is a cyclic group.
(D) Every cyclic group has cardinality the power of some prime.
(E) None of the statements is true.

4. If a finite commutative group has cardinality divisible by 8 then it has a subgroup of order
(A) 8
(B) 4
(C) 2
(D) 1.
(E) None of the numbers completes a true sentence.

EXERCISES

1. Find all commutative groups of cardinality 360.

2. Find all commutative groups of cardinality 1080.

3. Find all commutative groups of cardinality 144.

10.5 Similarity and canonical forms

In this section we study the matrix of an endomorphism of a vector space and produce by change of basis several canonical forms for the matrix.

We carry out our analysis of the matrix of an endomorphism by utilizing the invariant factor matrix. The invariant factors will provide us with the proper bases to produce our canonical matrices. Our results are not produced by immediate application; we must first construct from our given vector space a module over a principal domain to which we apply our invariant factor theory. From this will flow our canonical forms of similarity. Two matrices are similar if and only if they are matrices of the same endomorphism.

Theorem. *Let K be a field and M a finite m-dimensional vector space over K. Let f be an endomorphism of M. Then M with $K[X]$-exterior multiplication defined by $p(X)x = p(f)(x)$ for all x in M is a $K[X]$-module.*

PROOF. By $p(f)(x)$ we mean $a_0x + a_1f(x) + \cdots + a_kf^k(x)$ where $p(X)$ is the polynomial $a_0 + a_1X + \cdots + a_kX^k$. The necessary properties for a module are readily verified. □

We point out that M as a vector space is a module over the field K. The theorem introduces a new module (still M) in that exterior multiplication is defined not just for elements of K but also for polynomials over K. With respect to this new exterior multiplication M is not a vector space but is a module. Since $K[X]$ is a principal domain, M is a module over a principal domain which makes available the earlier work of this chapter. The definition given of $K[X]$-exterior multiplication depends upon the given endomorphism f and the character of the module will change with a change in f.

We now apply the theorem of the last section to the $K[X]$-module.

Theorem. *Let M be a K-vector space of finite dimension m and let f be an endomorphism of M. Then the $K[X]$-module defined by f is the direct sum of cyclic $K[X]$-modules, $[x_1] \oplus [x_2] \oplus \cdots \oplus [x_m]$, which have annihilating ideals $\langle\varphi_1(X)\rangle, \langle\varphi_2(X)\rangle, \ldots, \langle\varphi_m(X)\rangle$ in $K[X]$.*

PROOF. This theorem is a direct application of the fundamental resolution of a finitely generated module into cyclic submodules. M has a basis of m vectors as a K-vector space. This basis forms a generating set for the $K[X]$-module. The set, while independent over K, is not linearly independent over $K[X]$. No element x in M can have annihilating ideal $\langle\theta\rangle$ because it is impossible for all the vectors $x, Xx, X^2x, \ldots$ to be linearly independent in M.

Let $(y_1, y_2, \ldots, y_m)$ be the given basis for the K-vector space M. Let A be the matrix of f with respect to this basis. We define $\Phi: K[X]^m \to M$ such that $\Phi(r_1(X), r_2(X), \ldots, r_m(X)) = r_1(X)y_1 + r_2(X)y_2 + \cdots + r_m(X)y_m$. Φ is clearly a $K(X)$-module epimorphism. We further denote the standard basis of $K[X]^m$ by $(e_1, e_2, \ldots, e_m)$ and the kernel of Φ by N.

We now assert that a basis for N is $(z_1, z_2, \ldots, z_m)$ where $z_i = Xe_i - \sum_{j=1}^m A_{ji}e_j$. When we say basis here we are speaking of N as a $K[X]$-module, a submodule of the free $K[X]$-module $K[X]^m$. We must, of course, prove $(z_1, z_2, \ldots, z_m)$ to be linearly independent and that it generates N. Suppose then that $\sum_{i=1}^m b_i(X)\, z_i = \zeta$. Note the coefficients of this arbitrary linear combination lie in $K[X]$, not K in general. $\zeta = \sum_{i=1}^m b_i(X)(Xe_i - \sum_{j=1}^m A_{ji}e_j) = \sum_{i=1}^m (Xb_i(X) - \sum_{k=1}^m b_k(X)A_{ik})e_i$. Therefore $Xb_i(X) - \sum_{k=1}^m b_k(X)A_{ik} = \theta$ for all $i = 1, 2, \ldots, m$. If some $b_i(X)$ fails to be zero we choose one of maximum degree, say $b_l(X)$. The degree of $Xb_l(X)$ exceeds the degree of $\sum_{k=1}^m b_k(X)A_{ik}$ making it impossible for the expression to be zero. This contradiction proves all $b_i(X)$, $i = 1, 2, \ldots, m$, to be zero and the family $(z_1, z_2, \ldots, z_m)$ to be linearly independent.

We now prove $(z_1, z_2, \ldots, z_m)$ generates N. $\Phi(z_i) = \Phi(Xe_i - \sum_{j=1}^m A_{ji}e_j) = Xy_i - \sum_{j=1}^m A_{ji}y_j = f(y_i) - \sum_{j=1}^m A_{ji}y_j = \sum_{k=1}^m A_{ki}y_k - \sum_{j=1}^m A_{ji}y_j = \zeta$. $z_i \in N$ for each $i = 1, 2, \ldots, m$. $[z_1, z_2, \ldots, z_m] \subseteq N$. We next consider the set $\{r_1e_1 + r_2e_2 + \cdots + r_me_m | r_1, \ldots, r_m \in K\} + [z_1, z_2, \ldots, z_m]$. The set is an additive subgroup of the free module $K[X]^m$ and is closed under K-exterior multiplication. If it is furthermore closed under $K[X]$-exterior multiplication it will be a $K[X]$-submodule of $K[X]^m$. As it contains $e_1, e_2, \ldots, e_m$, it will have to be actually equal to $K[X]^m$ if it is such a $K[X]$-submodule. For any z in $[z_1, z_2, \ldots, z_m]$ we have $X(\sum_{i=1}^m r_ie_i + z) = f(\sum_{i=1}^m r_ie_i) + Xz = \sum_{i=1}^m r_i \sum_{k=1}^m A_{ji}e_j + Xz$ which is a member of $\{\sum_{i=1}^m r_ie_i | r_i \in K\} + [z_1, z_2, \ldots, z_m]$. Having proved the set to be closed under multiplication by X we can by induction demonstrate its closure under multiplication by arbitrary positive integral powers of X and then finally by polynomials in $K[X]$. We then have $K[X]^m = \{\sum_{i=1}^m r_ie_i | r_i \in K\} + [z_1, z_2, \ldots, z_m]$. We now use this equality to prove $(z_1, z_2, \ldots, z_m)$ generates N. Suppose $y \in N$. $y = z + \sum_{i=1}^m r_ie_i$ for some $z \in [z_1, z_2, \ldots, z_m]$ and some $r_i \in K$. $\zeta = \Phi(y) = \Phi(z) + \sum_{i=1}^m r_iy_i = \zeta + \sum_{i=1}^m r_iy_i$. But $(y_1, y_2, \ldots, y_m)$ is linearly independent in the K-vector space M showing $r_i = \theta$ for all

$i = 1, 2, \ldots, m$. $y = z$ and $y \in [z_1, z_2, \ldots, z_m]$. This completes showing that $(z_1, z_2, \ldots, z_m)$ generates N.

The change of basis matrix which expresses the $(e_1, e_2, \ldots, e_m)$ coordinates of a vector in N in terms of the $(z_1, z_2, \ldots, z_m)$ coordinates we can read off the defining equations

$$z_i = Xe_j - \sum_{j=1}^{m} A_{ji}e_j, \quad j = 1, 2, \ldots, m.$$

The matrix is

$$\begin{pmatrix} X - A_{11} & -A_{12} & \cdots & -A_{1m} \\ -A_{21} & X - A_{22} & \cdots & -A_{2m} \\ \cdots & \cdots & & \\ -A_{m1} & -A_{m2} & \cdots & X - A_{mm} \end{pmatrix}$$

or, more briefly, $X\delta - A$, where δ is the identity matrix. This matrix $X\delta - A$ can, step by step, be reduced to the invariant factor matrix

$$D = \begin{pmatrix} \varphi_1(X) & \theta & \cdots & \theta \\ \theta & \varphi_2(X) & \cdots & \theta \\ \cdots & & & \\ \theta & \theta & \cdots & \varphi_m(X) \end{pmatrix}$$

with $\varphi_i(X)$ a divisor of $\varphi_{i+1}(X)$, $i = 1, 2, \ldots, m - 1$. It is to be understood that the matrices $X\delta - A$ and D are matrices with entries in the principal domain $K[X]$. The elementary transformation or change of basis matrices used also have entries in $K[X]$.

Let us denote with $(u_1, u_2, \ldots, u_m)$ the new basis of $K[X]^m$ and by $(v_1, v_2, \ldots, v_m)$ the new basis of N producing the matrix D. We have from D the equations $v_i = \varphi_i(X)u_i$, $i = 1, 2, \ldots, m$, as the relation between the new bases. Denoting the change of basis matrix in N by P and the change of basis matrix in $K[X]^m$ by Q we have the equation $D = Q(X\delta - A)P^{-1}$. We offer the following diagram as a roadmap.

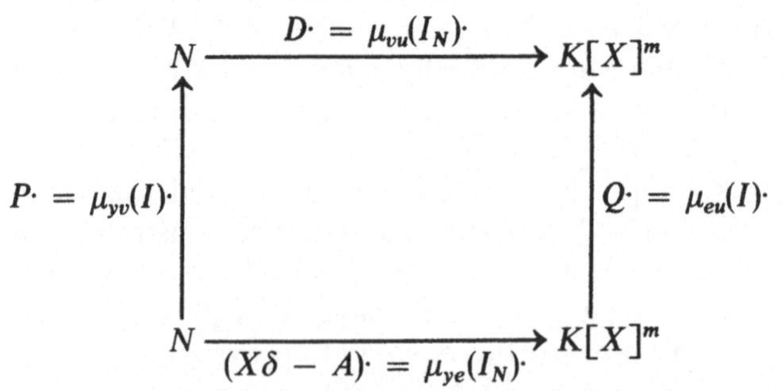

$$\begin{array}{ccc} N & \xrightarrow{\;D\cdot \,=\, \mu_{vu}(I_N)\cdot\;} & K[X]^m \\ \uparrow{\scriptstyle P\cdot \,=\, \mu_{yv}(I)\cdot} & & \uparrow{\scriptstyle Q\cdot \,=\, \mu_{eu}(I)\cdot} \\ N & \xrightarrow[\;(X\delta - A)\cdot \,=\, \mu_{ye}(I_N)\cdot\;]{} & K[X]^m \end{array}$$

Since N is the kernel of the epimorphism $\Phi: K[X]^m \to M$ we know that M and $K[X]^m/N$ are isomorphic. M is generated by the set $\{\Phi(u_1), \Phi(u_2), \ldots,$

$\Phi(u_m)\}$. An element $s_1(X)\Phi(u_1) + s_2(X)\Phi(u_2) + \cdots + s_m(X)\Phi(u_m)$ of M is zero if and only if $s_1(X)u_1 + s_2(X)u_2 + \cdots + s_m(X)u_m$ is in N if and only if $s_j(X)$ is a multiple of $\varphi_j(X)$ for all $j = 1, 2, \ldots, m$. M is then isomorphic to the product $K[X]/\langle\varphi_1(X)\rangle \times K[X]/\langle\varphi_2(X)\rangle \times \cdots \times K[X]/\langle\varphi_m(X)\rangle$. □

Corollary. *$\langle\varphi_m(X)\rangle$ is the annihilating ideal of the module M.*

Proof. $\varphi_1(X), \varphi_2(X), \ldots, \varphi_m(X)$ are all factors of $\varphi_m(X)$ and each element of M is annihilated by at least one of the invariant factors. No smaller degree polynomial can annihilate $[\Phi(u_m)] \subseteq M$. □

We remark that any one of the cyclic submodules in the direct sum resolution of M such as $[\Phi(u_i)]$ which is isomorphic to $K[X]/\langle\varphi_i(X)\rangle$ can be further decomposed into the direct sum of cyclic submodules according to the prime power factorization of the order $\varphi_i(X)$. If $\varphi_i(X) = p_1(X)^{\alpha_1}p_2(X)^{\alpha_2} \cdots p_k(X)^{\alpha_k}$ then

$$\begin{aligned}[\Phi(u_i)] &\approx K[X]/\langle\varphi_i(X)\rangle \\ &\approx K[X]/\langle p_1(X)^{\alpha_1}\rangle \times K[X]/\langle p_2(X)^{\alpha_2}\rangle \times \cdots \times K[X]/\langle p_k(X)^{\alpha_k}\rangle.\end{aligned}$$

Definition. Let M be a finite m-dimensional vector space over a field K and let f be an endormorphism of M with matrix A. Then $\varphi_m(X)$, the mth invariant factor of $X\delta - A$, is called a *minimal polynomial* of f or A. $\operatorname{Det}(X\delta - A)$ is called the *characteristic polynomial* of f or A.

Theorem. *Let M be a finite m-dimensional vector space over a field K and f be an endomorphism of M with matrix A. Then* $\det(X\delta - A) = u\varphi_1(X)\varphi_2(X) \cdots \varphi_m(X)$ *for some $u \in K$.* $\operatorname{Det}(X\delta - A)$ *annihilates M.*

Proof. A is the matrix of f with respect to some basis of M. $\mathscr{D}_m(X\delta - A)$ and $\mathscr{D}_m(D)$ are associates in $K[X]$ where D is the invariant factor matrix of $X\delta - A$. $\mathscr{D}_m(X\delta - A) = \det(X\delta - A)$ and $\mathscr{D}_m(D) = \det D = \varphi_1(X)\varphi_2(X) \cdots \varphi_m(X)$. Det D annihilates M because $\varphi_m(X)$ does. □

We now wish to illustrate the preceding theorems in an example.

Example. Let us consider the matrix

$$A = \begin{pmatrix} 7 & -2 & 1 \\ -2 & 10 & -2 \\ 1 & -2 & 7 \end{pmatrix}$$

as the matrix of an endomorphism of the $\mathbb{Q}$-vector space $\mathbb{Q}^3$. We will find the invariant factor matrix of $X\delta - A$ and simultaneously keep track of the transformations used to bring $X\delta - A$ to invariant factor form. We use a tabular form keeping record of the row transformations on the left and the column transformations on the right.

$$X\delta - A = \begin{pmatrix} X-7 & 2 & -1 \\ 2 & X-10 & 2 \\ -1 & 2 & X-7 \end{pmatrix}.$$

$$\begin{array}{ccc|ccc|ccc}
1 & 0 & 0 & X-7 & 2 & -1 & 1 & 0 & 0 \\
0 & 1 & 0 & 2 & X-10 & 2 & 0 & 1 & 0 \\
0 & 0 & 1 & -1 & 2 & X-7 & 0 & 0 & 1 \\
\hline
0 & 0 & 1 & -1 & 2 & X-7 & 0 & 0 & 1 \\
0 & 1 & 0 & 2 & X-10 & -1 & 0 & 1 & 0 \\
1 & 0 & 0 & X-7 & 2 & -1 & 0 & 0 & 1 \\
\hline
0 & 0 & 1 & -1 & 2 & X-7 & 1 & 0 & 0 \\
0 & 1 & 2 & 0 & X-6 & 2X-12 & 0 & 1 & 0 \\
1 & 0 & 0 & X-7 & 2 & -1 & 0 & 0 & 1 \\
\hline
0 & 0 & 1 & -1 & 2 & X-7 & 1 & 0 & 0 \\
0 & 1 & 2 & 0 & X-6 & 2(X-6) & 0 & 1 & 0 \\
1 & 0 & X-7 & 0 & 2(X-6) & (X-6)(X-8) & 0 & 0 & 1 \\
\hline
0 & 0 & 1 & -1 & 0 & 0 & 1 & 2 & X-7 \\
0 & 1 & 2 & 0 & X-6 & 2(X-6) & 0 & 1 & 0 \\
1 & 0 & X-7 & 0 & 2(X-6) & (X-6)(X-8) & 0 & 0 & 1 \\
\hline
0 & 0 & 1 & -1 & 0 & 0 & 1 & 2 & X-7 \\
0 & 1 & 2 & 0 & X-6 & 2(X-6) & 0 & 1 & 0 \\
1 & -2 & X-11 & 0 & 0 & (X-6)(X-12) & 0 & 0 & 1 \\
\hline
0 & 0 & 1 & -1 & 0 & 0 & 1 & 2 & X-11 \\
0 & 1 & 2 & 0 & X-6 & 0 & 0 & 1 & -2 \\
1 & -2 & X-11 & 0 & 0 & (X-6)(X-12) & 0 & 0 & 1
\end{array}$$

We can now write $Q(X\delta - A)P^{-1} = D$ as

$$\begin{pmatrix} 0 & 0 & 1 \\ 0 & 1 & 2 \\ 1 & -2 & X-11 \end{pmatrix}\begin{pmatrix} X-7 & 2 & -1 \\ 2 & X-10 & 2 \\ -1 & 2 & X-7 \end{pmatrix}\begin{pmatrix} 1 & 2 & X-11 \\ 0 & 1 & -2 \\ 0 & 0 & 1 \end{pmatrix} = \begin{pmatrix} -1 & 0 & 0 \\ 0 & X-6 & 0 \\ 0 & 0 & (X-6)(X-12) \end{pmatrix}.$$

From the invariant factor matrix D we read $\varphi_1(X) = -1$, $\varphi_2(X) = X - 6$, $\varphi_3(X) = (X - 6)(X - 12)$. From this we can conclude that a minimal polynomial of A is $(X - 6)(X - 12)$ and the characteristic polynomial is $-(X - 6)^2(X - 12)$.

$\mathbb{Q}^3$ is the direct sum of cyclic submodules $\Phi(u_1)$, $\Phi(u_2)$, $\Phi(u_3)$ with annihilating ideals $\langle -1\rangle$, $\langle X - 6\rangle$, $\langle (X - 6)(X - 12)\rangle$. In order to calculate these submodules we must find the new basis (u_1, u_2, u_3). We first find the inverse of the change of basis matrix Q. We use tabular form again to do this.

$$Q = \begin{pmatrix} 0 & 0 & 1 \\ 0 & 1 & 2 \\ 1 & -2 & X-11 \end{pmatrix}. \qquad \begin{array}{ccc|ccc} 0 & 0 & 1 & 1 & 0 & 0 \\ 0 & 1 & 2 & 0 & 1 & 0 \\ 1 & -2 & X-11 & 0 & 0 & 1 \\ \hline 1 & -2 & X-11 & 0 & 0 & 1 \\ 0 & 1 & 2 & 0 & 1 & 0 \\ 0 & 0 & 1 & 1 & 0 & 0 \\ \hline 1 & 0 & X-7 & 0 & 2 & 1 \\ 0 & 1 & 2 & 0 & 1 & 0 \\ 0 & 0 & 1 & 1 & 0 & 0 \\ \hline 1 & 0 & X-7 & 0 & 2 & 1 \\ 0 & 1 & 0 & -2 & 1 & 0 \\ 0 & 0 & 1 & 1 & 0 & 0 \\ \hline 1 & 0 & 0 & -(X-7) & 2 & 1 \\ 0 & 1 & 0 & -2 & 1 & 0 \\ 0 & 0 & 1 & 1 & 0 & 0 \end{array}$$

$$Q^{-1} = \begin{pmatrix} -(X-7) & 2 & 1 \\ -2 & 1 & 0 \\ 1 & 0 & 0 \end{pmatrix}.$$

From the equations $u_i = \sum_{h=1}^{3} Q_{hi}^{-1} e_h$, $i = 1, 2, 3$ we get $u_1 = -(X-7)e_1 - 2e_2 + e_3$, $u_2 = 2e_1 + e_2$, $u_3 = e_1$. Remembering Xe_1 means Ae_1 we obtain $u_1 = (0, 0, 0)$, $u_2 = (2, 1, 0)$, $u_3 = (1, 0, 0)$. Using the standard basis we have $\Phi(u_1) = (0, 0, 0)$, $\Phi(u_2) = (2, 1, 0)$, $\Phi(u_3) = (1, 0, 0)$. We have now $\mathbb{Q}^3 = [(0, 0, 0)] \oplus [(2, 1, 0)] \oplus [(1, 0, 0)]$ with annihilating ideals $\langle -1 \rangle$, $\langle X - 6 \rangle$, $\langle (X-6)(X-12) \rangle$. $[(0,0,0)] \approx \mathbb{Q}[X]/\langle -1 \rangle$, $[(2,1,0)] \approx \mathbb{Q}[X]/\langle X-6 \rangle$, $[(1, 0, 0)] \approx \mathbb{Q}[X]/\langle (X-6)(X-12) \rangle$. The submodule $[(1, 0, 0)]$ can be further resolved into submodules isomorphic to $\mathbb{Q}[X]/\langle X - 6 \rangle$ and $\mathbb{Q}[X]/\langle X - 12 \rangle$. Their generators can be calculated as $(X - 12)(1, 0, 0) = (-5, -2, 1)$ and $(X - 6)(1, 0, 0) = (1, -2, 1)$. $[(1, 0, 0)] = [(-5, -2, 1)] \oplus [(1, -2, 1)]$.

The component $[(0, 0, 0)]$ has annihilating ideal $\langle -1 \rangle$. -1 is a unit in $\mathbb{Q}[X]$ and therefore $\langle -1 \rangle$ is $\mathbb{Q}[X]$. That $[(0, 0, 0)]$ is annihilated by everything is consistent with its being a trivial submodule. As a component of the $\mathbb{Q}[X]$-module $\mathbb{Q}^3$ it can be omitted.

We now move on to explore how the $K[X]$-module M provides us with a basis for the K-vector space M. Our goal is to obtain a basis for the K-vector space M so that the matrix of the given endomorphism f has a simple form.

Definition. If $\{R_1, R_2, \ldots, R_k\}$ is a partition of the rows $\{1, 2, \ldots, m\}$ of a matrix A and $\{S_1, S_2, \ldots, S_k\}$ is a partition of the columns $\{1, 2, \ldots, m\}$ of A and $A_l = (A_{ij} | i \in R_l, j \in R_l)$, $l = 1, 2, \ldots, k$ and $A_{ij} = \theta$ if $(i, j) \notin R_l \times S_l$ for any $l = 1, 2, \ldots, k$ then A is said to be the *direct sum* of submatrices $A_1, A_2, \ldots, A_k$.

EXAMPLE

$$\begin{pmatrix} 6 & 0 & 0 & 0 & 0 \\ 0 & 4 & 1 & 0 & 0 \\ 0 & 3 & 2 & 0 & 0 \\ 0 & 0 & 0 & 0 & 2 \\ 0 & 0 & 0 & 1 & 0 \end{pmatrix}$$

is the direct sum of

$$(6), \quad \begin{pmatrix} 4 & 1 \\ 3 & 2 \end{pmatrix}, \quad \begin{pmatrix} 0 & 2 \\ 1 & 0 \end{pmatrix}.$$

One other direct sum possibility for this matrix is

$$\begin{pmatrix} 6 & 0 & 0 \\ 0 & 4 & 1 \\ 0 & 3 & 2 \end{pmatrix}, \quad \begin{pmatrix} 0 & 2 \\ 1 & 0 \end{pmatrix}.$$

We are interested only in submatrices formed from adjacent rows and columns of a matrix. Symbolically we represent a direct sum of a matrix by

$$A = \begin{pmatrix} A_1 & \theta & \cdots & \theta \\ \theta & A_2 & \cdots & \theta \\ \cdots & & & \\ \theta & \theta & \cdots & A_k \end{pmatrix}$$

where each A_i is a submatrix and each θ stands for a block of θ's.

Theorem. *Let M be an m-dimensional K-vector space and f an endomorphism of M. Then there exists a basis $(w_1, w_2, \ldots, w_m)$ of the K-vector space M such that the matrix of f with respect to this basis is the direct sum of submatrices $A_1, A_2, \ldots, A_m$ determined by the invariant factors $\varphi_1(X)$, $\varphi_2(X), \ldots, \varphi_m(X)$. If $\varphi_i(X) = a_{i0} + a_{i1}X + \cdots + a_{in_i-1}X^{n_i-1} + X^{n_i}$ then*

$$A_i = \begin{pmatrix} \theta & \theta & \theta & \cdots & \theta & -a_{i0} \\ v & \theta & \theta & \cdots & \theta & -a_{i1} \\ \theta & v & \theta & \cdots & \theta & -a_{i2} \\ \cdots & & & & & \\ \theta & \theta & \theta & \cdots & v & -a_{in_i-1} \end{pmatrix}.$$

If $\varphi_i(X)$ is unity in $K[X]$ then we say for counting convenience that the submatrix A_i is empty and it contributes no entries to the matrix. $n_1 + n_2 + \cdots + n_m = m$.

PROOF. We first notice that each invariant factor is presented with leading coefficient equal to unity. Since every nonzero constant is a unit in $K[X]$ we can always so normalize our invariant factors. For example, invariant factors of degree two and degree one, $a_0 + a_1X + X^2$, $a_0 + X$, lead to submatrices

$$\begin{pmatrix} \theta & -a_0 \\ v & -a_1 \end{pmatrix} \quad \text{and} \quad (-a_0).$$

As we state in the conclusion of the theorem a nonzero constant invariant factor gives rise to an empty submatrix.

The product of the invariant factors $\varphi_1(X)\varphi_2(X)\cdots\varphi_m(X)$ is an associate in $K[X]$ of the characteristic polynomial and is therefore of degree m. Thus $n_1 + n_2 + \cdots + n_m = m$. We have the cyclic $K[X]$-module direct sum of the cyclic submodules $[\Phi(u_1)], [\Phi(u_2)], \ldots, [\Phi(u_m)]$ which we now call $[t_1] \oplus [t_2] \oplus \cdots \oplus [t_m]$. If $t_i = \zeta$ and $\varphi_i(X)$ is unity then the submodule and corresponding submatrix is omitted. If $t_i \neq \zeta$ and $\varphi_i(X)$ is not unity then the vectors $t_i, Xt_i, X^2t_i, \ldots, X^{n_i-1}t_i$, are linearly independent over K and form a K-basis for the submodule $[t_i]$. The defining equations of the matrix of f on the submodule are

$$\begin{aligned}
f(t_i) &= Xt_i \\
f(Xt_i) &= X^2t_i \\
f(X^2t_i) &= X^3t_i \\
&\cdots \\
f(X^{n_i-1}t_i) &= -a_{i0}t_i - a_{i1}Xt_i - a_{i2}X^2t_i - \cdots - a_{in_i-1}X^{n_i-1}t_i.
\end{aligned}$$

These equations yield (remembering $Xx = f(x)$) a matrix A_i (as described in the statement of the theorem) with respect to the basis $(t_i, f(t_i), f^2(t_i), \ldots, f^{n_i-1}(t_i))$. Since each submodule $[t_i]$ is closed under f, $f([t_i]) \subseteq [t_i]$, the entire matrix A of f will be the direct sum of the separate A_i, $i = 1, 2, \ldots, m$. The complete basis for M is $t_1, f(t_1), \ldots, f^{n_1-1}(t_1), t_2, f(t_2), \ldots, f^{n_2-1}(t_2), \ldots, t_m, f(t_m), \ldots, f^{n_m-1}(t_m)$. □

We call the entire matrix obtained by the process described in the theorem the *first rational canonical form* while each separate submatrix A_i is called the *companion matrix* of the invariant factor $\varphi_i(X)$. Since two matrices are similar if and only if they are matrices of the same endomorphism the first rational canonical form is a canonical form for similarity.

EXAMPLE. We now continue the example for which we had obtained the direct sum $[(0, 0, 0)] \oplus [(2, 1, 0)] \oplus [(1, 0, 0)]$ and invariant factors -1, $X - 6$, $(X - 6)(X - 12)$. We rewrite the invariant factors in the normalized form: $1, -6 + X, 72 - 18X + X^2$. The trivial submodule $[(0, 0, 0)]$ makes no contribution to the first rational canonical form.

$$t_2 = (2, 1, 0) \quad\text{and}\quad A_2 = (6).$$

$$t_3 = (1, 0, 0) \quad\text{and}\quad A_3 = \begin{pmatrix} 0 & -72 \\ 1 & 18 \end{pmatrix}.$$

The basis for M is $(t_2, t_3, f(t_3))$ or $((2, 1, 0), (1, 0, 0), (7, -2, 1))$. With respect to this basis the matrix of f is

$$\begin{pmatrix} 6 & 0 & 0 \\ 0 & 0 & -72 \\ 0 & 1 & 18 \end{pmatrix}.$$

This is the first rational canonical form which is similar to the given matrix.

We can use the fact that $[(1, 0, 0)]$ splits into $[(-5, -2, 1)] \oplus [(1, -2, 1)]$ to obtain a further refinement. With respect to the basis $((2, 1, 0), (-5, -2, 1), (1, -2, 1))$ we will obtain a matrix

$$\begin{pmatrix} 6 & 0 & 0 \\ 0 & 6 & 0 \\ 0 & 0 & 12 \end{pmatrix}.$$

We now discuss, in general, how this further split can be obtained.

Theorem. *Let M be a finite dimensional vector space over a field K and f an endomorphism of M. Then there exists a basis for M such that the matrix of f is the direct sum of submatrices which are the companion matrices of powers of irreducible polynomials. The matrix associated with an invariant factor $\varphi_i(X)$ is the direct sum of matrices associated with the irreducible factors of $\varphi_i(X)$. If $p(X)^\alpha$ is the highest power of the irreducible factor $p(X)$ found in $\varphi_i(X)$ then the submatrix associated with $p(X)^\alpha$ (denoting $p(X)$ by $X^l + b_{l-1}X^{l-1} + \cdots + b_1X + b_0$) is*

$$\left(\begin{array}{ccccc|ccccc|ccccc|c|ccccc}
\theta&\theta&\cdots&\theta&-b_0 & \theta&&\cdots&&\theta & \theta&&\cdots&&\theta & & \theta&&\cdots&&\theta \\
v&\theta&\cdots&\theta&-b_1 & &&&& & &&&& & & &&&& \\
\theta&v&\cdots&\theta&-b_2 & \cdots&&&&\cdots & \cdots&&&&\cdots & \cdots & \cdots&&&&\cdots \\
&\cdots&&& & &&&& & &&&& & & &&&& \\
\theta&\theta&\cdots&v&-b_{l-1} & \theta&&\cdots&&\theta & \theta&&\cdots&&\theta & & \theta&&\cdots&&\theta \\
\hline
\theta&\theta&\cdots&\theta&v & \theta&\theta&\cdots&\theta&-b_0 & \theta&&\cdots&&\theta & & \theta&&\cdots&&\theta \\
\theta&\theta&\cdots&\theta&\theta & v&\theta&\cdots&\theta&-b_1 & &&&& & & &&&& \\
\theta&\theta&\cdots&\theta&\theta & \theta&v&\cdots&\theta&-b_2 & \cdots&&&&\cdots & \cdots & \cdots&&&&\cdots \\
&\cdots&&& & &\cdots&&& & &&&& & & &&&& \\
\theta&\theta&\cdots&\theta&\theta & \theta&\theta&\cdots&v&-b_{l-1} & \theta&&\cdots&&\theta & & \theta&&\cdots&&\theta \\
\hline
\theta&&\cdots&&\theta & \theta&\theta&\cdots&\theta&v & \theta&\theta&\cdots&\theta&-b_0 & & \theta&&\cdots&&\theta \\
&&&& & \theta&\theta&\cdots&\theta&\theta & v&\theta&\cdots&\theta&-b_1 & & &&&& \\
\cdots&&&&\cdots & \theta&\theta&\cdots&\theta&\theta & \theta&v&\cdots&\theta&-b_2 & \cdots & \cdots&&&&\cdots \\
&&&& & &\cdots&&& & &\cdots&&& & & &&&& \\
\theta&&\cdots&&\theta & \theta&\theta&\cdots&\theta&\theta & \theta&\theta&\cdots&v&-b_{l-1} & & \theta&&\cdots&&\theta \\
\hline
&\cdots&&& & &\cdots&&& & &\cdots&&& & & &\cdots&&& \\
\hline
\theta&&\cdots&&\theta & \theta&&\cdots&&\theta & \theta&&\cdots&&\theta & & \theta&\theta&\cdots&\theta&-b_0 \\
&&&& & &&&& & &&&& & & v&\theta&\cdots&\theta&-b_1 \\
\cdots&&&&\cdots & \cdots&&&&\cdots & \cdots&&&&\cdots & \cdots & \theta&v&\cdots&\theta&-b_2 \\
&&&& & &&&& & &&&& & & &\cdots&&& \\
\theta&&\cdots&&\theta & \theta&&\cdots&&\theta & \theta&&\cdots&&\theta & & \theta&\theta&\cdots&v&-b_{l-1}
\end{array}\right)$$

where the number of diagonal blocks is α.

PROOF. Suppose $\varphi_i(X) = p_1(X)^{\alpha_1}p_2(X)^{\alpha_2}\cdots p_r(X)^{\alpha_r}$. Then $[t_i]$, the cyclic submodule with annihilating ideal $\langle\varphi_i(X)\rangle$, is equal to a direct sum $[t_{i1}] \oplus$

$[t_{i2}] \oplus \cdots \oplus [t_{ir}]$ where $[t_{ij}]$ has annihilating ideal $\langle p_j(X)^{\alpha_j} \rangle$. Choose as a basis for $[t_{ij}]$ the vectors

$$\begin{array}{lll} t_{ij}, & Xt_{ij}, \ldots, & X^{l-1}t_{ij}, \\ p_j(X)t_{ij}, & Xp_j(X)t_{ij}, \ldots, & X^{l-1}p_j(X)t_{ij}, \\ & \ldots & \\ \multicolumn{3}{c}{p_j(X)^{\alpha_j-1}t_{ij},\ Xp_j(X)^{\alpha_j-1}t_{ij}, \ldots, X^{l-1}p_j(X)^{\alpha_j-1}t_{ij},} \end{array}$$

where we denote $p_j(X)$ by $b_0 + b_1X + \cdots + b_{l-1}X^{l-1} + X^l$. The defining equations for the matrix of f over the submodule $[t_{ij}]$ are

$$\begin{aligned} f(t_{ij}) &= Xt_{ij} \\ f(Xt_{ij}) &= X^2t_{ij} \\ &\ldots \\ f(X^{l-1}t_{ij}) &= -b_0t_{ij} - b_1Xt_{ij} - \cdots - b_{l-1}X^{l-1}t_{ij} + p_j(X)t_{ij} \\ f(p_j(X)t_{ij}) &= Xp_j(X)t_{ij} \\ f(Xp_j(X)t_{ij}) &= X^2p_j(X)t_{ij} \\ &\ldots \\ f(X^{l-1}p_j(X)t_{ij}) &= -b_0p_j(X)t_{ij} - b_1Xp_j(X)t_{ij} \\ &\quad - \cdots - b_{l-1}X^{l-1} + p_j(X)^2t_{ij} \\ &\ldots \\ f(p_j(X)^{\alpha_j-1}t_{ij}) &= Xp_j(X)^{\alpha_j-1}t_{ij} \\ f(Xp_j(X)^{\alpha_j-1}t_{ij}) &= X^2p_j(X)^{\alpha_j-1}t_{ij} \\ &\ldots \\ f(X^{l-1}p_j(X)^{\alpha_j-1}t_{ij}) &= -b_0p_j(X)^{\alpha_j-1}t_{ij} - b_1p_j(X)^{\alpha_j-1}Xt_{ij} \\ &\quad - \cdots - b_{l-1}X^{l-1}p_j(X)^{\alpha_j-1}t_{ij}. \end{aligned}$$

These equations yield a matrix as described in the theorem statement. □

EXAMPLE. Let an endomorphism f have a matrix

$$A = \begin{pmatrix} 0 & 1 & 0 & 2 \\ 5 & 1 & 0 & -2 \\ 4 & 1 & -1 & -4 \\ -4 & -2 & 0 & -1 \end{pmatrix}$$

with respect to the standard basis of the $\mathbb{Q}$-vector space $\mathbb{Q}^4$. We begin by finding the invariant factors for the matrix $X\delta - A$. We find by row and column transformations the invertible matrices Q and P such that $Q(X\delta - A)P^{-1} = D$, the invariant factor matrix. We list here only the results.

$$Q = \begin{pmatrix} 1 & 0 & 0 & 0 \\ -1 & 0 & 1 & 0 \\ -\frac{2}{3}(X-1) & 0 & \frac{2}{3}(X+2) & 1 \\ -\frac{1}{15}(X^2 - 4X - 22) & 1 & \frac{1}{15}(X^2 - X - 1) & -\frac{2}{5}(X-3) \end{pmatrix}$$

$$P^{-1} = \begin{pmatrix} 0 & 1 & 2 & \frac{1}{15}(X-3) \\ 1 & X & 2(X-1) & \frac{1}{15}(X^2+X+8) \\ 0 & 1 & 2 & \frac{1}{15}(X+12) \\ 0 & 0 & 1 & -\frac{2}{15}(X+2) \end{pmatrix}$$

$$D = \begin{pmatrix} -1 & 0 & 0 & 0 \\ 0 & -3 & 0 & 0 \\ 0 & 0 & 5(X+1) & 0 \\ 0 & 0 & 0 & \frac{1}{15}(X+1)(X^2-X-1) \end{pmatrix}.$$

The invariant factors are $-1, -3, 5(X+1), \frac{1}{15}(X+1)(X^2-X-1)$. Normalized to make each leading coefficient unity the invariant factors are $1, 1, 1+X, -1-2X+X^3$. In $\mathbb{Q}[X]$ we are simply choosing associates of the four polynomials.

To compute the basis (u_1, u_2, u_3, u_4) we must first find Q^{-1}. Again we do not show the work.

$$Q^{-1} = \begin{pmatrix} 1 & 0 & 0 & 0 \\ -X+1 & \frac{1}{3}(-X^2+X+5) & \frac{2}{5}(X-3) & 1 \\ 1 & 1 & 0 & 0 \\ -2 & -\frac{2}{3}(X+2) & 1 & 0 \end{pmatrix}.$$

$$u_1 = 1e_1 + (-X+1)e_2 + 1e_3 + (-2)e_4 = (0, 0, 0, 0).$$
$$u_2 = 0e_1 + \tfrac{1}{3}(-X^2+X+5)e_2 + e_3 + -\tfrac{2}{3}(X+2)e_4 = (0, 0, 0, 0).$$
$$u_3 = 0e_1 + \tfrac{2}{5}(X-3)e_2 + 0e_3 + 1e_4 = (\tfrac{2}{5}, -\tfrac{4}{5}, \tfrac{2}{5}, \tfrac{1}{5}).$$
$$u_4 = 0e_1 + 1e_2 + 0e_3 + 0e_4 = (0, 1, 0, 0).$$

Because we are using the standard basis, $\Phi(u_i) = t_i = u_i$, $i = 1, 2, 3, 4$. We then have our $\mathbb{Q}[X]$-module $\mathbb{Q}^4$ expressed as a direct sum. $\mathbb{Q}^4 = [(0,0,0,0)] \oplus [(0,0,0,0)] \oplus [(\frac{2}{5}, -\frac{4}{5}, \frac{2}{5}, \frac{1}{5})] \oplus [(0,1,0,0)]$. $[(0,0,0,0)] \approx \mathbb{Q}[X]/\langle -1\rangle$, $[(0,0,0,0)] \approx \mathbb{Q}[X]/\langle -3\rangle$, $[(\frac{2}{5}, -\frac{4}{5}, \frac{2}{5}, \frac{1}{5})] \approx \mathbb{Q}[X]/\langle 5(X+1)\rangle$, $[(0,1,0,0)] \approx \mathbb{Q}[X]/\langle \frac{1}{15}(X^3-2X-1)\rangle$. Our basis choice for the $\mathbb{Q}$-vector space $\mathbb{Q}^4$ to produce the rational canonical form is $((\frac{2}{5}, -\frac{4}{5}, \frac{2}{5}, \frac{1}{5}), (0,1,0,0), (1,1,1,-2), (-3,10,12,-4))$. With respect to this basis choice the matrix of the endomorphism is

$$\begin{pmatrix} -1 & 0 & 0 & 0 \\ 0 & 0 & 0 & 1 \\ 0 & 1 & 0 & 2 \\ 0 & 0 & 1 & 0 \end{pmatrix}.$$

This matrix is the direct sum of the two companion matrices to the polynomials $X + 1$ and $X^3 - 0X^2 - 2X - 1$, namely

$$(-1) \quad \text{and} \quad \begin{pmatrix} 0 & 0 & 1 \\ 1 & 0 & 2 \\ 0 & 1 & 0 \end{pmatrix}.$$

Now if we further split the direct summand $[(0, 1, 0, 0)]$ by means of the factorization $X^3 - 2X - 1 = (X + 1)(X^2 - X - 1)$ we have $[(0, 1, 0, 0)] = [(-4, 8, 11, -2)] \oplus [(1, 2, 1, -2)] \approx \mathbb{Q}[X]/\langle X + 1\rangle \times \mathbb{Q}[X]/\langle X^2 - X - 1\rangle$. We have obtained $(-4, 8, 11, -2)$ as $(A^2 - A - 1)(0, 1, 0, 0)$ and $(1, 2, 1, -2)$ as $(A + 1)(0, 1, 0, 0)$. With respect to the basis $((\frac{2}{5}, -\frac{4}{5}, \frac{2}{5}, \frac{1}{5}), (-4, 8, 11, -2), (1, 2, 1, -2), (-2, 11, 13, -6))$ the matrix of the endomorphism is

$$\begin{pmatrix} -1 & 0 & 0 & 0 \\ 0 & -1 & 0 & 0 \\ 0 & 0 & 0 & 1 \\ 0 & 0 & 1 & 1 \end{pmatrix}$$

which is the direct sum of the companion matrices of $X + 1$, $X + 1$, and $X^2 - X - 1$. Since $X^2 - X - 1$ is an irreducible polynomial in $\mathbb{Q}[X]$ we have obtained our rational form.

If this same problem is considered over $\mathbb{R}^4$ instead of $\mathbb{Q}^4$ then the polynomial $X^2 - X - 1$ is no longer irreducible but factors in $\mathbb{R}[X]$ into $(X - 1/2 - \sqrt{5}/2)(X - 1/2 + \sqrt{5}/2)$. We then have a matrix.

$$\begin{pmatrix} -1 & 0 & 0 & 0 \\ 0 & -1 & 0 & 0 \\ 0 & 0 & 1/2 + \sqrt{5}/2 & 0 \\ 0 & 0 & 0 & 1/2 - \sqrt{5}/2 \end{pmatrix}$$

for our rational canonical form.

It can be seen from the previous example that the canonical form as proposed here depends upon which polynomials are irreducible over the field of the given vector space. If all polynomials of positive degree over a given field can be written as a product of linear factors then the field is called *algebraically closed.* Alternatively said, a field K is algebraically closed if and only if every polynomial in $K[X]$ has at least one root in K. $\mathbb{C}$, the field of complex numbers, is algebraically closed while $\mathbb{R}$ and $\mathbb{Q}$ are not. This deficiency can be seen in the irreducibility of polynomials like $X^2 + 1$ and $X^2 - 3$, respectively. We now present a canonical form, called the Jordan canonical form, which shows what is possible when all the relevant polynomials split into linear factors. This Jordan form is actually a case of the previous theorem, but we separate it and call it the Jordan form for emphasis. This covers also the extreme case of an algebraically closed field where *all* polynomials completely factor into linear factors.

Corollary. *Let M be a finite dimensional vector space over a field K and let f be an endomorphism of M. Let A be the matrix of f with respect to a given basis of M and suppose all invariant factors of $X\delta - A$ have no irreducible factors of degree exceeding one. Then there exists a basis* *of M such that the matrix of f is the direct sum of matrices of the form*

$$\begin{pmatrix} \lambda & 0 & 0 & \cdots & 0 & 0 \\ \nu & \lambda & 0 & \cdots & 0 & 0 \\ 0 & \nu & \lambda & \cdots & 0 & 0 \\ \cdots \\ 0 & 0 & 0 & \cdots & \lambda & 0 \\ 0 & 0 & 0 & \cdots & \nu & \lambda \end{pmatrix}.$$

There will be one such submatrix with α rows and α columns for each irreducible factor power $(X - \lambda)^\alpha$ of $\varphi_i(X)$.

Proof. Suppose $(X - \lambda)^\alpha$ is one of the relatively prime factors of the invariant factor $\varphi_i(X)$ which is the order of the cyclic submodule $[t_{ij}]$. A basis for the submodule is $(t_{ij}, (X - \lambda)t_{ij}, (X - \lambda)^2 t_{ij}, \ldots, (X - \lambda)^{\alpha-1} t_{ij})$. The defining equations for the matrix are just as in the theorem with $p_j(X) = (X - \lambda)$ and $\alpha_j = \alpha$. □

Example. We begin with an endomorphism of the $\mathbb{Q}$-vector space $\mathbb{Q}^4$ which has a matrix

$$\begin{pmatrix} 1 & 0 & 0 & -1 \\ 0 & 2 & 0 & 0 \\ 0 & 0 & 1 & 1 \\ 0 & 0 & -1 & 3 \end{pmatrix}$$

with respect to the standard basis. We find the invariant factors of $X\delta - A$ by finding invertible matrices P^{-1}, Q such that $Q(X\delta - A)P^{-1} = D$.

$$P^{-1} = \begin{pmatrix} 0 & 0 & 0 & 1 \\ 0 & 0 & 1 & 0 \\ 0 & 1 & 0 & X^2 - 4X + 3 \\ 1 & 0 & 0 & -X + 1 \end{pmatrix}. \qquad Q = \begin{pmatrix} 1 & 0 & 0 & 0 \\ -X + 3 & 0 & 0 & 1 \\ 0 & 1 & 0 & 0 \\ X^2 - 4X + 4 & 0 & 1 & -X + 1 \end{pmatrix}.$$

$$X\delta - A = \begin{pmatrix} X - 1 & 0 & 0 & 1 \\ 0 & X - 2 & 0 & 0 \\ 0 & 0 & X - 1 & -1 \\ 0 & 0 & 1 & X - 3 \end{pmatrix}. \qquad D = \begin{pmatrix} 1 & 0 & 0 & 0 \\ 0 & 1 & 0 & 0 \\ 0 & 0 & X - 2 & 0 \\ 0 & 0 & 0 & (X - 1)(X - 2)^2 \end{pmatrix}.$$

We can determine the canonical forms directly from D without further work, but to know what bases will produce these canonical matrices requires further computation. The canonical matrices are

$$\begin{pmatrix} 2 & 0 & 0 & 0 \\ 0 & 0 & 0 & 4 \\ 0 & 1 & 0 & -8 \\ 0 & 0 & 1 & 5 \end{pmatrix} \quad \begin{pmatrix} 2 & 0 & 0 & 0 \\ 0 & 1 & 0 & 0 \\ 0 & 0 & 0 & -4 \\ 0 & 0 & 1 & 4 \end{pmatrix} \quad \begin{pmatrix} 2 & 0 & 0 & 0 \\ 0 & 1 & 0 & 0 \\ 0 & 0 & 2 & 1 \\ 0 & 0 & 0 & 2 \end{pmatrix}.$$

We now show how each of these forms are producced. To compute the bases we first find Q^{-1}.

$$Q^{-1} = \begin{pmatrix} 1 & 0 & 0 & 0 \\ 0 & 0 & 1 & 0 \\ -1 & X-1 & 0 & 1 \\ X-3 & 1 & 0 & 0 \end{pmatrix}.$$

Then

$$u_1 = e_1 - e_3 + (X - 3)e_4 = (0, 0, 0, 0).$$
$$u_2 = (X - 1)e_3 + e_4 = (0, 0, 0, 0).$$
$$u_3 = e_2 = (0, 1, 0, 0).$$
$$u_4 = e_3 = (0, 0, 1, 0).$$

From this we have the $Q[X]$ module $\mathbb{Q}^4 = [(0, 0, 0, 0)] \oplus [(0, 0, 0, 0)] \oplus [(0, 1, 0, 0)] \oplus [(0, 0, 1, 0)]$. $[(0, 0, 0, 0)] \approx \mathbb{Q}[X]/\langle 1\rangle$. $[(0, 1, 0, 0)] \approx Q[X]/\langle X - 2\rangle$ and $[(0, 0, 1, 0)] \approx \mathbb{Q}[X]/\langle (X - 1)(X - 2)^2\rangle$. We drop, of course, the trivial submodules leaving $\mathbb{Q}^4 = [(0, 1, 0, 0)] \oplus [(0, 0, 1, 0)]$. For the first submodule with annihilating ideal $\langle X - 2\rangle$ we use simply the basis $((0, 1, 0, 0))$. For the second submodule with annihilating ideal $\langle X^3 - 5X^2 + 8X - 4\rangle$ we use the basis $((0, 0, 1, 0), f(0, 0, 1, 0), f^2(0, 0, 1, 0))$ or $((0, 0, 1, 0), (0, 0, 1, -1), (1, 0, 0, -4))$. Altogether for the entire vector space we have the basis $((0, 1, 0, 0), (0, 0, 1, 0), (0, 0, 1, -1), (1, 0, 0, -4))$ with respect to which f has the matrix

$$\begin{pmatrix} 2 & 0 & 0 & 0 \\ 0 & 0 & 0 & 4 \\ 0 & 1 & 0 & -8 \\ 0 & 0 & 1 & 5 \end{pmatrix}.$$

This is the first rational canonical form.

To obtain the second rational canonical form we split the submodule $[(0, 0, 1, 0)]$ using the relatively prime factors of $X^3 - 5X^2 + 8X - 4 = (X - 1)(X - 2)^2$. We compute $(X - 2)^2(0, 0, 1, 0) = (f - 2)^2(0, 0, 1, 0) = (1, 0, 0, 0)$ and $(X - 1)(0, 0, 1, 0) = (f - 1)(0, 0, 1, 0) = (0, 0, 0, -1)$. Then $[(0, 0, 1, 0)] = [(1, 0, 0, 0)] \oplus [(0, 0, 0, -1)]$ with $[(1, 0, 0, 0)] \approx \mathbb{Q}[X]/\langle X - 1\rangle$ and $[(0, 0, 0, -1)] \approx \mathbb{Q}[X]/\langle (X - 2)^2\rangle$. For the first of these we have the basis $((1, 0, 0, 0))$ and for the second $((0, 0, 0, -1), X(0, 0, 0, -1)) = ((0, 0, 0, -1), f(0, 0, 0, -1)) = ((0, 0, 0, -1), (1, 0, -1, -3))$. Thus with respect to the basis $((0, 1, 0, 0), (1, 0, 0, 0), (0, 0, 0, -1), (1, 0, -1, -3))$ of $\mathbb{Q}^4$ the endomorphism has the matrix

$$\begin{pmatrix} 2 & 0 & 0 & 0 \\ 0 & 1 & 0 & 0 \\ 0 & 0 & 0 & -4 \\ 0 & 0 & 1 & 4 \end{pmatrix}.$$

Finally we come to the Jordan canonical form. We replace the basis $((0, 0, 0, -1), (1, 0, -1, -3))$ of the submodule $[(0, 0, 0, -1)]$ by the basis $((0, 0, 0, -1), (X - 2)(0, 0, 0, -1)) = ((0, 0, 0, -1), (1, 0, -1, -1))$. Thus with respect to the basis $((0, 1, 0, 0), (1, 0, 0, 0), (0, 0, 0, -1), (1, 0, -1, -1))$ we have the Jordan form

$$\begin{pmatrix} 2 & 0 & 0 & 0 \\ 0 & 1 & 0 & 0 \\ 0 & 0 & 2 & 0 \\ 0 & 0 & 1 & 2 \end{pmatrix}.$$

Questions

1. Let M be a vector space of finite dimension m over a field K and f be an endomorphism of M. Which of these statements are true?
   (A) The $K[X]$-module M defined by f is finite dimensional.
   (B) The $K[X]$-module M defined by f is finitely generated.
   (C) The $K[X]$-module $K[X]^m$ is finite dimensional.
   (D) The $K[X]$-module $K[X]^m$ is finitely generated.
   (E) None of the statements is true.

2. Let M be a vector space of finite dimension m over a field K, and f be an endomorphism of M. Which of these are true?
   (A) An f-invariant subspace of M, $f(S) \subseteq S$, is a submodule of the $K[X]$-module defined by f.
   (B) A subspace S of M is a submodule of the $K[X]$-module defined by f.
   (C) If the mth invariant factor of $X\delta - f$, $\varphi_m(X)$, is the product of distinct linear factors then so also are $\varphi_1(X), \varphi_2(X), \ldots, \varphi_{m-1}(X)$ products of distinct linear factors.
   (D) If the mth invariant factor $\varphi_m(X)$ is the product of distinct linear factors then $\mathscr{D}_m(D)$ is the product of distinct linear factors.
   (E) None of the statements is true.

3. Let M be a vector space of dimension 4 over the field $\mathbb{Q}$ and A be the matrix of an endomorphism of M. The first rational canonical form of A is calculated to be

$$\begin{pmatrix} 0 & -1 & 0 & 0 \\ 1 & 0 & 0 & 0 \\ 0 & 1 & 0 & -1 \\ 0 & 0 & 1 & 0 \end{pmatrix}.$$

   Which of these statements are true?
   (A) The invariant factors of $X\delta - A$ are $1, 1, X^2 + 1, X^2 + 1$.
   (B) The invariant factors of $X\delta - A$ are $1, X + 1, X + 1, (X + 1)^2$.
   (C) The invariant factors of $X\delta - A$ are $1, 1, 1, (X^2 + 1)^2$.
   (D) The invariant factors of $X\delta - A$ are $1, X - 1, X - 1, X^2 - 1$.
   (E) None of the statements is true.

4. Let A be the matrix of an endomorphism of $\mathbb{Q}^4$ for which $X\delta - A$ has invariant factors $1, 1, X + 1, (X + 1)(X^2 + 1)$. Which of the following matrices is the second rational canonical form for A?

(A) $\begin{pmatrix} 0 & 0 & 0 & 0 \\ 1 & 0 & 0 & 0 \\ 0 & 0 & 1 & 0 \\ 0 & 0 & 0 & 1 \end{pmatrix}$ (B) $\begin{pmatrix} -1 & 0 & 0 & 0 \\ 0 & 0 & 0 & -1 \\ 0 & 1 & 0 & -1 \\ 0 & 0 & 1 & -1 \end{pmatrix}$

(C) $\begin{pmatrix} -1 & 0 & 0 & 0 \\ 0 & -1 & 0 & 0 \\ 0 & 0 & 0 & -1 \\ 0 & 0 & 1 & 0 \end{pmatrix}$ (D) $\begin{pmatrix} -1 & 0 & 0 & 0 \\ 1 & -1 & 0 & 0 \\ 0 & 0 & 0 & -1 \\ 0 & 0 & 1 & 0 \end{pmatrix}$.

(E) None of the matrices is the second rational canonical form for A.

Exercises

1. A matrix and its transpose are similar. Show this.

2. For the matrix

$$\begin{pmatrix} 0 & 1 & 0 \\ 0 & 0 & 1 \\ 1 & 0 & 0 \end{pmatrix}$$

find the first rational canonical form in the vector space $\mathbb{Q}^3$. Find the Jordan canonical form in the vector space $\mathbb{C}^3$. Denote the three cube roots of 1 by ω, ω^2, and 1.

3. For the $\mathbb{Q}$-vector space $\mathbb{Q}^3$ given the Jordan canonical form possible for each of the following characteristic polynomials.
(a) $(X - 1)(X - 2)(X - 3)$
(b) $(X - 1)^2(X - 2)$
(c) $(X - 1)^3$.

4. For the matrix

$$\begin{pmatrix} 5 & -6 & -6 \\ -1 & 4 & 2 \\ 3 & -6 & -4 \end{pmatrix}$$

find the rational canonical forms and the Jordan canonical form over the vector space $\mathbb{Q}^3$. Include the bases which produce the canonical forms.

5. Suppose the endomorphism f of $\mathbb{Q}^3$ has the matrix

$$A = \begin{pmatrix} -1 & 3 & 0 \\ 0 & 2 & 0 \\ 2 & 1 & -1 \end{pmatrix}$$

with respect to the standard basis. Find the invariant factors of $X\delta - A$. What is the minimal polynomial of A? What is the characteristic polynomial of A? What is the second rational canonical form?

10.6 The characteristic polynomial and characteristic values

In this section we use the canonical forms to study a few of the properties of characteristic polynomials and characteristic values.

We will use the words characteristic value and characteristic root in preference to the word eigenvalue, which one often sees. By a *characteristic root* we mean a root of the characteristic polynomial $\det(X\delta - A)$ obtained from the matrix A. Characteristic value will be defined shortly. We begin this section with a theorem on the coefficients of the characteristic polynomial.

Theorem. *Let A be an m by m matrix with entries in a field K. If*

$$\det(X\delta - A) = \sum_{j=0}^{m} (-1)^{m-j} C_j X^j$$

then C_j is the sum of the principal $(m - j)$-minors of A (a principal minor is the determinant of a submatrix which has its diagonal lying on the diagonal of A).

PROOF. We denote the jth column of a matrix A by A_{*j}. $\operatorname{Det}(X\delta - A) = \det(X\delta_{*1} - A_{*1}, X\delta_{*2} - A_{*2}, \ldots, X\delta_{*m} - A_{*m}) = (-1)^m \det(A_{*1} - X\delta_{*1}, A_{*2} - X\delta_{*2}, \ldots, A_{*m} - X\delta_{*m}) = (-1)^m \sum \det(B_1, B_2, \ldots, B_m)$ where $B_j = A_{*j}$ or $-X\delta_{*j}$ for each $j = 1, 2, \ldots, m$ (2^m terms in the sum). Continuing, $\det(\delta X - A) = (-1)^m \sum_{j=0}^{m} (-X)^j \sum \det(B'_1, B'_2, \ldots, B'_m)$ where $B'_j = A_{*j}$ or δ_{*j} and the number of $\delta_{*1}, \delta_{*2}, \ldots, \delta_{*m}$ appearing is j. $\operatorname{Det}(\delta X - A) = (-1)^m \sum_{j=0}^{m} (-1)^j C_j X^j$ where $C_j =$ the sum of the principal $(m - j)$-minors of A. Note, for example, that

$$\det(\delta_{*1}, \delta_{*2}, \ldots, \delta_{*k}, A_{*k+1}, \ldots, A_{*m})$$

$$= \det\begin{pmatrix} 1 & \theta & \cdots & \theta & A_{1\,k+1} & \cdots & A_{1\,m} \\ \theta & 1 & \cdots & \theta & A_{2\,k+1} & \cdots & A_{2\,m} \\ \cdots \\ \theta & \theta & \cdots & 1 & A_{k\,k+1} & \cdots & A_{k\,m} \\ \theta & \theta & \cdots & \theta & A_{k+1\,k+1} & \cdots & A_{k+1\,m} \\ \cdots \\ \theta & \theta & \cdots & \theta & A_{m\,k+1} & \cdots & A_{m\,m} \end{pmatrix}$$

$$= \det\begin{pmatrix} A_{k+1\,k+1} & \cdots & A_{k+1\,m} \\ \cdots \\ A_{m\,k+1} & \cdots & A_{m\,m} \end{pmatrix},$$

a principal $(m - k)$-minor. □

We take note at this point of a theorem credited to W. R. Hamilton and A. Cayley. This theorem states that any matrix satisfies its own characteristic equation.

Theorem. *Let f be an endomorphism of a finite m-dimensional vector space M over a field K. Then $\sum_{j=0}^{m}(-1)^{m-j}C_j f^j = 0$, where $\sum_{j=0}^{m}(-1)^{m-j}C_j X^j$ is the characteristic polynomial of f.*

PROOF. That $\sum_{j=0}^{m}(-1)^{m-j}C_j f^j$ is zero means that it is the zero endomorphism of M or the zero mapping of M into M. Such a sum is the zero mapping if and only if it sends all vectors of M to the zero vector of M. Put another way we are asking whether or not in the $K[X]$-module M defined by f the polynomial $\sum_{j=0}^{m}(-1)^{m-j}C_j X^j$ annihilates M. Put in this way the answer is obvious. We know that not only does $\det(X\delta - A)$ annihilate M (using A as the matrix of f), but, in general, a polynomial of smaller degree, $\varphi_m(X)$, annihilates M. Not only does every matrix or endomorphism satisfy its own characteristic equation, but it satisfies its own minimal polynomial equation. □

Theorem. *If A and B are similar matrices with entries in a field K then A and B have the same characteristic polynomial.*

PROOF. $B = PAP^{-1}$ for some invertible matrix P. $\text{Det}(X\delta - B) = \det(X\delta - PAP^{-1}) = \det(PX\delta P^{-1} - PAP^{-1}) = \det P(X\delta - A)P^{-1} = \det P \det(X\delta - A) \det P^{-1} = \det P \det P^{-1} \det(X\delta - A) = \det(X\delta - A)$. □

We single out two of the coefficients of the characteristic polynomial for special attention: the sum of the principal 1-minors and the sum of the principal m-minors.

Definition. By the *trace* of the m by m matrix A we mean the sum of the diagonal elements of A, $\sum_{i=1}^{m} A_{ii}$.

Theorem. *If A and B are similar matrices with entries in a field K then trace A = trace B and* $\det A = \det B$.

PROOF. Because the characteristic polynomial is invariant under similarity all of its coefficients are also. □

Theorem. *If the characteristic polynomial* $\det(X\delta - A)$ *of the matrix A factors into linear factors only and $\lambda_1, \lambda_2, \ldots, \lambda_k$ are the characteristic roots with respective multiplicities $\alpha_1, \alpha_2, \ldots, \alpha_k$ then trace $A = \alpha_1\lambda_1 + \alpha_2\lambda_2 + \cdots + \alpha_k\lambda_k$ and* $\det A = \lambda_1^{\alpha_1}\lambda_2^{\alpha_2}\cdots\lambda_k^{\alpha_k}$.

PROOF. $(X - \lambda_1)^{\alpha_1}(X - \lambda_2)^{\alpha_2}\cdots(X - \lambda_k)^{\alpha_k} = \det(X\delta - A) = \varphi_1(X)\varphi_2(X)\cdots\varphi_m(X)$ where $\varphi_1(X), \varphi_2(X), \ldots, \varphi_m(X)$ are the invariant factors of $X\delta - A$ normalized (leading coefficient unity). The factor $(X - \lambda_1)$ appears α_1 times in the polynomials $\varphi_1(X), \varphi_2(X), \ldots, \varphi_m(X)$. In the Jordan canonical form for A, λ_1 appears on the diagonal α_1 times. So also for the other roots. The sum of the diagonal elements in the canonical form is then $\alpha_1\lambda_1 + \alpha_2\lambda_2 + \cdots + \alpha_k\lambda_k$. This is the trace of A. The canonical form for A contains the characteristic roots on the diagonal and elsewhere 0's and 1's. The determinant of such a matrix will be the product of the diagonal elements which can easily be seen from expanding the determinant first by column 1, then by column 2, etc. □

We now define characteristic value.

Definition. Let f be an endomorphism of a K-vector space M. λ in K is a *characteristic value* and x (nonzero) in M is a *characteristic vector* of f if and only if $f(x) = \lambda x$.

Given a finite basis of M with respect to which A is the matrix of f and X is the family of coordinates of the vector x we speak of λ and X as being characteristic value and characteristic vector of the matrix A. $AX = \lambda X$.

EXAMPLE

$$\begin{pmatrix} 1 & 0 & 0 & -1 \\ 0 & 2 & 0 & 0 \\ 0 & 0 & 1 & 1 \\ 0 & 0 & -1 & 3 \end{pmatrix} \begin{pmatrix} 0 \\ 1 \\ 0 \\ 0 \end{pmatrix} = \begin{pmatrix} 0 \\ 2 \\ 0 \\ 0 \end{pmatrix} = 2 \begin{pmatrix} 0 \\ 1 \\ 0 \\ 0 \end{pmatrix}.$$

2 is a characteristic value and (0, 1, 0, 0) is a characteristic vector.

Theorem. *Let A be an m by m matrix with entries in a field K. Then λ is a characteristic value of A if and only if λ is a characteristic root.*

PROOF. $AX = \lambda X$ and $X \neq 0$ if and only if $(\lambda\delta - A)X = 0$ and $X \neq 0$ if and only if nonzero X belongs to the kernel of $\lambda\delta - A$ if and only if $\det(\lambda\delta - A) = 0$ if and only if λ is a root of $\det(\lambda\delta - A)$. □

In multiplying a characteristic vector by a constant (the characteristic value) the matrix A sends the characteristic vector back into the vector subspace generated by the characteristic vector. This suggests the following theorem.

Theorem. *Let A be an m by m matrix with entries in a field K. Let λ be a characteristic value of A. Then $\{y \mid y \in M \text{ and } Ay = \lambda y\}$ is a subspace of M*

PROOF. $(\lambda\delta - A)(x + y) = (\lambda\delta - A)x + (\lambda\delta - A)y = \zeta$. $(\lambda\delta - A)(kx) = k(\lambda\delta - A)x = k\zeta = \zeta$. □

For a characteristic value λ of a matrix we now make two multiplicity definitions: one for the role of λ as a characteristic value and one for the role of λ as a characteristic root.

Definition. Let λ be a characteristic value of a matrix A with entries in a field K. By the *geometric multiplicity* of λ we mean $\dim\{x \mid Ax = \lambda x\}$ (from the characteristic value role of λ). By the *algebraic multiplicity* of λ we mean the multiplicity of λ as a root of the characteristic polynomial.

Theorem. *Let λ be a characteristic value of a matrix A with entries in a field K. Then the geometric multiplicity of λ is less than or equal to the algebraic multiplicity of λ.*

PROOF. Let the geometric multiplicity be n. $\text{Dim}\{x|(\lambda\delta - A)x = 0\} = n$. Rank $(\lambda\delta - A) = m - n$. D, we recall, we use for the matrix with the invariant factors $\varphi_1(X), \varphi_2(X), \ldots, \varphi_m(X)$ on the diagonal. By $D(\lambda)$ we will mean the matrix D with all X's replaced by λ. Rank $D(\lambda) = m - n$. $\varphi_{m-n+1}(\lambda) = \cdots = \varphi_m(\lambda) = 0$. $(X - \lambda)$ is a factor of $\varphi_{m-n+1}(X), \ldots, \varphi_m(X)$. $(X - \lambda)^n$ is a factor of the product $\varphi_1(X)\varphi_2(X)\cdots\varphi_m(X)$ which is equal to the characteristic polynomial. λ is a root of at least multiplicity n of the characteristic polynomial. □

EXAMPLE. Let the matrix A be

$$\begin{pmatrix} 1 & 0 \\ -1 & 1 \end{pmatrix}$$

and the vector space be $\mathbb{Q}^2$. The characteristic polynomial is

$$\det\begin{pmatrix} X - 1 & 0 \\ 1 & X - 1 \end{pmatrix} = (X - 1)^2$$

and so 1 has algebraic multiplicity 2.

$$\begin{aligned} \text{Ker}(1\delta - A) &= \left\{x \,\middle|\, \left(1\delta - \begin{pmatrix} 1 & 0 \\ -1 & 0 \end{pmatrix}\right)x = 0\right\} \\ &= \left\{x \,\middle|\, \begin{pmatrix} 0 & 0 \\ 1 & 0 \end{pmatrix}x = 0\right\} \\ &= \{(x_1, x_2)|x_1 = 0\} = \{(0, k)|k \in \mathbb{Q}\}. \end{aligned}$$

Dim $\ker(1\delta - A) = 1$ and therefore the geometric multiplicity is 1.

Theorem. *Let M be a K-vector space, f an endomorphism of M, and A a matrix of f with respect to some finite basis. The following statements are equivalent:*

1. *A is similar to some diagonal matrix*
2. *There exists a basis of characteristic vectors for M*
3. *The minimal polynomial of A is a product of distinct linear factors.*

PROOF. By a diagonal matrix we mean a matrix having only zero entries off the principal diagonal: $B_{ij} = \theta$ if $i \neq j$. Let A be similar to some diagonal matrix B. Each entry on the diagonal of B is a characteristic root with multiplicity equal to the number of times that root appears on the diagonal. Suppose $\lambda_1, \lambda_2, \ldots, \lambda_k$ are the distinct characteristic roots of B (and A). Dim $\ker(\lambda_j\delta - B) = m_j$, the number of zeros on the diagonal of the matrix $\lambda_j\delta - B$. Let $x_{j1}, x_{j2}, \ldots, x_{jm_j}$ be a basis for $\ker(\lambda_j\delta - B)$, $j = 1, 2, \ldots, k$. The vector space M is the direct sum of the subspaces $\ker(\lambda_j\delta - B)$ and hence has a basis $x_{11}, \ldots, x_{1m_1}, x_{21}, \ldots, x_{2m_2}, \ldots, x_{k1}, \ldots, x_{km_k}$. Each member of the basis is a characteristic vector of B and A. We have then a basis of characteristic vectors for M. Furthermore, still assuming the existence of the diagonal matrix B we can see that the entire $K[X]$-module M is annihilated by $(X - \lambda_1)(X - \lambda_2)\cdots(X - \lambda_k)$ since each basis element is annihilated by one of the factors. This shows that the minimal polynomial is a product of distinct linear factors.

Now if we assume that the minimal polynomial is a product of distinct linear factors then each annihilating polynomial (invariant factors of $X\delta - A$) $\varphi_1(X), \varphi_2(X), \ldots, \varphi_m(X)$ can have only distinct linear factors. From this we can conclude that the second rational canonical form is a matrix with all nondiagonal entries zero.

Assuming there exists a basis of characteristic vectors leads immediately to a diagonal matrix for that basis. □

Questions

1. The characteristic polynomial of the matrix

$$\begin{pmatrix} 1 & 1 & 1 \\ 1 & 1 & 1 \\ 1 & 1 & 1 \end{pmatrix}$$

is

(A) X^3
(B) $X^3 + X^2 + X + 1$
(C) $X^3 - 3X^2$
(D) $(X - 3)^3$.
(E) The correct characteristic polynomial is not listed.

2. Given the canonical form

$$\begin{pmatrix} 1 & 0 & 0 & 0 \\ 1 & 1 & 0 & 0 \\ 0 & 0 & 2 & 0 \\ 0 & 0 & 0 & 1 \end{pmatrix}$$

which one of the following choices describes the characteristic values and their multiplicities?

Choice	Characteristic values	Geometric multiplicity	Algebraic multiplicity
(A)	1	1	2
	2	1	2
(B)	1	2	3
	2	1	1
(C)	1	3	2
	2	1	1
(D)	1	3	3
	2	1	1

(E) None of the choices accurately describes the matrix.

3. Which of the following statements are true?
   (A) If the characteristic polynomial and the minimal polynomial are equal then all characteristic roots have algebraic multiplicity one.
   (B) If the characteristic polynomial and the minimal polynomial are equal then all characteristic values have geometric multiplicity one.
   (C) If λ is a characteristic root of geometric multiplicity two of a matrix A then λ must be a root of at least two separate invariant factors of $X\delta - A$.
   (D) Every matrix has at least one characteristic root even though the root may have zero geometric multiplicity.
   (E) None of the sentences is true.

Exercises

1. Let M be a finite dimensional vector space with dimension m. Let f be an endomorphism of M with m distinct characteristic values. Show that there is a basis for M so that the matrix of f is a diagonal matrix.

2. Show that some characteristic value of the matrix A is zero if and only if $\det A = \theta$.

3. Show that a matrix A has all its characteristic roots in the field K if and only if A is similar to a matrix in triangular form (B is in triangular form if and only if $B_{ij} = \theta$ for all $i > j$).

4. If $A^p = \theta$ for some square matrix A and $p \in \mathbb{N}$ then trace $A = \theta$.

5. For the matrix

$$\begin{pmatrix} 1 & 1 & 1 \\ 1 & 1 & 1 \\ 1 & 1 & 1 \end{pmatrix}$$

with entries in the field $\mathbb{Q}$ what are the characteristic values and their geometric and algebraic multiplicities?

6. Find the characteristic polynomial, the minimal polynomial, the characteristic values, their multiplicities, for the matrix

$$\begin{pmatrix} 1 & 0 & 1 \\ 0 & 1 & 0 \\ 1 & 0 & 1 \end{pmatrix}$$

with entries in $\mathbb{C}$.

Selected references

Comprehensive undergraduate algebras

1. G. Birkhoff and S. MacLane, *A Survey of Modern Algebra*, Third Edition. Macmillan: New York, 1965.
2. I. N. Herstein, *Topics in Algebra*. Blaisdell: Waltham, 1964.

Linear algebras

3. Paul R. Halmos, *Finite Dimensional Vector Spaces*, Second Edition. Springer: New York, 1974. Originally published by D. van Nostrand.
4. K. Hoffman and R. Kunze, *Linear Algebra*, Second Edition. Prentice-Hall: Englewood Cliffs, 1971.
5. Ivar Nering, *Linear Algebra and Matrix Theory*, Second Edition. Wiley: New York, 1970.

Comprehensive graduate texts in algebra

6. Nicholas Bourbaki, *Algèbre*. Hermann: Paris, 1951.
7. Nathan Jacobson, *Lectures in Abstract Algebra*. 3 volumes. Springer: New York, 1951, 1975, 1964. Originally published by D. van Nostrand.
8. Serge Lang, *Algebra*. Addison-Wesley: Reading, 1965.
9. Azriel Rosenfeld, *An Introduction to Algebraic Structures*. Holden: San Francisco, 1968.
10. B. L. van der Waerden, *Algebra*. [Translation of the Seventh Edition.] Ungar: New York, 1970.

Universal algebras

11. P. M. Cohn, *Universal Algebra*. Harper and Row: New York, 1965.
12. George Gratzer, *Universal Algebra*. Springer: New York, 1968. Originally published by D. van Nostrand.
13. Richard S. Pierce, *Introduction to the Theory of Abstract Algebras*. Holt: New York, 1968.

Set theory

14. Paul R. Halmos, *Naive Set Theory*. Springer: New York, 1974. Originally published by D. van Nostrand.
15. L. E. Sigler, *Exercises in Set Theory*. Springer: New York, 1975. Originally published by D. van Nostrand.

History of mathematics

16. *Encyclopaedia Britannica*. Encyc. Brit. Inc.: Chicago, 1973. Articles on Algebra, Algebra (History of), Algebra (Linear), Arithmetic, Complex Numbers, Groups, Mathematics (History of), Number, Polynomial, and many others.
17. E. T. Bell, *Men of Mathematics*. Shuster: New York, 1937.
18. Carl Boyer, *A History of Mathematics*. Wiley: New York, 1968.
19. Morris Kline, *Mathematical Thought from Ancient to Modern Times*. Oxford: New York, 1972.

Answers to questions

1.2 1D 2B 3B 4A 5C 6B 7C 8B 9ACEF
1.3 1BD 2E 3B 4D 5D
1.4 1C 2D 3A 4D 5C
1.5 1B 2E 3D 4C
1.6 1E 2ABCD 3C 4C 5B
1.7 1E 2C 3A 4D
1.8 1C 2A 3ABC 4D
2.1 1D 2BC 3ACD 4ACD 5D 6A 7D
2.2 1ABC 2AC 3ACD 4A
2.3 1ABCDE 2BD 3AB 4E 5ABD
2.4 1E 2BC 3E 4A 5E
2.5 1E 2AD 3BCD 4BD
2.6 1D 2BCD 3A 4A 5E
2.7 1BCDE 2ABC 3D 4ABCF 5ABCD
2.8 1CD 2AD 3E 4AB 5D
3.1 1AC 2B 3B 4E
3.2 1ABCD 2C 3AB 4C
3.3 1ABCD 2A
3.4 1AC 2BCDE
3.5 1ABC 2ABC 3ABCD
3.6 1ABC 2AC
3.7 1E 2BD 3BCD 4B
3.8 1BC 2BC 3E 4AC
4.1 1ABCD 2ABCD 3ACD 4ABCD 5A 6ABC
4.3 1BD 2BCD 3B 4ABC
4.4 1D 2B 3C 4BD 5ABCD
4.5 1BD 2ABD 3AB 4BD

4.6 1C 2C 3E 4C 5BC 6BD
5.1 1C 2A 3AC 4E 5BC 6A
5.2 1AD 2C 3AB 4BC
5.3 1C 2C 3E 4A 5A 6ABCD
5.4 1AB 2ABD 3C 4AB 5B 6A 7ABC 8AC 9E 10AC
5.5 1ABC 2E 3AC 4E
5.6 1D 2BCD 3A 4D
5.7 1B 2B 3ABD 4AB 5ABCD
5.8 1B 2AC 3CD 4ABE 5B 6D 7AB
6.1 1BCD 2AC 3ABC 4ABCD 5BCE 6BCD 7BD 8BC
6.2 1B 2BD 3BCD 4E 5B 6ABD 7ABC 8ACD
6.3 1A 2AC 3D 4BD 5CD 6D
6.4 1AB 2BCD 3BC 4AB 5E
6.5 1B 2E 3ABC 4ABC
6.6 1ABC 2ABCD 3AB 4ABCD 5ABCD
6.7 1E 2BCD 3AC 4AD 5A
6.8 1AC 2ABC 3C 4A
6.9 1E 2ABCD 3ABCD 4BCD
6.10 1BC 2BCD 3BCD 4D 5BD 6B 7A 8CD
7.1 1ABCD 2E 3D 4D
7.2 1BCD 2ABCD 3BCD 4AB
7.3 1ACD 2ABCD 3AD 4ABCD
7.4 1ABCD 2D 3ACD 4BD
7.5 1ABD 2BC 3B 4ABCD
7.6 1BCD 2AD 3CD 4BC 5BD 6ABD 7BD 8ACD
7.7 1AC 2BCD 3B 4BC 5ABC 6A 7B 8AD 9B 10B 11D 12C 13A
8.1 1AB 2AC 3ABD
8.2 1BC 2AD 3BC 4C
8.3 1BCD 2ABCD 3ABD 4ABCD
8.4 1D 2A 3ABCD 4AB
8.5 1B 2AC 3ABD
9.1 1BC 2C 3ABD 4ABCD 5CD 6BD 7AD 8ACD
9.2 1AD 2ABCD 3BD 4BD 5ABC 6C 7E 8D
9.3 1ABC 2ABCD 3D 4AD 5AC 6BD 7ABD 8ACD
9.4 1CDE 2ABC 3A 4ACD 5ACD 6BCD 7AB 8B
9.5 1B 2B 3CD 4ABD 5ABCD
10.1 1A 2E 3BC 4CD
10.2 1BCD 2AC 3ABC 4ABC 5BC
10.3 1AC 2AB 3AC 4AC
10.4 1ABC 2B 3AB 4ABCD
10.5 1BCD 2AC 3C 4C
10.6 1C 2B 3E

Index of symbols

Index

H

I

J

K

L

M

N

T

U

V

W

Z